Fortschritte der Chemie organischer Naturstoffe

Progress in the Chemistry of Organic Natural Products

49

Founded by L. Zechmeister
Edited by W. Herz, H. Grisebach, G.W. Kirby,
and Ch. Tamm

Authors:
R. A. Hill, H. Chr. Krebs, R. Verpoorte,
R. Wijnsma

Springer-Verlag
Wien New York 1986

Dr. W. HERZ, Professor of Chemistry, Department of Chemistry,
The Florida State University, Tallahassee, Florida, U.S.A.

Prof. Dr. H. GRISEBACH, Biologisches Institut II, Lehrstuhl für Biochemie der Pflanzen,
Albert-Ludwigs-Universität, Freiburg i. Br., Federal Republic of Germany

G.W. KIRBY, Sc. D., Regius Professor of Chemistry, Chemistry Department,
The University, Glasgow, Scotland

Prof. Dr. CH. TAMM, Institut für Organische Chemie der Universität Basel,
Basel, Switzerland

This work is subject to copyright.
All rights are reserved, whether the whole or part of the material is concerned, specifically those
of translation, reprinting, re-use of illustrations, broadcasting, reproduction by photocopying
machine or similar means, and storage in data banks.

© 1986 by Springer-Verlag/Wien
Softcover reprint of the hardcover 1st edition 1986

Library of Congress Catalog Card Number AC 39-1015

ISSN 0071-7886

ISBN-13: 978-3-7091-8848-4   e-ISBN-13: 978-3-7091-8846-0
DOI: 10.1007/978-3-7091-8846-0

Contents

List of Contributors	VIII
Naturally Occurring Isocoumarins. By R.A. HILL	1
I. Introduction	2
II. Nomenclature and Structural Types	2
III. Structure Determination	3
IV. Synthesis	5
V. Biosynthesis	5
VI. Biological Activity	10
VII. Introduction to the Tables	13
Table 1. Isocoumarins with no 8-Oxygenation	14
Table 2. Isocoumarins with 8-Oxygenation	17
(a) With no Carbon Substituent at C-3	17
(b) With a One-Carbon Substituent at C-3	18
(c) With a Substituent at C-3 Containing More than One Carbon.	23
Table 3. Isocoumarins with 6,8-Dioxygenation	25
(a) With no Carbon Substituent at C-3	25
(b) With a One-Carbon Substituent at C-3	26
(c) With a Substituent at C-3 Containing More than One Carbon.	32
Table 4. Isocoumarins with 6,7,8-Trioxygenation	37
(a) With a One-Carbon Substituent at C-3	37
(b) With a Substituent at C-3 Containing More than One Carbon.	39
Table 5. Isocoumarins with Fused Carbocyclic Rings	41
(a) 3,4-Fused	41
(b) 4,5-Fused	43
(c) 6,7-Fused	47
Table 6. Isocoumarins with Nitrogen-Containing Substituents	52
Formula Index	57
Trivial Name Index	58
Source Index	60
References	63

Anthraquinones in the Rubiaceae. By R. WIJNSMA and R. VERPOORTE 79

   I. Introduction 79
  II. Biological Activity 80
 III. Biosynthesis 81
 IV. Spectroscopy 83
     1. UV Spectroscopy 84
     2. IR Spectroscopy 84
     3. Mass Spectrometry 85
     4. ^{1}H-NMR Spectroscopy 85
     5. ^{13}C-NMR Spectroscopy 86
  V. Artifacts 87
 VI. Separation Methods 88
VII. Physical and Spectroscopic Properties of Anthraquinones (Table 7) 91
VIII. Rubiaceae Species Containing Anthraquinones (Table 8) 131
References 143

Recent Developments in the Field of Marine Natural Products with Emphasis on Biologically Active Compounds. By H.CHR. KREBS 151

   I. Introduction 152
  II. Porifera 153
     II.1. Steroids from Porifera 154
     II.2. Terpenoid Constituents from Porifera 157
     II.3. Amino Acid Derived Metabolites from Porifera 183
     II.4. Peptide Alkaloids, Peptides, and Proteins from Porifera 190
     II.5. Nucleosides from Porifera 193
     II.6. Alkaloids and Other Heterocyclic Compounds from Porifera . . . 194
     II.7. Macrolides from Porifera 199
     II.8. Phenols and Aromatic Ethers from Porifera 202
     II.9. Carboxylic Acids from Porifera 203
     II.10. Miscellaneous Other Compounds from Porifera 205
 III. Coelenterata (Cnidaria) 208
     III.1. Hydrozoa, Cubozoa, and Scyphozoa 208
     III.2. Hexacorallia: Sea Anemones 211
     III.3. Hexacorallia: Other Organisms 213
     III.4. Octocorallia 216
         III.4.1. Steroids from Octocorallia 216
         III.4.2. Terpenes from Octocorallia 222
         III.4.3. Miscellaneous Compounds from Octocorallia 248
 IV. Bryozoa 253

Contents

- V. Mollusca 257
  - V.1. Gastropoda 257
    - V.1.1. Nudibranchia 257
      - V.1.1.1. Steroids from Nudibranchia 261
      - V.1.1.2. Terpenes from Nudibranchia 261
      - V.1.1.3. Miscellaneous Compounds from Nudibranchia . . . 267
    - V.1.2. Aplysiidae 269
      - V.1.2.1. Terpenes from Aplysiidae 269
      - V.1.2.2. Miscellaneous Compounds from Aplysiidae 271
    - V.1.3. Conidae 273
    - V.1.4. Miscellaneous Compounds from Other Marine Snails . . . 275
  - V.2. Bivalvia 282
  - V.3. Cephalopoda 282
- VI. Echinodermata 282
  - VI.1. Saponins from Starfish 283
  - VI.2. Steroids from Starfish 295
  - VI.3. Miscellaneous Compounds from Starfish 298
  - VI.4. Saponins from Sea Cucumbers 299
  - VI.5. Steroids from Sea Cucumbers 307
  - VI.6. Crinoids and Ophiuroids 307
  - VI.7. Sea Urchins 308
- VII. Tunicata 309
- VIII. Miscellaneous Other Sources 317
- References 320
- **Author Index** 365
- **Subject Index** 385

List of Contributors

HILL, Dr. R.A., Chemistry Department, The University of Glasgow, Glasgow G12 8QQ, Scotland.

KREBS, Dr. H.CHR., Chemisches Institut der Tierärztlichen Hochschule Hannover, Bischofsholer Damm 15, D-3000 Hannover 1, Federal Republic of Germany.

VERPOORTE, Dr. R., Center for Bio-Pharmaceutical Sciences, Gorlaeus Laboratories, Leiden University, PO Box 9502, NL-2300 RA Leiden, The Netherlands.

WIJNSMA, Dr. R., Center for Bio-Pharmaceutical Sciences, Gorlaeus Laboratories, Leiden University, PO Box 9502, NL-2300 RA Leiden, The Netherlands.

Naturally Occurring Isocoumarins

By R.A. Hill, Chemistry Department, University of Glasgow, Scotland

Contents

I. Introduction	2
II. Nomenclature and Structural Types	2
III. Structure Determination	3
IV. Synthesis	5
V. Biosynthesis	5
VI. Biological Activity	10
VII. Introduction to the Tables	13
Table 1. Isocoumarins with no 8-Oxygenation	14
Table 2. Isocoumarins with 8-Oxygenation	17
(a) With no Carbon Substituent at C-3	17
(b) With a One-Carbon Substituent at C-3	18
(c) With a Substituent at C-3 Containing More than One Carbon	23
Table 3. Isocoumarins with 6,8-Dioxygenation	25
(a) With no Carbon Substituent at C-3	25
(b) With a One-Carbon Substituent at C-3	26
(c) With a Substituent at C-3 Containing More than One Carbon	32
Table 4. Isocoumarins with 6,7,8-Trioxygenation	37
(a) With a One-Carbon Substituent at C-3	37
(b) With a Substituent at C-3 Containing More than One Carbon	39
Table 5. Isocoumarins with Fused Carbocyclic Rings	41
(a) 3,4-Fused	41
(b) 4,5-Fused	43
(c) 6,7-Fused	47
Table 6. Isocoumarins with Nitrogen-Containing Substituents	52
Formula Index	57
Trivial Name Index	58
Source Index	60
References	63

I. Introduction

Isocoumarins have been isolated from a wide variety of microbial, plant and insect sources and have been shown to possess an impressive array of biological activities. Since the review by BARRY in 1963 (24), the number of known naturally occurring isocoumarins has increased dramatically. This increase is largely due to improvements in isolation procedures and structural analysis. Previous reviews have concentrated on fungal isocoumarins (293, 294) and mycotoxic isocoumarins (301). This review lists over 160 naturally occurring isocoumarins. Leading references on isolation, structure elucidation, biosynthesis and synthesis are given in the accompanying Tables.

The known natural isocoumarins are listed in the Tables according to the number and orientation of oxygen atoms on the benzenoid ring and by carbon substituents. For completeness, those isocoumarins bearing additional fused carbocyclic rings and those containing nitrogen substituents are included in separate Tables. It is hoped that by using these Tables in conjunction with the Formula Index, the Trivial Name Index and the Source Index the reader will be able to locate key references in the literature and gain an understanding of the fascinating chemistry and action of naturally occurring isocoumarins.

II. Nomenclature and Structural Types

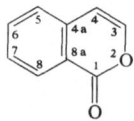

Isocoumarin

Many natural isocoumarins have been assigned trivial names and these are listed in the Tables and the Trivial Name Index. The numbering system of the isocoumarin ring system has led to some confusion in the past (323); the preferred system is given above and agrees with that used for the IUPAC and Chemical Abstracts preferred name 1H-2-benzopyran-1-one.

Isocoumarin itself has not been found to occur naturally but simple derivatives abound in nature; many of these have substituents at C-3. The Tables are arranged in increasing number of carbon atoms in the C-3 substituent. Isocoumarins which lack oxygenation at C-8 are

References, 63–78

found in plants and are not derived biogenetically from acetate. These are listed in Table 1. The acetate derived isocoumarins all have an oxygen atom at C-8 and some have retained the C-6 oxygen. Further oxygenation or alkylation may occur at the remaining positions.

Mellein (**19**) has been taken as the parent compound for simple isocoumarins. Thus the commonly occurring 3,4-dihydro-8-hydroxy-6-methoxy-3-methylisocoumarin (**53**) is known as 6-methoxymellein. This greatly aids discussion of these compounds.

3-Hydroxyisocoumarins such as ustic acid (**76**) exist in equilibrium with the corresponding keto acids. For this review they are treated as isocoumarins. Isocoumarins with 3,4-, 4,5- and 7,8-fused carbocyclic rings are listed in Table 5. 3,4-Dihydroisocoumarins are often found together with the corresponding isocoumarins, particularly for the 7,8-fused carbocyclic isocoumarins. These are listed together in the Tables.

III. Structure Determination

The range of spectroscopic techniques now available means that it is often a straightforward matter to arrive at a possible structure for a new isocoumarin. Key references are given in the Tables relating to the structure elucidation of the natural isocoumarins. Carbon-13 nuclear magnetic resonance spectroscopy has proved to be of use in determination of the structures of the more complex isocoumarins (*139*). The ^{13}C-NMR data for some simple isocoumarin derivatives have been listed (*227, 265*).

Crystal structure determinations have been used to confirm the structures of duclauxin (**127**) (*221*), cladosporin (**86**) (*266*), hydrangenol (**40**) (*249*), agrimolide (**93**) (*13*), dehydroaltenusin (**118**) (*238*), gilmaniellin (**124**) (*68*) and antibiotic AI-77-B (**152**) (*259*). The absolute configuration has been determined by crystal structure methods for "naphthalic anhydride" (**123**) (*143*) and actinobolin (**145**) (*302, 307*).

The absolute configurations of 5-methylmellein (**24**) (*20*), ochratoxin A (**158**) (*298*) and xanthomegnin (**138**) (*209a*) have been determined by degradation to 3-hydroxybutanoic acid of known configuration (Scheme 1).

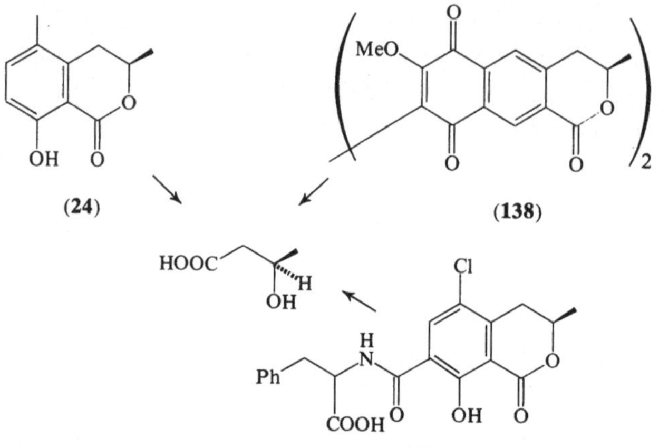

Scheme 1

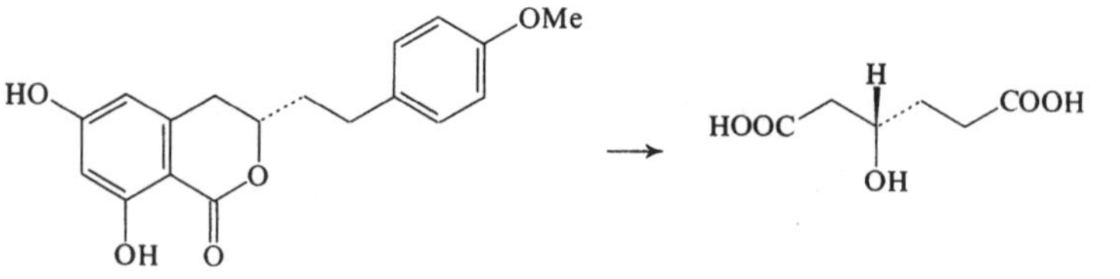

Scheme 2

Also, agrimolide (**93**) has been degraded to the known 3-hydroxyadipic acid to establish its absolute configuration (*13, 14*) (Scheme 2).

Circular dichroism has been used to correlate the absolute configuration at C-3 of dihydroisocoumarins, for example mellein (**19**), agrimolide (**93**) and phyllodulcin (**41**) (*12, 14*) and the fusarentins (*120*). However, some doubt on the validity of the correlation of Cotton effects with absolute configuration of isocoumarins has been expressed (*9*). Thus only for Cotton effects within the n→π* band is the helicity rule independent of the substitution of the benzenoid ring.

IV. Synthesis

The synthesis of natural isocoumarins has attracted considerable attention (*24, 316*). Key references to syntheses of specific isocoumarins are given in the Tables.

V. Biosynthesis

Most of the natural isocoumarins are derived biosynthetically from acetate *via* the acetate-polymalonate pathway (*293, 294*). Mellein (**19**) is formed from acetate and malonate, as in Scheme 3 (*140*). The early reduction of two of the carbonyl groups in the polyketone chain has been indicated by results of CD_3COOH feedings (*1*). Loss of the oxygen function at C-6 of an isocoumarin is quite common but loss of the hydroxyl group at C-8 never occurs in those isocoumarins derived from acetate, presumably as a consequence of the cyclization mechanism (*293*).

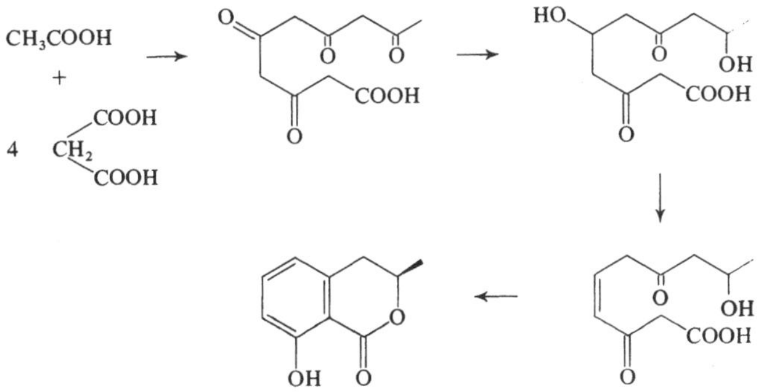

Scheme 3

6-Methoxymellein (**53**) (*272, 244*), reticulol (**98**) (*104*) and the chloroisocoumarin (**59**) (*141*) have been shown to arise by similar folding of the polyketone chain but without loss of the C-6 oxygen function. The O-methylation of these metabolites from the C-1 pool occurs at a late stage in the biosyntheses (*134*), whereas the introduction of chlorine appears to occur immediately after aromatization (*134*).

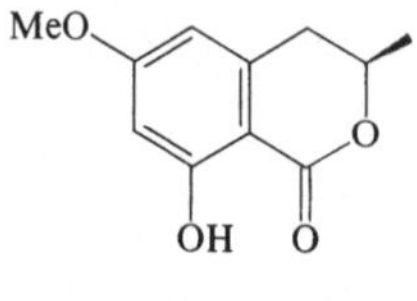

(**53**) 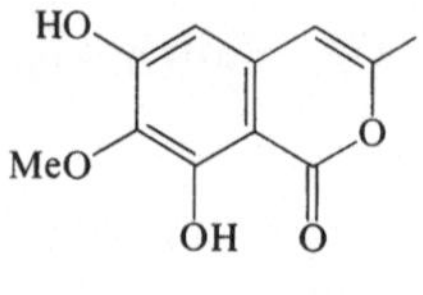   (**98**) 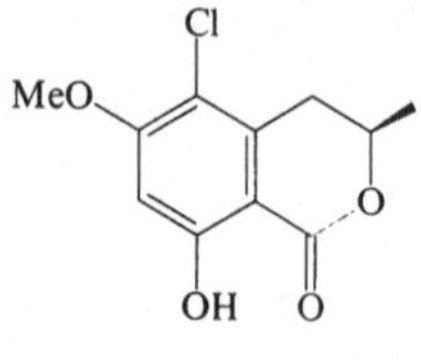   (**59**)

Longer polyketone chains are involved in the biosynthesis of monocerin (**109**), a heptaketide (*252a*) (Scheme 4), and asperentin (**86**), an octaketide (*57*) (Scheme 5).

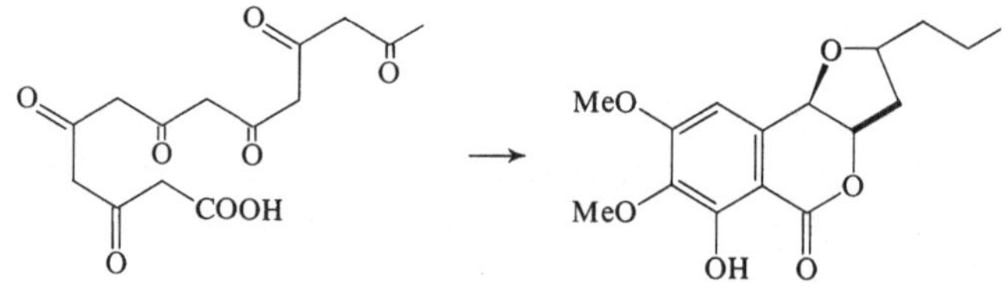

Scheme 4

(**109**)

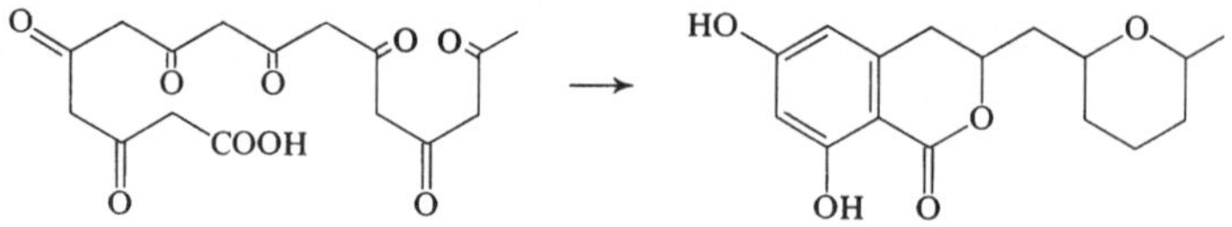

Scheme 5   (**86**)

Xanthomegnin (**138**) and the related metabolite viomellein (**135**) have been shown to be dimers of heptaketide precursors (*262*) (Scheme 6).

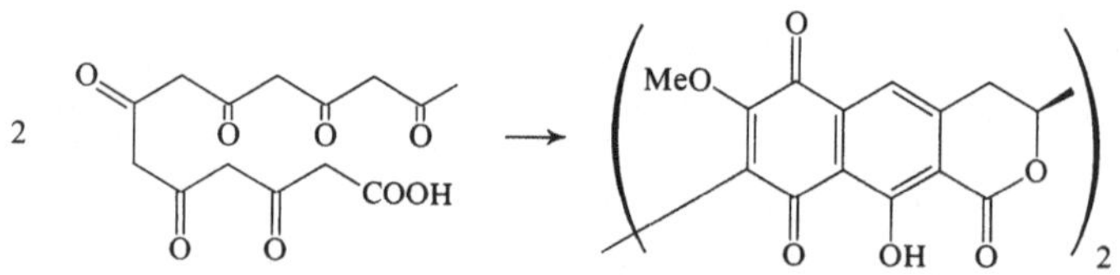

Scheme 6   (**138**)

References, 63–78

The biosynthesis of ochratoxin A (**158**) has attracted a great deal of attention (*88*, *99*, *180*, *275*, *307*, *309*, *321*). The additional carbon at C-7 (▲) is derived from the C-1 pool (*321*). The additional carbons (marked ▲) in oospolactone (**23**), oosponol (**16**) (*214*) and canescin (**67**) (*36*, *37*) are similarly derived. However the origin of the remaining 3 carbons in the side chain of canescin remains unclear (*37*).

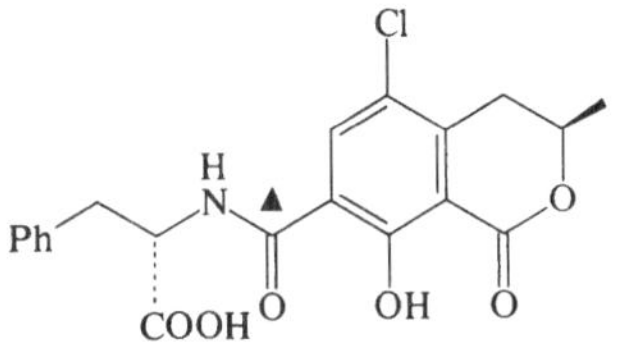

(**158**)

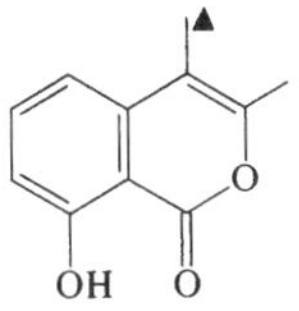

(**23**)

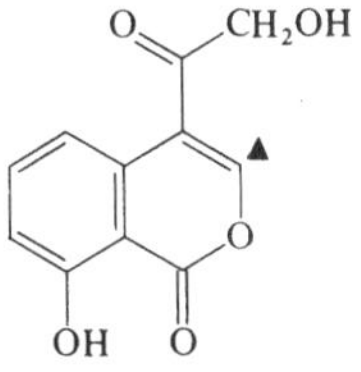

(**16**)

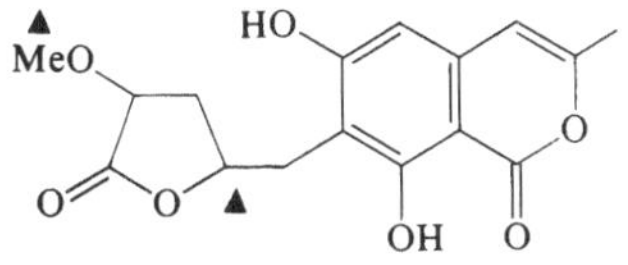

(**67**)

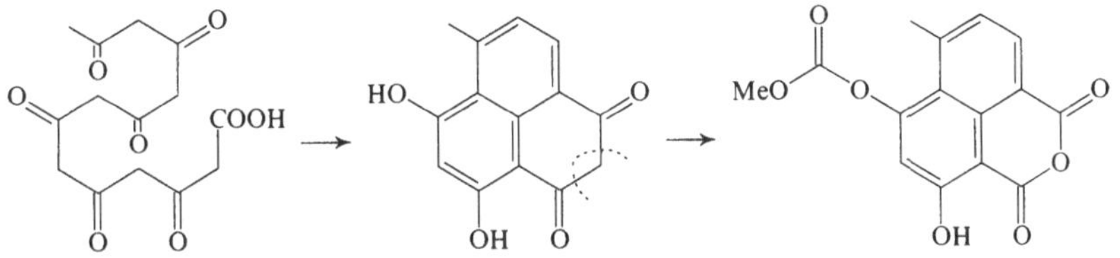

(**121**)

Scheme 7

Biosynthetic studies have indicated that 4-*O*-carbomethoxylamellicolic anhydride (**121**) may be derived from a phenalenone precursor (*176*) (Scheme 7). The origin of the carbomethoxy group is not certain (*105*, *176*, *289*). Duclauxin (**127**) is probably formed by dimerisation of an intermediate of similar origin (*243*).

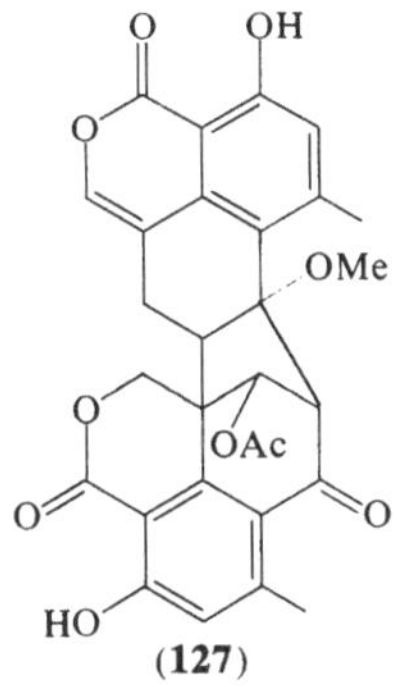

(**127**)

The incorporation of sclerotinin A (**72**) into sclerin (**17**) (*21, 106*) provides an interesting example of a biosynthetic rearrangement (Scheme 8).

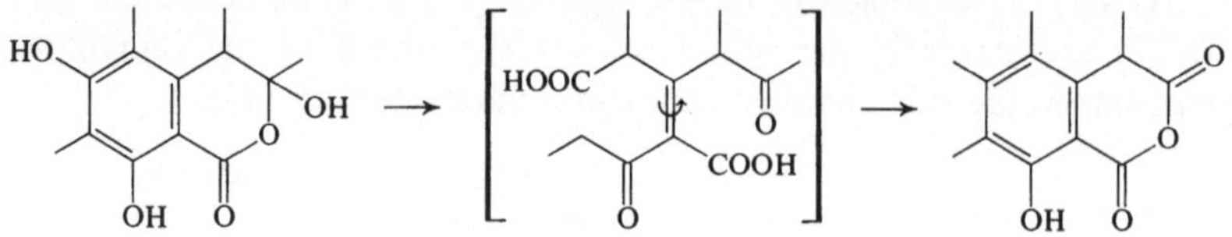

Scheme 8

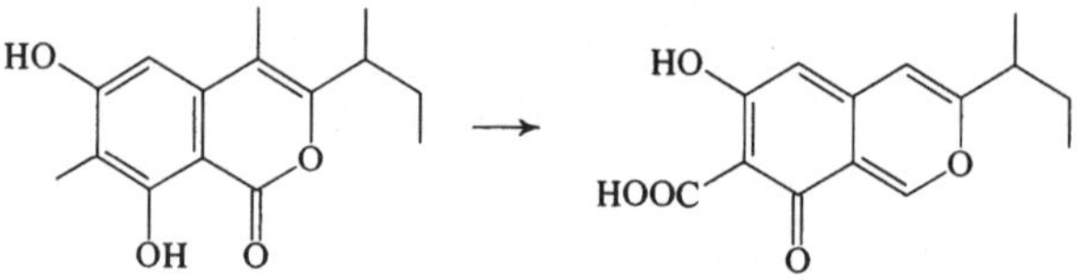

Scheme 9

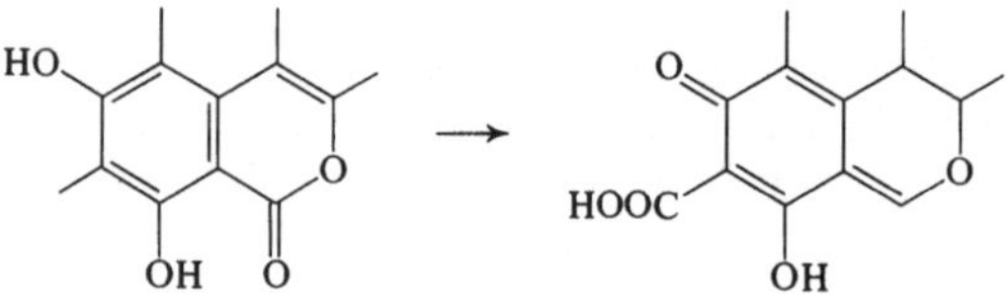

Ascochitine

Scheme 10

citrinin

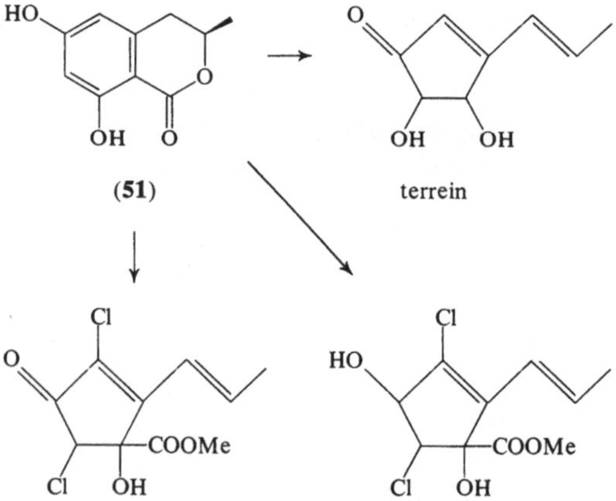

(**51**)

terrein

cryptosporiopsin

cryptosporiopsinol

Scheme 11

References, 63–78

The involvement of isocoumarins as biosynthetic intermediates of fungal metabolites has been well established. Ascochitine (Scheme 9) (*76, 77*) and citrinin (Scheme 10) (*22, 56, 75*) have been shown to be derived from the corresponding isocoumarin precursors or similar intermediates. Ring contraction of the dihydroxydihydroisocoumarin (**51**) in various fungi leads to terrein (*138, 323*), cryptosporiopsin (*323*) and cryptosporiopsinol (*134*) (Scheme 11).

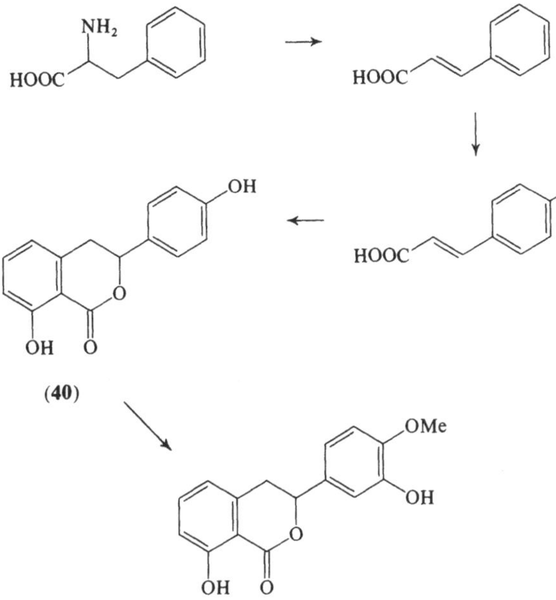

Scheme 12

The biosynthesis of plant derived isocoumarins has been studied to a lesser extent than fungal isocoumarins. Phyllodulcin (**41**) and hydrangenol (**40**) have been shown to be formed from phenylalanine *via* cinnamic acid and *p*-coumaric acid with the addition of three acetate units (Scheme 12) (*33, 34, 148, 312*).

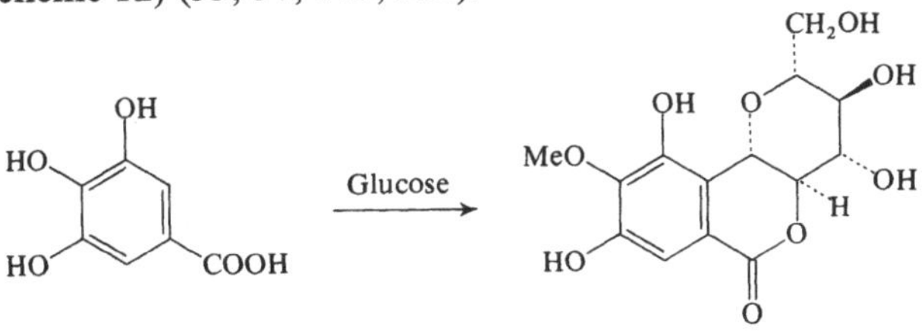

Scheme 13

Bergenin (**9**) is derived from *C*-glucosylation of gallic acid and subsequent lactone formation (*286*) (Scheme 13).

VI. Biological Activity

Isocoumarins display a very wide range of biological activities. Many fungal isocoumarins exhibit antifungal activity (*304*), particularly oospolactone (**23**) (*201*), cladosporin (**86**) (*253*) and 6-methoxymellein (**53**) (*85*).

(**23**)   (**86**)   (**53**)

This activity probably arises as a result of the competition between different fungal species, although, in common with many other isocoumarins, these antifungal agents also possess phytotoxic activity. Thus, cladosporin (**86**) has been shown to inhibit the growth of etiolated wheat coleoptiles (*266*). The role and origin of 6-methoxymellein (**53**) has been the subject of some debate as it is a metabolite of several fungi (*18, 92, 177, 252, 271, 294*) but is also a phytoalexin (*81*) produced by carrots under stress either by chemical reagents or infection by fungi (*78, 124, 194*).

Several 6,8-dihydroxyisocoumarins and 6,8-dihydroxy-3,4-dihydroisocoumarins are metabolites of phytopathogenic fungi. The hydroxydihydroisocoumarins (**70**), (**74**) and (**75**) were implicated as possible phytotoxins produced by *Ceratocystis ulmi*, the fungus responsible for Dutch Elm Disease (*71, 72, 135*). The hydroxydihydroisocoumarin (**74**) causes necrotic lesions on the leaves of pear trees and inhibits the growth of rice seedlings (*160*). The dihydroxydihydroisocoumarin (**51**) is a metabolite of *Pyricularium oryzae*, the rice blast fungus (*151*). The corresponding dihydroxyisocoumarin (**52**) and the hydroxy derivative (**57**) are metabolites of *Ceratocystis minor*, the fungus causing Southern Pine Disease (*178*). The methoxyisocoumarin (**54**) and its dihydro analogue (**53**) are metabolites of the phytopathogenic fungus *Ceratocystis fimbriata* (*81, 271*). Diaporthin (**78**) is a metabolite of *Endothia parasitica* that causes wilting of chestnut trees (*123*). Sclerin (**17**) and the sclerotinins A (**72**) and B (**71**), metabolites of *Sclerotinia sclerotiorum*, have plant growth regulatory effects (*246, 248*). Sclerotinin A (**72**) has a synergistic effect with gibberelin (*284*). The mode of action of these isocoumarins in phytopathogenic organisms is not clear although they are observed to build up in the trees and plants after infection by the fungi (*135*).

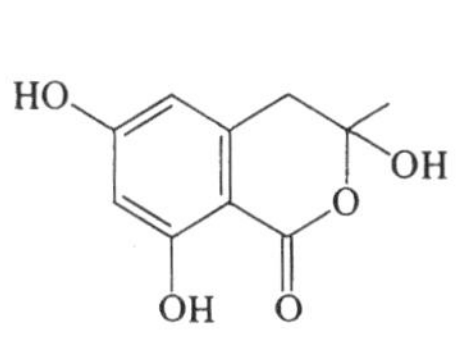

(70)

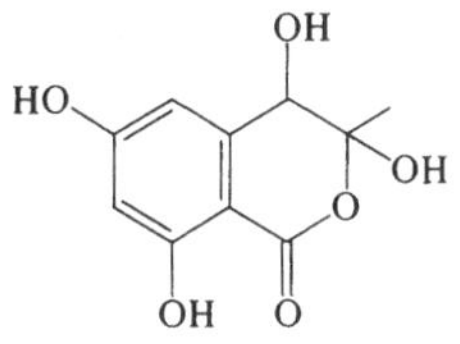

(74)

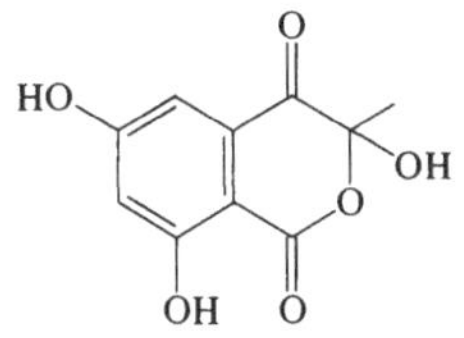

(75)

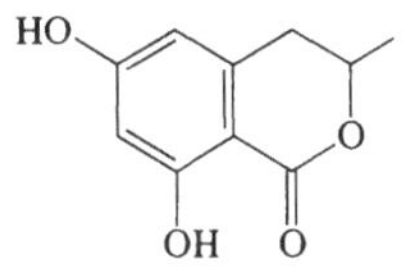

(51)

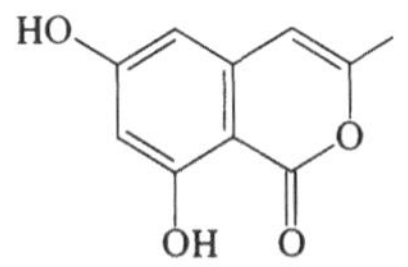

(52)

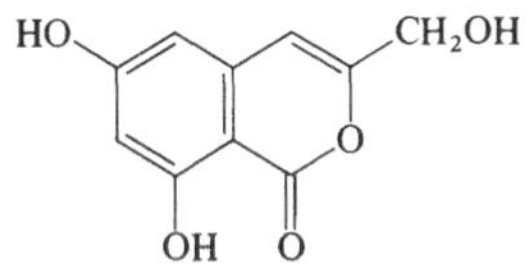

(57)

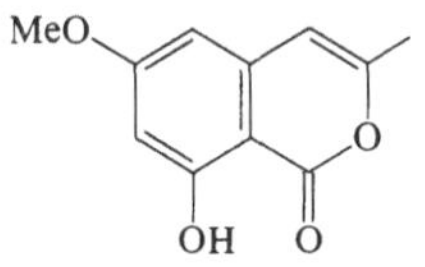

(54)

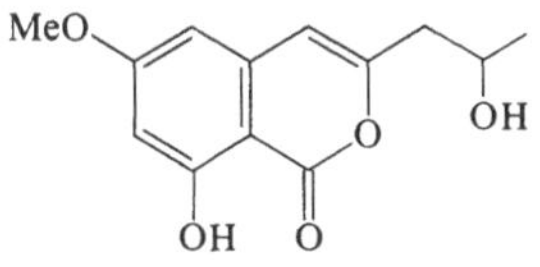

(78)

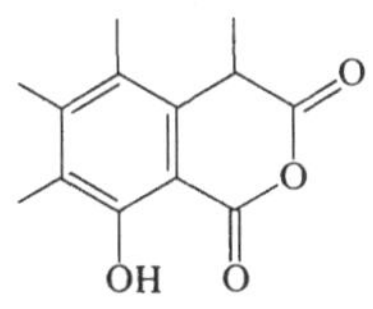

(17)

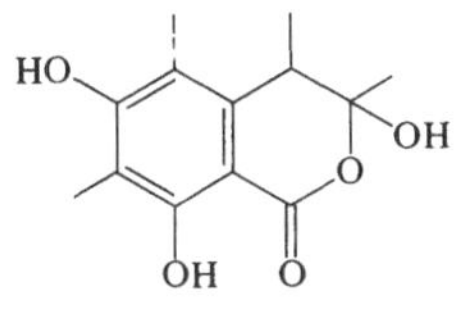

(72)

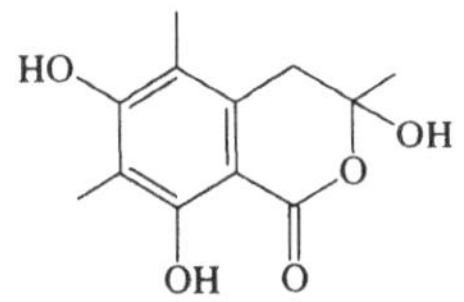

(71)

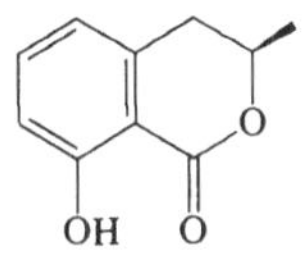

(19)

Mellein (**19**) causes inhibition of growth of corn seedlings (*90*) and has been found recently in several insects. The defensive secretion of termites (*41*) and Australian onerine ants (*52*), the mandibular gland secretion of carpenter ants (*48*) and the male hair pencils of the oriental fruit moth (*19*) all contain mellein (**19**). There is no evidence that mellein acts as a sex attractant; the insect irritant or insecticidal properties of mellein are implicated (*121*). 8-Hydroxyisocoumarin (**14**) and its dihydro derivative (**13**) are found in the defensive secretion of the tembrionid beetle, *Aspena pubescens* (*175*).

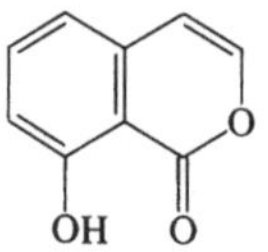

(**14**) 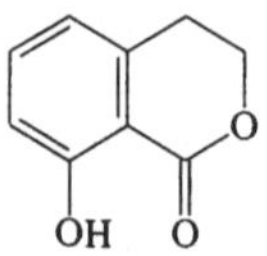 (**13**)

Some isocoumarins are toxic to man. Ochratoxin A (**158**) and ochratoxin B (**160**) are nephratoxic and hepatotoxic metabolites of several *Aspergillus* and *Penicillium* species (*273, 301*). Ochratoxin A (**158**) inhibits protein synthesis (*136*). Oosponol (**16**) inhibits dopamine β-hydroxylase and causes severe skin rash, bronchitis and pneumonia (*297*) and reticulol (**98**) inhibits cyclic AMPase (*103*). However, some isocoumarins have beneficial activities. Some show diuretic and antihypertensive activities (*144*) and some have been used in the treatment of lymphodema (*230*). The antitumour activity of duclauxin (**127**) has been demonstrated (*170*). Analogous of bactobolin A (**147**) have antileukemic activity (*196*) whereas bactobolin A (**147**) and its related metabolites are active against bacteria and viruses (*195*). Baciphelacin (*157*) is also an antibiotic and antiviral metabolite (*223*) and actinobolin (**145**) has antibiotic activity (*8*). The amicoumacins and AI-77s (**151**) to (**156**), besides being antibacterial, are gastroprotective substances against stress ulcers (*150, 259*). Phyllodulcin (**41**) is the sweet priciple found in hydrangea leaves (*91, 318, 319*) that also has antifungal properties (*219*).

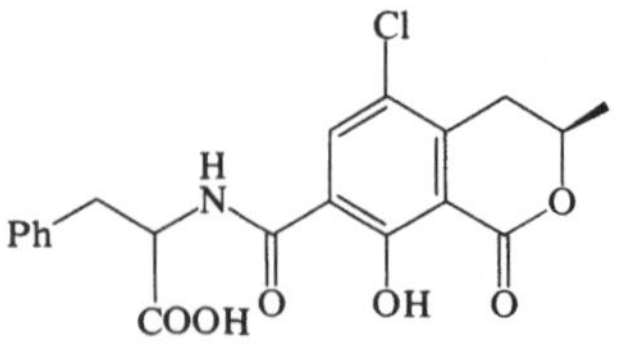

(**158**) 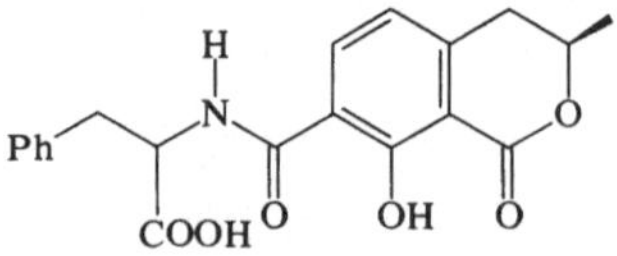 (**160**)

References, 63–78

VII. Introduction to the Tables

The naturally occurring isocoumarins are listed numerically in Tables 1 to 6. Their order is outlined in Sections I and II and is based on the occurrence of oxygen substituents at carbons 6 and 8, then upon the number of carbons in the substituents at C-3 and upon the oxidation level of the compound.

In each case the trivial name or names are given where appropriate. More than one melting point is given when the literature values vary greatly, otherwise the highest melting point is quoted. The melting points of some nitrogen compounds are those of the hydrochloride salts. This is indicated in these cases. The $[\alpha]_D^t$ (solvent) column refers to the specific rotation at t °C in the given solvent system at a wavelength of 589 nm. The reference column lists the sources of the isocoumarins and key references to biosynthetic, synthetic, crystal structure and carbon magnetic resonance studies.

Table 1. *Isocoumarins with no 8-Oxygenation*

Trivial Name(s)	Structure	Formula	M.p.	$[\alpha]_D^t$ (Solvent)	Source(s)	References
(1) Artemidinal		$C_{10}H_6O_3$	175–176		*Artemisia dracunculus* Synthesis	(*182*) (*31, 65*)
(2)		$C_{12}H_{12}O_3$	oil		*Felicia wrightii* Synthesis	(*44*) (*23, 198*)
(3) *trans*-Artemidin		$C_{13}H_{12}O_2$	49–50		*Anthemis fuscata* *Artemisia dracunculus* Synthesis	(*43*) (*112, 114, 181, 185*) (*23, 31, 60, 65*)
(3a) *cis*-Artemidin		$C_{13}H_{12}O_2$			*Artemisia dracunculus*	(*181*)
(4) Capillarin		$C_{13}H_{10}O_2$	120		*Anthemis fuscata* *Artemisia dracunculus* *Artemisia lamprocaules* Synthesis	(*43*) (*112*) (*296*) (*42*)

(5)	C$_{13}$H$_{12}$O$_3$	oil	*Anthemis fuscata*	(43)
(6) Artemidiol	C$_{13}$H$_{12}$O$_4$	131.5–133	*Artemisia dracunculus*	(112, 183)
(7) Artemidinol	C$_{13}$H$_{12}$O$_3$	180–181	*Artemisia dracunculus*	(112, 114, 184)
(8) Norbergenin	C$_{13}$H$_{14}$O$_9$	275–277 dec.	*Woodfordia fruticosa*	(158)

Table 1 (continued)

Trivial Name(s)	Structure	Formula	M.p.	$[\alpha]_D^t$(Solvent)	Source(s)	References
(9) Bergenin Ardisic acid B Corylopsin Peltaphorin Vakerin		$C_{14}H_{16}O_9$	238 220–226 140 (H_2O) 118	−45.3 (H_2O) −41.6 (MeOH) −37.7^{18} (EtOH)	Ardisia hortorum Astilbe macroflora Astilbe thunbergi Bergenia crassifolia Caesalpina digyna Corylopsis spp. Humiria balsamifera Mallotus japonicus Peltophorum interme Saxifragaceae Shorea leprosula Biosynthesis	(140) (145) (257) (102, 132, 232) (67) (129, 231) (87) (142) (153) (286, 287, 313) (55) (286)
(10)		$C_{15}H_{10}O_2$	89–90		Homalium laurifolium Synthesis	(58) (157, 236, 290)
(11) Dihydro-homalicine		$C_{21}H_{22}O_8$	200–202	−98.59 ($CHCl_3$)	Homalium zeylanicum Synthesis	(109) (208)
(12) Homalicine		$C_{21}H_{20}O_8$	241–243 dec.		Homalium zeylanicum	(109)

Table 2. Isocoumarins with 8-Oxygenation

Trivial Name(s)	Structure	Formula	M.p.	$[\alpha]_D^t$ (Solvent)	Source(s)	References
a) With no Carbon Substituent at C-3						
(13)		$C_9H_8O_3$	56–57		Apsena pubescens	(175)
(14)		$C_9H_6O_3$	123–124		Apsena pubescens	(175)
(15) Oospoglycol K-1		$C_{11}H_{10}O_5$	116	−40 (EtOH)	Oospora astringens Synthesis	(215, 295, 320) (295)
(16) Oosponol Lenzitin		$C_{11}H_8O_5$	176		Gleophyllum sepiarium Lenzites subferuginea Lenzites thermophylla Lenzites trabea Oospora astringens Biosynthesis Synthesis	(200) (294) (294) (188) (214, 315) (214) (213, 260, 295)

Table 2 (continued)

Trivial Name(s)	Structure	Formula	M.p.	$[\alpha]_D^1$ (Solvent)	Source(s)	References
(17) Sclerin		$C_{13}H_{14}O_4$	123	+7.85 (CHCl$_3$)	Aspergillus carneus	(69)
					Aspergillus cervinus	(210)
					Sclerotinia libertiana	(169, 291, 248)
					Sclerotinia sclerotiorum	(246)
					Biosynthesis	(21, 106, 292)

b) *With a One-Carbon Substituent at C-3*

Trivial Name(s)	Structure	Formula	M.p.	$[\alpha]_D^1$ (Solvent)	Source(s)	References
(18) Ramulosin		$C_{10}H_{14}O_3$	120–121	+18^{28} (EtOH)	Hypoxylon spp.	(7)
					Pestalotia ramulosa	(270)
					Truncatella hartigii	(294)
					Synthesis	(79a, 100)
(19) (−)-Mellein Ochracin		$C_{10}H_{10}O_3$	58	−108^{12} (CHCl$_3$)	Aspergillus melleus	(212)
					Aspergillus ochraceus	(193)
					Aspergillus oniki	(162)
					Camponotus spp.	(48)
					Cornitermes spp.	(41)
					Grapholithia molesta	(19, 211)
					Hypoxylon spp.	(7)
					Lasiodiplodia theobromae	(5)
					Marasmiellus ramealis	(152)
					Pestalotia ramulosa	(285)
					Rhytidoponera metallica	(52)
					Septoria nodorum	(89, 90)
					Biosynthesis	(1, 140)

Naturally Occurring Isocoumarins

Compound	Formula	mp	[α]	Source	Ref.
(20) (+)-Mellein				Synthesis	(10, 25, 63, 122, 126, 186, 206, 207, 235, 261)
				Biosynthesis	(1, 140)
				Abs. config.	(12, 14)
	$C_{10}H_{10}O_3$	58	$+88^{25}$ (MeOH) $+102^{25}$ (CHCl$_3$)	*Cercospora taiwanensis*	(17, 53)
				Fusarium larvarum	(120)
				Grignardia laricina	(247)
				Gyrostroma missouriense	(199)
				Unidentified fungus	(226)
(21)	$C_{10}H_8O_3$	99–100		*Marasmiellus ramealis*	(152)
				Synthesis	(25)
(22) 8-O-Methylmellein	$C_{11}H_{12}O_3$	82–84	-218^{20} (CHCl$_3$)	*Seporium nodorum*	(89)
(23) Oospolactone	$C_{11}H_{10}O_3$	129		*Gloephyllum sepiarium*	(201)
				Lenzites thermophila	(171)
				Oospora astringens	(214)
				Biosynthesis	(214)
				Synthesis	(66, 187)

Table 2 (continued)

Trivial Name(s)	Structure	Formula	M.p.	$[\alpha]_D^t$ (Solvent)	Source(s)	References
(24) 5-Methylmellein		$C_{11}H_{12}O_3$	131–133	-115^{20} (CHCl$_3$)	*Aspergillus caespitosus*	(252)
					Ceratocystis fimbriata	(271)
					Fusicoccum amygdali	(20)
					Hypoxylon spp.	(7)
					Semicarpus spp.	(54)
					Swartzia laevicarpa	(49)
					Tovomita brasiliensis	(50)
					Wood infected by fungi	(86)
					Synthesis	(19, 64, 59)
(25)	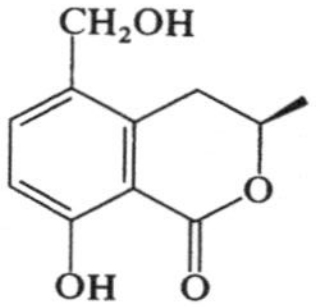	$C_{11}H_{12}O_4$	163–164		*Hypoxylon mammatum*	(7)
(26) 5-Formylmellein		$C_{11}H_{10}O_4$	127	-180^{20} (CHCl$_3$)	*Numularia* spp.	(7)
					Wood infected by fungi	(86)
(27) 5-Carboxymellein		$C_{11}H_{10}O_5$	247–249		*Numularia discreta*	(7)
					Hypoxylon spp.	(7)
					Virola venosa infected by fungi	(86)

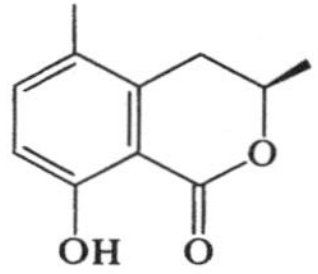

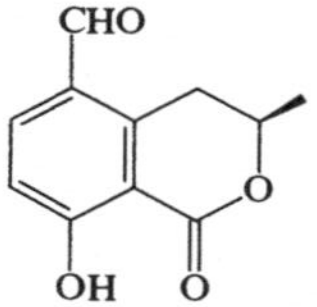

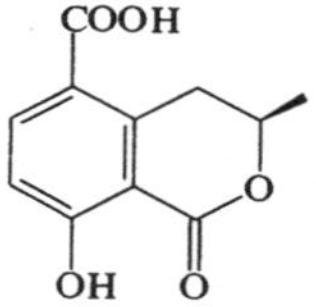

References, 63–78

(28)	$C_{12}H_{12}O_5$	65–66	-163^{22} (CHCl$_3$)	*Hypoxylon mammatum*	(7)
(29) 7-Methylmellein	$C_{11}H_{12}O_3$	100–102	-78^{21} (CHCl$_3$)	*Endlicheria servicea* infected by fungi	(86)
(30) 7-Carboxymellein	$C_{11}H_{10}O_5$	oil		*Penicillium viridicatum*	(147)
(31) 3-Hydroxymellein	$C_{10}H_{10}O_4$	109–109.5	0^{25} (CHCl$_3$)	*Aspergillus oniki*	(162)
(32) 4-Hydroxymellein	$C_{10}H_{10}O_4$	118–119	$+37.4^{20}$ (MeOH)	*Cercospora taiwanensis*	(17, 53)

Table 2 (continued)

Trivial Name(s)	Structure	Formula	M.p.	$[\alpha]_D^t$ (Solvent)	Source(s)	References
(33) 4-Hydroxymellein		$C_{10}H_{10}O_4$	112–117	-41^{20} (CHCl$_3$)	*Lasiodiplodia theobromae* *Septoria nodorum*	(5) (89)
(34) 4-Hydroxymellein		$C_{10}H_{10}O_4$			*Apiosporospora camptospora* 4-Hydroxymellein with unspecified stereochem. *Aspergillus ochraceus* *Aspergillus oniki* *Maringa oleifera*	(5) (74, 193) (162) (241)
(35) 5-Hydroxymellein		$C_{10}H_{10}O_4$	198–200		Wood infected by fungi Synthesis	(86) (126, 127)
(36) 5-Methoxymellein		$C_{11}H_{12}O_4$	71–72	-116^{21} (CHCl$_3$)	Wood infected by fungi	(86)

(37) 7-Hydroxymellein	$C_{10}H_{10}O_4$	100–101	-97° (CHCl$_3$)	*Septoria nodorum*	(89)

c) *With a Substituent at C-3 Containing More than One Carbon*

(38) 8-Hydroxyartemidin	$C_{13}H_{12}O_3$	oil		*Artemisia dracunculus*	(112, 114)
(39) 8-Hydroxy-capillarin	$C_{13}H_{10}O_3$	158		*Artemisia dracunculus* *Artemisia glauca*	(112, 113) (112)
(40) Hydrangenol	$C_{15}H_{12}O_4$	181		*Hydrangea macrophylla* var. *thunbergii* Synthesis Biosynthesis Crystal structure	(161) (26, 203, 218, 305, 306) (33, 34, 148) (249)
(41) Phyllodulcin	$C_{16}H_{14}O_5$	120	$+67.7$	*Hydrangea macrophylla* var. *thunbergii* Absolute configeration Biosynthesis Synthesis	(161) (12) (312) (16, 207, 279, 280, 281, 282, 305)

Table 2 (continued)

Trivial Name(s)	Structure	Formula	M.p.	$[\alpha]_D^t$ (Solvent)	Source(s)	References
(42)		$C_{17}H_{16}O_5$	115–116.5		*Hydrangea macrophylla* var. *thunbergii*	(278)
(43)		$C_{22}H_{24}O_{10}$	115–117	-60^{20} (EtOH)	*Hydrangea macrophylla* var. *thunbergii*	(277)
(44)		$C_{20}H_{30}O_3$	89–90	$+37.7^{21}$ (CHCl$_3$)	*Caulocystis cephalornthes*	(163)

Table 3. *Isocoumarins with 6,8-Dioxygenation*

Trivial Name(s)	Structure	Formula	M.p.	$[\alpha]_D^t$ (Solvent)	Source(s)	References
a) With no Carbon Substituent at C-3						
(**45**) Stellatin		$C_{11}H_{12}O_5$	126–128		*Aspergillus variecolor*	(263)
(**46**)		$C_{12}H_{12}O_5$			*Aspergillus viridinutans*	(294)
(**47**)		$C_{12}H_{10}O_5$	178–179		*Aspergillus viridinutans*	(4)
(**48**)		$C_{12}H_{12}O_6$			*Aspergillus viridinutans*	(294)

Table 3 (continued)

Trivial Name(s)	Structure	Formula	M.p.	$[\alpha]_D^t$ (Solvent)	Source(s)	References
(49)		$C_{13}H_{14}O_6$			*Aspergillus viridinutans*	(294)
b) *With a One-Carbon Substituent at C-3*						
(50) 6-Hydroxy-ramulosin		$C_{10}H_{14}O_4$	132–133	$+91.6^{25}$ (MeOH)	*Pestalotia ramulosa*	(285)
(51) 6-Hydroxy-mellein		$C_{10}H_{10}O_4$	214–215	-64^{20} (MeOH)	*Aspergillus terreus* *Daucus carota* *Gilmaniella humicola* *Pyricularia oryzae* Synthesis	(11, 83) (80) (68) (151) (133, 138, 264)
(52)		$C_{10}H_8O_4$	240–250 dec.		*Ceratocystis minor* *Delphinium* spp. Synthesis	(135, 178) (15) (82, 192)

Compound	Formula	M.p.	[α]	Source	Ref.
(53) 6-Methoxy-mellein	$C_{11}H_{12}O_4$	76–77	−60.71 (CHCl$_3$) −52.7^{27} (MeOH)	*Aspergillus caespitosus*	(252, 294)
				Aspergillus variecolor	(92)
				Carrots	(78, 80, 85, 124, 194)
				Ceratocystis fimbriata	(81)
				Kigelia pinnata	(111)
				Penicillium thomii	(294)
				Sporormia bipartis	(18)
				Sporormia affinis	(177)
				Biosynthesis	(244, 272)
				Synthesis	(28, 133, 217)
(54)	$C_{11}H_{10}O_4$	129		*Ceratocystis fimbriata*	(271)
				Streptomyces mobaraensis	(94)
(55)	$C_{16}H_{18}O_9$	238	−56^{20} (pyridine)	*Delphinium* spp.	(15)
(56)	$C_{17}H_{22}O_9$			Carrots	(194)

Structures:

(53) 6-Methoxymellein: 3-methyl-3,4-dihydroisocoumarin with MeO at 6-position and OH at 8-position

(54): 3-methylisocoumarin with MeO at 6-position and OH at 8-position

(55): 3-methylisocoumarin with OH at 6-position and OGlc at 8-position

(56): 3-methyl-3,4-dihydroisocoumarin with MeO at 6-position and OGlc at 8-position

Table 3 (continued)

Trivial Name(s)	Structure	Formula	M.p.	$[\alpha]_D^t$ (Solvent)	Source(s)	References
(57)	HO, CH₂OH, OH (isocoumarin)	$C_{10}H_8O_5$	220–225 dec.		*Ceratocystis minor*	(135, 178)
(58)	MeO, COOH, OH (isocoumarin)	$C_{11}H_8O_6$	>300 dec.		*Aspergillus ochraceus*	(322)
(59)	Cl, MeO, OH (dihydroisocoumarin with methyl)	$C_{11}H_{11}ClO_4$	123–124	-68^{25} (MeOH)	*Periconia macrospinosa* Biosynthesis Synthesis	(108) (134, 141) (61, 133, 217)
(60)	Cl, MeO, OH (isocoumarin with methyl)	$C_{11}H_9ClO_4$	155–157		*Swartzia laevicarpa* Synthesis	(49) (133)

(61)	$C_{11}H_{11}ClO_4$	170–171	-71.3^{25} (MeOH)	*Periconia macrospinosa* *Sporormia affinis* Synthesis	*(294)* *(177)* *(133)*
(62)	$C_{11}H_9ClO_4$	163–165		*Swartzia laevicarpa* *Tovimita brasiliensis* Synthesis	*(49)* *(50)* *(133)*
(63)	$C_{11}H_{10}Cl_2O_4$	225–226	-142^{25} (MeOH)	*Sporormia affinis* *Periconia macrospinosa* Synthesis	*(177)* *(294)* *(133)*
(64)	$C_{12}H_{14}O_4$	121	-99^{21} (CHCl$_3$)	*Hypoxylon atropunctatum* Wood infected by fungi	*(7)* *(86)*
(65)	$C_{12}H_{14}O_4$	216–218		*Penicillium citrinum*	*(84)*

Table 3 (continued)

Trivial Name(s)	Formula	M.p.	$[\alpha]_D^t$ (Solvent)	Source(s)	References
(66) Dihydrocitrone	$C_{13}H_{14}O_6$	137–138	$+104^{15}$ ($CHCl_3$)	Aspergillus carneus Aspergillus terreus Penicillium citrinum	(69) (142) (84)
(67) Canescin A	$C_{15}H_{14}O_7$	Mixture of A and B 200–202	$+17.8^{23}$ (Me_2CO)	Aspergillus malignus Penicillium canescens Biosynthesis	(35) (51) (36, 37)
(68) Canescin B	$C_{15}H_{14}O_7$			As for Canescin A	
(69) Leprocybin	$C_{24}H_{20}O_{13}$	240	-60^{20} (H_2O)	Cortinarius cotoneus	(167)

(**70**)	$C_{10}H_{10}O_5$	157–159		*Alternaria kikuchiana* (160)
				Ceratocystis ulmi (71, 72)
				Penicillium brevi-
				compactum (73, 275)
				Synthesis (119, 191)
(**71**) Sclerotinin B	$C_{12}H_{14}O_5$	192–195 dec.	0 (MeOH)	*Sclerotinia sclero-*
				tiorum (246)
				Synthesis (245)
(**72**) Sclerotinin A	$C_{13}H_{16}O_5$	205–208	$+36^{25}$ (MeOH)	*Penicillium citrinum* (84)
				Sclerotinia sclero-
				tiorum (246)
				Synthesis (283)
(**73**)	$C_{10}H_{10}O_5$	183–185		*Cercospora scirpicola* (17)
				Ceratocystis ulmi (135)
(**74**)	$C_{10}H_{10}O_6$	195 dec.	0	*Ceratocystis ulmi* (71, 72)
				Penicillium brevi-
				compactum (73, 275)
				Synthesis (119)

Table 3 (continued)

Trivial Name(s)	Structure	Formula	M.p.	$[\alpha]_D^t$ (Solvent)	Source(s)	References
(75)		$C_{10}H_8O_6$	188–190		*Ceratocystis ulmi* *Penicillium brevi-compactum* Synthesis	(71, 72) (73, 275) (119)
(76) Ustic Acid		$C_{11}H_{12}O_7$	169–170		*Aspergillus ustus* *Paecilomyces victoriae*	(233) (303)
(77) Dehydroustic Acid		$C_{11}H_{10}O_7$			*Aspergillus ustus* *Paecilomyces victoriae*	(233) (303)

c) *With a Substituent at C-3 Containing More than One Carbon*

(78) Diaporthin		$C_{13}H_{14}O_5$	90–92	+54 ($CHCl_3$)	*Endothia parasitica*	(123)

#	Compound	Formula	mp (°C)	Source	Ref
(79)	LL-Z1640-7	$C_{13}H_{10}O_6$	192–194	Lederle Culture Z1640	(96)
(80)	LL-Z1640-5	$C_{13}H_{13}ClO_6$	158–160	Lederle Culture Z1640	(96)
(81)	Lunatinin	$C_{13}H_{14}O_5$		Cochliobolus lunata	(220)
(82)	β-Alectoronic Acid	$C_{28}H_{22}O_9$	138–139	Parmelia birulae	(168)

Table 3 (continued)

Trivial Name(s)	Structure	Formula	M.p.	$[\alpha]_D^1$ (Solvent)	Source(s)	References
(83) β-Collatolic Acid		$C_{29}H_{24}O_9$	117		*Asahinea scholanderei*	(168)
(84) LL-Z1640-6		$C_{15}H_{16}O_6$	95–98	+4.4 (CHCl$_3$)	Lederle Culture Z1640	(96)
(85) Fusamarin		$C_{18}H_{20}O_4$	159	-11.39^{30} (MeOH)	*Fusarium* sp. Synthesis	(276) (2)
(86) Asperentin Cladosporin		$C_{16}H_{20}O_5$	187		*Aspergillus flavus* *Aspergillus proliferans* *Aspergillus repens* *Aspergillus* sp.	(117) (294) (266) (97)

(87)	[structure]	$C_{16}H_{20}O_5$	150–155		*Cladosporium cladosporioides* (253) Biosynthesis (57) CMR (57) *Aspergillus flavus* (117)
(88) 6-*O*-Methylasperentin	[structure]	$C_{17}H_{22}O_5$	98–98.5		*Aspergillus flavus* (117)
(89) 8-*O*-Methylasperentin	[structure]	$C_{17}H_{22}O_5$	225 and 235	$+72^{20}$ (CHCl$_3$)	*Aspergillus flavus* (117) *Aspergillus* sp. (294)
(90) 4'-Hydroxyasperentin	[structure]	$C_{16}H_{20}O_6$	195		*Aspergillus flavus* (118)
(91) 5'-Hydroxyasperentin	[structure]	$C_{16}H_{20}O_6$	229–230	-30^{20} (MeOH)	*Aspergillus flavus* (117) *Aspergillus repens* (294) *Penicillium* sp. (294)

Table 3 (continued)

Trivial Name(s)	Structure	Formula	M.p.	$[\alpha]_D^t$ (Solvent)	Source(s)	References
(92) 5'-Hydroxy-asperentin 8-O-Methyl ether		$C_{17}H_{22}O_6$	246-248 dec.		*Aspergillus flavus*	(118)
(93) Agrimolide		$C_{18}H_{18}O_5$	175.5-176.5	+8.1^{18} (Me$_2$CO)	*Agrimonia pilosa* Absolute configuration Synthesis	(13, 317) (12, 13, 14) (317)
(94)		$C_{21}H_{32}O_4$	98	−23.2 (CHCl$_3$)	*Ononis natrix*	(242)
(95) Peniolactol		$C_{24}H_{38}O_5$	150 dec.		Wood infected by *Peniophora sanguinea*	(115)

Table 4. *Isocoumarins with 6,7,8-Trioxygenation*

Trivial Name(s)	Structure	Formula	M.p.	$[\alpha]_D^t$ (Solvent)	Source(s)	References
a) With a One-Carbon Substituent at C-3						
(**96**)		$C_{10}H_8O_5$	234		*Streptomyces mobaraensis*	(94)
(**97**) 6-Demethyl-kigelin		$C_{11}H_{12}O_5$	148	−80.78 (CHCl$_3$)	*Kigelia pinnata*	(111)
(**98**) Reticulol		$C_{11}H_{10}O_5$	194–195		*Streptomyces mobaraensis* *Streptomyces rubrireticulae* CMR Biosynthesis Synthesis	(94, 103, 172) (189) (104) (104) (59, 64)
(**99**) Kigelin		$C_{12}H_{14}O_5$	144	−79.91 (CHCl$_3$)	*Kigelia pinnata* Synthesis	(111, 154) (29, 30, 110, 205, 261)

Table 4 (continued)

Trivial Name(s)	Structure	Formula	M.p.	$[\alpha]_D^t$ (Solvent)	Source(s)	References
(100)	MeO, MeO, OH	$C_{12}H_{12}O_5$	199		Streptomyces mobaraensis Synthesis	(94, 103, 172) (173)
(101)	HO, MeO, CH$_2$OH, OH	$C_{11}H_{10}O_6$	186–188		Streptomyces mobaraensis	(103)
(102)	MeO, MeO, CH$_2$OH, OH	$C_{12}H_{12}O_6$	158.5–160		Streptomyces mobaraensis	(103)
(103)	MeO, MeO, OH	$C_{13}H_{16}O_5$	oil		Wood infected by fungi	(86)
(104)	MeO, MeO, OH, OH	$C_{12}H_{14}O_6$	190–192		Aspergillus terreus	(11)

b) With a Substituent at C-3 Containing More than One Carbon

(105) Monocerolide		$C_{13}H_{12}O_7$	196–199		*Helminthosporium monoceras* (6)
(106) 6-*O*-Methylfusarentin		$C_{15}H_{20}O_6$	137	-30^{20} (MeOH)	*Fusarium larvarum* (120)
(107) 6,7-Dimethyl ether of fusarentin		$C_{16}H_{22}O_6$	99	-29^{20} (MeOH)	*Fusarium larvarum* (120)
(108) 6,8-Dimethyl ether of fusarentin		$C_{16}H_{22}O_6$	Gum		*Fusarium larvarum* (120)

Table 4 (continued)

Trivial Name(s)	Structure	Formula	M.p.	$[\alpha]_D^1$ (Solvent)	Source(s)	References
(109) Monocerin		$C_{16}H_{20}O_6$	64–66	$+53^{24}$ (MeOH)	*Drechslera ravenelii* *Fusarium larvarum* *Helminthosporum monoceras* Biosynthesis	(252a) (120) (6) (252a)
(110) 7-O-Demethyl-monocerin		$C_{15}H_{18}O_6$	172–175		*Fusarium larvarum*	(120)
(111) Monocerone		$C_{16}H_{18}O_7$	112–113		*Helminthosporium monoceras*	(6)
(112) 9-Hydroxy-monocerin		$C_{16}H_{20}O_7$	103–104		*Helminthosporium monoceras*	(6)

References, 63–78

Table 5. *Isocoumarins with Fused Carbocyclic Rings*

Trivial Name(s)	Structure	Formula	M.p.	$[\alpha]_D^t$ (Solvent)	Source(s)	References
a) 3,4-Fused						
(113) Brevifolin-carboxylic Acid		$C_{13}H_8O_8$	>250 dec.		*Algarobilla* spp. Synthesis	(250, 251) (27, 116, 131, 250)
(114) Altenuisol		$C_{14}H_{10}O_6$	277–282		*Alternaria tenuis*	(228)
(115) Alternariol		$C_{14}H_{10}O_5$	350 dec.		*Alternaria dauci* *Alternaria tenuis* Biosynthesis	(101) (234) (1, 107, 288)

Table 5 (*continued*)

Trivial Name(s)	Structure	Formula	M.p.	$[\alpha]_D^t$ (Solvent)	Source(s)	References
(**116**) Alternariol methyl ether		$C_{15}H_{12}O_5$	266–267 330–340 dec.		*Alternaria dauci* *Alternaria tenuis*	(*101*) (*234*)
(**117**) Altenuene		$C_{15}H_{16}O_6$	190–191		*Alternaria tenuis*	(*179, 229*)
(**118**) Dehydro-altenusin		$C_{15}H_{12}O_6$	190–193		*Alternaria tenuis* Crystal structure	(*79, 238, 239*) (*238*)

Naturally Occurring Isocoumarins

(119) Botrallin	$C_{16}H_{14}O_7$	165–185 dec.	*Botrytis allii*	(159, 224)
b) 4,5-Fused				
(120) Lamellicolic anhydride	$C_{13}H_8O_6$	>300 dec.	*Cephalosporium* sp. *Verticillium lamellicola*	(294) (176a)
(121) 4-O-carbomethoxylamellicolic anhydride	$C_{15}H_{10}O_8$	202–204	*Verticillium lamellicola* Biosynthesis	(176a) (176, 286)
(122) 3-Chlorolamellicolic anhydride	$C_{13}H_7ClO_6$	>250	*Scolecobasiella avellanea* Synthesis	(294) (176a)

Table 5 (continued)

Trivial Name(s)	Structure	Formula	M.p.	$[\alpha]_D$ (Solvent)	Source(s)	References
(123) "Naphthalic anhydride"		$C_{18}H_{18}O_6$	259–261		Fusicoccum putrefaciens Penicillium herquei Roesleria pallida Crystal structure	(240) (204) (299) (143)
(124) Gilmaniellin		$C_{26}H_{19}ClO_{10}$	> 350	-132^{20} (CHCl$_3$/MeOH)	Gilmaniella humicola Crystal structure	(68) (68)

(125) Dechlorogilmaniellin	$C_{26}H_{20}O_{10}$	>330 dec.	-130^{20} (CHCl$_3$/MeOH) -128 (Me$_2$CO)	*Gilmaniella humicola* (68)
(126) Xenoclauxin	$C_{28}H_{18}O_{11}$	>300 dec.	$+310^{26}$ (THF)	*Penicillium duclauxi* (222)

Table 5 (continued)

Trivial Name(s)	Formula	M.p.	$[\alpha]_D^t$ (Solvent)	Source(s)	References
(127) Duclauxin	$C_{29}H_{22}O_{11}$	230 dec.	+272.5³⁰ (CHCl$_3$)	*Penicillium duclauxi* *Penicillium stipitatum* Biosynthesis Crystal structure	(256) (170) (243) (221)
(128) Cryptoclauxin	$C_{29}H_{22}O_{12}$	>300		*Penicillium duclauxi*	(222)

c) 6,7-Fused

(129) Semi-vioxanthin	$C_{15}H_{14}O_5$	185	*Penicillium citreo-viride* (324)
(130) Xanthoviridicatin D	$C_{26}H_{20}O_9$	133–136	*Penicillium viridicatum* (268)
(131) Xanthoviridicatin G	$C_{25}H_{16}O_8$	318–320	*Penicillium viridicatum* (268)
(132) Rubrosulphin	$C_{29}H_{20}O_{10}$	280	*Aspergillus sulphureus* (93) *Penicillium viridicatum* (267) CMR (139)

Table 5 (continued)

Trivial Name(s)	Structure	Formula	M.p.	$[\alpha]_D^t$ (Solvent)	Source(s)	References
(133) Viopurpurin		$C_{29}H_{20}O_{11}$	310 dec.		Aspergillus melleus Aspergillus sulphureus Trichophyton violaceum CMR	(93) (93) (43, 209a) (139)
(134) Vioxanthin		$C_{30}H_{26}O_{10}$	190–195 dec.		Trichophyton violaceum	(43, 209a)
(135) Viomellein		$C_{30}H_{24}O_{11}$	275 dec.		Aspergillus melleus Aspergillus ochraceus Aspergillus sulphureus Nannizzia cajetani Penicillium citreo- viride Penicillium cyclopium Penicillium viridicatum Biosynthesis CMR	(93) (296) (93) (255) (324) (269) (267, 269) (262) (139, 262)

Naturally Occurring Isocoumarins

(136) 3,4-Dehydroviomellein $C_{30}H_{22}O_{11}$ >150 dec.

Nannizzia cajetani	(255)

(137) Luteosporin $C_{28}H_{18}O_{12}$

Microsporium cookei	(3)

(138) Xanthomegnin $C_{30}H_{22}O_{12}$ 264 dec. 340 dec. -156^{22} (CHCl$_3$)

Aspergillus melleus	(93)
Aspergillus ochraceus	(269)
Aspergillus sulphureus	(93)
Microsporium cookei	(3)
Nannizzia cajetani	(255)
Penicillium citreo-viride	(324)
Penicillium cyclopium	(269)
Penicillium viridicatum	(267, 269)
Trichophyton rubrum	(311)
Trichophyton violaceum	(43, 209 a)
Trychophyton megnini	(156)
Biosynthesis	(262)
CMR	(139, 262)

Table 5 (continued)

Trivial Name(s)	Structure	Formula	M.p.	$[\alpha]_D^t$ (Solvent)	Source(s)	References
(139) 3,4-Dehydroxanthomegnin		$C_{30}H_{20}O_{12}$	150 dec.		Nannizzia cajetani Penicillium citreoviride	(255) (324)
(140) 3,4,3',4'-Bisdehydroxanthomegnin		$C_{30}H_{18}O_{12}$			Nannizzia cajetani	(255)
(141) Xylindein		$C_{32}H_{20}O_{10}$	300 dec.		Chlorociboria aeruginosa (Chloroplenium aeruginosum)	(38, 39, 95)

(142) SC-30532	$C_{34}H_{30}O_{12}$	224–227		Spicaria divaricata	(190)
(143) SC-28763	$C_{34}H_{30}O_{13}$	238–242 dec.	-200.5^{28} $(CHCl_3)$	Spicaria divaricata	(155)
(144) Viriditoxin SC-28762	$C_{34}H_{30}O_{14}$	242–245 dec. 263–267 dec.	-202^{28} $(CHCl_3)$ 1210.5^{28} $(CHCl_3)$	Aspergillus viridinutans Spicaria divaricata	(174, 308) (155)

Table 6. *Isocoumarins with Nitrogen-Containing Substituents*

Trivial Name(s)	Structure	Formula	M.p.	$[\alpha]_D^t$ (Solvent)	Source(s)	References
(145) Actinobolin		$C_{13}H_{20}N_2O_6$		$+59^{18}$ (H_2O)	*Streptomyces griseoviridus* var. *atrofaciens* Crystal structure Synthesis	(8, 197, 209) (302, 310) (322a)
(146) Bactobolin C		$C_{14}H_{20}Cl_2N_2O_5$ (HCl)	235–236 dec.	-23.9^{25} (H_2O)	*Pseudomonas yoshitomiensis*	(195)
(147) Bactobolin A		$C_{14}H_{20}Cl_2N_2O_6$ (HCl)	220–222	-6.3^{25} (H_2O)	*Pseudomonas yoshitomiensis*	(195)

Compound	Formula	mp (°C)	[α]	Source	Ref.
(148) Bactobolin B N-L-Alanyl-bactobolin	$C_{16}H_{25}Cl_2N_3O_7$ (HCl)	231.2 dec.	-27^{25} (H_2O)	*Pseudomonas yoshitomiensis*	*(195)*
(149) Hippeastrine	$C_{17}H_{17}NO_5$	214–215	$+160^{22}$ ($CHCl_3$)	*Hyppeastrum vittatum*	*(46, 47, 130)*
(150) Homolycorine Narcipoetine	$C_{18}H_{21}NO_4$	175	$+86^{22}$ (EtOH)	*Leucoium vernum* *Lycoris radiata* *Narcissus poeticus*	*(45)* *(130, 164, 165)* *(45, 166)*
(151) AI-77-F	$C_{20}H_{23}NO_7$	182–183		*Bacillus pumilus*	*(259)*

Table 6 (continued)

Trivial Name(s)	Formula	M.p.	$[\alpha]_D^t$ (Solvent)	Source(s)	References
(152) Amicoumacin B, AI-77-B	$C_{20}H_{28}N_2O_8$	139.5–140 dec.	-106.1^{23} (MeOH)	*Bacillus pumilus*	*(149, 150, 259)*
(153) Amicoumacin C	$C_{20}H_{26}N_2O_7$	131–133 dec.	-81.6^{23} (MeOH)	*Bacillus pumilus*	*(149, 150)*
(154) AI-77-C	$C_{22}H_{28}N_2O_8$	210–211 dec.		*Bacillus pumilus*	*(259)*

(155) AI-77-D $C_{23}H_{30}N_2O_8$ *Bacillus pumilus*   (259)

(156) Amicoumacin A $C_{20}H_{29}N_3O_7$ (HCl)   132–135 dec. -97.2^{25} (MeOH)   *Bacillus pumilus*   (149, 150)

(157) Baciphelacin $C_{22}H_{34}N_2O_6$ *Bacillus thiaminolyticus*   (223)

Table 6 (continued)

Trivial Name(s)	Structure	Formula	M.p.	$[\alpha]_D^t$ (Solvent)	Source(s)	References
(158) Ochratoxin A		$C_{20}H_{18}ClNO_6$	169	−118 (CHCl$_3$)	Aspergillus ochraceus Penicillium cyclopium Penicillium viridicatum Biosynthesis Synthesis CMR	(216, 298) (216) (125, 147, 216, 254, 300) (88, 99, 180, 275, 307, 309, 321) (237, 274) (88, 309)
(159) 4-Hydroxyochratoxin A		$C_{20}H_{18}ClNO_7$	216–218		Penicillium viridicatin	(147)
(160) Ochratoxin B		$C_{20}H_{19}NO_6$	221	−35 (EtOH)	Aspergillus ochraceus Penicillium viridicatum Synthesis	(298) (147) (274)
(161) Ochratoxin C		$C_{22}H_{22}ClNO_6$	Gum	−100 (EtOH)	Aspergillus ochraceus	(298)

Naturally Occurring Isocoumarins

Formula Index

Formula	Compound number	Formula	Compound number
$C_9H_6O_3$	(14)	$C_{13}H_{12}O_7$	(105)
$C_9H_8O_3$	(13)	$C_{13}H_{13}ClO_6$	(80)
		$C_{13}H_{14}O_4$	(17)
$C_{10}H_6O_3$	(1)	$C_{13}H_{14}O_5$	(78), (81)
$C_{10}H_8O_3$	(21)	$C_{13}H_{14}O_6$	(49), (66)
$C_{10}H_8O_4$	(52)	$C_{13}H_{14}O_9$	(8)
$C_{10}H_8O_5$	(57), (96)	$C_{13}H_{16}O_5$	(72), (103)
$C_{10}H_8O_6$	(75)	$C_{13}H_{20}N_2O_6$	(145)
$C_{10}H_{10}O_3$	(19), (20)		
$C_{10}H_{10}O_4$	(31), (32), (33), (34), (35), (37), (51)	$C_{14}H_{10}O_5$	(115)
		$C_{14}H_{10}O_6$	(114)
$C_{10}H_{10}O_5$	(70), (73)	$C_{14}H_{16}O_9$	(9)
$C_{10}H_{10}O_6$	(74)	$C_{14}H_{20}Cl_2N_2O_5$	(146)
$C_{10}H_{14}O_3$	(18)	$C_{14}H_{20}Cl_2N_2O_6$	(147)
$C_{10}H_{14}O_4$	(50)		
		$C_{15}H_{10}O_2$	(10)
$C_{11}H_8O_5$	(16)	$C_{15}H_{10}O_8$	(121)
$C_{11}H_8O_6$	(58)	$C_{15}H_{12}O_4$	(40)
$C_{11}H_9ClO_4$	(60), (61)	$C_{15}H_{12}O_5$	(116)
$C_{11}H_{10}Cl_2O_4$	(63)	$C_{15}H_{12}O_6$	(118)
$C_{11}H_{10}O_3$	(23)	$C_{15}H_{14}O_5$	(129)
$C_{11}H_{10}O_4$	(26), (54)	$C_{15}H_{14}O_7$	(67), (68)
$C_{11}H_{10}O_5$	(15), (27), (30), (98)	$C_{15}H_{16}O_6$	(84), (117)
$C_{11}H_{10}O_6$	(101)	$C_{15}H_{18}O_6$	(110)
$C_{11}H_{10}O_7$	(77)	$C_{15}H_{20}O_6$	(106)
$C_{11}H_{11}ClO_4$	(59), (62)		
$C_{11}H_{12}O_3$	(22), (24), (29)	$C_{16}H_{14}O_5$	(41)
$C_{11}H_{12}O_4$	(25), (36), (53)	$C_{16}H_{14}O_7$	(119)
$C_{11}H_{12}O_5$	(45), (97)	$C_{16}H_{18}O_7$	(111)
$C_{11}H_{12}O_7$	(76)	$C_{16}H_{18}O_9$	(55)
		$C_{16}H_{20}O_5$	(86), (87)
$C_{12}H_{10}O_5$	(47)	$C_{16}H_{20}O_6$	(90), (91), (119)
$C_{12}H_{12}O_2$	(2)	$C_{16}H_{20}O_7$	(112)
$C_{12}H_{12}O_5$	(28), (46), (100)	$C_{16}H_{22}O_6$	(107), (108)
$C_{12}H_{12}O_6$	(48), (102)	$C_{16}H_{25}Cl_2N_3O_7$	(148)
$C_{12}H_{14}O_4$	(64), (65)		
$C_{12}H_{14}O_5$	(71), (99)	$C_{17}H_{16}O_5$	(42)
$C_{12}H_{14}O_6$	(104)	$C_{17}H_{17}NO_5$	(149)
		$C_{17}H_{22}O_5$	(88), (89)
		$C_{17}H_{22}O_6$	(92)
$C_{13}H_7ClO_6$	(122)	$C_{17}H_{22}O_9$	(56)
$C_{13}H_8O_6$	(120)		
$C_{13}H_8O_8$	(113)	$C_{18}H_{18}O_5$	(93)
$C_{13}H_{10}O_2$	(4)	$C_{18}H_{18}O_6$	(123)
$C_{13}H_{10}O_3$	(39)	$C_{18}H_{20}O_4$	(85)
$C_{13}H_{10}O_6$	(79)	$C_{18}H_{21}NO_4$	(150)
$C_{13}H_{12}O_2$	(3), (4)		
$C_{13}H_{12}O_3$	(5), (7), (38)	$C_{20}H_{18}ClNO_6$	(158)
$C_{13}H_{12}O_4$	(6)	$C_{20}H_{18}ClNO_7$	(159)

Formula Index *(continued)*

Formula	Compound number	Formula	Compound number
$C_{20}H_{19}NO_6$	(160)	$C_{26}H_{20}O_{10}$	(125)
$C_{20}H_{23}NO_7$	(151)		
$C_{20}H_{26}N_2O_7$	(153)	$C_{28}H_{18}O_{11}$	(126)
$C_{20}H_{28}N_2O_8$	(152)	$C_{28}H_{18}O_{12}$	(137)
$C_{20}H_{29}N_3O_7$	(156)	$C_{28}H_{22}O_9$	(82)
$C_{20}H_{30}O_3$	(44)		
		$C_{29}H_{20}O_{10}$	(132)
$C_{21}H_{20}O_8$	(12)	$C_{29}H_{20}O_{11}$	(133)
$C_{21}H_{22}O_8$	(11)	$C_{29}H_{22}O_{11}$	(127)
$C_{21}H_{32}O_4$	(94)	$C_{29}H_{22}O_{12}$	(128)
		$C_{29}H_{24}O_9$	(83)
$C_{22}H_{24}O_{10}$	(43)		
$C_{22}H_{28}N_2O_8$	(154)	$C_{30}H_{18}O_{12}$	(140)
$C_{22}H_{34}N_2O_6$	(157)	$C_{30}H_{20}O_{12}$	(139)
		$C_{30}H_{22}O_{11}$	(136)
$C_{23}H_{30}N_2O_8$	(155)	$C_{30}H_{22}O_{12}$	(138)
		$C_{30}H_{24}O_{11}$	(135)
$C_{24}H_{20}O_{13}$	(69)	$C_{30}H_{26}O_{10}$	(134)
$C_{24}H_{38}O_5$	(95)		
		$C_{32}H_{20}O_{10}$	(141)
$C_{25}H_{16}O_8$	(131)		
		$C_{34}H_{30}O_{12}$	(142)
$C_{26}H_{19}ClO_{10}$	(124)	$C_{34}H_{30}O_{13}$	(143)
$C_{26}H_{20}O_9$	(130)	$C_{34}H_{30}O_{14}$	(144)

Trivial Name Index

Name	Compound number	Name	Compound number
Actinobolin	(145)	Artemidinal	(1)
Agrimolide	(93)	Artemidinol	(7)
AI-77-B	(152)	Artemidiol	(6)
AI-77-C	(154)	Asperentin	(86)
AI-77-D	(155)	Baciphelacin	(157)
AI-77-F	(151)	Bactobolin A	(147)
Altenuene	(117)	Bactobolin B	(148)
Altenuisol	(114)	Bactobolin C	(146)
Alternariol	(115)	Bergenin	(9)
β-Alectoronic acid	(82)	Botrallin	(119)
Amicoumacin A	(156)	Brevifolincarboxylic acid	(113)
Amicoumacin B	(152)		
Amicoumacin C	(153)		
Ardisic acid B	(9)	Canescin A	(67)
Artemidin	(3), (4)	Canescin B	(68)

References, 63–78

Trivial Name Index *(continued)*

Name	Compound number	Name	Compound number
Capillarin	(4)	LL-Z1640-5	(80)
4-*O*-Carbomethoxylamellicolic anhydride	(121)	LL-Z1640-6	(84)
		LL-Z1640-7	(79)
5-Carboxymellein	(27)	Lunatinin	(81)
7-Carboxymellein	(30)	Luteosporin	(136)
3-Chlorolamellicolic anhydride	(122)	Mellein	(19), (20)
Cladosporin	(86)	5-Methoxymellein	(36)
β-Collatolic acid	(83)	6-Methoxymellein	(53)
Corylopsin	(9)	6-*O*-Methylasperentin	(88)
Cryptoclauxin	(128)	8-*O*-Methylasperentin	(89)
		6-*O*-Methylfusarentin	(106)
Dechlorogilmaniellin	(125)	5-Methylmellein	(24)
Dehydroaltenusin	(118)	7-Methylmellein	(29)
Dehydroustic acid	(77)	Monocerin	(109)
7-*O*-Demethylmonocerin	(110)	Monocerolide	(105)
Diaporthin	(78)	Monocerone	(111)
Dihydrocitrone	(66)		
Duclauxin	(127)	"Naphthalic anhydride"	(123)
		Narcipoetine	(150)
5-Formylmellein	(26)	Norbergenin	(8)
Fusamarin	(85)		
		Ochracin	(19), (20)
Gilmaniellin	(124)	Ochratoxin A	(158)
		Ochratoxin B	(160)
Hippeastrine	(149)	Ochratoxin C	(161)
Homalicine	(12)	Oospoglycol	(15)
Homolycorine	(150)	Oospolactone	(23)
Hydrangenol	(40)	Oosponol	(16)
8-Hydroxyartemidin	(38)		
4'-Hydroxyasperentin	(90)	Peltaphorin	(9)
5'-Hydroxyasperentin	(91)	Peniolactol	(95)
8-Hydroxycapillarin	(39)	Phyllodulcin	(41)
3-Hydroxymellein	(31)		
4-Hydroxymellein	(32), (33), (34)	Ramulosin	(18)
5-Hydroxymellein	(35)	Reticulol	(98)
6-Hydroxymellein	(51)	Rubrosulphin	(132)
7-Hydroxymellein	(37)		
9-Hydroxymonocerin	(112)	SC-28762	(144)
4-Hydroxyochratoxin	(159)	SC-28763	(143)
6-Hydroxyramulosin	(50)	SC-30532	(142)
		Sclerin	(17)
Kigelin	(99)	Sclerotinin A	(72)
		Sclerotinin B	(71)
Lamellicolic anhydride	(120)	Semi-vioxanthin	(129)
Lenzitin	(16)	Stellatin	(45)
Leprocybin	(69)		
		Ustic acid	(76)

Trivial Name Index *(continued)*

Name	Compound number	Name	Compound number
Vakerin	(9)	Xanthomegnin	(138)
Viomellein	(135)	Xanthoviridicatin D	(130)
Viopurpurin	(133)	Xanthoviridicatin G	(131)
Vioxanthin	(134)	Xenoclauxin	(126)
Viriditoxin	(144)	Xylindein	(141)

Source Index

Source	Compound number
Agrimonia pilosa	(93)
Algarobilla spp.	(113)
Alternaria dauci	(115), (116)
Alternaria kikuchiana	(70)
Alternaria tenuis	(114), (115), (116), (117), (118)
Anthemis fuscata	(3), (4), (5)
Apiosporosa camptospora	(34)
Apsena pubescens	(13), (14)
Ardisia hortorum	(9)
Artemisia dracunculus	(1), (3a), (4), (6), (7), (38), (39)
Artemisia glauca	(39)
Artemisia lamprocaules	(4)
Asahinae scholanderei	(83)
Aspergillus caespitosus	(24), (53)
Aspergillus carneus	(17), (66)
Aspergillus cervinus	(17)
Aspergillus flavus	(86), (87), (88), (89), (90), (91), (92)
Aspergillus malignus	(67), (68)
Aspergillus melleus	(19), (133), (135)
Aspergillus ochraceus	(19), (58), (135), (158), (160), (161)
Aspergillus oniki	(19), (31)
Aspergillus proliferans	(86)
Aspergillus repens	(86), (91)
Aspergillus sulphureus	(132), (133), (135)
Aspergillus terreus	(51), (104), (142)
Aspergillus ustus	(76), (77)
Aspergillus variecolor	(45), (53)
Aspergillus viridinutans	(46), (47), (48), (49), (144)
Astilbe macroflora	(9)
Astilbe thunbergi	(9)
Bacillus pumilus	(151), (152), (153), (154), (155), (156)
Bacillus thiaminolyticus	(157)
Bergenia crassifolia	(9)
Botrytis allii	(119)

References, 63–78

Source Index *(continued)*

Source	Compound number
Caesalpina digyna	(9)
Camponotus spp.	(19)
Caulocystis cephalornthes	(44)
Cephalosporum sp.	(120)
Cercospora scirpicola	(73)
Cercospora taiwanensis	(20), (32)
Ceratocystis fimbriata	(24), (53), (54)
Ceratocystis minor	(52), (57)
Ceratocystis ulmi	(70), (73), (74), (75)
Chlorociboria aeruginosa	(141)
Chloroplenium aeruginosum	(141)
Cladosporium cladosporioides	(86)
Cornitermes spp.	(19)
Corylopsis spp.	(19)
Daucus carota	(51), (53), (56)
Delphinium spp.	(52), (55)
Drechslera ravenelii	(109)
Endlicheria servicea	(29)
Endothia parasitica	(78)
Felicia wrightii	(2)
Fusarium larvarum	(20), (85), (106), (107), (108), (109), (110)
Fusicoccum amygdali	(24)
Fusicoccum putrefaciens	(123)
Gilmaniella humicola	(51), (124), (125)
Gleophyllum separium	(16), (23)
Grapholithia molesta	(19)
Grignardia laricina	(20)
Gyrostroma missouriense	(20)
Helminthosporium monoceras	(105), (109), (111), (112)
Homalium laurifolium	(10)
Homalium zeylanicum	(11), (12)
Humiria balsamifera	(9)
Hydrangea macrophylla	(40), (41), (42), (43)
Hypoxylon spp.	(18), (19), (24), (25), (27), (28), (64)
Hyppeastrium vittatum	(149)
Kigelia fimbriata	(53)
Kigelia pinnata	(97), (99)
Lasiodiplodia theobromae	(19), (33)
Lederle culture Z1640	(79), (80), (84)
Lenzites subferuginea	(16)
Lenzites thermophylla	(16), (23)

Source Index *(continued)*

Source	Compound number
Lenzites trabea	(16)
Leucoium vernum	(150)
Lycoris radiata	(150)
Mallotus japonicus	(9)
Marasmiellus ramealis	(19), (21)
Microsporium cookei	(137)
Nannizzia cajetani	(135), (136), (138), (139), (140)
Narcissus poeticus	(150)
Numularia spp.	(26), (27)
Ononis natrix	(94)
Oospora astringens	(15), (16), (23)
Paecilomyces victoria	(76), (77)
Parmelia birulae	(82)
Peltophorum interme	(9)
Penicillium brevicompactum	(70), (74), (75)
Penicillium canescens	(67), (68)
Penicillium citreo-viride	(129), (135), (138), (139)
Penicillium citrinum	(65), (72)
Penicillium cyclopium	(135), (138), (158)
Penicillium duclauxi	(126), (127), (128)
Penicillium herquei	(123)
Penicillium stipitatum	(127)
Penicillium thomii	(53)
Penicillium viridicatum	(30), (130), (131), (132), (135), (138), (158), (159), (160)
Peniophora sanguinea	(95)
Periconia macrospinosa	(59), (61), (63)
Pestalotia ramula	(18), (19), (50)
Pseudomonas yoshitomiensis	(146), (147), (148)
Pyricularia oryzae	(51)
Rhytidoponera metallica	(19)
Roesleria pallida	(123)
Sclerotinia libertiana	(17)
Sclerotinia sclerotiorum	(17), (71), (72)
Scolecobasiella avellanea	(122)
Semicarpus spp.	(24)
Septoria nodorum	(19), (22), (33), (37)
Shorea leprosula	(9)
Spicaria divaricata	(142), (143), (144)
Sporormia affinis	(53), (61), (63)
Sporormia bipartis	(53)
Sporormia griseoviridus	(145)
Streptomyces mobaraensis	(54), (96), (98), (100), (101), (102)

References, 63–78

Source Index *(continued)*

Source	Compound number
Streptomyces rubireticulae	(**98**)
Swartzia laevicarpa	(**24**), (**60**), (**62**)
Tovomita brasiliensis	(**24**)
Trichophyton megnini	(**138**)
Trichophyton rubrus	(**138**)
Trichophyton violaceum	(**134**), (**138**)
Truncatella hartigii	(**18**)
Verticillium lamellicola	(**120**), (**121**)
Woodfordia fruticosa	(**8**)

References

1. ABELL, C., M.J. GARSON, F.J. LEEPER, and J. STAUTON: Biosynthesis of the Fungal Metabolites Alternariol, Mellein, Rubrofusarin, and 6-Methylsalicylic Acid from Acetic Acid-2,2,2-d_3. Chem. Commun. **1982**, 1011.
2. AFZAL, S.M., R. PIKE, N.H. RAMA, L.R. SMITH, E.S. TURNER, and W.B. WHALLEY: The Chemistry of Fungi – Part 74. Synthesis of ($\pm$)-5-Butyl-6,8-dihydroxy-3-pentyl-3,4-dihydroisocoumarin. J. Chem. Soc. Perkin Trans. I **1978**, 81.
3. AKITA, T., K. KAWAI, H. SHIMONAKA, Y. NOZAWA, Y. ITO, S. NISHIBE, and Y. OGIHARA: Biochemical Studies on Pigments from the Dermatophyte *Microsporum cookei* II. Effects of Luteosporin on the Oxidative Phosphorylation of Rat Liver Mitochondria. Shinkin to Shinkinsho. **16**, 177 (1975); Chem. Abstr. **88**, 69634 (1978).
4. ALDRIDGE, D.C., J.F. GROVE, and W.B. TURNER: 4-Acetyl-6,8-dihydroxy-5-methyl-2-benzopyran-1-one, a Metabolite of *Aspergillus viridinutans*. J. Chem. Soc. (C) **1966**, 126.
5. ALDRIDGE, D.C., S. GALT, D. GILES, and W.B. TURNER: Metabolites of *Lasiodiplodia theobromae*. J. Chem. Soc. (C) **1971**, 1623.
6. ALDRIDGE, D.C., and W.B. TURNER: Metabolites of *Helminthosporium monoceras*, Structure of Monocerin and Related Benzopyrans. J. Chem. Soc. (C) **1970**, 2598.
7. ANDERSON, J.R., R.L. EDWARDS and A.J.S. WHALLEY: Metabolites of the Higher Fungi – Part 21. 3-Methyl-3,4-dihydroisocoumarins and Related Compounds from the Ascomycete Family Xylariaceae. J. Chem. Soc. Perkin Trans. I **1983**, 2185.
8. ANTOSZ, F.J., D.B. NELSON, D.L. HERALD, JR, and M.E. MUNK: The Structure and Chemistry of Actinobolin. J. Amer. Chem. Soc. **92**, 4933 (1970).
9. ANTUS, S., G. SNATZKE, and I. STEINHE: Synthesis and Circular Dichroism of Steroids with an Isochromanone chromophore. Liebigs Ann. Chem. **1983**, 2247.
10. ARAI, Y., T. KAMIKAWA, and T. KUBOTA: A Facile Synthesis of Mellein. Bull. Chem. Soc. Japan **46**, 3311 (1973).
11. ARAY, K., T. YOSHIMURA, Y. ITATANI, and Y. YAMAMOTO: Metabolic Products of *Aspergillus terreus*. VIII Astepyrone, a Novel Metabolite of the Strain IFO 4100. Chem. Pharm. Bull. (Japan) **31**, 925 (1983).
12. ARAKAWA, H.: Absolute Configuration of Mellein. Bull. Chem. Soc. (Japan) **41**, 2541 (1968).

13. ARAKAWA, H., N. TORIMOTO, and Y. MASUI: The Absolute Configuration of (−)-β-Tetralol and Agrimolide. Tetrahedron Letters **1968**, 4115.
14. − − − Absolute Configuration of Optically Active Naturally Occuring Isocoumarins. II Determination of the Absolute Configuration of Agrimonolide and Mellein. Liebigs Ann. Chem. **728**, 152 (1969).
15. ARAZASHVILI, A.I., G.K. NIKONOV, and E.P. KEMERTELIDZE: The Structure of Delphoside, a New Isocoumarin Glycoside. Khim. Prior. Soedin. **1974**, 705.
16. ARAHINA, H., and J. ASANO: Synthesis of Phyllodulcin Dimethyl Ether. Chem. Ber. **64**, 1252 (1931).
17. ASSANTE, G., R. LOCCI, L. CAMARDA, L. MERLINI, and G. NASINI: Metabolites of *Cercospora*. Part 4. Screening of the Genus *Cercospora* of secondary Metabolites. Phytochem. **16**, 243 (1977).
18. AUE, R., R. MAULI, and H.P. SIGG: Production of 6-Methoxymellein by *Sporormia bipartes*. Experientia **22**, 575 (1966).
19. BAKER, T.C., R. NISHIDA, and W. ROELOFS: Close-range Attraction of Female Oriental Fruit Moths to Herbal Scent of Male Hairpencils. Science **214**, 1359 (1981).
20. BALLIO, A., S. BARCELLONA, and B. SANTURBANO: 5-Methylmellein, a New Natural Dihydroisocoumarin. Tetrahedron Letters **1966**, 3723.
21. BARBER, J., M.J. GARSON, and J. STAUTON: The Biosynthesis of Fungal Metabolites: Sclerin, a Plant Growth Hormone from *Sclerotinia sclerotiorum*. J. Chem. Soc. Perkin I **1981**, 2584.
22. BARBER, J., and J. STAUTON: Protium as a Tracer in Polyketide Biosynthesis: Incorporation of $^{13}CH_3^{13}CO_2H$ into Citrinin Produced on a Medium Based on D_2O. Chem. Commun. **1979**, 1098.
23. BATU, G., and R. STEVENSON: Synthesis of Natural Isocoumarins, Artemidin and 3-Propylisocoumarin. J. Organ. Chem. (USA) **45**, 1532 (1980).
24. BARRY, R.D.: Isocoumarins. Developments since 1950. Chem. Rev. **64**, 229 (1964).
25. BELGAONKAR, V.H., and R.N. USGAONKAR: Isocoumarins Part XIX. Synthesis of 8-Hydroy-3-methylisocoumarin and (±)-Mellein from *m*-Dinitrobenzene and a Convenient Synthesis of 3-Methoxyhomophthalic Acid. Indian J. Chem. **17B**, 430 (1979).
26. BELLINGER, G.C.A., W.E. CAMPBELL, R.G.F. GILES, and J.D. TOBIAS: Formation of Some 3-Aryl-3,4-dihydroisocoumarins by Thermal Ring Closure of Stilbene-2-carboxylic Acids. J. Chem. Soc. Perkin Trans. I **1982**, 2819.
27. BERNAUER, K., and O.T. SCHMIDT: The Synthesis of Brevifolin Trimethyl Ether. Liebigs Ann. Chem. **591**, 153 (1955).
28. BHIDE, B.H., and D.I. BRAHMBHATT: Isocoumarins Part 4. Synthesis of 5,6-dimethoxy-, 6,7-dimethoxy-, 7,8-dimethoxy-, 5,7-dimethoxy-, 5,8-dimethoxy-3-methylisocoumarins and a New Synthesis of (±)-6-Methoxymellein. Proc. Indian Acad. Sci. **89**, 525 (1980).
29. BHIDE, B.H., and V.P. GUPTA: Isocoumarins Part I. Synthesis of 5,6,7-Trimethoxyisocoumarin and a Novel Synthesis of Kigelin. Indian J. Chem. **15B**, 512 (1977).
30. BHIDE, B.H., V.P. GUPTA, N.S. NARASIMHAN, and R.S. MALI: Novel Synthesis of (±)-Kigelin. Chem. and Ind. **1975**, 519.
31. BHIDE, B.H., V.P. GUPTA, and K.K. SHAM: Synthesis of Artemidin. Chem. and Ind. **1980**, 84.
32. BHIDE, B.H., and K.K. SHAM: Isocoumarins Part III. Synthesis of Methyl and Methylenedioxy-dihydroisocoumarins and a New Synthesis of (±)-5-Methylmellein. Indian J. Chem. **19B**, 9 (1980).
33. BILLEK, G., and H. KINDL: Phenolic Content of the Saxifragaceae Family. Monatsh. Chem. **93**, 85 (1962).
34. − − Biosynthesis of Plant Stilbenes Part II. Formation of Ring A of Hydrangenol. Monatsh. Chem. **93**, 814 (1962).

35. BIRCH, A.J., J.H. BIRKINSHAW, P. CHAPLEN, L. MO, A.H. MANCHANDA, A. PELTER, and M. RIANO-MARTIN: The Structure of Canescin A and B. Austral. J. Chem. **22**, 1933 (1969).
36. BIRCH, A.J., F. GAGER, L. MO, A. PELTER, and J.J. WRIGHT: Studies in Relation to Biosynthesis Part XLI. Canescin. Austral. J. Chem. **22**, 2429 (1969).
37. BIRCH, A.J., J.J. WRIGHT, F. GAGER, L.MO, and A. PELTER: Biosynthesis of Canescin, a C_1 Unit in a Chain. Tetrahedron Letters **1969**, 1519.
38. BLACKBURN, G.M., D.E.U. EKONG, A.H. NEILSON, and LORD TODD: Xylindein from *Chlorociboria aeruginosa*. Chimia **19**, 208 (1965).
39. BLACKBURN, G.M., A.H. NEILSON, and LORD TODD: The Structure of Xylindein. Proc. Chem. Soc. (London) **1962**, 327.
40. BLANK, F., A.S. NG, and G. JUST: Isolation and Tentative Structures of Vioxanthin and Viopurpurin, Two Coloured Metabolites of *Trichophyton violaceum*. Canad. J. Chem. **44**, 2873 (1966).
41. BLUM, M.S., T.H. JONES, D.F. HOWARD, and W.L. OVERAL: Biochemistry of Termite Defences: *Coptotermes*, *Rhinotermes* and *Cornitermes* Species. Comp. Biochem. Physiol. **71B**, 731 (1982).
42. BOHLMANN, F., and K. DIETER: Polyacetylene Compounds Part CIX. Synthesis of Naturally Occurring Aromatic Substituted Acetylene Compounds. Chem. Ber. **99**, 2822 (1966).
43. BOHLMANN, F., and C. ZDERO: Polyacetylenic Compounds Part 184. Constituents of *Anthemis fuscata*. Chem. Ber. **103**, 2856. (1970).
44. - - New Constituents from *Felicia* Species. Phytochem. **15**, 1318 (1976).
45. BOIT, H.-G.: The Alkaloids of *Leucoium vernum* and *Narcissus poeticus* var. *ornatus*. Chem. Ber. **87**, 681 (1954).
46. - Alkaloids from *Chidanthus fragram*, *Vallota purpurea*, *Nerine undulata*, and *Hippeastrum vittatum*. Chem. Ber. **89**, 1129 (1956).
47. BOIT, H.-G., and H. EHMKE: Alkaloids from *Nerine bowdenii*, *Crinum powellii*, *Amaryllis belladonna*, and *Pancratium maritimus*. Chem. Ber. **89**, 2093 (1956).
48. BRAND, J.M., H.M. FALES, E.A. SOKOLOSKI, J.G. MACCONNELL, M.S. BLUM, and R.M. DUFFIELD: Identification of Mellein in the Mandibular Gland Secretion of Carpenter Ants. Life Science **13**, 201 (1973).
49. BRAZ FILHO, R., M.P.L. DE MORAES, and O.R. GOTTLIEB: Pterocarpans from *Swartzia laevicarpa*. Phytochem. **19**, 2003 (1980).
50. BRAZ FILHO, R., C.A.S. MIRANDA, O.R. GOTTLIEB, and M.T. MAGALHAES: Chemical Constituents of *Tovomita brasiliensis*. Acta Amazonica **12**, 801 (1982); Chem. Abstr. **99**, 155230 (1983).
51. BRIAN, P.W., H.G. HEMMING, J.S. MOFFAT, and C.H. UNWIN: Canescin, an Antibiotic Produced by *Penicillium canescens*. Trans. Brit. Mycol. Soc. **36**, 243 (1953).
52. BROPHY, J.J., G.W.K. CAVIL, and W.D. PLANT: Volatile Constituents of an Australian Ponerine Ant, *Rhytidoponera metallica*. Insect Biochem. **11**, 307 (1981).
53. CAMARDA, L., L. MERLINI, and G. NASINI: Metabolites of *Cercospora*, Taiwapyrone, and α-Pyrone of Unusual Structure from *Cercospora taiwanensis*. Phytochem. **15**, 537 (1976).
54. CARPENTER, R.C., S. SOTHEESWARAN, M.U.S. SULTANBAWA, and S. BALUSUBRAMANIAM: (−)-5-Methylmellein and Catechol Derivatives from Four *Semicarpus* species. Phytochem. **19**, 445 (1980).
55. CARRUTHERS, W.R., J.E. HAY, and L.J. HAYNES: Isolation of Bergenin from *Shorea leprosula*. The Identity of Vakerin and Bergenin. Chem. and Ind. **1957**, 76.
56. CARTER, R.H., M.J. GARSON, and J. STAUNTON: Biosynthesis of Citrinin: Incorporation Studies with Advanced Precursors. Chem. Commun. **1979**, 1097.

57. CATTEL, L., J.F. GROVE, and D. SHAW: New Metabolic Products of *Aspergillus flavus*, III Biosynthesis of Asperentin. J. Chem. Soc. Perkin Trans I **1973**, 2626.
58. CHARUBALA, R., A. GUGGISBERG, M. HESSE, and H. SCHMID: Natural Occurrence of 3-Phenylisocoumarin. Helv. Chim. Acta **57**, 1096 (1974).
59. CHATTERJEA, J.N., B.K. BANERJEE, C. BHAKTA, and I. MUKERJI: Synthesis of *O,O*-Dimethylreticulol and (±)-5-Methylmellein. J. Indian Chem. Soc. **49**, 797 (1972).
60. CHATTERJEA, J.N., C. BHAKTA, and S.K. MUKHERJEE: Synthesis of *cis*- and *trans*-Artemidins. Indian J. Chem. **20B**, 992 (1981).
61. − − − On the Synthesis of Some Chlorinated Isocoumarins J. Indian Chem. Soc. **58**, 888 (1981).
62. CHATTERJEA, J.N., C. BHAKTA, and N.D. SINHA: A Synthesis of *O*-Methylkigelin. J. Indian Chem. Soc. **52**, 158 (1975).
63. CHATTERJEA, J.N., C. BHAKTA, and T.R. VAKULA: Synthesis of Some 3-Methylisocoumarins. J. Indian Chem. Soc. **49**, 1161 (1972).
64. CHATTERJEA, J.N., J. MUKRJI, C. BHAKTA, and B.K. BANERJEE: Synthesis of the *O,O*-Dimethyl Ether of Reticulol and (±)-5-Methylmellein. Current Sci. (India) **38**, 493 (1969).
65. CHATTERJEA, J.N., S.K. MUKHERJEE, and C. BHAKTA: New Synthesis of *trans*-Artemidin, Artemidinal, and Tetrahydrocapillarin. Indian J. Chem. **20B**, 359 (1981).
66. CHATTERJEA, J.N., S.K. MUKHERJEE, C. BHAKTA, H.C. JHA, and F. ZILLIKEN: A New Synthesis of Tetrahydrocapillarin, *O*-Methylglomellin, and Oospolactone. Chem. Ber. **113**, 3927 (1980).
67. CHAUDHURY, G.R.: Chemical Examination of the Roots of *Caesalpinia digyna*. Identy of Vakerin with Bergenin. J. Sci. Indust. Res. (India) **16B**, 511 (1957).
68. CHEXAL, K.K., CH. TAMM, K. HIROTSU, and J. CLARDY: Gilmaniellin and Dechlorogilmeniellin, Two Novel Dimeric Oxaphenalenones. Helv. Chim. Acta **62**, 1785 (1979).
69. CHIEN, M.M., P.L. SCHIFF, JR, D.J. SLATHIN, and J.E. KNAPP: The Isolation of Citrinin, Dihydrocitrinin and Sclerin from *Aspergillus carneus*. Lloydia **40**, 301 (1977).
70. CIEGLER, A., D.J. FENNELL, H.-J. MINTZLAFF, and L. LEISTNER: Ochratoxin Synthesis by *Penicillium* Species. Naturwiss. **59**, 365 (1972).
71. CLAYDON, N., J.F. GROVE, and M. HOSHEN: Phenolic Metabolic Products of *Ceratocystis ulmi*. Chem. and Ind. **1974**, 344.
72. − − − Phenolic Metabolites of *Ceratocystis ulmi*. Phytochem. **13**, 2567 (1974).
73. CLUTTERBUCK, P.W., A.E. OXFORD, H. RAISTRICK, and G. SMITH: The Metabolic Products of the *Penicillium brevi-compactum* Series. Biochem. J. **26**, 1441 (1932).
74. COLE, R.J., J.H. MOORE, N.D. DAVIS, J.W. KIRKSEY, and U.L. DIENER: 4-Hydroxymellein, a New Metabolite of *Aspergillus ochraceus*. J. Agric. Food Chem. **19**, 909 (1971).
75. COLOMBO, L., C. GENNARI, D. POTENZA, C. SCOLASTICO, F. ARAGOZZINI, and C. MERENDI: Biosynthesis of Citrinin and Synthesis of its Biogenetic Precursors. J. Chem. Soc. Perkin Trans. I **1981**, 2594.
76. COLOMBO, L., C. GENNARI, C. SCOLASTICO, F. ARAGOZZINI, and C. MERENDI: Biosynthesis of Ascochitine: Incorporation Studies with Advanced Precursors. Chem. Commun. **1979**, 492.
77. − − − − − Biosynthesis of Aschochitine and Synthesis of its Biogenetic Precursors. J. Chem. Soc. Perkin Trans. I **1980**, 2549.
78. CONDON, P., J. KUC., and H.N. DRAUDT: Production of 3-Methyl-6-methoxy-8-hydroxy-3,4-dihydroisocoumarin by Carrot Root Tissue. Phytopathology **53**, 1244 (1963).
79. COOMBE, R.G., J.J. JACOBS, and T.R. WATSON: Metabolites of Some *Alternaria* spe-

cies. The Structures of Altenusin and Dehydroaltenusin. Austral. J. Chem. **23**, 2343 (970).
79a. CORDOVA, R., and B.B. SNIDER: A Synthetic Approach to Actinobolin. Total synthesis of ($\pm$)-Ramulosin. Tetrahedron Letters **25**, 2945 (1984).
80. COXON, D.T., R.F. CURTIS, K.R. PRICE, and G. LEVETT: Abnormal Metabolites Produced by *Daucus carota* roots stored under Conditions of Stress. Phytochem. **12**, 1881 (1973).
81. CURTIS, R.F.: 6-Methoxymellein as a Phytoalexin. Experientia **24**, 1187 (1968).
82. CURTIS, R.F., P.C. HARRIES, and C.H. HASSAL: The Synthesis of (2-Carboxy-3,5-dihydroxyphenyl)propan-2-one (*C*-Acetyl-*o*-orsellinic Acid). J. Chem. Soc. (London) **1964**, 5382.
83. CURTIS, R.F., P.C. HARRIES, C.H. HASSALL, and J.D. LEVI: The Relationship of Some Phenolic Metabolites of Mutants of *Aspergillus terreus* Thom I.M.I. 16043. Biochem. J. **90**, 43 (1964).
84. CURTIS, R.F., C.H. HASSAL, and M. NAZAR: The Biosynthesis of Phenols Part XV. Some Metabolites of *Penicillium citrinum* related to Citrinin. J. Chem. Soc. (C) **1968**, 85.
85. DAVIES, W.P., and B.G. LEWIS: Antifungal Activity in Carrot Roots in Relation to Storage Infection by *Mycocentrospora acerina* (Hartig) Deighton. New Phytol. **89**, 109 (1981).
86. DE ALVARENGA, M.A., BRAZ FO, O.R. GOTTLIEB, J.P. DE P. DIAS, A.F. MAGALHAES, E.G. MAGALHAES, G.C. DE MAGALHAES, M.T. MAGALHAES, J.G.S. MAIA, R. MARQUES, A.J. MARSAIOLI, A.A.L. MESQUITA, A.A. DE MORAES, A.B. DE OLIVEIRA, G.G. DE OLIVEIRA, G. PEDREIRA, S.A. PEREIRA, S.L.V. PINHO, A.E.G. SANT'ANA, and C.C. SANTOS: Dihydroisocoumarins and Phalides from Wood Samples Infested by Fungi. Phytochem. **17**, 511 (1978).
87. DEAN, B.M., and J. WALKER: A New Source of Bergenin. Chem. and Ind. **1958**, 1696.
88. DE JESUS, A.E., P.S. STEYN, R. VLEGGAR, and P.L. WESSELS: Carbon-13 Nuclear Magnetic Resonance Assignments and Biosynthesis of the Mycotoxin Ochratoxin A. J. Chem. Soc. Perkin Trans. I **1980**, 52.
89. DEVYS, M., J.F. BOUSQUET, A. KOLLMANN, and M. BARBIER: Dihydroisocoumarins and Mycophenolic Acid from the Culture Medium of the Mushroom Phytopathogen, *Septoria nodorum*. Phytochem. **19**, 2221 (1980).
90. DEVYS, M., J.F. BOUSQUET, M. SKAJENNIKOFF, and M. BARBIER: Ochracin (Mellein), a Phytotoxin from the Culture Medium of *Septoria nodorum*. Phytopathol. Z. **81**, 92 (1974).
91. DICK, JR, W.E.: Structure-taste Correlations for Flavans and Flavones Conformationally Equivalent to Phyllodulcin. J. Agric. Food Chem. **29**, 305 (1981).
92. DUNN, A.W., R.A.W. JOHNSTONE, T.J. KING, L. LESSINGER, and B. SKLARZ: Structures of C_{25} Compounds from *Aspergillus variecolor*. J. Chem. Soc. Perkin Trans. I **1979**, 2113.
93. DURLEY, R.C., J. MACMILLAN, T.J. SIMPSON, A.T. GLEN, and W.B. TURNER: Xanthomegnin, Viomellein, Rubrosulphin, and Viopurpurin, Pigments from *Aspergillus sulphureus* and *Aspergillus melleus*. J. Chem. Soc. Perkin Trans. I **1975**, 163.
94. EATON, M.A.W., and D.W. HUTCHINSON: Isocoumarins from *Streptomyces mobaraensis*. Tetrahedron Letters **1971**, 1337.
95. EDWARDS, R.L., and N. KALE: The Structure of Xylindein. Tetrahedron **21**, 2095 (1965).
96. ELLESTAD, G.A., F.M. LOVELL, N.A. PERKINSON, R.T. HARGREAVES, and W.J. MCGAHREN: New Zearalenone Related Macrolides and Isocoumarins from an Unidentified Fungus. J. Organ. Chem. (USA) **43**, 2339 (1978).

97. ELLESTAD, G.A., P. MIRANDO, and M.P. KUNSTMANN: Structure of the Metabolite LL-S4908 from an Unidentified *Aspergillus* species. J. Organ. Chem. (USA) **38**, 4204 (1973).
98. FARRELL, I.W., T.G. HALSALL, V. THALLER, A.P.W. BRADSHAW, and J.R. HANSON: Structures of Some New Sesquiterpenoid Metabolites of *Marasmius alliaceus*. J. Chem. Soc. Perkin Trans. I **1981**, 1790.
99. FERREIRA, N.P., and M.J. PITOUT: The Biogenesis of Ochratoxin A.J.S. Afr. Chem. Inst. **22**, 1S (1969).
100. FINDLAY, J.A., J.M. MATSOUKAS, and J. KREPINSKY: Synthesis of *dl*-Epiramulosin and the Configuration of Ramulosin. Canad. J. Chem. **54**, 3419 (1976).
101. FREEMAN, G.G.: Isolation of Alternariol and Alternariol Monomethyl Ether from *Alternaria dauci* (Khuhn) Groves and Sholko. Phytochem. **5**, 719 (1966).
102. FUJISE, S., M. SUZUKI, Y. WATANABE, and S. MATSUEDA: The Structure of Bergenin. Bull. Chem. Soc. Japan **32**, 97 (1959).
103. FURUTANI, Y., H. NAGANAWA, T. TAKEUCHI, and H. UMEZAWA: Isolation and Structure of New Isocoumarins. Agric. Biol. Chem. **41**, 1179 (1977).
104. FURUTANI, Y., I. TSUCHIYA, H. NAGANAWA, T. TAKEUCHI, and H. UMEZAWA: Biosynthetic Studies of Reticulol, an Isocoumarin, by Carbon-13 NMR Spectroscopy. Agric. Biol. Chem. **41**, 1581 (1977).
105. GANDHI, R.N.: Biosynthesis of Methyl Carbonate Unit in 4-*O*-Carbomethoxylamellicolic Anhydride. Indian J. Chem. **15B**, 482 (1977).
106. GARSON, M.J., and J. STAUNTON: Biosynthesis of Sclerin: Incorporation Studies with Advanced Precursors. Chem. Commun. **1978**, 158.
107. GATENBECK, S., and S. HERMODSSON: Enzymatic synthesis of the Aromatic Product Alternariol. Acta Chem. Scand. **19**, 65 (1965).
108. GILES, D., and W.B. TURNER: Chlorine-containing Metabolites of *Periconia macrospinosa*. J. Chem. Soc. (C) **1969**, 2187.
109. GOVINDACHARI, T.R., P.C. PARTHASARATHY, H.K. DESAI, and K.S. RAMACHANDRAN: Homalicine and (−)-Dihydrohomalicine, Two New Isocoumarin Glucosides from *Homalium zeylanicum* Benth. Indian J. Chem. **13**, 537 (1975).
110. GOVINDACHARI, T.R., S.J. PATANKAR, and N. VISWANATHAN: Synthesis of *dl*-Kigelin. Indian J. Chem. **9**, 507 (1971).
111. − − − Isolation and Structure of Two New Dihydroisocoumarins from *Kigelia pinnata*. Phytochem. **10**, 1603 (1971).
112. GREGER, H.: Aromatic Acetylenes and Dehydrofalcurinone Derivatives from the *Artemisia dracunculus* group. Phytochem. **18**, 1319 (1979).
113. GREGER, H., and F. BOHLMANN: 8-Hydroxycapillarin. A new Isocoumarin from *Artemisia dracunculus*. Phytochem. **18**, 1244 (1979).
114. GREGER, H., F. BOHLMANN, and C. ZDERO: New Isocoumarins from *Artemisia dracunculus*. Phytochem. **16**, 795 (1977).
115. GRIDENBERG, J.: Peniolactol Obtained from Wood Attacked by the Fungus *Peniophora sanguinea*. Acta Chem. Scand. **28B**, 505 (1974).
116. GRIMSHAW, J., and R.D. HAWORTH: The Position of the Carboxyl Group in Isogalloflavin and a Synthesis of Trimethylbrevifolin. J. Chem. Soc. (London) **1956**, 418.
117. GROVE, J.F.: Asperentin, its Methyl Ethers and 5′-Hydroxyasperentin, New Metabolic Products of *Aspergillus flavus*. J. Chem. Soc. Perkin Trans. I **1972**, 2400.
118. − 4′-Hydroxyasperentin and 5′-Hydroxyasperentin 8-Methyl Ether, New Metabolic Products of *Aspergillus flavus*. J. Chem. Soc. Perkin Trans. I **1973**, 2704.
119. GROVE, J.F., and M. POPLE: The Synthesis and Reactions of Some Derivatives of *C*-Acetylorsellinic Acid. J. Chem. Soc. Perkin Trans. I **1979**, 337.
120. − − Metabolic Products of *Fusarium larvarum* Fuckel. The Fusarentins and the Absolute Configuration of Monocerin. J. Chem. Soc. Perkin I **1979**, 2048.

121. – – The Insecticidal Activity of Some Fungal Dihydroisocoumarins. Mycopathologia **76**, 65 (1981).
122. GUYUT, M., and D. MOLHO: New Method of Homophthalic Acid and Homophthalimide Synthesis by an Aryne Route. Application to the Synthesis of Mellein. Tetrahedron Letters **1973**, 3433.
123. HARDEGGER, E., W. RIEDER, A. WALSER, and F. KUGLER: Structure of the Diaporthins and the Synthesis of Diaporthic Acid. Helv. Chim. Acta **49**, 1283 (1966).
124. HARDING, V.K., and J.B. HEALE: Isolation and Identification of the Antifungal Compounds Accumulation in the Induced Resistance Response to Carrot Root Slices of *Botrytis cinerea*. Physiol. Plant. Pathol. **17**, 277 (1980).
125. HARWIG, J., and Y.-K. CHEN: Some Conditions Favouring Production of Ochratoxin A and Citrinin by *Penicillium viridicatum* in Wheat and Barley. Canad. J. Plant Sci. **54**, 17 (1974).
126. HARWOOD, L.M.: Trifluoroacetic Acid-Catalyzed Claisen Rearrangement of 5-Allyloxy-2-hydroxybenzoic Acid and Esters: An Efficient Synthesis of Mellein. Chem. Commun. **1982**, 1120.
127. – An Investigation into the Regioselectivity of the Acid Catalyzed Claisen Rearrangement of Methyl 4- and 5-Allyloxy-2-hydroxybenzoate and derivatives. Chem. Commun. **1983**, 530.
128. HASSAL, C.H., and D.W. JONES: The Biosynthesis of Phenols Part IV. A New Metabolic Product of *Aspergillus terreus* Thom. J. Chem. Soc. (London) **1962**, 4189.
129. HATTORI, S.: Corylopsin, a crystalline Constituent of the Bark of *Corylopsis spicata*. Acta Phytochim. (Tokyo) **4**, 327 (1929); Chem. Abstr. **24**, 1862 (1930).
130. HAWKSWORTH, W.A., P.W. JEFFS, B.K. TIDD, and T.P. TOUBE: The Aromatic Oxygenation Patterns and Stereochemistry of Some Trioxyaryl Alkaloids of the Hemiacetal and Lactone Series. J. Chem. Soc. (London) **1965**, 1991.
131. HAWORTH, R.D., and J. GRIMSHAW: Synthesis of Trimethylbrevifolin. Chem. and Ind. **1955**, 199.
132. HAY, J.E., and L.J. HAYNES: Bergenin, a *C*-Glycopyranosyl Derivative of 4-*O*-Methylgallic Acid. J. Chem. Soc. (London) **1958**, 2231.
133. HENDERSON, G.B., and R.A. HILL: Synthesis of Chlorinated Isocoumarin Derivatives. J. Chem. Soc. Perkin Trans. I **1982**, 1111.
134. – – The Biosynthesis of Chlorine-containing Metabolites of *Periconia macrospinosa*. J. Chem. Soc. Perkin Trans. I **1982**, 3037.
135. HEMINGWAY, R.W., G.W. MCGRAW, and S.J. BARRAS: Polyphenols in *Ceratocystis minor* infected *Pinus taeda*: Fungal Metabolites, Phloem and Xylem Phenols. J. Agric. Food Chem. **25**, 717 (1977).
136. HELLER, K., and R. ROSCHENTHALER: Inhibition of Protein Synthesis in *Streptococcus faecalis* by Ochratoxin A. Canad. J. Microbiol. **24**, 466 (1978).
137. HILL, R.A., R.H. CARTER, and J. STAUNTON: Biosynthesis of Terrein, a Metabolite of *Aspergillus terreus* Thom. Chem. Commun. **1975**, 380.
138. – – – Biosynthesis of Fungal Metabolites. Terrein, a Metabolite of *Aspergillus terreus* Thom. J. Chem. Soc. Perkin Trans. I **1981**, 2570.
139. HOFLE, G., and K. ROSER: Structure of Xanthomegnin and Related Pigments: Reinvestigation by ^{13}C Nuclear Magnetic Resonance Spectroscopy. Chem. Commun. **1978**, 611.
140. HOLKER, J.S.E. and T.J. SIMPSON: Studies on Fungal Metabolites Part 2. Carbon-13 Nuclear Magnetic Resonance Studies on Pentaketide Metabolites of *Aspergillus melleus*: 3-(1′,2′-Epoxypropyl)-5,6-dihydro-5-hydroxy-6-methylpyran-2-one and Mellein. J. Chem. Soc. Perkin Trans. I **1981**, 1397.
141. HOLKER, J.S.E., and K. YOUNG: Biosynthesis of Metabolites of *Periconia macrospinosa* from $[1-^{13}C]$-, $[2-^{13}C]$- and $[1,2-^{13}C_2]$acetate. Chem. Commun. **1975**, 525.

142. Homma, K.: Chemical Constituents of *Mallotus japonicus* J. Mueller I. J. Agric. Chem. Soc. Japan **15**, 394 (1939); Chem. Abstr. **33**, 6303 (1939).
143. Homma, K., K. Fukuyama, Y. Katsube, Y. Kimura, and T. Hamasaki: Structure and Absolute Configuration of an Atrovenetin-like Metabolite from *Aspergillus silvaticus*. Agric. Biol. Chem. **44**, 1333 (1980).
144. Houlchein, W.J., and J. Nadelson: British Patent 1374337 (1974); Chem. Abstr. **83**, 43196 (1975).
145. Hsu, H.-Y., and M.-C. Liau: Constituents of *Astilbe macroflora*. J. Taiwan Pharm. Assoc. **11**, 2 (1959); Chem. Abstr. **54**, 13556 (1960).
146. Hung, S.-H., and J.-H. Chu: The Constituents of the Chinese Drug, Kai-Ho-Chien, *Ardisia hortorum* II. Identification of Ardisic Acid B as Bergenin. Hua Hsueh Hsueh Pao **23**, 255 (1957); Chem. Abstr. **52**, 15827 (1958).
147. Hutchinson, R.D., and P.S. Steyn: The Isolation and Structure of 4-Hydroxyochratoxin A and 7-Carboxy-3,4-dihydro-8-hydroxy-3-methylisocoumarin from *Penicillium viridicatum*. Tetrahedron Letters **1971**, 4033.
148. Ibrahim, R.K., and G.H.N. Towers: Studies in Hydrangenol in *Hydrangea macrophylla* Part II. Biosynthesis of Hydrangenol from ^{14}C-Labelled Compounds. Canad. J. Biochem. Physiol. **40**, 449 (1962).
149. Itoh, J., S. Omoto, N. Nishzawa, Y. Kodama, and S. Inoye: Chemical Structures of Amicoumacins Produced by *Bacillus pumilus*. Agric. Biol. Chem. **46**, 2659 (1982).
150. Itoh, J., T. Shomura, S. Omoto, S. Miyato, Y. Yuda, U. Shibata, and S. Inouye: Isolation, Physio-chemical Properties and Biological Activities of Amicoumacins Produced by *Bacillus pumilus*. Agric. Biol. Chem. **46**, 1255 (1982).
151. Iwasaki, S., H. Muro, K. Sasaki, S. Nozoe, S. Okuda, and Z. Sato: Isolation of Phytotoxic Substances Produced by *Pyricularia oryzae*. Tetrahedron Letters **1973**, 3537.
152. Jarrah, M.Y., and V. Thaller: Isolation and Partial Synthesis of 3-Methoxycarbonyl-7-formyl-1-benzoxepin-5(2H)-one, the Ester of a Metabolite from Shake Cultures of the Fungus *Marasmiellus ramealis* (Bull. ex Fr.) Singer. J.Chem. Soc. Perkin Trans. I **1983**, 1719.
153. Joshi, B.S., and V.N. Kamat: Identity of Peltophorin with Bergenin. Naturwiss. **56**, 89 (1969).
154. Joshi, K.C., P. Singh, and S. Taneja: Crystalline Components of the Stem Heartwood of *Randia dumatorium* and the Roots Heartwood of *Kigelia pinnata*. J. Indian Chem. Soc. **58**, 825 (1981).
155. Jui, J., and S. Mizuba: Metabolic Products from *Spiricaria divaricata*. J. Antibiotics **27**, 760 (1974).
156. Just, G., W.C. Day, and F. Blank: The Structure of Xanthomegnin. Canad. J. Chem. **41**, 74 (1963).
157. Kaiser, P., and J. Schnehenburger: Acyl Derivatives of Methylene Active Dicarbonyl Compounds Part 12. 3-Substituted Isocoumarins from 4-Acylhomphthalic Anhydrides. Z. Naturforsch. **25 B**, 1190 (1970).
158. Kalidhar, S.B., M.R. Parthasathy, and P. Sharma: Norbergenin, a New C-Glycoside from *Woodfordia fruticosa* Kurz. Indian J. Chem. **20**, 720 (1981).
159. Kameda, K., H. Aoki, M. Namiki, and J.C. Overeem: An Alternative Structure for Botrallin, A Metabolite of *Botrytis allii*. Tetrahedron Letters **1974**, 103.
160. Kameda, K., H. Aoki, H. Tanaka, and M. Namiki: Metabolites of *Alternaria kikuchiana*, a Phytopathogenic Fungus of Japanese Pear. Agric. Biol. Chem. **37**, 2137 (1973).
161. Kaneko, H., T. Fujimori, H. Matsushita, and M. Nogcuhi: Chemical Constituents in Amacha Extract. Nippon Nogei Kagaku Zasshi **47**, 605; Chem. Abstr. **80**, 93153 (1974).

162. KANEKO, Y, K. OSHITA, H. TAKAMATSU, Y. ASAO, and T. YOKOTSUKA: Compounds Produced by Moulds Part VII. Isolation of Isocoumarin Compounds. Agric. Biol. Chem. **34**, 1296 (1970).
163. KAZLAUSKAS, R., J. MULDER, P.T. MURPHY, and R.J. WELLS: New Metabolites from the Brown Alga *Caulocystis cephalornithos*. Austral. J. Chem. **33**, 2097 (1980).
164. KIAGAWA, H., W.I. TAYLOR, S. UYEO, and H. YAJIMA: The Constitution of Homolycorine and Lycorenine. J. Chem. Soc. (London) **1955**, 1066.
165. KITAGAWA, T., S. UYEO, and N. YOKOYAMA: The Stereochemistry of Lycorenine, Homolycorine, Pluviine, and Their Hydrogenation Products. J. Chem. Soc. (London) **1959**, 3741.
166. KOLLE, F., and K.E. GLOPPE: Alkaloids of *Narcissus poeticus* L. Pharm. Zentralhalle **75**, 237 (1934).
167. KOPANSKI, L., M. KLAAR, and W. STEGLICH: Leprocybin, the Fluorescent Principle of *Cortinarius coteneus* and Related Leprocybes (Agaricales). Liebigs Ann. Chem. **1982**, 1280.
168. KRIVOSHCHEKOVA, O.E., N.P. MISHCHENKO, L.S. STEPANENKO, and O.B. MAKSIMOV: Aromatic Metabolites of Parmeliaceae Lichens. Part I Depsidones. Khim. Prior. Soedin **19**, 13 (1983).
169. KUBOTA, T., T. TOKOROYAMA, T. KAMIKAWA, and Y. SATOMURA: Structures of Sclerin and Sclerolide, Metabolites of *Sclerotinia libertiana*. Tetrahedron Letters **1966**, 5205.
170. KUHR, I., J. FUSKA, P. SEDMERA, M. PODOJIL, J. VOKOUN, and Z. VANEK: Antitumour Antibiotic Produced by *Penicillium stipitatum*. Its Identity with Duclauxin. J. Antibiotics **26**, 535 (1973).
171. LE BLANC, G.D., and L.M. BABINEAU: 2,5-Dimethoxybenzoquinone, 3,4-Dimethyl-8-hydroxyisocoumarin and Eburicoic Acid Isolated from *Lenzites thermophila*. Canad. J. Microbiol. **18**, 261 (1972).
172. LIN, J.-Y., S. YOSHIDA, and N. TAKAHASHI: Metabolites Produced by *Streptomyes mobaraensis*. Agric. Biol. Chem. **35**, 363 (1971).
173. – – – Synthesis of Some Isocoumarin Derivatives. Agric. Biol. Chem. **36**, 506 (1972).
174. LILLEHOJ, E.B., and M.S. MILBURN: Viriditoxin Production by *Aspergillus viridinutans* and Related Species. Appl. Microbiol. **26**, 202 (1973).
175. LLOYD, H.A., S.L. EVANS, A.H. KHAN, W.R. TSCHINKEL, and M.S. BLUM: 8-Hydroxyisocoumarin and 3,4-Dihydro-8-hydroxyisocoumarin in the Defensive Secretion of the Tenbrionid Beetle *Apsena pubescens*. Insect Biochem. **8**, 333 (1978).
176. MCCORKINDALE, N.J.: Personal Communication..
176a. – Lamellicolic Anhydride, 4-O-Carbomethoxylamellicolic anhydride and Monomethyl 3-Chlorlamellicolate. Metabolites of *Verticillium lamellicola*. Tetrahedron **39**, 2283 (1983).
177. MCGAHREN, W.J., and L.A. MITSCHER: Dihydroisocoumarins from a *Sporormia* Fungus. J. Organ. Chem. (USA) **33**, 1577 (1968).
178. MCGRAW, G.W., and R.W. HEMMINGWAY: 6,8-Dihydroxy-3-methylisocoumarin and Other Phenolic Metabolites of *Ceratocystis minor*. Phytochem. **16**, 1315 (1977).
179. MCPHAIL, A.T., R.W. MILLER, D. HARVAN, and R.W. PERO: X-Ray Crystal Structure Revision for the Fungal Metabolite ($\pm$)-Altenuene. Chem. Commun. **1973**, 682.
180. MAEBAYASHI, Y., K. MIYAKI, and M. YAMAZAKI: Application of ^{13}C-NMR to Biosynthetic Investigations. I. Biosynthesis of Ochratoxin A. Chem. Pharm. Bull. (Japan) **20**, 2172 (1972).
181. MALLABAEV, A., I.M. SAIBAEVA, and G.P. SIDYAKIN: Structure of the Isocoumarin Artemidin. Khim. Prior. Spedin **6**, 531 (1970).
182. – – – Artemidinal, an Isocoumarin from *Artemisia dracunculus*. Khim. Prior. Soedin **7**, 257 (1971).

183. MALABAEV, A., and G.P. SIDYAKIN: Artemidiol, a New Isocoumarin from *Artemisia dracunculus*. Khim. Prior. Soedin **1974**, 720.
184. – – Artemidinol, a New Isocoumarin from *Artemisia dracunculus*. Khim. Prior. Soedin **1976**, 811.
185. MALLABAEV, A., M.R. YAGUDAEV, I.M. SAITBAEVA, and G.P. SIDYAKIN: Isocoumarin Artemidin from *Artemisia dracunculus*. Khim. Prior. Soedin **6**, 467 (1970).
186. MATSUI, M., K. MORI, and S. ARASAKI: Synthesis of Isocoumarins I ($\pm$)-Mellein. Agric. Biol. Chem. **28**, 896 (1964).
187. MATSUI, M., K. MORI, and Y. OZAWA: Synthesis of Isocoumarins II Oospolactone. Agric. Biol. Chem. **30**, 193 (1966).
188. MIR, I., S. AHMED, and A. HAMID: Chemical Investigation on *Lenzites trabea*. Pak. J. Sci. Inf. Res. **14**, 479 (1971).
189. MITSCHER, L.A., W.W. ANDRES, and W. MCCRAE: Reticulol, a New Metabolic Isocoumarin. Experientia **20**, 258 (1964).
190. MIZUBA, S.S., C.F.J. HSU, and J. JIU: A Third Metabolite from *Spicaria divaricata* NRRL 5771. J. Antibiotics **30**, 670 (1977).
191. MONEY, T., F.W. COMER, G.R.B. WEBSTER, I.G. WRIGHT, and A.I. SCOTT: Pyrone Studies. I. Biogenetic Type Synthesis of Phenolic Compounds. Tetrahedron **23**, 3435 (1967).
192. MONEY, T., I.H. QURESHI, G.B. WEBSTER, and A.I. SCOTT: Chemistry of Polypyrones. A Model for Acetogenin Biosynthesis. J. Amer. Chem. Soc. **87**, 3004 (1965).
193. MOORE, J.H., N.D. DAVIS, and U.L. DIENER: Mellein and 4-Hydroxymellein Production by *Aspergillus ochraceus*. Appl. Microbiol. **23**, 1067 (1972).
194. MUELLER, H.: Studies on the Formation of 3-Methyl-6-methoxy-8-hydroxy-3,4-dihydroisocoumarin during Carrot Storage. Phytopathol. Z. **93**, 241 (1978).
195. MUNAKATA, T.: Bactobolins, Antitumour Antibiotics from Pseudomonas. Part 2. Synthesis and Microbial Activity of Related Compounds. Yakugaku Zasshi **101**, 138 (1981); Chem. Abstr. **95**, 42826 (1981).
196. MUNAKATA, T., and T. OKUMOTO: Some Structure-activity Relationships for Bactobolin Analogs in the Treatment of Mouse Leukemia P388. Chem. Parm. Bull. (Japan) **29**, 891 (1981).
197. MUNK, M.E., D.B. NELSON, F.J. ANTOSZ, D.L. HERALD, JR, and T.H. HASKELL: The Structure of Actinobolin. J. Amer. Chem. Soc. **90**, 1087 (1968).
198. NADKARNI, D.R., and R.N. USGAONKAR: Convenient Syntheses of Naturally Occurring 3-Propylisocoumarin and 3-Propyl-1(2H)-isoquinoline and Related Compounds. Indian J. Chem. **16B**, 320 (1978).
199. NAIR, M.S.R., and S.T. CAREY: Metabolites of Pyrenomycetes: XII. Polyketides from Hypocreales. Mycologia **71**, 1089 (1979).
200. NAKAJIMA, S., K. KAWAI, and S. YAMADA: The Identification of Lenzitin as Oosponol. Pytochem. **15**, 327 (1976).
201. NAKAJIMA, S., K. KAWAI, S. YAMADA, and Y. SAWAI: Isolation of Oospolactone as the Antifungal Principle of *Gloephyllum sepiarum*. Agric. Biol. Chem. **40**, 811 (1976).
202. NAOI, Y., S. HIGUCHI, H. ITO, T. NAKANO, K. SAKAI, T. MATSUI, S. WAGATSUMA, A. NISHI, and S. SANO: Total Synthesis of *dl*-Phyllodulcin. Org Prep. Proceed. Int. **7**, 129 (1975).
203. NAOI, Y., S. HIGUCHI, T. NAKANO, K. SAKAI, A. NISHI, and S. SANO: New Synthesis of Hydrangenol. Synth. Commun. **5**, 387 (1975).
204. NARASIMHACHARI, N., and L.C. VINING: Studies on the Pigments of *Penicillium herquei*. Canad. J. Chem. **41**, 641 (1963).
205. NARASIMHAN, N.S., and C.P. BAPAT: A New Synthesis of Kigelin. Conversion of Elemicin into Kigelin. J. Chem. Soc. Perkin Trans. I **1982**, 2099.

206. NARASIMHAN, N.S., and B.H. BHIDE: A Novel Synthesis of (±)-Mellein. Chem. Commun. **1970**, 1552.
207. – – Synthetic Applications of Lithiation Reactions. V. Novel Synthesis of Methoxyisocoumarin. Synthesis of (±)-Mellein. Tetrahedron **27**, 6171 (1971).
208. NARASIMHAN, N.S., R.S. MALI, and B.K. KULKARNT: Synthesis of (±)-3,4-Dihydro-3-phenylisocoumarin and Aglycone of (±)-Dihydrohomalicine. Indian J. Chem. **22B**, 850 (1983).
209. NELSON, D.B., and M.E. MUNK: Alanylactinobone. A Basic Hydrolisis Product of the Antibiotic Actinobolin. J. Organ. Chem. (USA) **35**, 3832 (1970).
209a. NG, A.S., G. JUST, and F. BLANK: Metabolites of Pathogenic Fungi. VII. On the Structure and Stereochemistry of Xanthomegnin, Vioxanthin and Viopurpurin, Pigments from *Trichophyton violaceus*. Canad. J. Chem. **47**, 1223 (1969).
210. NIAZI, H.M.: Isolation and Characterization of the Metabolites of *Aspergillus cervinus* Massee (Monoliaceae) and Isolation and Characterization of Certain Constituents of *Piper guineese* Schum and Thom. (Piperaceae). Univ. of Pittsburgh PhD. Thesis; Diss. Abstr. Int. **37B**, 1644 (1976).
211. NISHIDA, R., T.C. BAKER, and W.L. ROELOFS: Hairpencil Pheromone Components of Male Oriental Fruit Moths, *Grapholitha molesta*. J. Chem. Ecol. **8**, 947 (1982).
212. NISHIKAWA, W.: A Metabolic Product of *Aspergillus melleus* Yukawa. J. Agric. Chem. Soc. Japan **9**, 772 (1933); Chem. Abstr. **28**, 2751 (1934).
213. NITTA, K., J. IMAI, I. YAMAMOTO, and Y. YAMAMOTO: Determination of the Structure of Oosponol by Synthesis. Agric. Biol. Chem. **27**, 817 (1963).
214. NITTA, K., Y. YAMAMOTO, T. INOUE, and T. HYADO: Biogenesis of Oospolactone and Oosponol. Chem. Parm. Bull. (Japan) **14**, 363 (1966).
215. NITTA, K., Y. YAMAMOTO, I. YAMAMOTO, and S. YAMATODANI: Chemical Structure of Oospoglycol and its Formation from Ooponol by Fungus. Agric. Biol. Chem. **27**, 822 (1963).
216. NORTHOLT, M.D., H.P. VAN EGMOND, and W.E. PAULSCH: Ochratoxin A Production by Some Fungal Species in Relation to Water Activity and Temperature. J. Food Prot. **42**, 485 (1979).
217. NOZAWA, K., S. NAKAJIMA, M. YAMADA, and K.-I. KAWAI.: Synthesis of Two Microbial Metabolites, 5-Chloro-3,4-dihydro-8-hydroxy-6-methoxy-3-methyl-isocoumarin and 8-Hydroxy-6-methoxy-3-methylisocoumarin. Chem. Pharm. Bull. (Japan) **28**, 1622 (1980).
218. NOZAWA, K., M. YAMADA, Y. TSUDA, K. KAWAI, and S. NAKAJIMA: Synthesis and Antifungal Activity of 3-Substituted Isocoumarins. Chem. Pharm. Bull. (Japan) **29**, 2491 (1981).
219. – – – – – Antifungal Activity of Oosponol, Oospolactone, Phylloducin, Hydrangenol and Some Related Compounds. Chem. Pharm. Bull. (Japan) **29**, 2689 (1981).
220. NUKINA, M., T. SASSA, and S. MARUMO: Chemical Studies on Aversion-antagonism among Different Strains of the Same Fungal Species. Aversion Factor and New Metabolites of *Cochliobolus lunata*. Tennen. Yuki Kagobutsu Toronkai Koen. Yoshishu **1978**, 141; Chem. Abstr. **90**, 99737 (1979).
221. OGIHARA, Y., Y. IITAKA, and S. SHIBATA: The X-Ray Study of Monobromduclauxin. Tetrahedron Letters **1965**, 1289.
222. OGIHARA, Y., O. TANAKA, and S. SHIBATA: The Reactions of Duclauxin with Ammonia and Primary Amines. The Structures of Desacetylduclauxin, Neoduclauxin, Xenoclauxin and Cryptoclauxin. Tetrahedron Letters **1966**, 2867.
223. OKAZAKI, H., T. KISHI, T. BEPPU, and K. ARIMA: A New Antibiotic, Baciphelacin. J. Antibiotics **28**, 717 (1975).
224. OVEREEM, J.C., and A VON DIJKMAN: Botrallin, a Novel Quinone Produced by *Botrytis allii*. Rec. trav. chim. Pays-Bas **87**, 940 (1968).

225. OXFORD, A.E., and H. RAISTRICK: The Molecular Constitution of the Metabolic Products of *Penicillium brevi-compactum* Dierckx and Related Species. Biochem. J. **27**, 634 (1933).
226. PATTERSON, E.L., W.W. ANDRES, and N. BOHONOS: Isolation of The Optical Antipode of Mellein from an Unidentified Fungus. Experientia **22**, 209 (1966).
227. PELTER, A., R.S. WARD, and T.I. GRAY: The Carbon-13 Nuclear Magnetic Resonance Spectra of Flavonoids and Related Compounds. J. Chem. Soc. Perkin Trans. I **1976**, 2475.
228. PERO, R.W., D. HARVON, and M.-C. BLOIS: Isolation of the Toxin, Altenuisol, from the Fungus *Alternaria tenuis* Auct. Tetrahedron Letters **1973**, 945.
229. PERO, R.W., R.G. OWENS, S.W. DALE, and D. HARVON: Isolation and Identification of a New Toxin, Altenuene, from the Fungus *Alternaria tenuis*. Biochem. Biophys. Acta **230**, 170 (1971).
230. PILLER, N.B.: Conservative Treatment of Acute and Chronic Lymphoedema with Benzopyrones. Lymphology **9**, 132 (1976).
231. PLOUVIER, V.: The Study of Quinic and Shikimic Acids, Bergenin and Heterosides from Some Hamamelideae. C.R. hebd, seances Acad. Sci. **252**, 599 (1961).
232. POSTERNAK, T., and K. DURR: The Structure of Bergenin. Helv. Chim. Acta **41**, 1159 (1958).
233. RAISTRICK, H., and C.E. STICKINGS: Ustic Acid, a Metabolic Product of *Aspergillus ustus* (Bainier) Thom and Church. Biochem. J. **48**, 53 (1951).
234. RAISTRICK, H., C.E. STICKINGS, and R. THOMAS: Alternariol and Alternariol Monomethyl Ether. Metabolic Products of *Alternaria tenuis*. Biochem. J. **55**, 421 (1953).
235. REGAN, A.C., and J. STAUNTON: Asymmetric Synthesis of Mellein Methyl Ether: Use of *ortho*-Toluate Carbanions Generated by Chiral Bases. Chem. Commun. **1983**, 764.
236. RENSON, M., and L. CHRISTAENS: Preparation of 1-Isochromanones from Chlorides of *o*-(α,α-Dimethylacetonyl)- and *o*-(α,α-Dimethylphenacyl)-benzoic Acids. Bull soc. chim. Belges **71**, 394 (1962).
237. ROBERTS, J.C., and P. WOOLVEN: Synthesis of Ochratoxin A, a Metabolite of *Aspergillus ochraceus*. J. Chem. Soc. (C) **1970**, 278.
238. ROGERS, D., D.J. WILLIAMS, and R. THOMAS: The Crystal Structure of ($\pm$)-Dehydroaltenusin. Chem Commun. **1971**, 393.
239. ROSETT, T., R.H. SANKHALA, C.E. STICKINGS, M.E.U. TAYLOR, and R. THOMAS: Biochem. J. **67**, 390 (1957).
240. ROSSI, C., and R. UBALDI: Characterization of a Pigment Produced by *Fusicoccum putrifaciens*. Ann. 1st. Super. Sanita **9**, 320 (1973): Chem. Abstr. **82**, 82634 (1975).
241. SALUJA, M.P., R.S. KAPIL, and S.P. POPLI: Chemical Constituents of *Moringa oleifera* Lamk. (Hybrid Variety) and Isolation of 4-Hydroxymellein. Indian J. Chem. **16B**, 1044 (1978).
242. SAN FELICIANO, A., A.F. BARRERO, M. MEDARDE, J.M. MIGUEL DEL CARROL, and M.V. CALLE: An Isocoumarin and Other Phenolic Components of *Ononis natrix*. Phytochem. **22**, 2031 (1983).
243. SANKAWA, U., H. TAGUCHI, Y. OGIHARA, and S. SHIBITA: The Biosynthesis of Duclauxin. Tetrahedron Letters **1966**, 2883.
244. SARKAR, S.K., and C.T. PHAN: Biosynthesis of 8-Hydroxy-6-methoxy-3-methyl-3,4-dihydroisocoumarin and 5-Hydroxy-7-methoxy-2-methylchromone in Carrot Root Tissues Treated with Ethylene. Physiol. Plant **33**, 108 (1975).
245. SASSA, T., H. AOKI, and K. MUNAKATA: Synthesis of Sclerotinin B. Tetrahedron Letters **1968**, 5703.
246. SASSA, T., H. AOKI, M. NAMIKI, and K. MUNAKATA: Isolation and Structures of Sclerotinins A and B. Agric. Biol. Chem. **32**, 1432 (1968).

247. SATO, H., T. TAKASHIMA, N. OTOMO, and S. SAKAMURA: Phytotoxins Produced by the Fungus of the Larch Shoot Blight. Nippon Nogei Kagaku Kaishi **56**, 649 (1982); Chem. Abstr. **98**, 84628 (1983).
248. SATOMURA, Y., and A. SATO: Isolation and Physiological Activity of Sclerin, a Metabolite of *Sclerotinia* Fungus. Agric. Biol. Chem. **29**, 337 (1965).
249. SCHMALLE, H.W., O.H. JARCHAW, B.M. HAUSEN, and K.H. SCHULZ: 3,4-Dihydro-8-hydroxy-3-(4-Hydroxyphenyl)isocoumarin, Hydrangenol. Acta Crystallogr. **38B**, 2938 (1982).
250. SCHMIDT, O.T., and K. BERNAUER: Brevifolin and Brevifolin Carboxylic Acid. Liebigs Ann. Chem. **588**, 211 (1954).
251. SCHMIDT, O.T., and R. ECKERT: The Structure of Brevifolin Carboxylic Acid. Liebigs Ann. Chem. **618**, 71 (1958).
252. SCHROEDER, H.W., and R.D. STIPANOVIC: Production of 3-Methyl-6-methoxy-8-hydroxy-3,4-dihydroisocoumarin by *Aspergillus caespitosus*. Appl. Microbiol. **29**, 706 (1975).
252a. SCOTT, F.E., T.J. SIMPSON, L.A. TRIMBLE, and J.C. VEDERAS: Biosynthesis of Monocerin. Incorporation of ^{2}H-, ^{13}C- and ^{18}O-Labelled Acetates. Chem. Commun. **1984**, 756.
253. SCOTT, P.M., and W. VAN WALBEEK: Cladosporin, a New Antifungal Metabolite from *Cladosporium cladospioides*. J. Antibiotics **24**, 747 (1971).
254. SCOTT, P.M., W. VAN WALBEEK, J. HARWIG, and D.I. FENNELL: Occurrence of a Mycotoxin, Ochratoxin A, in Wheat and Isolation of Ochratoxin A and Citrinin Producing Strains of *Penicillium viridicatum*. Canad. J. Plant Sci. **50**, 583 (1970).
255. SEDMERA, P., J. VOLC, J. WEIJER, J. VOKOUN, and V. MUSILEK: Xanthomegnin and Viomellein Derivatives from Submerged Cultures of the Ascomycete *Nannizzia cajetani*. Collect. Czech. Chem. Comm. **46**, 1210 (1981).
256. SHIBATU, S., Y. OGIHARA, N. TOKUTAHE, and O. TANAKA: Duclauxin, a Metabolite of *Penicillium duclauxi* (Delacroix). Tetrahedron Letters **1965**, 1287.
257. SHIMADA, H., T. SAWADA, and S. FUKUDA: Constituents of *Astilbe thunbergi*. J. Pharm. Soc. Japan **72**, 578 (1952): Chem. Abstr. **46**, 8810 (1952).
258. SHIMOJIMA, Y., and H. HAYASHI: 1H-2-Benzopyran-1-one derivatives, Microbial Products with Pharmacological Activity. Relationships between Structure and Activity in 6-[[1S-(3S,4-Dihydro-8-hydroxy-1-oxo-1H-2-benzopyranon-3-yl-3-methylbutyl]amino]-4S,5S-dihydroxy-6-oxo-3S-ammoniohexanate. J. Med. Chem. **26**, 1370 (1983).
259. SHIMOJIMA, Y., H. HAYASHI, T. OOKA, M. SHIBKAWA, and Y. ITTAKA: Studies on AI-77s, Microbial Products with Gastroprotective Activity: Structures and the Chemical Nature of AI-77s. Tetrahedron **40**, 2519 (1984).
260. SHIOZAKI, M., K. MORI, and M. MATSUI: Synthesis of Isocoumarins III Oosponol Diacetate. Agric. Biol. Chem. **32**, 42 (1968).
261. SIBI, M.P., M.A. MIAH, and S.V. JALIL: Ortho-lithiated Tertiary Benzamides. Magnesium Transmetalation and Synthesis of Phthalides and Isocoumarins including Mellein and Kigelin. J. Organ. Chem. (USA) **49**, 737 (1984).
262. SIMPSON, T.J.: ^{13}C Nuclear Magnetic Resonance Spectra and Biosynthetic Studies of Xanthomegnin and Related Pigments from *Aspergillus sulphureus* and *melleus*. J. Chem. Soc. Perkin Trans. I **1977**, 592.
263. - Use of Long Range Proton-Carbon-13 Couplings in Structure Determination. Stellatin, a Novel Dihydrisocoumarin from *Aspergillus variecolor*. Chem. Commun. **1978**, 627.
264. SLATES, H.L., S. WEBER, and N.L. WENDLER: 3,5-Dimethoxyhomophthalic Acid and its Transformation into Intermediates in Mold Metabolite Synthesis. Chimia **21**, 468 (1967).

265. SPASOR, S., I. ATANOSOVA, and M. HAIMOVA: Carbon-13 NMR Spectra of Isocoumarins, N-Methyl-1($2H$)-isoquinolines and Related Compounds. Org. Magn. Resonan. **22**, 194 (1984).
266. SPRINGER, J.P., H.G. CUTLER, F.G. CRUMLEY, R.H. COX, E.E. DAVIS, and J.E. THEAN: Plant Growth Regulatory Effects and Steroechemistry of Cladosporin, J. Agric. Food Chem. **29**, 853 (1981).
267. STACK, M.E., R.M. EPPLEY, P.A. DREIFUSS, and A.E. POHLAND: Isolation and Identification of Xanthomegnin, Viomellein, Rubrosulphin, and Viopurpurin as Metabolites of *Penicillium viridicatum*. Appl. Environ. Microbiol. **33**, 351 (1977).
268. STACK, M.E., E.P. MAZZOLA, and R.M. EPPLEY: Structures of Xanthoviridicatin D and Xanthoviridicatin G. Metabolites of *Penicillium viridicatum*: Application of Proton and Carbon-13 NMR Spectroscopy. Tetrahedron Letters **1979**, 4989.
269. STACK, M.E., and P.B. MISLIVEC: Production of Xanthomegnin and Viomellein by Isolates of *Aspergillus ochraceus, Penicillium cyclopium* and *Penicillium viridicatum*. Appl. Environ. Microbiol. **36**, 552 (1978).
270. STODOLA, F.H., C. CABOT, and C.R. BENJAMIN: Structure of Ramulosin, a Metabolic product of the Fungus *Pestalotia ramulosa*. Biochem. J. **93**, 92 (1964).
271. STOESSL, A.: 8-Hydroxy-6-methoxy-3-methylisocoumarin and Other Metabolites of *Ceratocystis fimbriata*. Biochem. Biophys. Res. Comm. **35**, 186 (1969).
272. STOESSL, A., and J.B. STOTHERS: Postinfectional Inhibitors from Plants. Part XXXII. A Carbon-13 Biosynthetic Study of Stress Metabolites from Carrot Roots: Eugenin and 6-Methoxymellein. Canad. J. Bot. **56**, 2589 (1978).
273. STEYN, P.S.: Ochratoxin and Other Dihydroisocoumarins. Microbial Toxins **6**, 179 (1971).
274. STEYN, P.S., and C.W. HOLZAPFEL: Synthesis of Ochratoxin A and B. Metabolites *Aspergillus ochraceus*. Tetrahedron **23**, 4449 (1967).
275. STEYN, P.S., C.W. HOLZAPFEL, and N.P. FERREIRA: The Biosynthesis of the Ochratoxins, Metabolites of *Aspergillus ochraceus*. Phytochem. **9**, 1977 (1970).
276. SUZUKI, Y.: Fusamarin, a New Metabolite from a Species of *Fusarium*. Agric. Biol. Chem. **34**, 760 (1970).
277. SUZUKI, H., T. IKEDA, T. MATSUMOTO, and M. NOGUCHI: Isolation and Identification of a New Glycoside, Phyllodulcin-8-O-β-D-glucose from the Cultured Cells and Fresh Leaves of Amacha (*Hydrangea macropylla* Seringe var. *Thunbergii* Makino). Agric. Biol. Chem. **41**, 1815 (1977).
278. SUZUKI, H., T. MATSUMOTO, and M. NOGUCHI: Polyphenol Components in the Leaves of Amacha (*Hydrangea macrophylla* Seringe var *Thunbergii* Makino). Agric. Biol. Chem. **43**, 653 (1979).
279. TAKEUCHI, N., M. MASAYUKI, K. OCHI, and S. TOBINAGA: Biogenetic-type Synthesis of ($\pm$)-Phylloducin, a Sweet Principe of *Hydrangea serrata* Seringe var. *Thunbergii* Sugimoto. Chem. Commun. **1976**, 820.
280. TAKEUCHI, N., M. MURASE, K. OCHI, and S. TOBINAGO: Biogenetic-type Synthesis of Phyllodulcin, a Sweet Principle of *Hydrangea serrata* Seringe var. *Thunbergii* Sugimoto. Chem. Pharm. Bull. (Japan) **28**, 3013 (1980).
281. TAKEUCHI, N., K. OCHI, M. MURASE, and S. TOBINAGA: An Easy Two Synthon Synthesis of a Sweet Dihydroisocoumarin ($\pm$)-Phyllodulcin. Chem. Commun. **1980**, 593.
282. ---- An Easy Two-Synthon Synthesis of a Sweet Dihydroisocoumarin ($\pm$)-Phyllodulcin. Chem. Pharm. Bull. (Japan) **31**, 4360 (1983).
283. TANAKA, A.K., A. KOBAYASHI, and K. YAMASHITA: A Total Synthesis of Sclerotinin A. Agric. Biol. Chem. **37**, 669 (1973).
284. TANAKA, A.K., C. SATO, Y. SHIBATA, A. KOBAYASHI, and K. YAMASHITA: Growth Promoting Activities of Sclerotinin A and its Analogs. Agric. Biol. Chem. **38**, 1311 (1974).

285. TANENBAUM, S.W., S.C. AGARWAL, T. WILLIAMS, and R.G. PITCHER: 6-Hydroxyramulosin – a New Metabolite from *Pestalotia ramulosa*. Tetrahedron Letters **1970**, 2377.
286. TANEYAMA, M., and S. YOSHIDA: Incorporation of Glucose-^{14}C into Bergenin and Arbutin in *Saxifraga stolonifera*. Botan. Mag. **92**, 69 (1979).
287. TATSUMAKI, T., and T. INOUE: Distribution of Bergenin in Saxifragaceae. Sigenkagaku Kenkyusyo Iko **10**, 13 (1946); Chem. Abstr. **42**, 3034 (1948).
288. THOMAS, R.: The Biosynthesis of Alternariol and its Relation to Other Fungal Phenols. Biochem. J. **78**, 748 (1961).
289. – The Biosynthesis of Phenalenones. Pure Appl. Chem. **34**, 515 (1973).
290. TIRODKAR, R.B., and R.N. USGAONKAR: Isocoumarins. New Findings about the Condensation of Homophthalic Anhydride with Benzoic Anhydride and Benzyl Chloride. New Synthesis for Some 3-Phenylisocoumarins. Current Sci. (India) **41**, 701 (1972).
291. TOKOROYAMA, T., T. KAMIKAWA, and T. KUBOTA: The Structure of Sclerin, a Metabolite of *Sclerotinia libertiana*. Tetrahedron **24**, 2345 (1968).
292. TOKOROYAMA, T., and T. KUBOTA: Biosynthesis of Sclerin. J. Chem. Soc. (C) **1971**, 2703.
293. TURNER, W.B.: Fungal Metabolites. London: Academic Press 1971.
294. TURNER, W.B., and D.C. ALDRIDGE: Fungal Metabolites II. London: Academic Press 1983.
295. UEMURA, M., and T. SAKAN: The Synthesis of Oosponol and Oospoglycol. Chem. Commun. **1971**, 921.
296. ULUBELEN, A., S. OKSUZ, Y. AYNEHCHI, S.M.H. SALEKI, and A. SOURI: Capillarin and Scaparone from *Artemisia lamprocaules*. J. Nat. Prod. **47**, 170 (1984).
297. UMEZAWA, H.: Chemistry of Enzyme Inhibitors of Microbial Origin. Pure Appl. Chem. **33**, 129 (1973).
298. VAN DER MERWE, K.J., P.S. STEYN, and L. FOURIE: The Constitution of Ochratoxins A, B, and C. Metabolites of *Aspergillus ochraceus* Wilh. J. Chem. Soc. (London) **1965**, 7083.
299. VAN EIJK, G.W.: A Naphtho[1,2-b]furan Derivative from the Fungus *Roesleria pallida*. Phytochem. **10**, 3203 (1971).
300. VAN WALBEEK, W., P.M. SCOTT, J. HARWIG, and J.W. LAURENCE: Canad. J. Microbiol. **15**, 1281 (1969).
301. VLEGGAAR, R., and P.S. STEYN: The Biosynthesis of Some Miscellaneous Mycotoxins. In: Steyn, P.S.: The Biosynthesis of Mycotoxins, a Study in Secondary Metabolism. New York: Academic Press 1980.
302. VON DREELE, R.B.: The Crystal and Molecular Structure of *N*-Acetylactinobolin: the α-Helix. Acta Crystalogr. **32B**, 2852 (1976).
303. VORA, V.C.: Metabolic Products of *Paecilomyces victoriae*. J. Sci. Indust. Res. (India) **13B**, 842 (1954).
304. WAITZ, J.A., and C.G. DRUBE: Antifungal Agents. Annu. Rep. Med. Chem. **7**, 109 (1972).
305. WATANABE, M., M. SAHARA, S. FURUKAWA, R. BILLEDEAU, and V. SNIECKUS: Directed Metalation of Tertiary Benzamides. Short Syntheses of Hydrangenol and Phyllodulcin. Tetrahedron Letters **23**, 1647 (1982).
306. WATANABE, M., M. SAHARA, M. KUBO, S. FURKAWA, R.J. BILLEDEAU, and V. SNIECKUS: Ortho-lithiated Tertiary Benzamides. Chain Extension via *o*-Toluamide Anion and General Synthesis of Isocoumarins Including Hydrangenol and Phyllodulcin. J. Organ. Chem. (USA) **49**, 742 (1984).
307. WEI, R.-D., F.M. STRONG, and E.B. SMALLEY: Incorporation of Chlorine-36 into Ochratoxin A. Appl. Microbiol. **22**, 276 (1971).

308. WEISLEDER, D., and E.B. LILLEHOJ: Structure of Viriditoxin, a Toxic Metabolite of *Aspergillus viridi-nutans*. Tetrahedron Letters **1971**, 4705.
309. – – Carbon-13 Magnetic Resonance Assignments and Biosynthesis of Ochratoxin A. Tetrahedron Letters **1980**, 993.
310. WETHERINGTON, J.B., and J.W. MONCRIEF: The Crystal Structure and Absolute Configuration of the Antibiotic Actinobolin. Acta Crystallogr. **31 B**, 501 (1975).
311. WIRTH, J.C., T.E. BEESLEY, and S.R. ANAND: The Isolation of Xanthomegnin from Several Strains of the Dermatophyte *Trichophyton rubrus*. Phytochem. **4**, 505 (1965).
312. YAGI, A., Y. OGATA, T. YAMAUCHI, and I. NISHIOKA: Metabolism of Phenylpropanoids in *Hydrangea serrata* var *thunbergii* and the Biosynthesis of Phyllodulcin. Phytochem. **16**, 1098 (1977).
313. YAMAKI, T., and T. INOUE: Isolation of Bergenin from the Roots of Several Saxifragaceae. Acta Phytochim. (Tokyo) **14**, 93 (1944); Chem. Abstr. **45**, 4308 (1951).
314. – – Distribution of Bergenin in Saxifragaceae. Misc. Rep. Res. Inst. Nat. Resources (Tokyo) **10**, 13 (1946); Chem. Abstr. **44**, 9013 (1950).
315. YAMAMOTO, I., K. NITTA, and Y. YAMAMOTO: Chemical Structure of Oosponol. Agric. Biol.Chem. **26**, 486 (1962).
316. YAMATO, M.: Synthesis and Reactrions of 3,4-Dihydroisocoumarins and Isochromans. Yuki Gosei Kagaku Kyokaishi **41**, 958 (1983); Chem. Abstr. **100**, 68041 (1984).
317. YAMATO, M., and K. HASHIGAKI: Synthesis of *dl*-Agrimonolide (Constituent of the Rhizome of *Agrimonia pilosa* Leder). Chem. Pharm. Bull. (Japan) **24**, 200 (1976).
318. YAMATO, M., K. HASHIGAKI, E. HONDA, K. SATO, and T. KOYAMA: Chemical Structure and Sweet Taste of Isocoumarin and Related Compounds. Chem. Pharm. Bull. (Japan) **25**, 695 (1977).
319. YAMATO, M., K. SATO, K. HASHIGAKI, M. OKI, and T. KOYAMA: Chemical Structures and Sweet Taste of Isocoumarins and Related Compounds. Chem. Pharm. Bull. (Japan). **22**, 475 (1974).
320. YAMATODANI, S., T. YAMANO, Y. KOZU, and M. ABE: Isolation of a New Isocoumarin Derivative, K-1, from the Saprophytic Culture of *Oospora astringens*. Nippon Nogei Kagaku Kaishi **37**, 240 (1963); Chem. Abstr. **63**, 570 (1965).
321. YAMAZAKI, Y., Y. MAEBAYASHI, and K. MIYAKI: Biosynthesis of Ochratoxin A. Tetrahedron Letters **1971**, 2301.
322. – – – Isolation of a New Metabolite, 6-Methoxy-8-hydroxyisocoumarin-3-carboxylic Acid from *Aspergillus ochraceus* Wilk. Chem. Pharm. Bull. (Japan) **20**, 2276 (1972).
322a. YOSHIOKA, M., H. NAKAI, and M. OHNO: Stereocontrolled Total Synthesis of (+) Actinobolin by an Intramolecular Diels-Aber Reaction of a chiral Z Diene from L-Threonine. J. Amer. Chem. Soc. **106**, 1133 (1984).
323. ZAMIR, L.O., and C.C. CHIN: Aromatic Origin of Cyclopentenoid Metabolites. Bioorganic Chem. **11**, 338 (1982).
324. ZEECK, A., P. RUB, H. LAATSH, W. LOEFFLER, H. WEHOLE, H. ZAHNER, and H. HOLST: Isolation of the Antibiotiv Semi-vioxanthin from *Penicillium citreo-viride* and Synthesis of Xanthomegnin. Chem. Ber. **112**, 957 (1979).

(Received August 13, 1985)

Anthraquinones in the Rubiaceae [1]

R. WIJNSMA and R. VERPOORTE, Biotechnology Delft Leiden, Center for Biopharmaceutical Sciences, Division of Pharmacognosy, State University of Leiden, Gorlaeus Laboratories, Leiden, The Netherlands

Contents

I. Introduction 79
II. Biological Activity 80
III. Biosynthesis 81
IV. Spectroscopy 83
    1. UV Spectroscopy 84
    2. IR Spectroscopy 84
    3. Mass Spectrometry 85
    4. ^{1}H-NMR Spectroscopy 85
    5. ^{13}C-NMR Spectroscopy 86
V. Artifacts 87
VI. Separation Methods 88
VII. Physical and Spectroscopic Properties of Anthraquinones (Table 7) 91
VIII. Rubiaceae Species Containing Anthraquinones (Table 8) 131
References 143

I. Introduction

This review has been written in order to update the literature on anthraquinones occurring in the Rubiaceae. Since appearance of the excellent book on naturally-occurring quinones by R.H. THOMSON (*120*) in 1971 about 50 new anthraquinones have been isolated from members

---

[1] This article is dedicated to Prof. Dr. A. BAERHEIM SVENDSEN on the occasion of his 65th birthday.

of the Rubiaceae. Also several new methods have been used for structure analysis and separation of the anthraquinones which have not been subject to review before. For this review we have confined ourselves to the naturally occurring anthraquinones and therefore the synthesis of anthraquinones is not included. The article covers the literature from 1969 to 1984.

Plants in the family Rubiaceae, and specifically in the genera *Rubia*, *Galium* and *Morinda*, have long been known to contain substantial amounts of anthraquinones (*120*), with the roots being especially rich sources of anthraquinones. Tissue cultures of Rubiaceae also produce large amounts of anthraquinones (*112, 129*) and in some cases the percentage found in tissue culture even exceeds that of the parent plant (*112*). The anthraquinones are most often present as aglycones and sometimes in the form of glycosides; in the latter case the sugar moiety is most often primeveroside (glucose-xylose).

Because of their anthraquinone content many plants of the Rubiaceae haven been used for the preparation of natural dyes all over the world. The best known for this purpose is madder, i.e. the ground root of *Rubia tinctorum* L., which produces alizarin (*13*). The colours most frequently observed when alizarin is used with an aluminum mordant are various shades of red. But as alizarin forms lakes with many metal oxides many other colours can be obtained as well. The colours for which madder has been used most frequently are red, pink, violet, lilac, dark-brown and black (*50*).

The cultivation of madder for dyeing purposes was once of great economic importance as can be seen from the value of the traditional madder industry in France which was around 1865 worth about 16 million US dollars per year. The synthesis of alizarin in 1869 by Graebe and Liebermann wiped out the madder industry in about ten years, with the revenues being transferred completely from France to Germany (*50*).

II. Biological Activity

The naturally occurring anthraquinones can be divided into two groups on the basis of their biosynthetic pathways, a division which parallels therapeutic use. The first of the two groups is the acetate-derived class of anthraquinones (*86*). This group of anthraquinones which occurs mostly in the form of glycosides is found in the Leguminosae (*Cassia* sp.), Rhamnaceae (*Rhamnus* sp.) and Polygonaceae (*Rheum*

References, pp. 143–149

sp.) and is widely used in medicine because of their laxative action (*7, 51*).

The second group of anthraquinones found in the Rubiaceae and some related families, e.g. the Bignoniaceae, Scrophulariaceae and Verbenaceae, has been used in traditional medicine (*101*), for example in Japan, for the coagulation of blood (*117*). In West and East Germany extracts of the roots of *Rubia tinctorum* L. have been used for the treatment of kidney stones (*16, 111*). To our knowledge no other therapeutic uses have been reported for rubiaceous anthraquinones in western medicine and they do not have any laxative action (*111*). Some of the rubiaceous anthraquinones, however, exhibit very interesting biological *in vitro* activities: antimicrobial (*101, 127*), hypotensive (*92*) and antileukemic (*41, 42*) properties have been reported.

III. Biosynthesis

Much work on the biosynthesis of anthraquinones in the Rubiaceae has been performed by LEISTNER and co-workers during the past twenty years. Earlier work of LEISTNER (*78, 79, 81*) has shown that the rubiaceous anthraquinones are formed through the shikimate-mevalonate pathway. In these studies the specific incorporation of radioactive labeled precursors such as o-succinylbenzoic acid and shikimic acid into anthraquinones was demonstrated. In a more recent study INOUE and co-workers (*72*) have also shown that shikimic acid and o-succinylbenzoic acid are involved in anthraquinone biosynthesis in *Galium mollugo* L. cell suspension cultures and that 1,4-dihydroxynaphtoic acid and 1,4-dihydroxy-3-prenyl-2-naphtoic acid are important intermediates in this biosynthetic pathway. LEISTNER and co-workers recently have shown how the shikimate pathway is linked to the vitamin K_2 biosynthetic pathway in *E. coli*. (*134*). With crude enzyme preparations o-succinylbenzoic acid was formed from iso-chorismic acid and α-ketoglutaric acid and thiamine pyrophosphate as cofactor. Chorismic acid which was previously (*135*) believed to be a precursor for o-succinylbenzoic acid was not converted to any measurable extent to o-succinylbenzoic acid by these enzyme preparations. KOLKMANN and LEISTNER (*136*) have further demonstrated that the activated o-succinylbenzoic acid is the "aliphatic" acetyl coenzyme ester of o-succinylbenzoic acid which cyclizes to 1,4-dihydroxy-2-naphtoic acid. It seems thus that there are major similarities in the biosynthesis of anthraquinones in the Rubiaceae (see Scheme 1) and of menaquinone in bacteria (*135*), with 1,4-di-

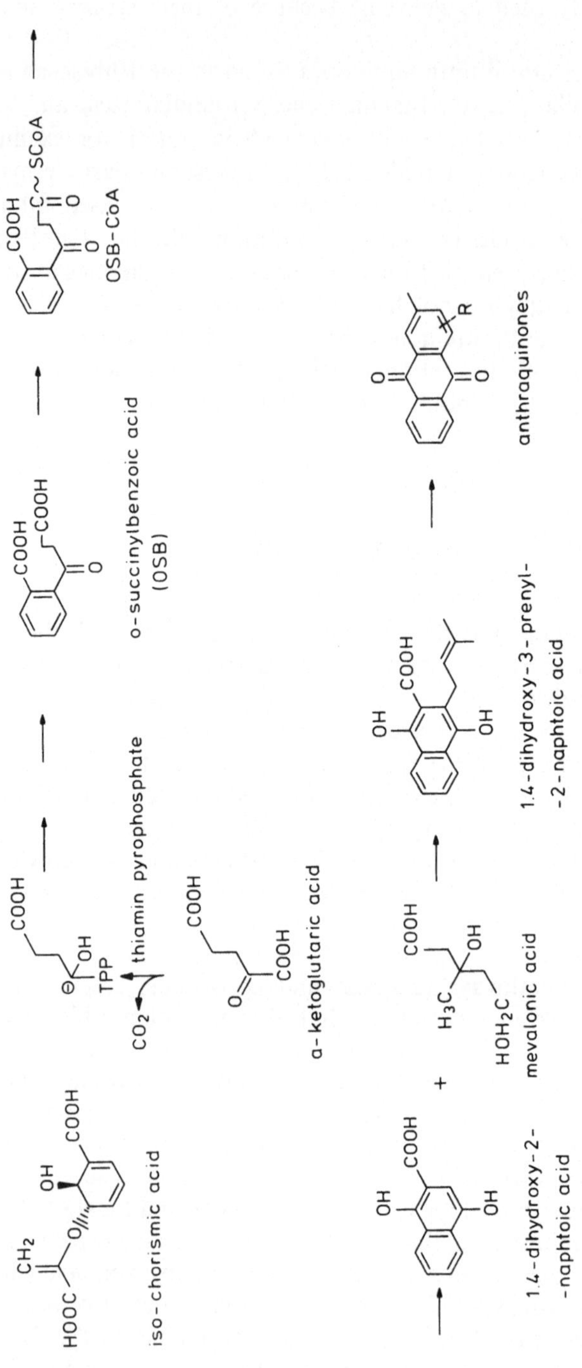

Scheme 1. Biosynthetic Pathway leading to Anthraquinones in the Rubiaceae

hydroxy-2-naphtoic acid being the branchpoint between bacterial vitamin K biosynthesis and anthraquinone biosynthesis in higher plants.

From this biosynthetic route it is obvious that a methyl group or an oxidized methyl group is always found in the 2- or in the 6-position which are equivalent from a biogenetic point of view. It is therefore proposed to use a biogenetic numbering for the anthraquinones found in the Rubiaceae (Fig. 1). Some authors have given names to anthraquinones isolated from rubiaceous plants which at first sight do not obey this rule. These compounds, however, can be easily given names which do fit into the general biogenetic nomenclature, i.e. compounds in which the (oxidized) methyl group is present in the 2-position. A good example of this procedure is 8-hydroxysubspinosin (*89*), which if the biogenetic numbering were followed, would become 3,5-dihydroxy-2-ethoxymethyl-/4-methoxyanthraquinone. This structure is in another aspect also more plausible for an anthraquinone found in a member of the Rubiaceae because 5- and/or 6- and/or 7-substituents are much more frequently encountered in the rubiaceous anthraquinones than 8-substituents. In this review we will consequently follow the proposed biogenetic numbering and the original numbering (or name) will be given in parentheses.

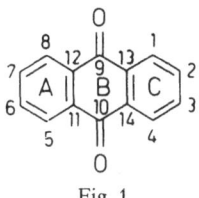

Fig. 1

IV. Spectroscopy

The anthraquinones found in the Rubiaceae constitute a homogeneous group of compounds, all being derivatives from the basic tricyclic structure given in Fig. 1. They differ only in the nature of substituents and the substitution pattern of ring A and C. The substituents found most frequently are, in addition to the already mentioned (oxidized) methyl group in the 2-position, hydroxyl and methoxyl groups. These substituents are mostly found in ring C at the 1-, 2-, 3- and 4-positions and in ring A at the 5- and 6-positions. To establish the nature and number of substituents as well as the substitution pattern spectroscopic methods have proved extremely valuable. In this review we will consecutively discuss the merits of UV-, IR-, mass-, ^{1}H-NMR- and ^{13}C-NMR spectroscopy.

1. UV Spectroscopy

As can be seen from Table 1 the UV-VIS-spectrum recorded in alkaline methanol or ethanol gives us information on the location of hydroxyl groups (*19*). The values for the UV maximum in alkaline ethanol in Table 1 are recorded for pure hydroxy-anthraquinones not bearing any other substituents. In the case of naturally occurring anthraquinones which contain other substituents as well, either positive or negative effects due to these substituents on the bathochromic shift are possible.

Table 1. *Longest Wavelength Maxima in the UV Spectra of Hydroxyanthraquinones Recorded in Alkaline Ethanol, Useful for the Determination of Hydroxyl Group Substitution Patterns*

	λ_{max} (nm)
1-hydroxyanthraquinone	500
2-hydroxyanthraquinone	478
1,2-dihydroxyanthraquinone	576
1,3-dihydroxyanthraquinone	485
1,4-dihydroxyanthraquinone	560
1,5-dihydroxyanthraquinone	510
1,8-dihydroxyanthraquinone	515
1,4,5,8-tetrahydroxyanthraquinone	630

2. IR Spectroscopy

The IR spectrum of an anthraquinone gives information on the environment of the carbonyl groups in the 9- and 10-positions. If a *peri* hydroxyl group is present, the carbonyl absorption band will show a shift to smaller wave numbers due to hydrogen bonding between the *peri* hydroxyl and the adjacent carbonyl group (*120*). The presence

Table 2. *Absorption Bands of Primary Diagnostic Value in the IR Spectra of Anthraquinones*

	wave number (cm^{-1})
Free hydroxyl group	3400
Not-chelated carbonyl group	1675
Chelated carbonyl group	1630

of a free (or 2-, 3-, 6-, 7-positioned) hydroxyl group can also be readily deduced from the IR spectrum (see Table 2). For a more detailed discussion of the possibilities for structure elucidation of anthraquinones by means of IR spectroscopy the reader is referred to (*120*).

3. Mass Spectrometry

Mass spectrometry gives information on the molecular weight of a compound, the M^+ peak often being the base peak in the EI mass spectra of the anthraquinone aglycones. From the mass spectrum also information on the location of methoxyl groups can be obtained. In the case of a *peri* positioned methoxyl group loss of water from the molecular ion will occur, giving rise to an M^+-18 peak (*20*); however, this information should be viewed with caution if in the same molecule a hydroxyl group is present. The loss of water from the molecular ion will also occur when a hydroxyl group is positioned *ortho* to a methoxyl group (*25*).

4. ¹H NMR Spectroscopy

As is generally the case in natural products chemistry ¹H/NMR/ spectroscopy is an indispensable tool for structure elucidation of anthraquinones. In many instances the ¹H NMR spectrum allows deduction of the nature and the number of substituents, because all the substituents (hydroxyl groups in 2-, 3-, 6- or 7-position are not always readily observable) will give rise to signals with a characteristic chemical shift (see Table 3). By means of high resolution ¹H NMR spectrometry the substitution pattern for both ring A and C can also be deduced from the splitting patterns observed in the aromatic region between 7.0 and 8.5 ppm. The usual coupling constants for the aromatic protons are 7–8 Hz for *ortho* coupling and ca. 1.5 Hz for *meta* coupling. Also from the chemical shifts observed for the aromatic protons information on their position can be obtained. In the case of a proton at C-1, C-4, C-5 or C-8 a signal above 7.5 ppm will be seen whereas a proton at C-2, C-3, C-6 or C-7 will give rise to a signal below 7.5 ppm. Because the two carbonyl groups in the molecule exert a relatively strong deshielding effect on substituents in the *peri* position this can be of help in the location of substituents. A *peri* positioned methoxyl group will give rise to a signal at ca. 4.05 ppm; when a methoxyl group is in the 2-, 3-, 6- or 7-position, the signal will be found at ca. 3.9 ppm. The deshielding effect of and hydrogen bonding with the carbonyl

Table 3. *Characteristic Features in the ^1H NMR Spectra of Anthraquinones ($CDCl_3$)*

Substituents	Chemical shift (ppm)
CH_3	2.1–2.4
OCH_3 (in 2,3,6,7-position)	3.8–3.95
OCH_3 (in peri-position)	3.9–4.15
CH_2O-	4.4–4.8
$O-CH_2-O$	6.2
H-6, H-7 (of an unsubstituted A-ring)	7.7–7.8
H-5, H-8 (of an unsubstituted A-ring)	8.1–8.3
Aldehydic H	ca. 10.5
Chelated OH	ca. 13

A signal due to a unchelated hydroxyl group is usually not observed in the ^1H NMR spectra, but according to some authors it is seen at ca. 10 ppm.

groups helps in discriminating between a free hydroxyl group and a hydroxyl group in the *peri* position. The signal due to a *peri* positioned hydroxyl group will be found between 12 and 13 ppm; the signal due to a free hydroxyl group – if it is seen at all – will be found at ca. 10 ppm. All chemical shift values given in Table 3 or in the text are derived from spectra recorded in $CDCl_3$. When spectra of anthraquinones are recorded in $DMSOd_6$ or CD_3OD, the values observed for the chemical shifts are lower, usually by ca. 0.10–0.20 ppm.

5. ^{13}C NMR Spectroscopy

^{13}C NMR data for both synthetic and naturally occurring anthraquinones have been reported in recent years and attempts have been made to determine substituent induced chemical shift parameters. For monosubstituted anthraquinones the SCS values were roughly consistent with those of monosubstituted benzene. However, hydroxyl groups in the *peri* position cause a considerable paramagnetic shift in the C-9 (C-10) signal, ca. 5 ppm due to hydrogen bonding. This also results in rather different SCS values for the two *ortho* carbons. Typical shifts for C-9 and C-10 are 183 ppm; in the case of hydrogen bonding to one hydroxyl group this is shifted to 188 ppm, and, if hydrogen bonding to two hydroxyl group is possible, to 192 ppm. Methoxyl signals are usually observed at ca. 56 ppm; however, in the case of a 1-methoxyl group in combination with a substituent at C-2 the methoxyl signal is shifted downfield to ca. 61–64 ppm, due to steric hindrance (*18*).

References, pp. 143–149

From an experimental point of view it is to be noted that the signals due to the carbons 11, 12, 13, 14 are difficult to observe due to their very long relaxation times. By using a pulse delay of 10 sec. (*17*) or adding of a relaxation agent such as chromium tris(acetylacetonate) (*5, 18*) the signals of these carbons can be made visible. Spectra have been recorded in $CDCl_3$ and $DMSO_{d6}$, shift differences due to the solvent being in the range of 0–1 ppm (*18*).

HÖFLE (*67*) found assignment of the various signals in the anthraquinone spectra quite difficult due to the small chemical shift differences. The anthraquinones were converted into the corresponding acetoxy derivatives to improve their solubility in organic solvents. He concluded that the use of long-range C-H couplings is useful in some cases for making assignments and thus to establish the sites of substitution.

ITOKAWA et al. (*74*) reported the ^{13}C NMR spectra of some anthraquinones isolated from *Rubia* species. For one of the compounds a 2-methyl-1,3,6 (or 7)-trihydroxy-9,10-anthraquinone structure was assumed based on various spectral data. Eventually it was stated that the hydroxyl group was in the 6-position rather than in the 7-position because of a 0.6–0.8 ppm upfield shift observed for C-9 if compared with compounds having a 1,3-dihydroxy-2-hydroxymethyl substitution pattern. For a better insight in the validity of using ^{13}C NMR spectrometry for distinguishing 6/7 or 2/3 substitution in anthraquinones bearing several substituents, further ^{13}C NMR studies of such multiply-substituted anthraquinones seem desirable.

In conclusion it can be said that for structure elucidation of anthraquinones ^{13}C NMR spectroscopy does not yield new information that cannot also be obtained from other spectral data such as UV, IR, MS and ^1H NMR; however, for multiply-substituted anthraquinones ^{13}C NMR data might conceivably lead to more definitive structure assignments than are now possible using the other spectroscopic methods.

V. Artifacts

Some anthraquinones which have been isolated are believed to be artifacts formed during extraction with methanol or ethanol (*80*). These anthraquinones all show the presence of 2-methoxymethyl or 2-ethoxymethyl groups, respectively. The procedures for isolation of these compounds generally employed boiling methanol or ethanol as extraction solvents. From this one could conclude that for the extraction of anthraquinones from plant material the use of hot methanol or ethanol should be avoided; whether these solvents can be used in the cold, however, remains to be investigated. Use of ethanol is to be preferred to methanol because ethanol derivatives are not likely to occur in natural compounds.

VI. Separation Methods

For separation of the naturally occurring anthraquinones several chromatographic procedures have been used which include paper chromatography, thin layer chromatography, gas-liquid chromatography, high-performance liquid chromatography and droplet counter-current chromatography. Although it has been extensively used in the past, very little use has recently been made of paper chromatography. FORMANEK (52) and FORMANEK and RACZ (55) employed paper chromatography for separation of anthraquinones from a number of Rubiaceae, using n-butanol:acetic acid:water (4:1:5) as eluent for both the glycosides and the aglycones.

Thin layer chromatography (TLC) has been used extensively for separation of anthraquinones, the most frequently used adsorbents being silica and silica impregnated with tartaric acid. Especially noteworthy is the extensive use of the last-named system by BAUCH and LEISTNER (11) for TLC of anthraquinones from *Galium mollugo* cell cultures. Since almost every research group has used different solvent systems for the separation of anthraquinones, we will give only a few examples (see Table 4), although a large number of other solvent systems have been used in combination with silica.

Table 4. *TLC-Systems Used for the Separation of Rubiaceous Anthraquinone Aglycones and Glycosides*

Adsorbent	Eluent		Refs.
SiO_2	toluene:MeOH = 9:1	aglycones	(75, 128)
SiO_2	C_6H_6:HCO_2Et:HCO_2H = 75:24:1	aglycones	(116)
SiO_2	$CHCl_3$:MeOH:25% NH_4OH = 85:14:1	aglycones	(128)
SiO_2	Hexane:EtOAc = 1:1	aglycones	(119)
SiO_2	C_6H_6	aglycones	(38)
SiO_2	EtOAc:MeOH:H_2O = 100:16.5:13.5	glycosides	(106)
SiO_2 (acid washed)	$CHCl_3$:MeOH:H_2O = 64:36:8	glycosides	(11)
Polyamide (acetylated)	80% EtOH	aglycones	(11)
Cellulose	EtOAc:Py:H_2O = 4:2:2	glycosides	(11)

References, pp. 143–149

The detection of anthraquinones on TLC plates is very simple due to their colour. However, use of other detection methods gives much more information from one TLC plate. Some of the anthraquinones show a colour when observed in UV_{254} light and most of them show colours when observed in UV_{366} light. A change in colour is also observed when hydroxyanthraquinones are sprayed with a solution of KOH or NaOH in methanol (5% w/v) (Bornträger reaction) or after exposure of the chromatograms to NH_3 vapour. As the colour after spraying with KOH in methanol and especially after exposure to NH_3 vapour is not very stable, it must be evaluated as quickly as possible after treatment. The colours observed in daylight, UV light and after spraying and treatment with NH_3 vapour may be of help in the identification of an unknown compound.

Column chromatography has been employed frequently, using either silica or alumina stationary phases, for isolation of individual anthraquinones. Eluents used for column chromatography usually consist of a series of solvents of increasing polarity. Examples of such separations are found in references (*41, 47, 119*).

Gas-liquid chromatography (GLC) for analytical separations has proved to be a very valuable technique when applied to anthraquinones. Because the anthraquinones as such are not sufficiently volatile, more volatile derivatives have to be prepared prior to GLC. INOUE *et al.* (*70*) used trimethylsilylation for the derivatization of the anthraquinones from cell cultures of *Morinda citrifolia*. VAN EIJK *et al.* (*123, 124*) also prepared trimethylsilyl ethers of the anthraquinones prior to GLC. VAN EIJK and ROEIJMANS (*123*) have effected an excellent separation of 18 naturally occurring anthraquinones using a glass column (1.7 m × 4 mm ID) packed with 3% SE-30 on Gas-Chrom Q (125–150 µm). In a more recent publication (*124*) they described the separation of a large number (37) of naturally occurring anthraquinones by means of capillary gas chromatography. The method was also used in combination with mass spectrometry. HENRIKSEN *et al.* (*63*) claimed that reductive silylation of hydroxylated naturally occurring anthraquinones is the method of choice for derivatization of such compounds due to the excessive tailing observed when other derivatization methods were used. For the GLC analysis they used a 36 m long, all-glass capillary SE-30 SCOT column.

Until now there has only been one report on the use of reversed-phase liquid chromatography (LC) for the isolation of anthraquinones from rubiaceous plants. DEMAGOS and co-workers (*44*) used reversed-phase liquid chromatography for the separation of anthraquinone glycosides from *Morinda lucida* heartwood. For the separation of anthraquinone glycosides reversed-phase high performance (HP) LC seems

Table 5. *HPLC Systems Used for Separation of Anthraquinone Aglycones and Glycosides*

Stationary phase	Eluent		Refs.
Lichrosorb RP18 10 µm	MeOH:H_2O:HAc (40:60:1)	glycosides	*(44)*
Spherisorb ODS 5	MeOH:2.5% HCOOH in H_2O (81.5:18.5)	aglycones	*(14)*
Permaphase ODS	MeOH:H_2O	aglycones/glycosides	*(103)*
Micropack Si-10	MeOH:n-pentane	aglycones/glycosides	*(103)*
Corasil II Si	grad. cyclohexane→EtOAc	aglycones	*(104)*
Nucleosil 5 N($CH_3)_2$	THF:H_2O:HAc (8:2:1)	glycosides	*(99)*

to be very suitable, as in the separation of the anthraquinones from *Senna, Rheum* or *Rhamnus* sp. good results were obtained with a reversed-phase HPLC system (see Table 5). Also for the separation of anthracycline antibiotics which are comparable to anthraquinone glycosides several reversed-phase HPLC systems have been described (*6*). The eluents used for the separations of the glycosides generally consisted of mixtures of water and acetonitrile or of water and methanol. Also for the separation of anthraquinone aglycones the use of reversed-phase stationary phases in combination with eluents consisting of water-methanol mixtures with a pH of 2–3 seems very useful. Thus mixtures of anthraquinone aglycones from suspension cultured cells of *Cinchona ledgeriana* could be separated by means of Lichrosorb RP 18 5 µm and water:methanol:acetic acid (30:70:1) as an eluent *(127)*.

A new separation technique which has been applied successfully to the isolation of anthraquinones is droplet countercurrent chromatography (DCCC). Thus INOUE and co-workers have used DCCC for the isolation of anthraquinones (*70, 72*). For the solvent systems used see Table 6.

Table 7 which follows lists the anthraquinones isolated from Rubiaceae in alphabetical order, together with their formulas, physical prop-

Table 6. *DCCC Systems Used for the Separation of Antraquinone Aglycones and Glycosides*

Solvent system	Method		Refs.
n-Hexane:EtOH:H_2O:EtOAc (5:4:1:2)	ascending	aglycones	*(70)*
EtOAc:n-PrOH:H_2O (7:3:9)	ascending	glycosides	*(70)*
$CHCl_3$:MeOH:H_2O (5:5:3)	ascending	glycosides	*(70, 72)*

References, pp. 143–149

erties, such spectral properties as are recorded in the literature and their plant sources. Table 8 (p. 131) lists species of Rubiaceae reported to contain anthraquinones and the anthraquinones isolated from them.

Table 7. *Spectroscopic and Physical Properties of Anthraquinones Isolated from the Rubiaceae*[a]

[Structure of Alizarin with ^{13}C-NMR chemical shifts labeled on carbons: 126.4, 133.5, 116.1, 150.8, 152.8, 125.9, 121.1, 123.7, 132.7, 126.6, 180.4, 188.7, 133.9, 135.0; ^{1}H-NMR shifts in parentheses: (8.20), (7.93), (7.93), (8.20), (7.70), (7.12); OH groups and two C=O groups shown]

Alizarin

$C_{14}H_8O_4$
1: red-purple; 2, 3: dark; 4: purple-red; 5: purple *(75)*
mp: 288–289° *(38)*; 198–205° *(74)*
UV (EtOH): 248, 264, 277, 330 (sh.), 432 *(75)*
IR (KBr): 3340, 1660, 1625, 1585 *(75)*
MS (50 eV): 240 (M$^+$, 100), 212 (8), 184 (5), 138 (6), 92 (6), 77 (6), *(20)*
^{1}H-NMR (90 MHz, DMSO$_{d6}$) *(75, 74)*
^{13}C-NMR (22.63 MHz, DMSO$_{d6}$): *(18, 67, 74)*

Sources:

Asperula azura L. *(55)*
A. besseriana Klok. *(21)*
A. galioides M.B. *(55)*
A. odorata L. *(37, 55)*
A. tinctoria L. *(55)*
Cinchona ledgeriana Moens. *(94)*
C. pubescens Vahl. *(95)*
Galium album Mill. *(75)*
G. aparine L. *(37, 55)*
G. cruciata (L.) Scop. *(55)*
G. dasypodum Klok. *(131)*
G. fleuroti Jord. *(37)*
G. mollugo L. *(11, 12, 37, 55, 125)*
G. normani Dahl. *(37)*
G. polonicum Br. *(55)*
G. pumilum (L.) Murr. *(37, 55)*
G. purpureum L. *(55)*

G. rubioides L. *(55)*
G. ruthenicum Willd. *(21)*
G. saxatile L. *(37)*
G. saxatile × *G. sterneri* *(37)*
G. schultesii Vest. *(55)*
G. sterneri Ehrend *(37)*
G. verum L. *(37)*
G. vulgare S.F. Gray *(55)*
Morinda citrifolia L. *(79, 80)*
M. umbellata L. *(39)*
Rubia cordifolia L. *(74, 96, 118)*
R. iberica C. Koch. *(98)*
R. oliveri L. *(55)*
R. petiolaris L. *(55)*
R. tinctorum L. *(15, 16, 36, 38, 52, 55, 78, 82, 83, 84, 87, 97)*

---

[a] 1: colour of the anthraquinone in daylight; 2: colour of the anthraquinone in UV 254 nm light; 3: colour of the anthraquinone in UV 366 nm light; 4: colour of the anthraquinone after treatment with NH$_3$ vapour; 5: colour of the anthraquinone after treatment with KOH/NaOH.

UV spectral data given in nm (log ε); IR spectral data in cm^{-1}; ^{1}H- and ^{13}C-NMR spectra from first reference given, in most instances this is the most recent one or the spectrum recorded at highest resolution

Table 7 (*continued*)

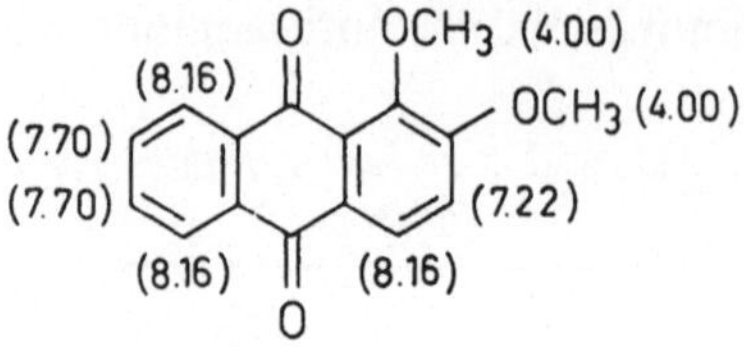

Alizarin dimethyl ether

$C_{16}H_{12}O_4$
1: yellow (*38*)
mp: 214–216° (EtOH) (*38*); 220–221° (*45*)
IR 1668, 1654 (*45*)
^{1}H-NMR (CDCl$_3$) (*100*)
Sources: *Rubia tinctorum* L. (*38*)

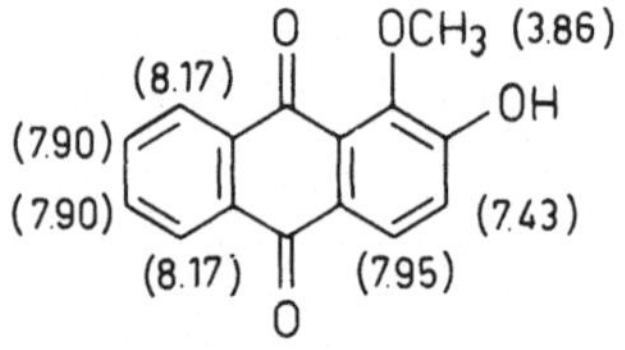

Alizarin 1-methyl ether

$C_{15}H_{10}O_4$
1: orange-yellow; 2, 3: browngreen; 4: orange; 5: orange (*75*)
mp: 178–179° (EtOH) (*38*); 182–184° (CH$_2$Cl$_2$/P.E.) (*26*)
UV (EtOH): 242 (sh., 4.35), 249 (4.37), 270 (4.36), 284 (4.24), 385 (3.65) (*58*)
IR (KBr): 3540, 1670, 1590 (*75*)
MS: 254 (M$^+$, 77), 237 (12), 236 (43), 225 (14), 211 (14), 209 (18), 208 (100), 183 (14), 180 (10), 152 (14) (*75*)
^{1}H-NMR (90 MHz, DMSO$_{d6}$): (*75, 41, 58*)

Sources:

Asperula odorata L. (*38*)
Cinchona ledgeriana Moens. (*94*)
C. pubescens Vahl. (*95*)
Damnacanthus major Sieb. & Zucc. (*26*)
Galium album Mill. (*75*)
G. aparine L. (*37*)
G. mollugo L. (*37*)
G. pumilum L. (*37*)
G. saxatile L. (*37*)
G. saxatile × *G. sterneri* (*37*)

G. sterneri Ehrend. (*37*)
G. verum L. (*37*)
Morinda lucida Benth. (*2*)
M. parvifolia Bartl. (*41, 42*)
M. umbellata L. (*39*)
Oldenlandia umbellata L. (*102*)
Ploclama pendula Ait. (*58*)
Relbunium hypocarpium Hemsl. (*113*)
Rubia tinctorum L. (*38*)

Table 7 (*continued*)

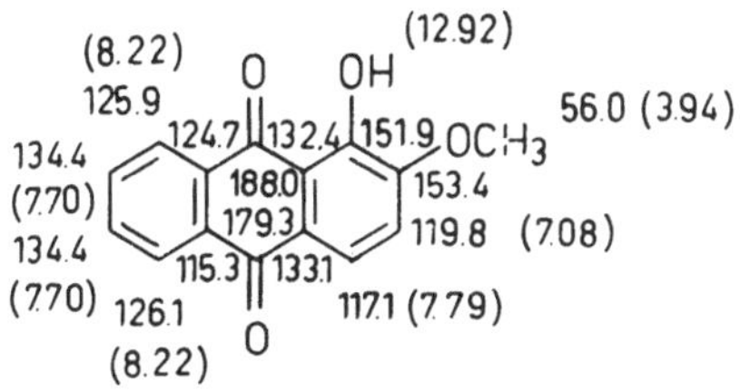

Alizarin 2-methyl ether (with NMR shifts labeled on structure):
(8.22) 125.9; 134.4 (7.70); 134.4 (7.70); 126.1 (8.22); 124.7; 188.0; 179.3; 115.3; 133.1; 132.4; 151.9; 153.4; 119.8 (7.08); 117.1 (7.79); OCH₃ 56.0 (3.94); OH (12.92)

$C_{15}H_{10}O_4$
1: yellow-orange; 3: brown; 5: purple (*47*)
mp: 220° (CH_2Cl_2) (*47*); 230–232° (EtOH) (*38*)
UV (EtOH): 223 (4.21), 250 (3.99), 425 (3.60) (EtOH/OH^-): 232, 515 (*47*)
IR (KBr): 1670, 1640, 1600, 1460, 1440, 1375, 1300, 1270, 1080, 1050, 980, 720 (*47*)
MS (70 eV, 200°): 254 (M^+, 100), 236 (10), 225 (46), 211 (21), 208 (14), 127 (17), 111, 97, 85, 83, 71, 57 (*47*)
^{1}H-NMR (270 MHz, $CDCl_3$): (*47*)
^{13}C-NMR (22.63 MHz, $CDCl_3$): (*18*)

Sources:

Asperula odorata L. (*37*)   *G. sterneri* Ehrend (*37*)
Galium aparine L. (*37*)   *G. verum* L. (*37*)
G. mollugo L. (*37*)   *Morinda umbellata* L. (*39*)
G. pumilum L. (*37*)   *Rubia cordifolia* L. (*47*)
G. saxatile L. (*37*)   *R. tinctorum* L. (*38*)

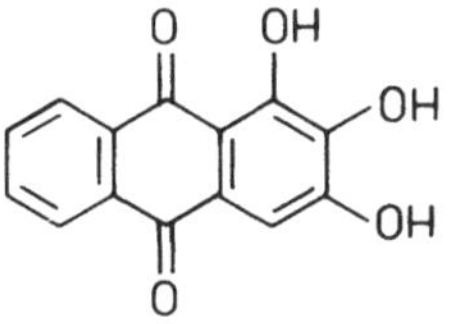

Anthragallol

$C_{14}H_8O_5$
1: orange (*38*)
mp: 309–310° (nitrobenzene) (*38*)
UV (EtOH): 239 (sh., 4.08), 245 (4.12), 286 (4.33), 410 (3.80) (*108*)
IR (nujol): 3450, 3370, 1653, 1630, 1584 (*108*)

Sources:

Hymenodictyon excelsum Wall. (*26*)
Rubia tinctorum L. (*38*)

Table 7 *(continued)*

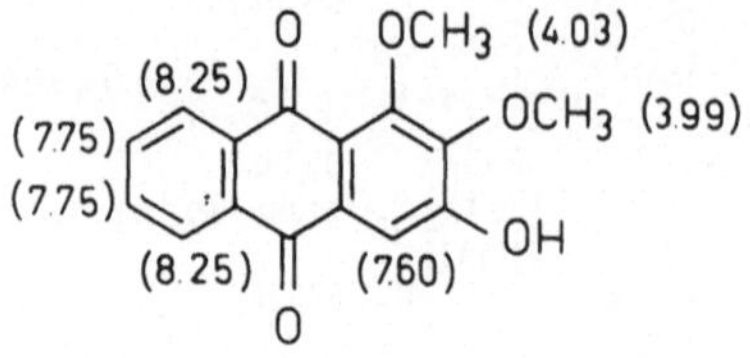

Anthragallol 1,2-dimethyl ether

$C_{16}H_{12}O_5$
1: yellow; 3: red; 4: orange; 5: orange *(128)*
mp: 225–227° (MeOH) *(102)*; 228–232° *(128)*
UV (MeOH): 238, 242 (sh.), 279, 309 (MeOH/OH⁻): 244, 311, 466 *(128)*
IR (CHCl₃): 3420, 2980, 1660, 1575 *(128)*
MS (70 eV, 200°): 284 (M⁺, 95), 269 (80), 266 (5), 256 (25), 83 (63), 55 (*85*), 43 (100) *(128)*
¹H-NMR (100 MHz, CDCl₃): *(128)*

Sources:

Cinchona ledgeriana Moens. *(128)*   *Relbunium hypocarpium* Hemsl. *(113)*
Oldenlandia umbellata L. *(102)*

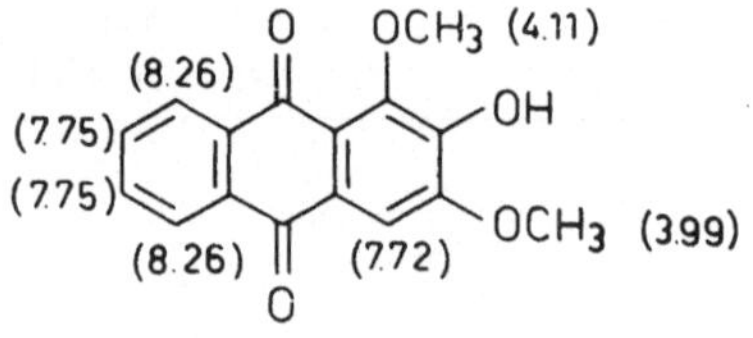

Anthragallol 1,3-dimethyl ether

$C_{16}H_{12}O_5$
1: dark-yellow; 3: orange; 4: orange; 5: red *(128)*
mp: 212° (MeOH) *(102)*
UV (MeOH): 220 (sh.), 242 (sh.), 278, 310 (sh.) (MeOH/OH⁻): 230 (sh.), 246, 312, 490 *(128)*
IR (KBr): 3420, 1675, 1570 *(128)*
MS (70 eV, 200°): 284 (M⁺, 39), 269 (30), 266 (5), 71 (60), 57 (100), 43 (90) *(128)*
¹H-NMR (300 MHz, CDCl₃): *(128)*

Sources:

Cinchona ledgeriana Moens. *(128)*   *Relbunium hypocarpium* Hemsl. *(113)*
Oldenlandia umbellata L. *(102)*

References, pp. 143–149

Table 7 (*continued*)

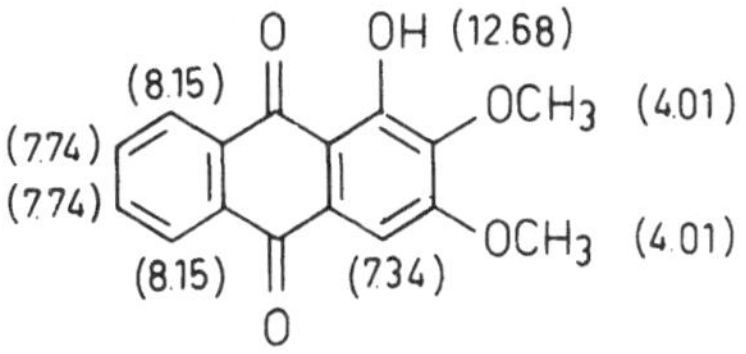

Anthragallol 2,3-dimethyl ether

Structure with NMR shifts: O (8.15), OH (12.68), OCH₃ (4.01), (7.74), (7.74), (8.15), (7.34), OCH₃ (4.01)

$C_{16}H_{12}O_5$
1: yellow (*38*)
mp: 166–167° (EtOH) (*38*); 162–163° (EtOH) (*109*)
IR (KBr): 1667, 1634, 1591, 1576, 1275 (*109*)
^{1}H-NMR (90 MHz, CDCl$_3$): (*109*)

Sources:
Rubia tinctorum L. (*38*)

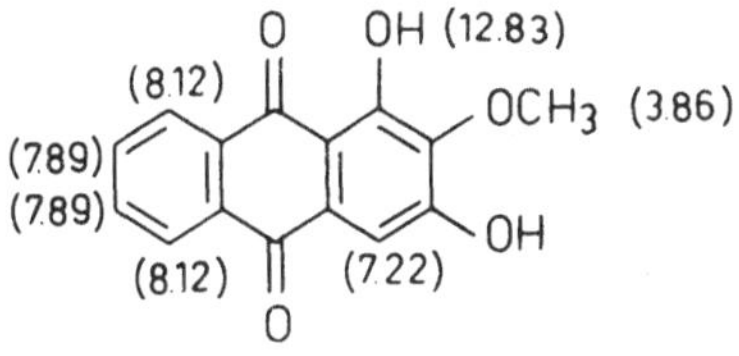

Anthragallol 2-methyl ether

Structure with NMR shifts: O (8.12), OH (12.83), OCH₃ (3.86), (7.89), (7.89), (8.12), (7.22), OH

$C_{15}H_{10}O_5$
1: yellow (*27*)
mp: 216° (*27*); 219° (dioxane) (*109*)
IR (KBr): 3450, 3420, 1666, 1634, 1592, 1582, 1262 (*109*)
^{1}H-NMR (90 MHz, DMSO$_{d6}$): (*109*)

Sources: *Coprosma linariifolia* Hook.f. (*27*)

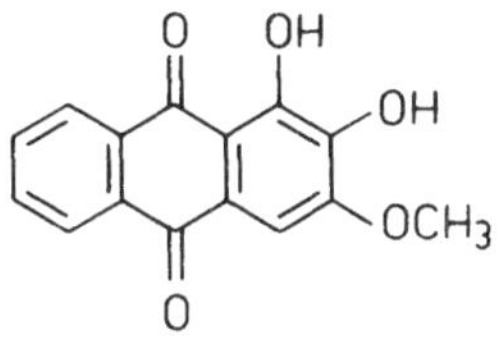

Anthragallol 3-methyl ether

$C_{15}H_{10}O_5$
1: orange (*38*)
mp: 240–242° (acetone) (*38*)

Sources: *Rubia tinctorum* L. (*38*)

Table 7 (*continued*)

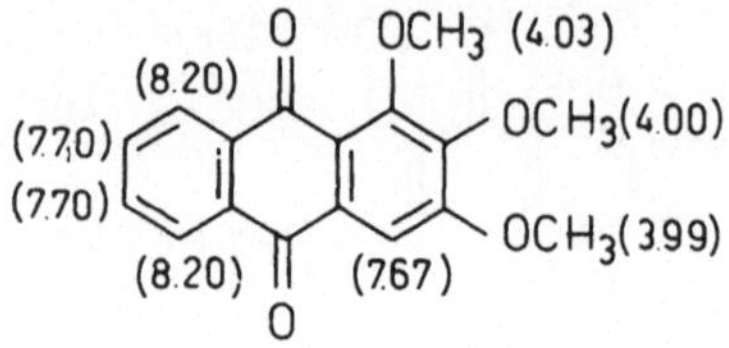

Anthragallol 1,2,3-trimethyl ether

Ring positions with NMR shifts: (8.20), (7.70), (7.70), (8.20), (7.67), OCH₃ (4.03), OCH₃ (4.00), OCH₃ (3.99)

$C_{17}H_{14}O_5$
1: pale-yellow (*102*)
mp: 164–166° (acetone: hexane) (*102*); 171–172° (acetone) (*108*)
UV (EtOH): 239 (4.11), 245 (4.08), 276 (4.64), 350 (3.92) (*108*)
IR (KBr): 1665, 1590, 1574 (*108*)
^{1}H-NMR (90 MHz, CDCl₃): (*108*)

Sources: *Oldenlandia umbellata* L. (*102*)

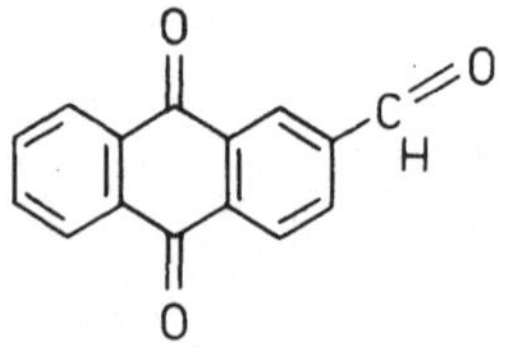

Anthraquinone-2-carbaldehyde

$C_{15}H_8O_3$

Sources: *Morinda lucida* Benth. (*2*)

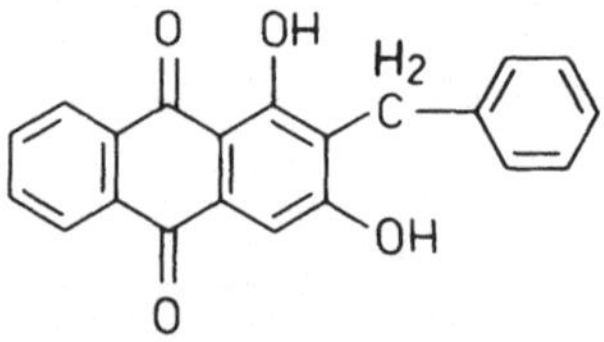

2-Benzylxanthopurpurin

$C_{21}H_{14}O_4$
1: yellow (*26*)
mp: 300° (EtOEt: P.E.) (*26*)
UV (EtOH): 241 (4.40), 246 (4.45), 280 (4.51), 335 (3.43), 413 (3.85) (*26*)
IR (KBr): 3390, 1665, 1630, 1588, (*26*)
MS (70 eV): 330 (M$^+$, 100), 329 (13), 254 (10), 253 (13), 252 (29), 225 (12), 165 (10), 139 (13), 115 (14), 113 (13), 105 (11), 103 (11), 91 (80), 76 (13), 74 (25), 51 (12) (*26*)

Sources:
Damnacanthus major Sieb. & Zucc. (*26*)   *Hymenodictyon excelsum* Wall. (*26*)

Table 7 (*continued*)

Copareolatin 1(or-5),6-dimethyl ether

$C_{17}H_{24}O_6$
mp: 271–273° (*101*)
UV (EtOH): 216 (4.34), 279 (4.46), 407 (3.82), (EtOH/OH⁻): 250 (4.25), 310 (4.47), 407 (3.82), 480 (sh., 3.25) (*101*)
IR (KBr): 3295, 1658, 1630, 1590, 1565, 1423, 1332, 1282, 1123, 985, 840 (*101*)
MS (70 eV): 314 (M⁺, 87), 300 (20), 299 (100), 297 (20), 296 (5), 285 (20), 283 (100), 281 (22), 271 (18), 255 (10) (*101*)
^{1}H-NMR (90 MHz, DMSO$_{d6}$): (*109*; copareolatin 1,6-dimethyl ether) (*101*)

Sources: *Morinda roioc* L. (*101*)

Copareolatin 6-methyl ether

$C_{16}H_{12}O_6$
mp: 238–240° (CHCl$_3$) (*101*)
UV (EtOH): 215 (4.30), 256 (sh., 4.16), 279 (4.35), 307 (3.98), 435 (3.99). (EtOH/OH⁻): 217 (4.31), 255 (4.24), 311 (4.36), 442 (3.91), 510 (sh., 3.79) (*101*)
IR (KBr): 3444, 1630, 1587, 1450, 1440, 1390, 1302, 1249, 1166, 1010, 968, 940, 775 (*101*)
MS (70 eV): 300 (M⁺, 100), 285 (15), 283 (15), 282 (75), 271 (23), 269 (7), 257 (37), 254 (26), 242 (9), 241 (6), 229 (9), 226 (15), 139 (9), 135 (18), 107 (23) (*101*)
^{1}H-NMR (60 MHz, CDCl$_3$): (*101*)

Sources: *Morinda roioc* L. (*101*)

Table 7 (*continued*)

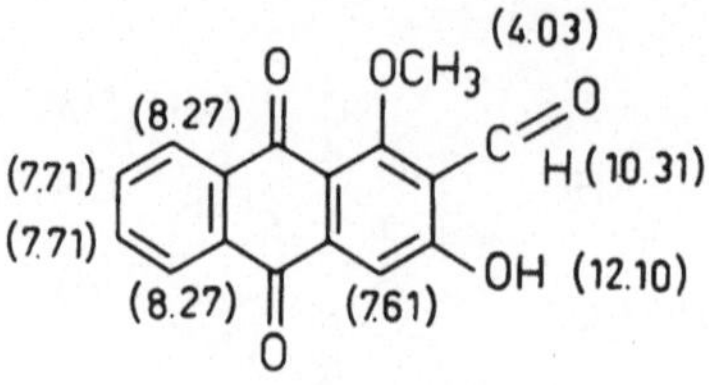

Damnacanthal

$C_{16}H_{10}O_5$
mp: 212° (EtOEt:P.E.) (*26*)
UV (EtOH): 247 (4.23), 282 (4.20), 375 (3.65). (EtOH/OH⁻): 247 (4.36), 262 (sh., 4.26), 307 (4.20), 375 (3.65) (*101*)
IR (KBr): 1678, 1659, 1649, 1590, 1567, 1340, 1255, 970, 710 (*101*)
MS (70 eV): 282 (M^+, 45), 267 (15), 264 (18), 254 (100), 253 (23), 226 (11), 225 (42), 224 (11), 208 (17), 196 (20), 168 (20) 148 (23), 147 (23), 139 (30) (*101*)
^{1}H-NMR (60 MHz, CDCl$_3$): (*101*)

Sources:

Damnacanthus major Sieb. & Zucc. (*26*)
Lasianthus chinensis Benth. (*68*)
Morinda lucida Benth. (*1, 2*)
M riooc L. (*101*)
M. tinctoria Roxb. (*49*)
Prismatomeris tetrandra (Roxb.) K. Schum. (*122*)

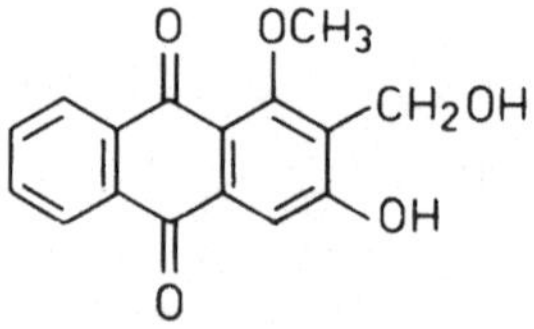

Damnacanthol

$C_{16}H_{12}O_5$
1: yellow (*27*)
mp: >300° (*27*)

Sources: *Coprosma rotundifolia* A. Cunn. (*27*)

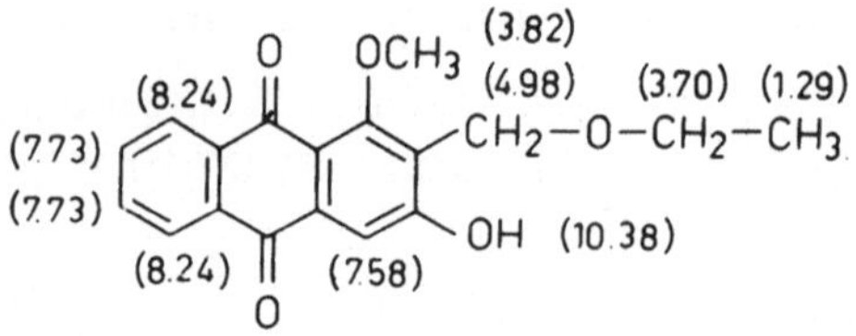

Damnacanthol ω-ethyl ether

Table 7 (*continued*)

$C_{18}H_{16}O_5$
1: red (*58*)
mp: 197–198° (*58*)
UV (EtOH): 240 (4.22), 246 (sh., 4.21), 280 (4.42), 330 (3.60), 375 (3.46). (EtOH/OH⁻): 480 (*58*)
IR (KBr): 3300, 1645, 1590, 1570 (*58*)
^{1}H-NMR (60 MHz, CDCl$_3$): (*58*)

Sources:
Ploclama pendula Ait. (*58*) *Putoria calabrica* Perss. (*57*)

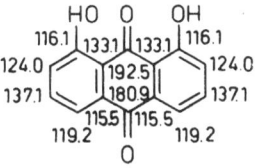

Damnacanthol ω-methyl ether

$C_{17}H_{14}O_5$
1: yellow (*44*)
mp: 204° (*44*)
UV (MeOH): 238 (4.45), 242 (4.45), 274 (4.56), 308 (4.04), 380 (3.53). (MeOH/OH⁻): 242 (4.62), 308 (4.50), 458 (3.87) (*44*)
IR (KBr): 1675, 1650, 710 (*44*)
MS (70 eV): 298 (M$^+$, 60), 283 (54), 266 (100), 265 (67), 253 (23), 251 (65), 238 (39), 237 (17), 225 (4), 223 (8), 210 (16), 182 (5), 181 (11), 154 (7), 153 (16), 138 (21) (*44*)
^{1}H-NMR (CDCl$_3$: DMSO$_{d6}$, 8:2) (*44*)

Sources: *Morinda lucida* Benth. (*44*)

1,8-Dihydroxyanthraquinone

$C_{14}H_8O_4$
MS (50 eV): 240 (M$^+$, 100), 223 (5), 212 (10), 184 (8), 138 (5), 120 (5), 92 (7) (*20*)
^{13}C-NMR (22.63 MHz, DMSO$_{d6}$): (*17*)

Sources:
Cinchona ledgeriana Moens. (*94*) *C. pubescens* Vahl. (*95*)

Table 7 (*continued*)

1,3-Dihydroxy-2,5-dimethoxyanthraquinone

$C_{16}H_{12}O_6$
1: orange-red; 2: red-brown; 3: orange; 4: red-brown; 5: purple (*128*)
UV (MeOH): 220 (sh.), 280, 310 (sh.), 390. (MeOH/OH⁻): 248, 313, 505 (*128*)
IR (KBr): 3420, 1670, 1630 (*128*)
MS (70 eV, 200°): 300 (M^+, 82), 285 (88), 282 (24), 270 (40), 239 (35), 227 (43), 69 (45), 57 (70), 43 (100) (*128*)
^{1}H-NMR (300 MHz, DMSO$_{d6}$): (*128*)

Sources: *Cinchona ledgeriana* Moens. (*128*)

1,4-Dihydroxy-2-ethoxymethylanthraquinone

$C_{17}H_{14}O_5$
1: yellow-orange; 4: red (*16*)
mp: 176–178° (EtOEt) (*16*); 183–185° (MeOH) (*74*)
UV (MeOH): 240 (4.07), 244 (4.08), 278 (4.00), 406 (3.78) (*74*)
MS (70 eV, 120°): 298 (M^+, 60), 269, 252 (100), 224, 196, 168, 139, 138 (*16*)
^{1}H-NMR (100 MHz, DMSO$_{d6}$): (*74, 16*)

Sources:
Rubia cordifolia L. (*74*)   *R. tinctoria* L. (*16*)

3,5-Dihydroxy-2-ethoxymethyl-4-methoxyanthraquinone (8-hydroxysubspinosin)

Table 7 (*continued*)

$C_{18}H_{16}O_6$
mp: 184.5–186.5° (*89*)
UV (MeOH): 206 (sh., 4.38), 224 (4.49), 278 (4.52), 431 (sh., 3.56) (*89*)
IR (nujol): 3310, 1660, 1640, 1580, (*89*)
MS: 328 (M$^+$), 300, 299, 285, 283, 282, 281 (*89*)
^{1}H-NMR (100 MHz, CDCl$_3$): (*89*)

Sources: *Damnacanthus subspinosus* Hand-Mazz. (*89*)

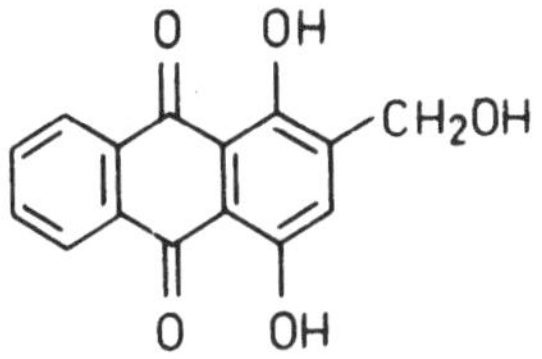

1,4-Dihydroxy-2-hydroxymethylanthraquinone

$C_{15}H_{10}O_5$
1: yellow; 4: yellow-red (*15*)
mp: >300° (*15*)
MS (70 eV, 140°): 270 (M$^+$, 20), 268 (8), 254 (14), 252 (100), 240 (7), 224 (10), 223 (5), 212 (3), 196 (21), 168 (11), 140 (6), 139 (15), 105 (8) (*15*)

Sources: *Rubia tinctorum* L. (*15*)

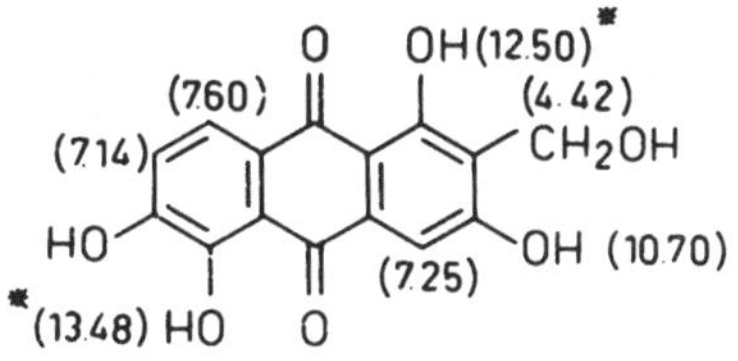

5,6-Dihydroxylucidin

$C_{15}H_{10}O_7$
1: red (*70*)
mp: >300° (*70*)
UV (MeOH): 233 (4.45), 268 (4.45), 288 (4.12), 315 (sh., 3.95), 446 (3.93) (*70*)
IR (KBr): 3350, 1600 (*70*)
^{1}H-NMR (60 MHz, DMSO$_{d6}$): (*70*)

Sources: *Morinda citrifolia* L. (*70*)

Table 7 (*continued*)

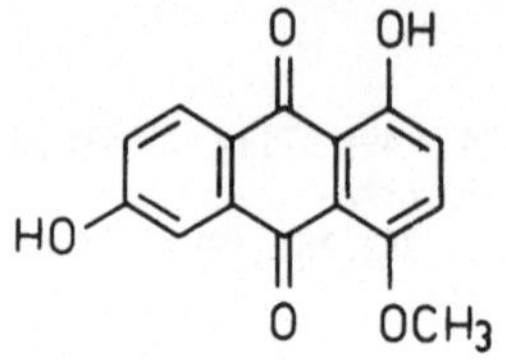

1,6- (or 1,7-)Dihydroxy-4-methoxyanthraquinone

$C_{15}H_{10}O_5$
1: orange-yellow; 2,3: orange; 4: purple-red; 5: purple-red (*75*)
UV (EtOH): 250, 268, 275 (sh.), 320, 410 (*75*)
IR (KBr): 3310, 3110, 1665, 1625, 1585 (*75*)
MS: 270 (M$^+$, 100), 253 (17), 252 (47), 241 (21), 227 (13), 224 (86), 213 (5), 199 (9), 196 (7), 171 (11), 168 (10) (*75*)
^{1}H-NMR (90 MHz, DMSO$_{d6}$): (*75*)

Sources: *Galium album* Mill. (*75*)

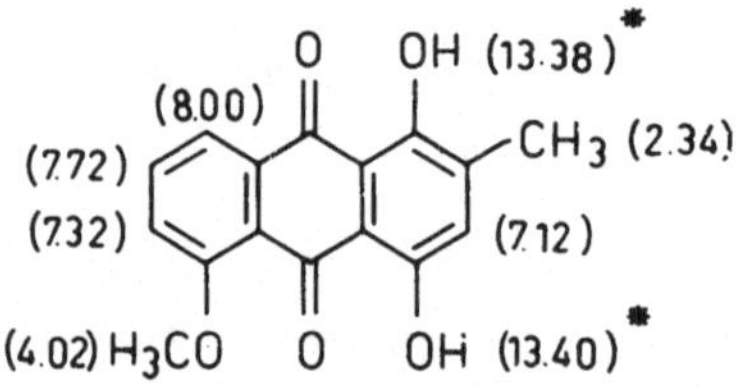

1,4-Dihydroxy-5-(or -8-)methoxy-2-methylanthraquinone

$C_{16}H_{12}O_5$
1: orange; 3: yellow-orange; 5: red (*47*)
mp: 165° (CH$_2$Cl$_2$) (*47*)
UV (EtOH): 212 (sh.), 232 (4.04), 250 (3.88), 286 (3.66), 480 (3.52), 530 (3.32). (EtOH/ OH$^-$): 225, 260, 320, 550, 594 (*47*)
IR (KBr): 1625, 1580, 1450, 1440, 1405, 1300, 1290, 1235, 1045, 965, 810, 790 (*47*)
MS (70 eV, 200°): 284 (M$^+$, 100), 266 (62), 238 (40), 210 (8), 181 (8), 86, 84, 49 (*47*)
^{1}H-NMR (270 MHz, CDCl$_3$): (*47*)

Sources: *Rubia cordifolia* L. (*47*)

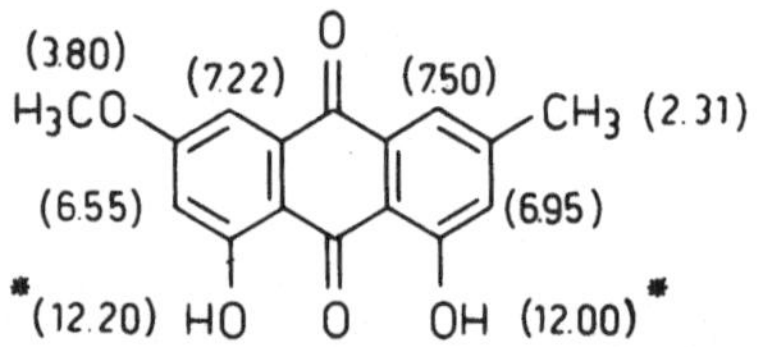

4,5-Dihydroxy-7-methoxy-2-methylanthraquinone (physcion)

$C_{16}H_{12}O_5$
1: yellow-orange; 3: orange; 5: red (*119*)
mp: 201–202° (CH$_2$Cl$_2$) (*119*); 207° (*60*)

References, pp. 143–149

Table 7 *(continued)*

UV (EtOH): 207, 225, 249 (sh.), 255, 265, 287, 435, 450 (sh.). (EtOH/OH⁻): 210, 230, 237, 255, 310, 512 *(119)*
IR (KBr): 2995, 1685, 1640, 1488, 1395, 1375, 1332, 1305, 1280, 1170, 1110, 1045, 990, 905, 880, 765 *(119)*
MS (70 eV, 200°): 284 (M⁺, 100), 255 (8), 241 (8), 213 (5), 128 (8), 43 (11) *(119)*
^{1}H-NMR (270 MHz, CDCl$_3$): *(119)*
Sources: *Rubia cordifolia* L. *(119)*

1,4-Dihydroxy-2-methylanthraquinone

$C_{15}H_{10}O_4$
1: orange-red; 3: orange-brown; 5: red-purple *(46)*
mp: 179° (MeOH) *(46)*; 183–185° (MeOH) *(65)*
UV (EtOH): 207 (4.16), 230 (4.16), 251 (4.29), 257 (4.30), 286 (3.85), 456 (3.66), 482 (3.60), 516 (3.55). (EtOH/OH⁻): 212, 249, 255, 446, 580 *(46)*
IR (KBr): 1630, 1590, 1465, 1400, 1380, 1280, 1165, 790, 750, 730 *(46)*
MS (70 eV, 200°): 254 (M⁺, 100), 240 (5), 239 (5), 74 (6), 59 (11) *(46)*
^{1}H-NMR (270 MHz, CDCl$_3$): *(46, 59)*
Sources: *Rubia cordifolia* L. *(46)*

1,5-Dihydroxy-2-methylanthraquinone

$C_{15}H_{10}O_4$
1: yellow-orange; 3: yellow-orange; 5: red *(46)*
mp: 194° (MeOH) *(46)*
UV (EtOH): 205 (sh.), 229 (4.25), 255 (4.04), 280 (3.70), 290 (3.70), 396 (sh.), 420 (4.07), 442 (4.07). (EtOH/OH⁻): 211, 235, 287, 506 *(46)*
IR (KBr): 1630, 1615, 1585, 1480, 1460, 1435, 1380, 1320, 1300, 1270, 1165, 1155, 1090, 1055, 935, 900, 840, 800, 730 *(46)*
MS (70 eV, 200°): 254 (M⁺, 100), 236 (7), 225 (5), 196 (7), 169 (8), 151 (6), 141 (5), 127 (9), 115 (7), 76 (7), 63 (6) *(46)*
^{1}H-NMR (270 MHz, CDCl$_3$): *(46, 58)*
Sources:
Ploclama pendula Ait. *(58)*   *Rubia cordifolia* L. *(46)*

Table 7 (*continued*)

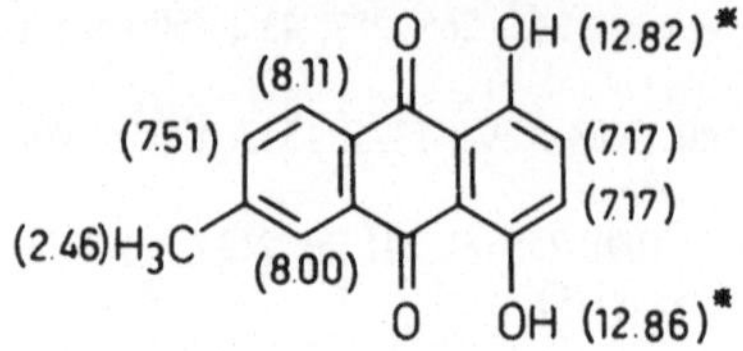

1,4-Dihydroxy-6-methylanthraquinone

(8.11), (7.51), (2.46)H₃C, (8.00), OH (12.82)*, (7.17), (7.17), OH (12.86)*

$C_{15}H_{10}O_4$
1: red; 3: orange-brown; 5: red-purple (*119*)
mp: 167–168° (MeOH) (*119*)
UV (EtOH): 211, 225, 255, 260, 283, 291 (sh.), 328, 468, 500 (sh.), 514. (EtOH/OH⁻): 209, 230, 257, 288, 325 (sh.), 542, 580 (*119*)
IR (KBr): 1640, 1605, 1598, 1460, 1360, 1340, 1282, 1230, 1210, 1162, 800 (*119*)
MS (70 eV, 200°): 254 (M⁺, 100), 197 (5), 169 (3), 127 (5), 115 (7), 96 (5), 70 (42) (*119*)
¹H-NMR (270 MHz, CDCl₃): (*119*)

Sources: *Rubia cordifolia* L. (*119*)

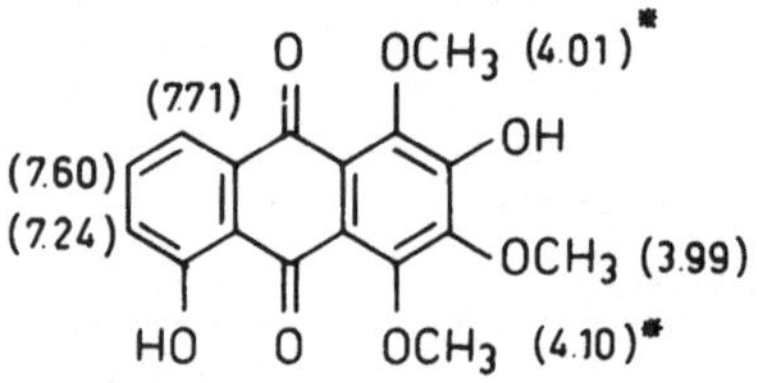

2,5- (or 3,5-)Dihydroxy-1,3,4- (or-1,2,4-)trimethoxyanthraquinone

(7.71), (7.60), (7.24), OCH₃ (4.01)*, OH, OCH₃ (3.99), HO, OCH₃ (4.10)*

$C_{17}H_{14}O_7$
1: orange; 3: orange; 4: red; 5: red (*128*)
UV (MeOH): 218, 276, 410. (MeOH/OH⁻): 253, 319, 500 (*128*)
IR (KBr): 3400, 2920, 2840, 1660, 1630, 1540 (*128*)
MS (70 eV, 110°): 330 (M⁺, 100), 315 (60), 312 (5), 297 (20), 287 (22), 272 (24), 227 (20), 58 (23) (*128*)
¹H-NMR (300 MHz, CDCl₃): (*128*)

Sources: *Cinchona ledgeriana* Moens. (*128*)

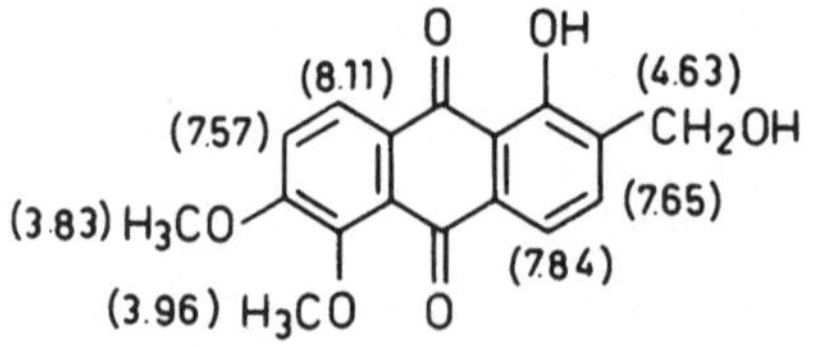

5,6-Dimethoxy-1- (or -4-)hydroxy-2- (or -3-)hydroxymethylanthraquinone

(8.11), (7.57), (3.83)H₃CO, (3.96)H₃CO, OH, (4.63), CH₂OH, (7.65), (7.84)

References, pp. 143–149

Table 7 (*continued*)

$C_{17}H_{14}O_6$
1: orange-yellow; 2: red; 3: orange; 4: red; 5: red *(128)*
UV (MeOH): 223, 240 (sh.), 275 (sh.), 395. (MeOH/OH$^-$): 235, 480 *(128)*
MS (70 eV, 115°): 314 (M$^+$, 42), 296 (8), 283 (100), 209 (52), 152 (34) *(128)*
^{1}H-NMR (300 MHz, CDCl$_3$): *(128)*

Sources: *Cinchona ledgeriana* Moens. *(128)*

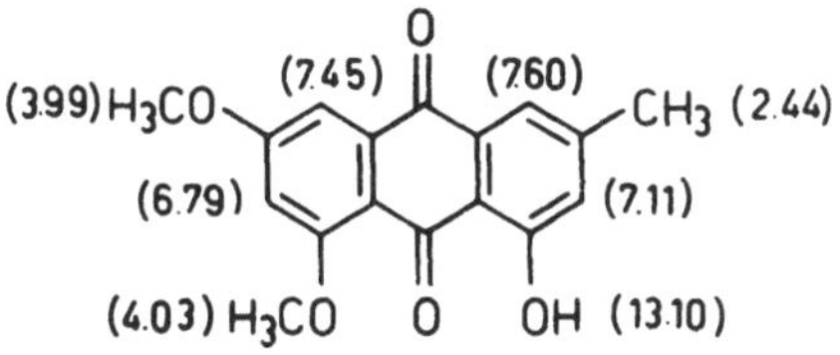

5,7-Dimethoxy-4-hydroxy-2-methylanthraquinone   (6,8-dimethoxy-1-hydroxy-3-methyl-anthraquinone)

$C_{17}H_{14}O_5$
1: orange *(121)*
mp: 210° *(121)*
UV (EtOH): 225 (4.48), 270 (4.21), 281 (4.21), 430 (3.83) *(40)*
IR: 2890, 1678, 1630 *(121)*
MS: 298 (M$^+$, 100), 281 (70), 280 (97), 269 (84), 252 (78) *(40)*
^{1}H-NMR (60 MHz, CDCl$_3$): *(8)*

Sources: *Morinda citrifolia* L. *(121)*

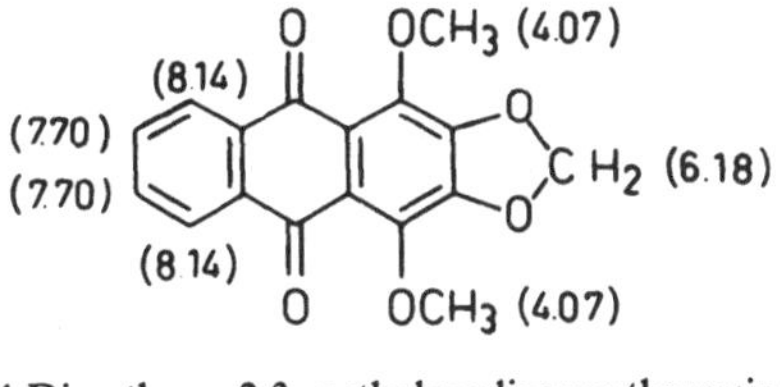

1,4-Dimethoxy-2,3-methylenedioxyanthraquinone

$C_{17}H_{12}O_6$
1: bright-yellow; 3: bright-yellow; 4: yellow; 5: yellow *(128)*
UV (MeOH): 220 (sh.), 241 (sh.), 276, 375^a. (MeOH/OH$^-$): 220 (sh.), 241 (sh.), 276, 375^b *(128)*
MS (70 eV, 250°): 312 (M$^+$, 48), 297 (9), 294 (5), 269 (12), 254 (14), 85 (40), 71 (62), 57 (90), 43 (100) *(128)*
^{1}H-NMR (100 MHz, CDCl$_3$): *(128)*

Sources: *Cinchona ledgeriana* Moens. *(128)*

---

^{a} In ref. *128* erroneously published as 355 nm
^{b} In ref. *128* erroneously published as 386 nm

Table 7 (*continued*)

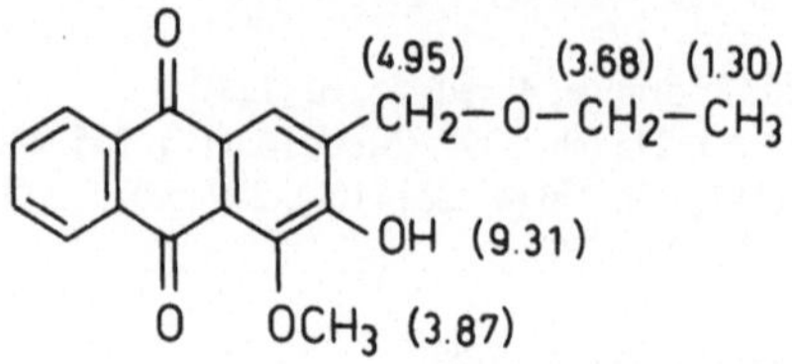

[Structure: 2-Ethoxymethyl-3-hydroxy-4-methoxyanthraquinone with NMR shifts: CH₂–O–CH₂–CH₃ (4.95)(3.68)(1.30), OH (9.31), OCH₃ (3.87)]

2-Ethoxymethyl-3-hydroxy-4-methoxyanthraquinone (subspinosin)

$C_{18}H_{16}O_5$
mp: 197–198.5° (*89*)
UV (MeOH): 207 (4.45), 223 (4.33), 240 (4.38), 246 (4.37), 278 (4.59), 334 (3.57), 370 (3.53) (*89*)
IR (nujol): 3300, 1674, 1653, 717 (*89*)
MS: 312 (M⁺), 297, 284, 283, 269, 268, 267, 266 (*89*)
¹H-NMR (100 MHz, CDCl₃): (*89*)

Sources: *Damnacanthus subspinosus* Hand-Mazz. (*89*)

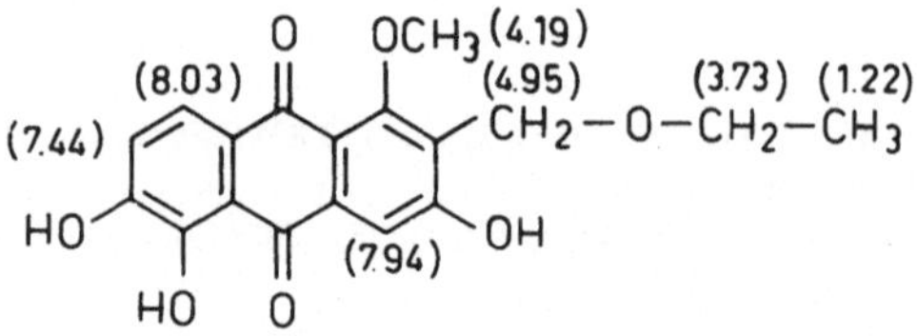

[Structure: 2-Ethoxymethyl-1-methoxy-3,5,6-trihydroxyanthraquinone with NMR shifts: (8.03), (7.44), OCH₃ (4.19), CH₂–O–CH₂–CH₃ (4.95)(3.73)(1.22), HO, HO, OH (7.94)]

2-Ethoxymethyl-1-methoxy-3,5,6-trihydroxyanthraquinone

$C_{18}H_{16}O_7$
1: red (*57*)
mp: >300° (EtOEt) (*57*)
UV (EtOH): 225 (4.39), 277 (4.50), 310 (4.27), 350 (3.93), 430 (3.72) (*57*)
IR (KBr): 3445, 3275, 1648, 1628, 1565 (*57*)
¹H-NMR (60 MHz, C₅D₅N): (*57*)

Sources: *Putoria calabrica* Perss. (*57*)

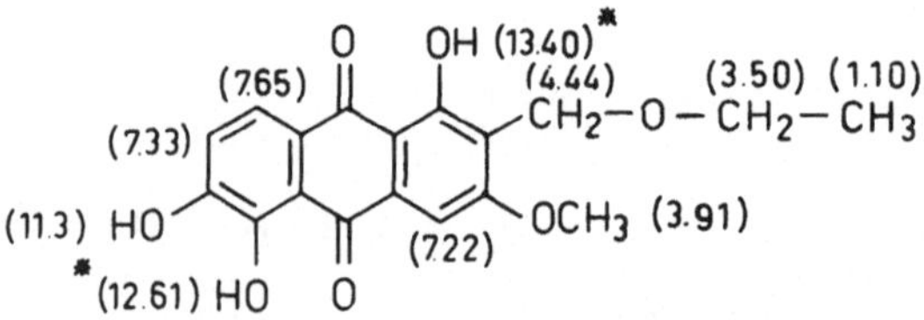

[Structure: 2-Ethoxymethyl-3-methoxy-1,5,6-trihydroxyanthraquinone with NMR shifts: (7.65), (7.33), OH (13.40)*, CH₂–O–CH₂–CH₃ (4.44)(3.50)(1.10), (11.3) HO*, (12.61) HO, (7.22), OCH₃ (3.91)]

2-Ethoxymethyl-3-methoxy-1,5,6-trihydroxyanthraquinone

$C_{18}H_{16}O_7$
1: red (*80*)
mp: >345° (*80*)

References, pp. 143–149

Table 7 (*continued*)

UV (EtOH): 229 (4.36), 267 (4.26), 290 (4.26), 452 (3.89) *(80)*
IR (KBr): 3400–3500, 3230, 1603, 1590, 1388, 1262 *(80)*
MS: 344 (M$^+$, 55), 315 (25), 300 (45), 299 (57), 298 (100), 269 (25), 227 (17), 199 (9) *(80)*
^{1}H-NMR (DMSO$_{d6}$): *(80)*

Sources: *Morinda citrifolia* L. *(80)*

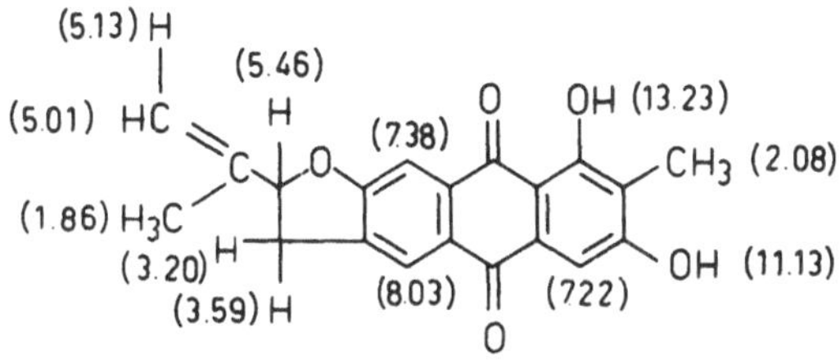

Galiprenylin

$C_{20}H_{16}O_5$
1: yellow; 2, 3: orange; 4: orange; 5: orange *(75)*
mp: 216° (EtOEt) *(76)*
UV (MeOH): 218 (3.67), 284 (4.25), 340 (sh., 3.71), 422 (3.31) *(76)*
IR (KBr): 3390, 1653, 1620, 1592, 1095 *(76)*
MS (60 eV): 336 (M$^+$, 100), 321 (92), 318 (15), 307 (11) *(76)*
^{1}H-NMR (200 MHz, DMSO$_{d6}$): *(76)*

Sources: *Galium album* Mill. *(76)*

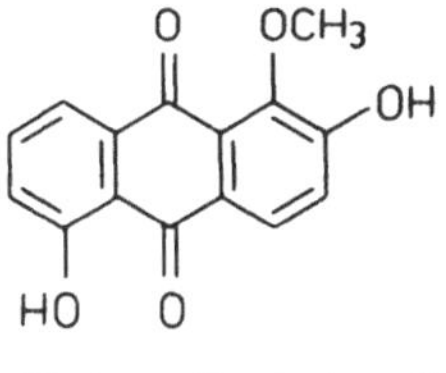

5 (or 8)-Hydroxyalizarin 1-methyl ether

$C_{15}H_{10}O_5$
1: orange-yellow *(26)*
mp: 235–236° (CH$_2$Cl$_2$: PE) *(26)*
UV (EtOH): 247 (4.23), 260 (sh., 4.28), 266 (4.30), 276 (sh., 4.28), 285 (sh., 4.24), 397 (sh., 3.91), 410 (3.94), 424 (sh., 3.91) *(26)*
IR (KBr): 3325, 1670, 1632, 1578 *(26)*
MS (70 eV): 270 (M$^+$, 65), 252 (33), 241 (13), 225 (19), 224 (100), 199 (20), 196 (12), 184 (12), 171 (21), 168 (22), 115 (25), 88 (11), 69 (11), 63 (17), 55 (13) *(26)*

Sources: *Damnacanthus major* Sieb. & Zucc. *(26)*

Table 7 (*continued*)

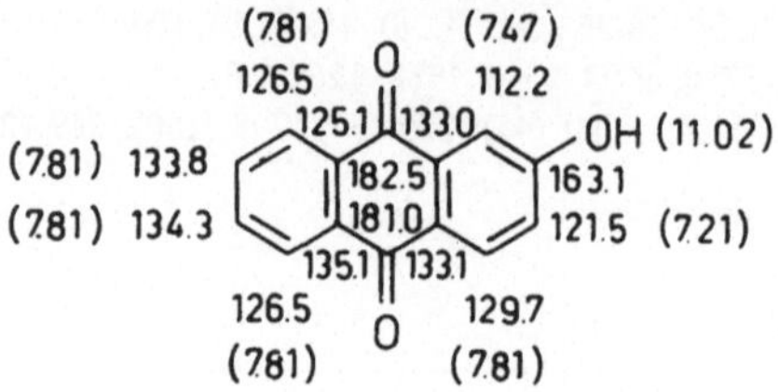
(7.81)
(7.47)
O
126.5
112.2
125.1
133.0
OH (11.02)
(7.81) 133.8
182.5
163.1
(7.81) 134.3
181.0
121.5 (7.21)
135.1
133.1
126.5
129.7
O
(7.81)
(7.81)

2-Hydroxyanthraquinone

$C_{14}H_8O_3$
1: yellow (*38*)
mp: 302–304° (EtOH) (*38*)
UV (H_2SO_4): 250, 295, 410, 485 (*91*)
IR (KBr): 3370, 1680, 1300 (*91*)
MS (50 eV): 224 (M^+, 100), 196 (27), 195 (9), 168 (29), 139 (30), 84 (9) (*20*)
^{1}H-NMR (90 MHz, $DMSO_{d6}$): (*35*)
^{13}C-NMR (22.63 MHz, $DMSO_{d6}$): (*17*)

Sources:

Asperula odorata L. (*38*)
Galium mollugo L. (*37*)
G. pumilum L. (*37*)
G. saxatile L. (*37*)
G. verum L. (*37*)
Morinda umbellata L. (*39*)
Rubia tinctorum L. (*38*)

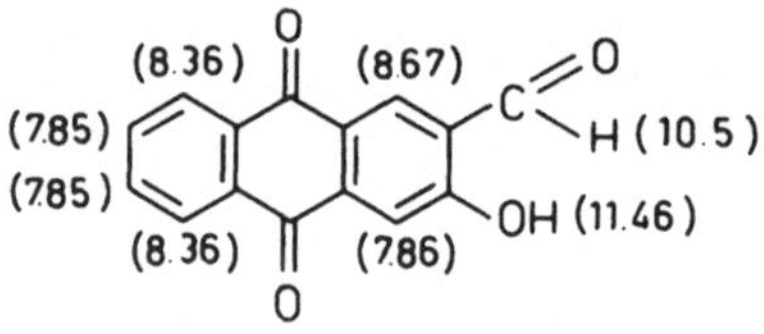
O
(8.36)
(8.67)
O
C
H (10.5)
(7.85)
(7.85)
OH (11.46)
(8.36)
(7.86)
O

3-Hydroxyanthraquinone-2-carbaldehyde

$C_{15}H_8O_4$
1: pale-yellow (*44*)
mp: 259–261° (subl.) (*101*)
UV (EtOH): 207 (4.15), 245 (4.32), 272 (4.43), 314 (sh., 3.81), 380 (3.73), 460 (3.46). (EtOH/OH^-): 206 (4.34), 257 (4.42), 277 (sh., 4.35), 305 (4.10), 385 (3.81), 460 (3.56) (*101*)
IR (KBr): 1683, 1668, 1595, 1580, 1495, 1390, 1335, 1200, 1140, 1110, 968, 890, 788, 760, 720 (*101*)
MS (70 eV): 252 (M^+, 100), 251 (54), 234 (15), 224 (9), 223 (9), 206 (11), 196 (5), 195 (3), 178 (4), 168 (4), 167 (4), 150 (4), 139 (49) (*101*)
^{1}H-NMR ($CDCl_3$): (*44*)

Sources:

Morinda lucida Benth. (*44*)
M. roioc L. (*101*)

Table 7 *(continued)*

8-Hydroxydamnacanthol ω-ethyl ether

$C_{18}H_{16}O_6$
mp: 174.5–176.5° *(88)*
UV (MeOH): 208 (sh., 4.44), 222 (4.50), 278 (4.55), 382–436 (4.10–3.71) *(88)*
IR (nujol): 3290, 1653, 1632, 1568, 897 *(88)*
MS: 328 (M$^+$, 59), 313 (50), 300 (12), 299 (71), 285 (17), 283 (31), 282 (95), 281 (99), 254 (100) *(88)*
^{1}H-NMR (100 MHz, CDCl$_3$): *(88)*

Sources: *Damnacanthus subspinosus* Hand-Mazz. *(87)*

1-Hydroxy-2-hydroxyethyl-3-methoxyanthraquinone

$C_{17}H_{14}O_5$
1: yellow; 2, 3: white-yellow; 4: orange; 5: orange *(75)*
UV (EtOH): 247, 282, 418 *(75)*
IR (KBr): 3380, 1655, 1625, 1570 *(75)*
MS: 298 (M$^+$, 52), 267 (29), 266 (100), 238 (38), 210 (6), 182 (12) *(75)*
^{1}H-NMR (90 MHz, DMSO$_{d6}$): *(75)*

Sources: *Galium album* Mill. *(75)*

1-Hydroxy-2-hydroxymethylanthraquinone

$C_{15}H_{10}O_4$
1: bright-yellow; 2: red; 3: orange; 4: red; 5: red *(128)*
UV (MeOH): 220, 252, 270 (sh.), 280 (sh.), 320, 385. (MeOH/OH$^-$): 243, 270 (sh.), 310, 468 *(128)*

Table 7 (*continued*)

IR (KBr): 3360, 1670, 1635, 1595 (*128*)
MS (70 eV, 150°): 254 (M$^+$, 60), 226 (18), 225 (60), 152 (32), 71 (67), 57 (100), 43 (90) (*128*)
^{1}H-NMR (100 MHz, CDCl$_3$): (*128, 42, 69, 75*)

Sources:
Cinchona ledgeriana Moens. (*94, 128*)   *Morinda parvifolia* Bartl. (*42*)
C. pubescens Vahl. (*95*)   *Rubia tinctorum* L. (*75*)
Galium album Mill. (*75*)

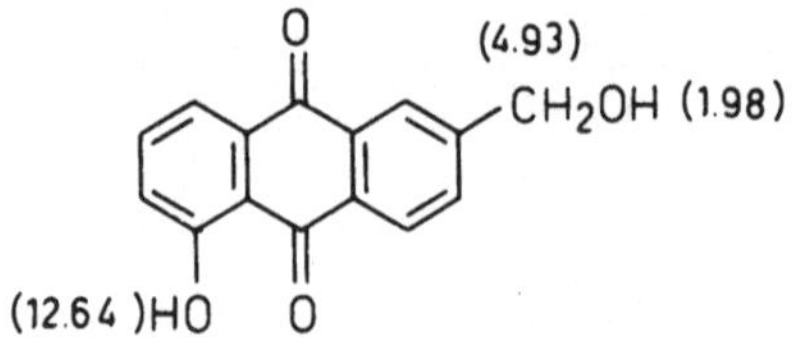

5-Hydroxy-2-hydroxymethylanthraquinone (1-hydroxy-6 (or -7)-hydroxymethylanthraquinone)

$C_{15}H_{10}O_4$
1: orange-yellow; 5: red (*42*)
mp: 172° (*42*)
UV (EtOH): 210 (4.67), 256 (4.78), 280 (sh., 4.36), 335 (4.32), 398 (4.58) (*42*)
IR: 3535, 1662, 1631, 1595 (*42*)
MS (70 eV): 254 (M$^+$, 100), 226 (13), 225 (50), 208 (7), 197 (9), 180 (6), 169 (5), 168 (4), 152 (8), 139 (9) (*42*)
^{1}H-NMR (60 MHz, CDCl$_3$): (*42*)

Sources: *Morinda parvifolia* Bartl. (*42*)

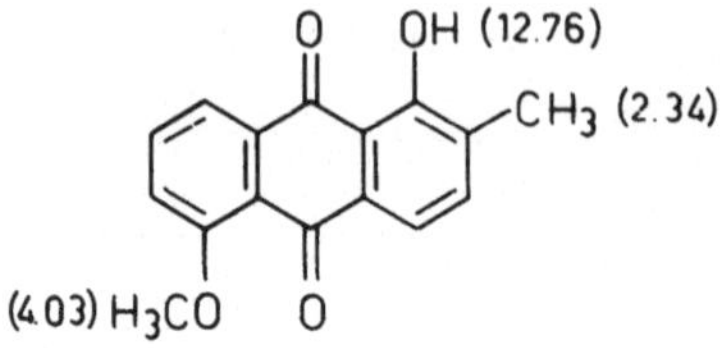

1-Hydroxy-5-methoxy-2-methylanthraquinone

$C_{16}H_{12}O_4$
1: red-orange (*58*)
mp: 189–191° (EtOH) (*58*)
UV (EtOH): 226, 255, 285, 420. (EtOH/OH$^-$): 520 (*58*)
IR (KBr): 1665, 1635, 1585 (*58*)
^{1}H-NMR (60 MHz, CDCl$_3$): (*58*)

Sources: *Ploclama pendula* Ait. (*58*)

Table 7 (continued)

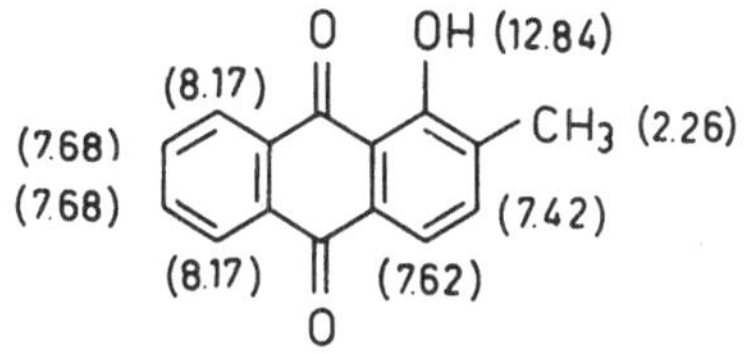

```
         O   OH (12.84)
  (8.17) ||
(7.68)   /\/\/\--CH₃ (2.26)
(7.68)   \/\/\/
  (8.17) ||  (7.42)
         O   (7.62)
```

1-Hydroxy-2-methylanthraquinone

$C_{15}H_{10}O_3$
1: yellow; 2: red-brown; 3: orange; 4: yellow; 5: red (128)
mp: 180–182° (MeOH) (119)
UV (MeOH): 226 (4.24), 247 (4.40), 254 (4.42), 279 (4.02), 325 (3.43), 408 (3.74) (73)
(EtOH/OH⁻): 210, 240 (sh.), 252, 270 (sh.), 317, 510 (119)
IR (KBr): 1678, 1642, 1600, 1435, 1368, 1300, 1270, 718 (119)
MS (70 eV, 200°): 238 (M⁺, 100), 181 (16), 153 (10), 152 (15), 76 (12.5) (119)
¹H-NMR (270 MHz, CDCl₃): (119, 57, 58, 73, 74, 110)

Sources:

Cinchona ledgeriana Moens. (128)   *Morinda lucida* Benth. (2)
Galium aparine L. (37)   *M. umbellata* L. (39)
G. mollugo L. (37)   *Ploclama pendula* Ait. (58)
G. pumilum L. (37)   *Putoria calabrica* Perss. (57)
G. saxatile L. (37)   *Rubia cordifolia* L. (74, 119)
G. verum L. (37)   *R. tinctorum* L. (38)

```
              O    (4.91)
       (8.31) || (8.31)
  (7.82)    /\/\/\--CH₂OH (1.99)
  (7.82)    \/\/\/
       (8.31) || (8.31) (7.82)
              O
```

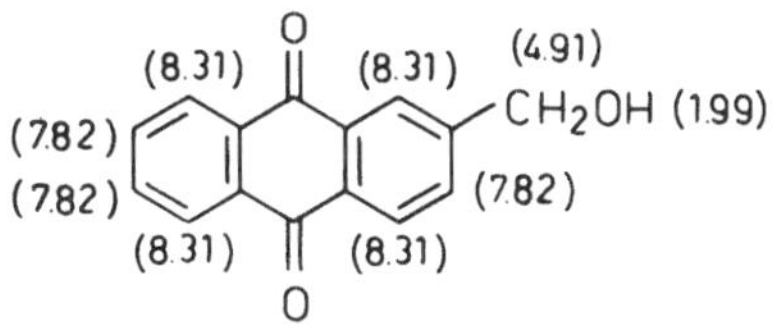

2-Hydroxymethylanthraquinone

$C_{15}H_{10}O_3$
1: pale-yellow; 5: pale-yellow (42)
mp: 194° (42)
IR: 3520, 1670, 1586 (42)
MS (70 eV): 238 (M⁺, 97), 210 (27), 209 (100), 192 (11), 181 (20), 164 (9), 152 (21) (42)
¹H-NMR (60 MHz, CDCl₃): (42)

Sources: *Morinda parvifolia* Bartl. (42)

Table 7 (*continued*)

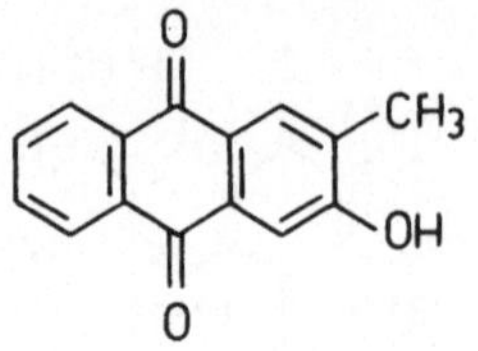

3-Hydroxy-2-methylanthraquinone

$C_{15}H_{10}O_3$
mp: 293–296° (HAc) (*27*)

Sources: *Coprosma tenuicaulis* Hook f. (*27*)

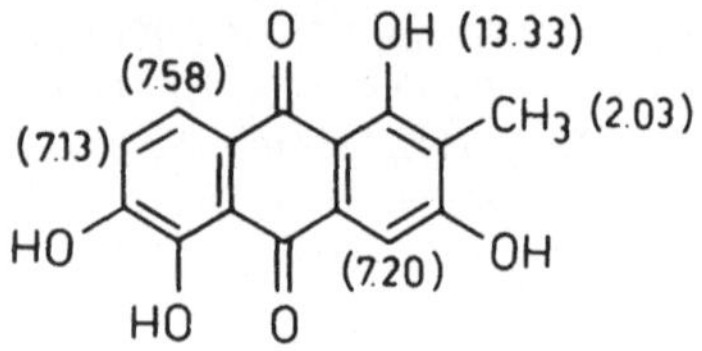

3-Hydroxymorindone

$C_{15}H_{10}O_6$
1: red (*70*)
mp: >300° (*70*)
UV (MeOH): 227 (4.27), 276 (4.43), 316 (3.95), 450 (3.90) (*70*)
IR (KBr): 3420, 1595 (*70*)
^{1}H-NMR (60 MHz, DMSO$_{d6}$): (*70*)

Sources: *Morinda citrifolia* L. (*70*)

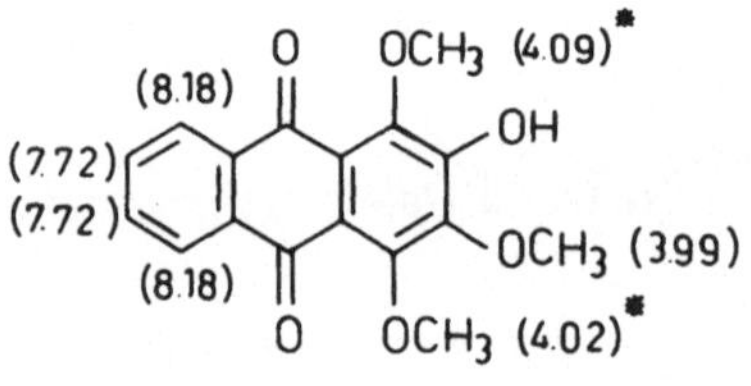

2-Hydroxy-1,3,4-trimethoxyanthraquinone

$C_{17}H_{14}O_6$
1: dark-yellow; 3: red; 4: orange; 5: orange (*128*)
UV (MeOH): 205, 240, 276, 366. (MeOH/OH$^-$): 251, 315, 476 (*128*)
IR (KBr): 3400, 1668, 1590 (*128*)
MS (70 eV, 95°): 314 (M$^+$, 100), 299 (60), 296 (8), 281 (20), 271 (20), 256 (30), 211 (23), 157 (23) (*128*)
^{1}H-NMR (300 MHz, CDCl$_3$): (*128*)

Sources: *Cinchona ledgeriana* Moens. (*128*)

References, pp. 143–149

Table 7 (*continued*)

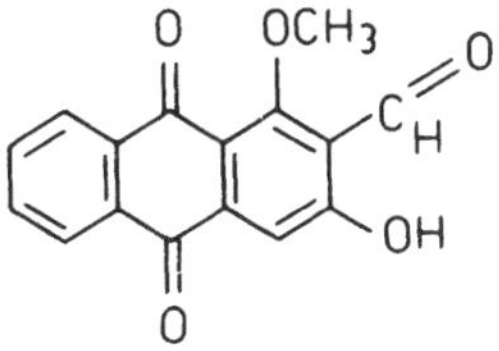

Juzunal

$C_{16}H_{10}O_6$
1: yellow (*26*)
mp: 247–249° (*26*)

Sources: *Damnacanthus major* Sieb. & Zucc. (*26*)

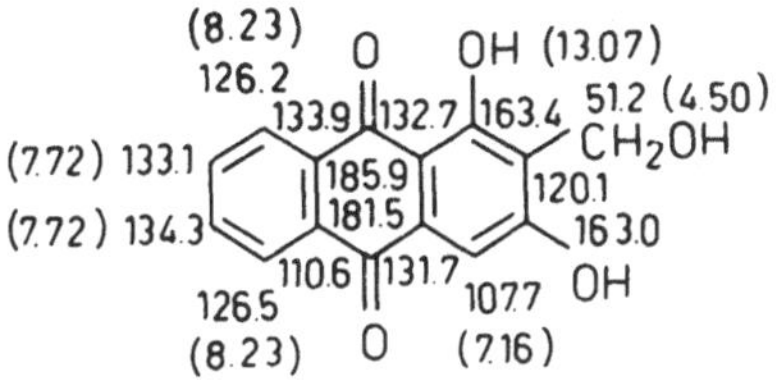

Lucidin

$C_{15}H_{10}O_5$
1: yellow (*70*)
mp: >330° (dioxane) (*38*)
UV (EtOH): 242 (4.38), 246 (4.39), 280 (4.35), 330 (3.54), 4.15 (3.75) (*80*)
IR (KBr): 3410, 1663, 1620, 1593, 1375, 1287 (*80*)
MS: 270 (M$^+$, 22), 252 (100), 224 (11), 196 (25), 168 (12), 139 (17) (*80*)
^{1}H-NMR (200 MHz, DMSO$_{d6}$): (*72, 27, 44, 66, 70, 74, 80*)
^{13}C-NMR (50.1 MHz, CDCl$_3$): (*72, 44, 71, 74*)

Sources:

Asperula besseriana Klok. (*21*)
A. odorata L. (*37*)
Commitheca liebrechtsiana Brem. (*66*)
Coprosma lucida J.R. et G. Forst. (*27*)
C. rotundifolia A. Cunn. (*27*)
Galium aparine L. (*37*)
G. dasypodum Klok. (*131, 133*)
G. mollugo L. (*11, 12, 37, 71, 72, 126*)
G. pumilum L. (*37*)
G. ruthenicum Willd. (*21*)
G. saxatile L. (*37*)
G. saxatile × *G. sterneri* (*37*)

G. semiamictum Klok. (*130*)
G. sterneri Ehrend. (*37*)
G. verum L. (*37*)
Hymenodictyon excelsum Wall. (*26*)
Morinda citrifolia L. (*70, 80*)
M. lucida Benth. (*44*)
M. roioc L. (*101*)
M. umbellata L. (*39*)
Rubia cordifolia L. (*74*)
R. iberica C. Koch. (*96*)
R. tinctorum L. (*38, 97*)

Table 7 (*continued*)

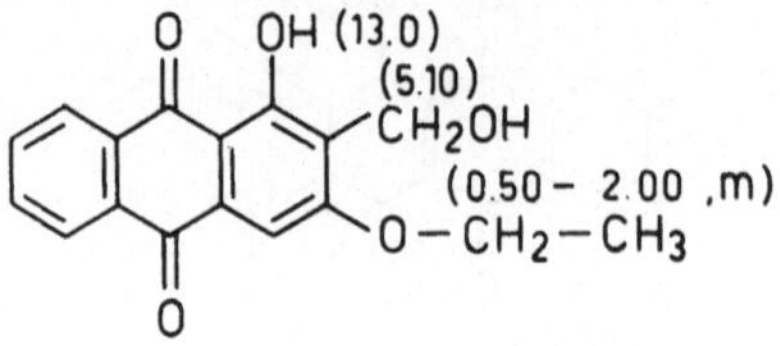

Lucidin 3-ethyl ether

structure: anthraquinone with OH (13.0), CH₂OH (5.10), O–CH₂–CH₃ (0.50–2.00, m)

$C_{17}H_{14}O_5$
1: yellow (*118*)
mp: 180–181° (*118*)
UV: 206, 286, 425 (*118*)
IR (KBr): 3400, 1690, 1580, 1280 (*118*)
^{1}H-NMR (100 MHz, CDCl$_3$): (*118*)

Sources: *Rubia cordifolia* L. (*118*)

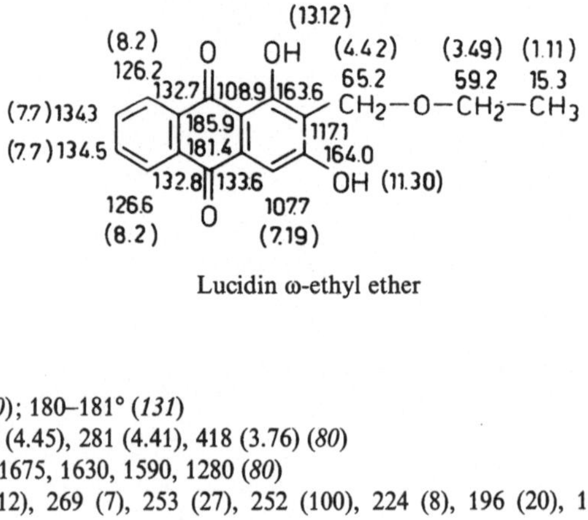

Lucidin ω-ethyl ether

$C_{17}H_{14}O_5$
1: yellow (*80*)
mp: 168–170° (*80*); 180–181° (*131*)
UV (EtOH): 246 (4.45), 281 (4.41), 418 (3.76) (*80*)
IR (KBr): 3210, 1675, 1630, 1590, 1280 (*80*)
MS: 298 (M$^+$, 12), 269 (7), 253 (27), 252 (100), 224 (8), 196 (20), 168 (8), 139 (12) (*80*)
^{1}H-NMR (DMSO$_{d6}$): (*80, 42, 74*)
^{13}C-NMR (DMSO$_{d6}$): (*74*)

Sources:
Galium dasypodum Klok. (*131, 133*)   *M. parvifolia* Bartl. (*42*)
G. mollugo L. (*11*)   *Rubia cordifolia* L. (*74*)
Morinda citrifolia L. (*80*)

Table 7 (continued)

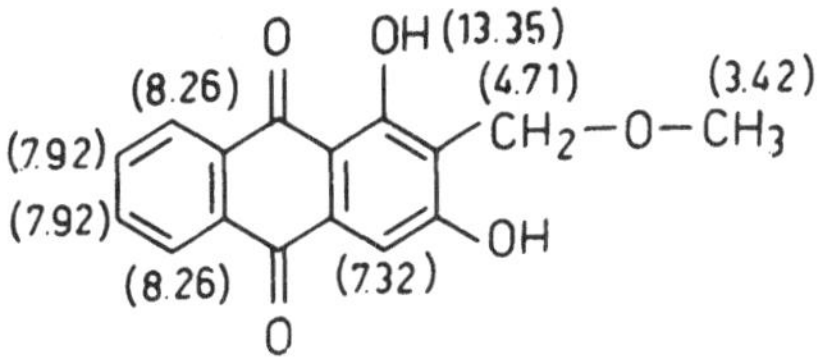

Lucidin ω-methyl ether

$C_{16}H_{12}O_5$
1: yellow-orange; 5: red (42)
mp: 170° (42)
UV (EtOH): 246 (4.51), 282 (4.40), 315 (sh., 4.07), 417 (3.72) (42)
IR: 3180, 1673, 1625, 1592 (42)
MS (70 eV): 284 (M$^+$, 30), 269 (10), 254 (10), 253 (32), 252 (100), 224 (10), 196 (31), 168 (9), 139 (11) (42)
^{1}H-NMR (250 MHz, acetone$_{d6}$): (42, 75)

Sources:
Galium album Mill. (75)   *M. parvifolia* Bartl. (42)
Morinda citrifolia L. (80)

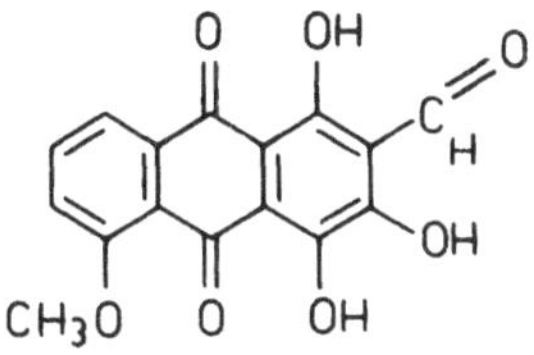

Majoranal

$C_{16}H_{10}O_7$
1: red (26)
mp: 267–269° (CHCl$_3$) (26)
UV (CHCl$_3$): 252.5 (4.48), 262 (sh., 4.36), 290 (4.45), 295 (sh., 4.45), 335 (sh., 3.76), 490 (3.99), 535 (sh., 3.78) (26)
IR (KBr): 1652, 1624, 1590 (26)
MS (70 eV): 314 (M$^+$, 100), 299 (12), 296 (13), 286 (34), 285 (19), 269 (12), 268 (48), 256 (14), 240 (22), 111 (11), 109 (10), 97 (18), 95 (16), 93 (34), 85 (18), 83 (21), 81 (19), 71 (32), 70 (12), 69 (29), 67 (15), 63 (34), 57 (56), 56 (15), 53 (43) (26)

Sources: *Damnacanthus major* Sieb. & Zucc. (26)

Table 7 (continued)

[Structure of 2-Methoxyanthraquinone with ^{13}C-NMR chemical shifts labeled: 127.1, 127.3, 133.8, 110.1, 55.8 (4.04), 134.0, 183.1, OCH₃, 133.5, 181.8, 121.1, 135.7, 133.7, 127.1, 129.7]

2-Methoxyanthraquinone

$C_{15}H_{10}O_3$
1: pale-yellow (38)
mp: 194–195° (38)
MS (50 eV): 238 (M⁺, 100), 210 (8), 209 (15), 208 (14), 195 (9), 167 (9), 152 (8), 139 (21) (20)
¹H-NMR (CDCl₃): (45)
¹³C-NMR (22.63 MHz, CDCl₃): (17)

Sources:

Asperula odorata L. (37)
Galium aparine L. (37)
G. mollugo L. (37)
G. saxatile L. (37)

G. saxatile × *G. sterneri* (37)
G. verum L. (37)
Morinda umbellata L. (39)
Rubia tinctorum L. (38)

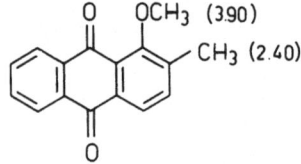

1-Methoxy-2-methylanthraquinone

$C_{16}H_{12}O_3$
1: pale-yellow (38)
mp: 153–154° (38)
UV (EtOH): 211 (4.21), 256 (4.37), 275 (sh., 4.00), 336 (3.48), 402 (3.20) (110)
IR (CHCl₃): 1670, 1590, 1320, 1270 (110)
¹H-NMR (60 MHz, CDCl₃): (110)

Sources:

Asperula odorata L. (37)
Galium aparine L. (37)
G. saxatile L. (37)
G. verum L. (37)

Morinda lucida Benth. (2)
M. umbellata L. (39)
Rubia tinctorum L. (38)

1-Methoxy-2-methyl-3,5,6-trihydroxyanthraquinone

$C_{16}H_{12}O_6$
1: orange (58)

Sources: *Putoria calabrica* Perss. (57)

5-Methoxy-2 (or -3)-methyl-1,4,6-trihydroxyanthraquinone

$C_{16}H_{12}O_6$
1: orange-red; 2: red-brown; 3: yellow-orange; 4: red-purple; 5: red-purple (128)
UV (MeOH): 222, 272, 457. (MeOH/OH$^-$): 240, 302, 525 (128)
IR (CHCl$_3$): 3420, 1610, 1565 (128)
MS (70 eV, 200°): 300 (M$^+$, 40), 282 (39), 254 (20), 135 (25), 97 (31), 83 (63), 71 (60), 57 (100) (128)
^{1}H-NMR (100 MHz, CDCl$_3$): (128)

Sources: *Cinchona ledgeriana* Moens. (128)

4-Methoxy-1,3,5-trihydroxyanthraquinone

$C_{15}H_{10}O_6$
1: orange; 3: red; 4: red-brown; 5: red-purple (128)
UV (MeOH): 279, 320, 425, 470 (sh.), 485 (sh.). (MeOH/OH$^-$): 283, 315 (sh.), 485 (128)
IR (KBr): 3420, 2920, 2860, 1720, 1630, 1470 (128)
MS (70 eV, 110°): 286 (M$^+$, 100), 268 (87), 257 (10), 243 (38), 212 (27), 180 (30) (128)
^{1}H-NMR (100 MHz, CDCl$_3$): (128)

Sources: *Cinchona ledgeriana* Moens. (128)

Table 7 (*continued*)

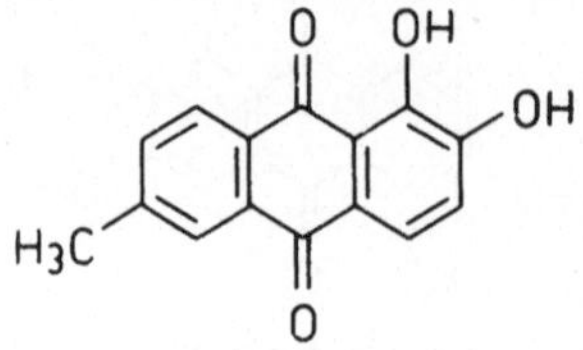

6-Methylalizarin

$C_{15}H_{10}O_4$
1: orange (*27*)
mp: 222–224 (*26*)
UV (EtOH): 233 (4.25), 262.5 (4.49), 278 (4.34), 330 (sh., 3.55), 429 (3.75) (*26*)
IR (KBr): 3480, 1670, 1647, 1608 (*26*)

Sources:

Coprosma tenuicaulis Hook f. (*27*) *Hymenodictyon excelsum* Wall. (*26*)

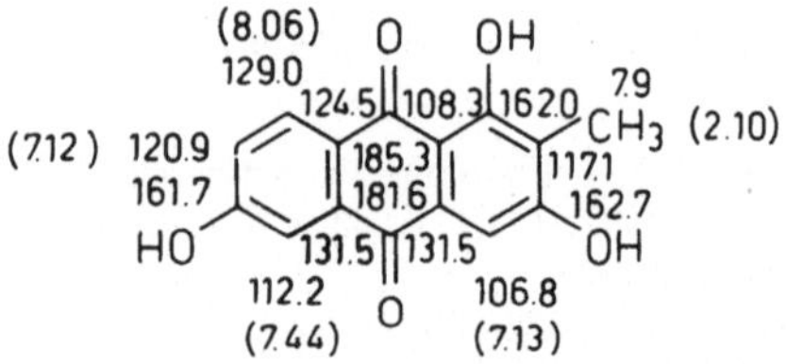

2-Methyl-1,3,6-trihydroxyanthraquinone

$C_{15}H_{10}O_5$
1: yellow (*74*)
mp: 236–238° (MeOH) (*74*)
UV (MeOH): 284 (4.63), 304 (sh., 4.22), 321 (3.82), 424 (3.83) (*74*)
IR (KBr): 3400, 1660, 1620, 1590 (*74*)
^{1}H-NMR (100 MHz, CD_3OD): (*74*)
^{13}C-NMR ($DMSO_{d6}$): (*74*)

Sources: *Rubia cordifolia* L. (*74*)

2-Methyl-3,5,6-trihydroxyanthraquinone

References, pp. 143–149

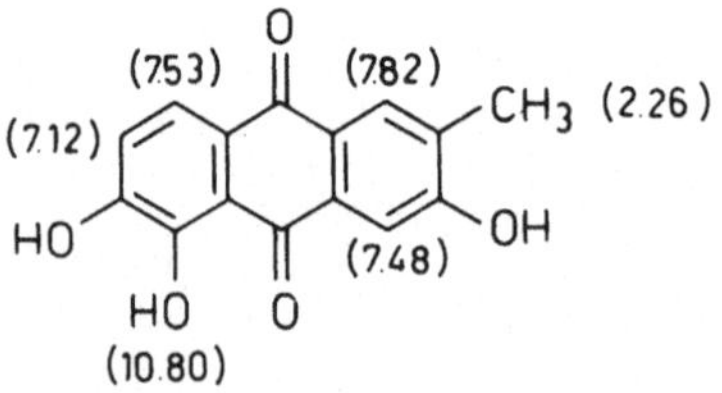

Table 7 (*continued*)

$C_{15}H_{10}O_5$
1: orange-red *(70)*
mp: >300° *(70)*
UV (MeOH): 222 (4.30), 279 (4.65), 338 (3.71), 414 (3.74) *(70)*
IR (KBr): 3440, 1630, 1570 *(70)*
^{1}H-NMR (100 MHz, DMSO$_{d6}$): *(70)*

Sources: *Morinda citrifolia* L. *(70)*

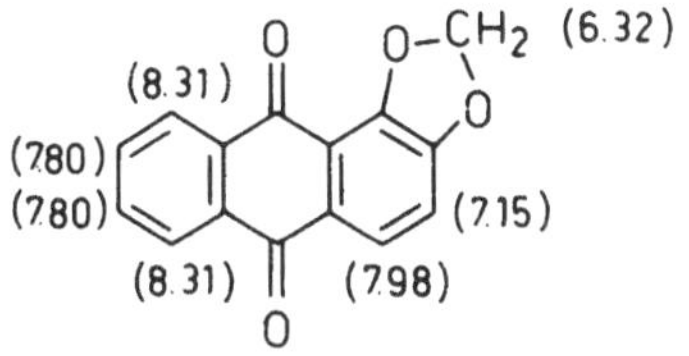

Morindaparvin-A

$C_{15}H_8O_4$
1: yellow; 2: yellow; 3: yellow *(41)*
mp: 257° (CHCl$_3$) *(41)*
UV: 248, 273 *(41)*
IR: 1675, 1587, 1455, 1295, 1260, 1075, 1040, 994, 934 *(41)*
MS (70 eV): 252 (M$^+$), 223, 196 *(41)*
^{1}H-NMR (100 MHz, CDCl$_3$): *(41)*

Sources: *Morinda parvifolia* Bartl. *(41, 42)*

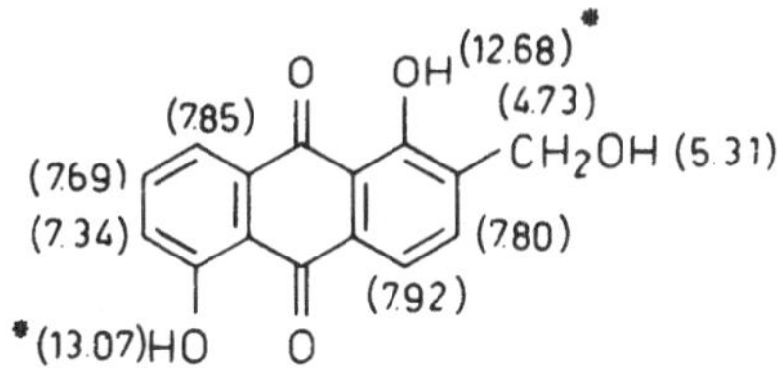

Morindaparvin-B

$C_{15}H_{10}O_5$
1: orange-yellow; 5: orange-red *(42)*
mp: 208.5–209.5° *(42)*
UV (EtOH): 227 (4.32), 255 (4.13), 277 (sh., 3.64), 287.5 (3.65), 420 (3.64), 430 (3.65) *(42)*
MS (70 eV): 270 (M$^+$, 100), 252 (9), 241 (51), 224 (13), 213 (4), 196 (5), 168 (6), 139 (9), 121 (8) *(42)*
^{1}H-NMR: *(42)*

Sources: *Morinda parvifolia* Bartl. *(42)*

Table 7 (*continued*)

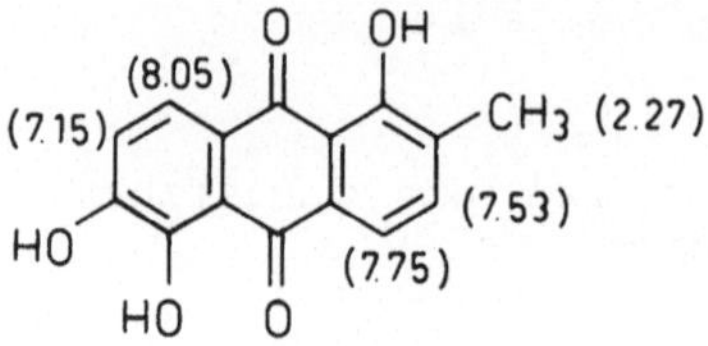

Morindone

$C_{15}H_{10}O_5$
1: red (*80*)
mp: 248–249.5° (*80*); 282–284° (toluene) (*125*)
UV (MeOH): 232 (4.49), 260 (4.50), 293 (4.18), 335 (3.42), 432 (3.78) (*70*)
IR (KBr): 3440, 1630, 1605 (*70*)
MS: 270 (M^+, 100), 253 (7), 242 (25), 213 (8), 185 (5), 168 (5), 157 (3), 139 (13), 135 (24) (*80*)
^{1}H-NMR (100 MHz, $DMSO_{d6}$): (*70, 110, 125*)

Sources:

Coprosma australis Robinson (*110*)
Hymenodictyon excelsum Wall. (*26*)
Morinda angustifolia Roxb. (*105*)
M. citrifolia L. (*70, 79, 80*)

M. rioc L. (*101*)
M. tinctoria Roxb. (*49, 92*)
M. tinctoria var. *tomentosa* (*106*)

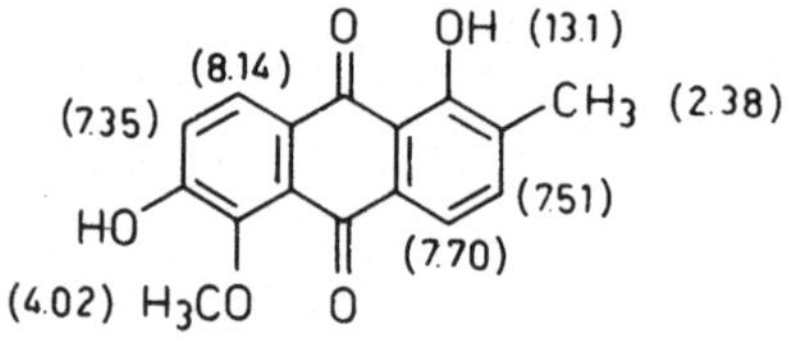

Morindone 5-methyl ether

$C_{16}H_{12}O_5$
1: dark yellow; 2: red-brown; 3: yellow-orange; 4: red-purple; 5: red-purple (*128*)
mp: 223° (*44*)
UV (EtOH): 223 (4.45), 249 (sh., 4.24), 266 (4.36), 283 (4.18), 292 (4.12), 415 (3.82).
(EtOH/OH^-): 210 (4.47), 235 (4.28), 251 (4.39), 266 (sh., 4.28), 313 (4.20), 503 (3.97) (*101*)
IR (KBr): 3420, 2920, 2850, 1660, 1625, (*128*)
MS (70 eV, 110°): 284 (M^+, 100), 266 (45), 238 (48), 60 (30) (*128*)
^{1}H-NMR (300 MHz, $CDCl_3$): (*128, 44, 58*)

Sources:

Cinchona ledgeriana Moens. (*128*)
Morinda lucida Benth. (*44*)

M. rioc L. (*101*)
Ploclama pendula Ait. (*58*)

Table 7 (*continued*)

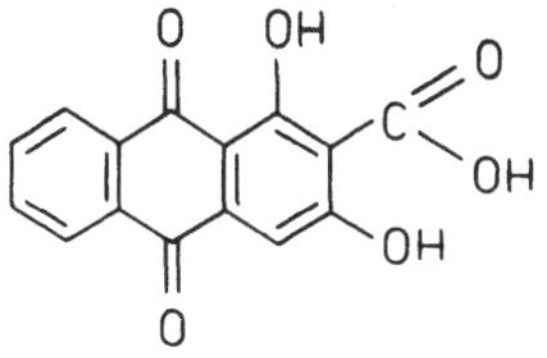

Munjistin

$C_{15}H_8O_6$
mp: 129–131° (dil. HAc) (*38*); 232° (1% HCl) (*65*)
IR: 3250, 3080, 1688, 1626 (*65*)

Sources:

Asperula azura L. (*55*)
A. galioides M.B. (*55*)
A. tinctoria L. (*55*)
Galium aparine L. (*55*)
G. cruciata (L.) Scop. (*55*)
G. mollugo L. (*55*)
G. polonicum Br. (*55*)
G. pumilum Murr. (*55*)
G. purpureum L. (*55*)

G. rubioides L. (*55*)
G. schultesii Vest. (*55*)
G. vulgare S.F. Gray (*55*)
Morinda umbellata L. (*39*)
Relbunium hypocarpium Hemsl. (*9*)
Rubia oliveri L. (*55*)
R. petiolaris L. (*55*)
R. tinctorum L. (*38, 52, 55*)

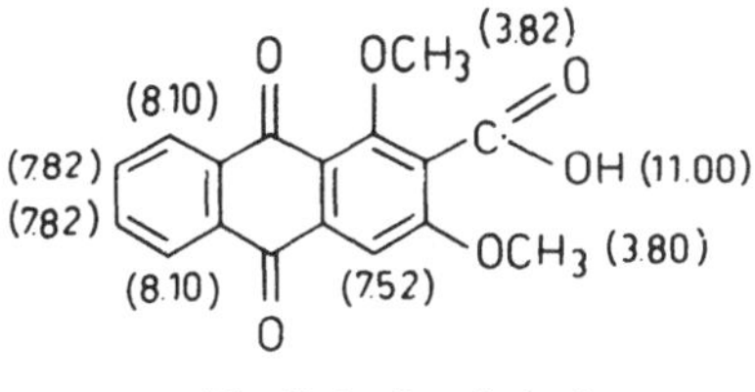

Munjistin dimethyl ether

$C_{17}H_{12}O_6$
1: yellow; 3: brown; 5: orange-red (*47*)
mp: 243° (CH_2Cl_2) (*47*); 212–213° (MeOH) (*65*)
UV (EtOH): 207 (4.13), 248 (4.22), 318 (4.09), 450 (3.49). (EtOH/OH^-): 213, 245, 312, 450 (*47*)
IR (KBr): 3330, 1745, 1680, 1650, 1570, 1460, 1420, 1370, 1310, 1280, 1245, 1225, 1090, 960, 890, 720 (*47*)
MS (70 eV, 200°): 312 (M^+, 7), 294 (10), 280 (46), 267 (53), 144 (100), 115, 74, 59, 43, 31 (*47*)
^{1}H-NMR (270 MHz, $DMSO_{d6}$): (*47*)

Sources: *Rubia cordifolia* L. (*47*)

Table 7 (*continued*)

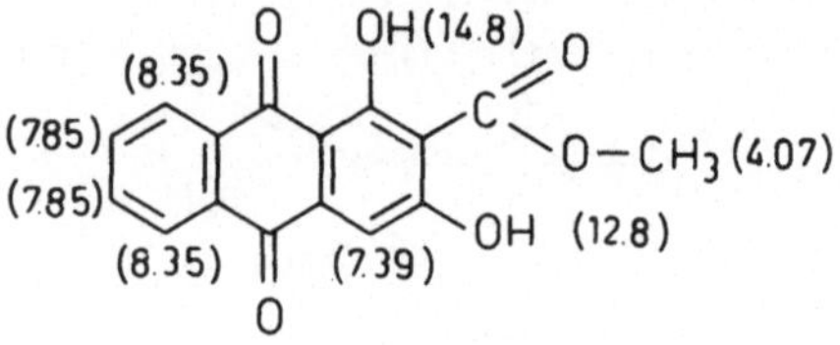

Munjistin methyl ester

C$_{16}$H$_{10}$O$_6$
1: yellow (*44*)
mp: 202° (*44*)
UV (MeOH): 242 (4.52), 280 (4.31), 408 (3.71). (MeOH/OH$^-$): 224 (4.38), 246 (4.48), 310, (4.36), 468 (3.90) (*44*)
IR (KBr): 1670, 1660, 1625, 708 (*44*)
MS (70 eV): 298 (M$^+$, 49), 266 (100), 238 (37) (*44*)
^{1}H-NMR (CDCl$_3$): (*44*)

Sources: *Morinda lucida* Benth. (*44*)

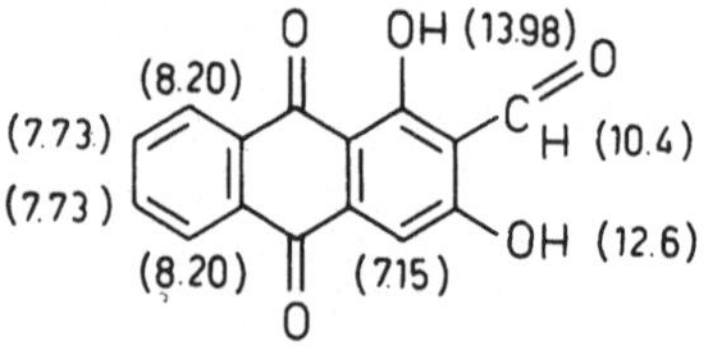

Nordamnacanthal

C$_{15}$H$_8$O$_5$
1: brown-orange; 3: bright yellow; 5: red (*119*)
mp: 220–222° (MeOH) (*119*); 200–202° (C$_6$H$_6$) (*80*)
UV (EtOH): 248, 272, 348, 450. (EtOH/OH$^-$): 233 (sh.), 250, 287, 362, 524 (*119*)
IR (KBr): 1690, 1660, 1645, 1605, 1585, 1398, 1310, 1295, 1280, 1205, 730 (*119*)
MS (70 eV, 200°): 268 (M$^+$, 51), 240 (100), 212 (22), 184 (15), 138 (16) (*119*)
^{1}H-NMR (270 MHz, CDCl$_3$): (*119, 80*)

Sources:

Coprosma linariifolia Hook f. (*27*)   *M. lucida* Benth. (*2*)
C. rotundifolia A. Cunn. (*27*)   *M. rooic* L. (*101*)
Damnacanthus major Sieb. & Zucc. (*26*)   *M. tinctoria* Roxb. (*49, 92*)
Hymenodictyon excelsum Wall. (*26*)   *Rubia cordifolia* L. (*119*)
Morinda citrifolia L. (*80*)   *R. iberica* C. Koch. (*98*)

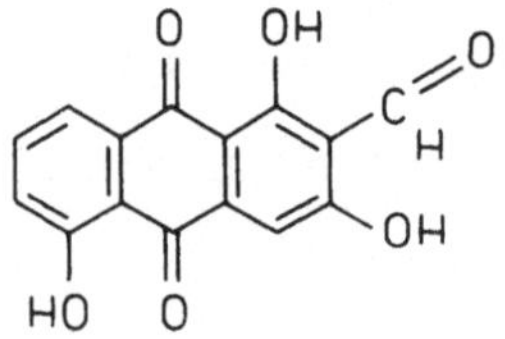

Norjuzunal

$C_{15}H_8O_6$
1: orange (26)
mp: 265° (EtOEt/P.E.) (26)
UV (hexane): 232.5 (4.50), 245 (4.49), 252 (4.51), 260 (sh., 4.48), 293 (sh., 4.30), 302 (4.33), 432 (4.13), 474 (sh., 3.42) (26)
IR (KBr): 1662, 1631, 1598 (26)

Sources: *Damnacanthus major* Sieb. & Zucc. (26)

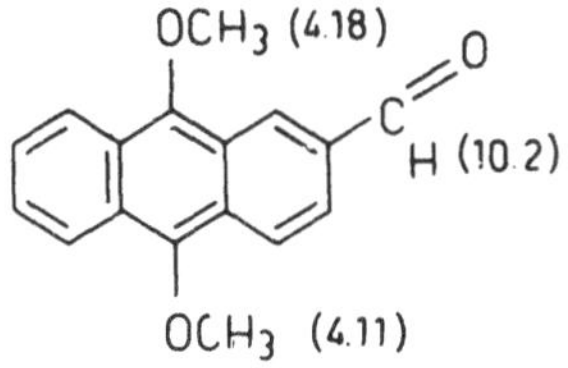

OCH$_3$ (4.18)
H (10.2)
OCH$_3$ (4.11)

Oruwal

$C_{17}H_{14}O_3$
1: bright yellow (1)
mp: 157–158° (MeOH) (2)
UV (cyclohex.): 238 (sh., 4.41), 243 (4.43), 272 (sh., 4.76), 277 (4.81), 360 (3.63), 375 (3.75) (2)
IR (nujol): 1695, 1680, 1620, 695 (2)
MS: 266 (M$^+$), 251, 235, 223, 207, 193, 179, 165, 151 (2)
^{1}H-NMR (60 MHz, CDCl$_3$): (2)

Sources: *Morinda lucida* Benth. (1, 2)

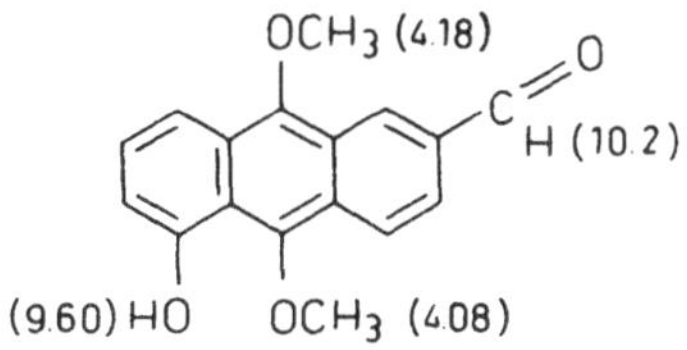

OCH$_3$ (4.18)
H (10.2)
(9.60) HO   OCH$_3$ (4.08)

Oruwalol

$C_{17}H_{14}O_4$
mp: 198–201° (2)
IR (nujol): 3320, 1695, 1680, 1620, 695 (2)
MS: 282 (M$^+$, 100), 267, 251, 223, 195, 167, 139 (2)
^{1}H-NMR (60 MHz, CDCl$_3$): (2)

Sources: *Morinda lucida* Benth. (2)

Table 7 *(continued)*

Pseudopurpurin

$C_{15}H_8O_7$
1: yellow; 2, 3: yellow; 5: red *(55)*
mp: 221–223° (EtOH) *(38)*; 229–230° (CH$_2$Cl$_2$) *(118)*
UV: 208, 284, 496 *(118)*
IR: 3000, 1700, 1570 *(118)*

Sources:

Asperula azura L. *(55)*
A. galioides M.B. *(55)*
A. odorata L. *(37, 55)*
A. tinctoria L. *(55)*
Galium aparine L. *(37, 55)*
G. cruciata (L.) Scop. *(55)*
G. fleuroti Jord. *(37)*
G. mollugo L. *(11, 12, 37, 55, 126)*
G. normani Dahl. *(37)*
G. polonicum Br. *(55)*
G. pumilum (L.) Murr. *(37, 55)*
G. purpureum L. *(55)*

G. rubioides L. *(55)*
G. saxatile L. *(37)*
G. saxatile × *G. sterneri* *(37)*
G. schultesii Vest. *(55)*
G. sterneri Ehrend *(55)*
G. verum L. *(37)*
G. vulgare S.F. Gray *(55)*
Relbunium hypocarpium Hemsl. *(9)*
Rubia cordifolia L. *(118)*
R. oliveri L. *(55)*
R. petiolaris L. *(55)*
R. tinctorum L. *(15, 36, 38, 52, 55, 88)*

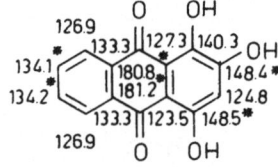

Purpurin

$C_{14}H_8O_5$
1: red-purple; 3: red; 4: red-purple; 5: red-purple *(128)*
mp: 262–263° (EtOH) *(38)*
UV (MeOH): 218, 260, 280 (sh.), 515 *(128)*
MS (70 eV, 90°): 256 (M$^+$, 30), 111 (35), 97 (50), 83 (50), 71 (75), 57 (100) *(128)*
^{13}C-NMR (20 MHz, CDCl$_3$, as triacetate derivative): *(67)*

Sources:

Asperula odorata L. *(37)*
Cinchona ledgeriana Moens. *(128)*
Galium album Mill. *(75)*

G. aparine L. *(37)*
G. fleuroti Jord. *(37)*
G. mollugo L. *(11, 12, 37, 72)*

Table 7 (*continued*)

G. *normani* Dahl. (*37*)
G. *pumilum* L. (*37*)
G. *saxatile* L. (*37*)
G. *saxatile* × G. *sterneri* (*37*)
G. *sterneri* Ehrend. (*37*)

G. *verum* L. (*37*)
Relbunium hypocarpium Hemsl. (*9, 113*)
Rubia cordifolia L. (*96, 118*)
R. tinctorum L. (*15, 16, 36, 38, 52, 55, 82, 97*)

Purpurin 1-methyl ether

$C_{15}H_{10}O_5$
1: orange; 3: red; 4: red-brown; 5: red-purple (*128*)
mp: 219–221° (hexane/CHCl$_3$) (*26*)
UV (MeOH): 209, 240, 245 (sh.), 284, 312, 415, 480 (sh.). (MeOH/OH$^-$): 240, 245 (sh.), 312, 485 (*128*)
IR (KBr): 3400, 2920, 2840, 1660, 1630, 1590 (*128*)
MS (70 eV, 95°): 270 (M$^+$, 100), 252 (63), 241 (8), 227 (32), 196 (23), 85 (20), 71 (33), 57 (44) (*128*)
^{1}H-NMR (100 MHz, CDCl$_3$): (*128*)

Sources: *Cinchona ledgeriana* Moens. (*128*)

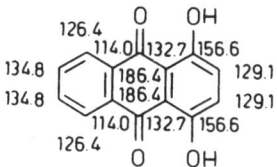

Quinizarin

$C_{14}H_8O_4$
UV (H$_2$O): 228, 256, 262, 288, 332, 464 (*48*)
MS (50 eV): 240 (M$^+$, 100), 183 (5), 102 (6), 77 (5), 51 (6) (*20*)
^{13}C-NMR (22.63 MHz, DMSO$_{d6}$): (*17*)

Sources: *Rubia tinctorum* L. (*16*)

Table 7 (*continued*)

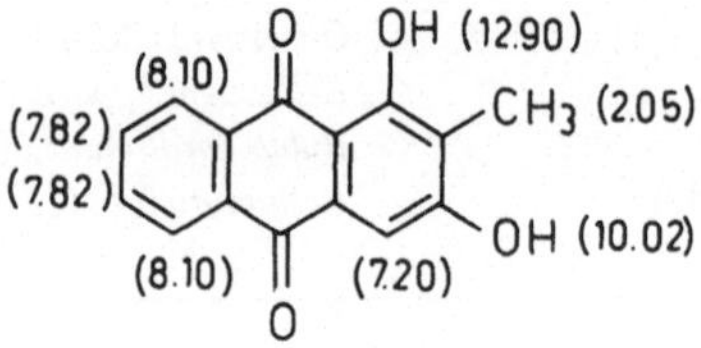

Rubiadin

$C_{15}H_{10}O_4$
1: yellow; 3: brown; 4: brown; 5: red (*128*)
mp: 300° (MeOH) (*47*); 280–283° (*80*)
UV (MeOH): 242 (sh.), 243, 260 (sh.), 279, 308, 410. (MeOH/OH$^-$): 245, 300 (sh.), 310, 484 (*128*)
IR (KBr): 3410, 1655, 1620, 1590, 1335, 1310, 1120, 710 (*128*)
MS (70 eV, 200°): 254 (M$^+$, 100), 226 (8), 225 (6), 197 (8), 169 (3), 152, 141, 115, 105, 76 (*47*)
^{1}H-NMR (270 MHz, DMSO$_{d6}$): (*47, 4, 57, 58, 66, 70, 75, 128*)

Sources:

Asperula besseriana Klok. (*21*)
A. odorata L. (*37*)
Cinchona ledgeriana Moens. (*94, 128*)
C. pubescens Vahl. (*95*)
Commitheca liebrechtsiana Brem. (*66*)
Coprosma linariifolia Hook. f. (*27*)
C. robusta Raoul. (*27*)
C. rotundifolia A. Cunn. (*27*)
C. tenuicaulis Hook. f. (*27*)
Damnacanthus major Sieb. & Zucc. (*26*)
Danais fragrans Gaertn. (*4*)
Galium album Mill. (*75*)
G. aparine L. (*37, 55*)
G. dasypodum Klok. (*131, 132, 133*)
G. mollugo L. (*55*)
G. normani Dahl. (*37*)
G. polonicum Br. (*55*)
G. pumilum L. (*37, 55*)
G. purpureum L. (*55*)
G. rubioides L. (*55*)

G. ruthenicum Willd. (*21*)
G. saxatile L. (*37*)
G. saxatile × *G. sterneri* (*37*)
G. schultesii Vest. (*55*)
G. semiamictum Klok. (*130*)
G. sterneri Ehrend. (*37*)
G. verum L. (*37*)
G. vulgare S.F. Gray (*55*)
Hymenodictyon excelsum Wall. (*26*)
Morinda citrifolia L. (*70, 80*)
M. lucida Benth. (*2*)
M. umbellata L. (*39*)
Ploclama pendula Ait. (*58*)
Prismatomeris malayana Ridley (*77*)
P. tetrandra (Roxb.) K. Schum. (*122*)
Putoria calabrica Perss. (*57*)
Rubia cordifolia L. (*47*)
R. oliveri L. (*55*)
R. petiolaris L. (*55*)
R. tinctorum L. (*36, 38, 52, 55*)

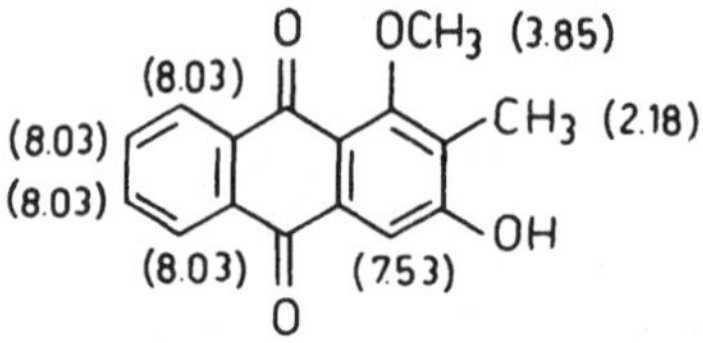

Rubiadin 1-methyl ether

References, pp. 143–149

Table 7 (*continued*)

$C_{16}H_{12}O_4$
1: yellow (*27*)
mp: 282-284° (dioxane) (*27*); 300° (acetone) (*110*)
UV (MeOH): 208 (4.32), 240 (4.28), 245 (4.26), 279 (4.55), 366 (3.65) (*89*)
IR (nujol): 3300, 1673, 1653, 1569, 710 (*89*)
MS (70 eV): 268 (M$^+$, 100), 254 (11), 253 (47), 251 (23), 250 (28), 240 (7), 239 (30), 237 (5), 225 (7), 222 (15), 211 (7), 209 (5), 181 (13), 139 (10) (*101*)
^{1}H-NMR (60 MHz, DMSO$_{d_6}$): (*101, 110*)

Sources:

Asperula besseriana Klok. (*21*)	*Galium dasypodum* Klok. (*131, 132, 133*)
Coprosma australis Robinson (*110*)	*Hymenodictyon excelsum* Wall. (*26*)
C. linariifolia Hook. f. (*27*)	*Morinda lucida* Benth. (*1, 2*)
C. robusta Raoul. (*27*)	*M. roioc* L. (*101*)
C. rotundifolia A. Cunn. (*27*)	*M. umbellata* L. (*39*)
C. tenuicaulis Hook f. (*27*)	*Prismatomeris malayana* Ridley (*77*)
Damnacanthus subspinosus Hand-Mazz. (*89*)	*P. tetrandra* (Roxb.) K. Schum. (*122*)

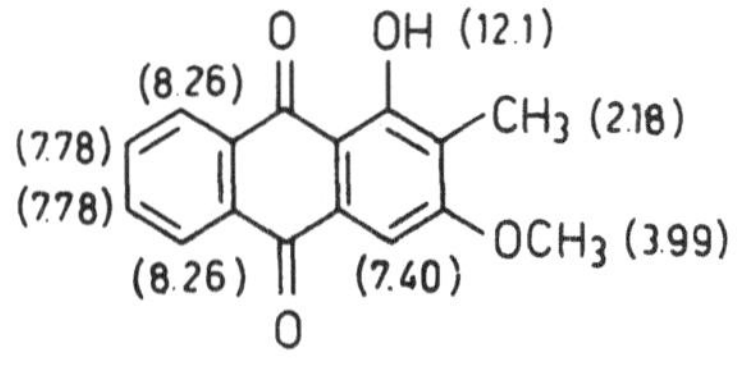

Rubiadin 3-methyl ether

$C_{16}H_{12}O_4$
1: yellow; 2, 3: red; 4: yellow-orange; 5: red-orange (*75*)
mp: 190-191° (EtOH) (*58*)
UV (EtOH): 240 (sh., 4.41), 246 (4.45), 276 (4.64), 337 (3.54), 410 (3.93) (*58*)
IR (KBr): 3260, 1663, 1645, 1560 (*75*)
MS: 268 (M$^+$, 100), 253 (42), 251 (22), 250 (26), 239 (26), 225 (6), 223 (7), 222 (12), 211 (5) (*75*)
^{1}H-NMR (60 MHz, CDCl$_3$): (*58, 75*)

Sources: *Galium album* Mill. (*75*)

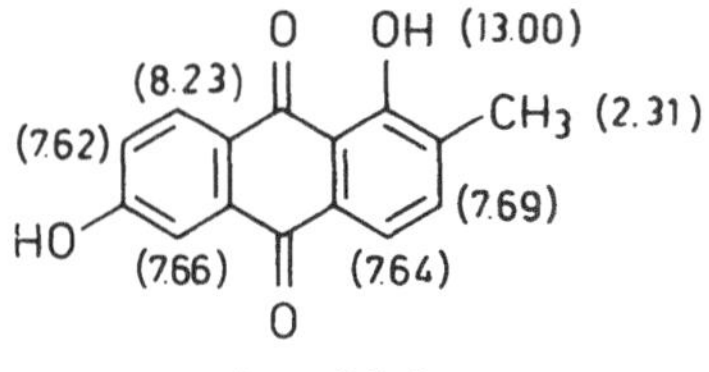

Soranjidiol

$C_{15}H_{10}O_4$
1: yellow (*26*)
mp: 271-273° (*2*); 286-288° (subl.) (*101*)

Table 7 (*continued*)

UV (EtOH): 217 (4.44), 267 (4.45), 282 (sh., 4.28), 292 (4.19), 392 (sh., 3.86), 408 (3.91), 430 (sh., 3.82). (EtOH/OH⁻): 220 (4.37), 244 (4.29), 304 (4.37), 385 (sh., 3.66), 480 (3.88) (*101*)
IR (KBr): 3415, 1672, 1637, 1600, 1437, 1360, 1279, 1256, 1028, 869, 827, 745 (*101*)
MS (70 eV): 254 (M⁺, 100), 253 (10), 237 (8), 226 (8), 225 (8), 209 (9), 197 (9), 180 (38), 149 (24), 145 (33), 120 (45) (*101*)
^{1}H-NMR (270 MHz, DMSO$_{d6}$): (*44, 35, 101*)

Sources:
Hymenodictyon excelsum Wall. (*26*)   *M. rioic* L. (*101*)
Morinda lucida Benth. (*1, 2, 44*)

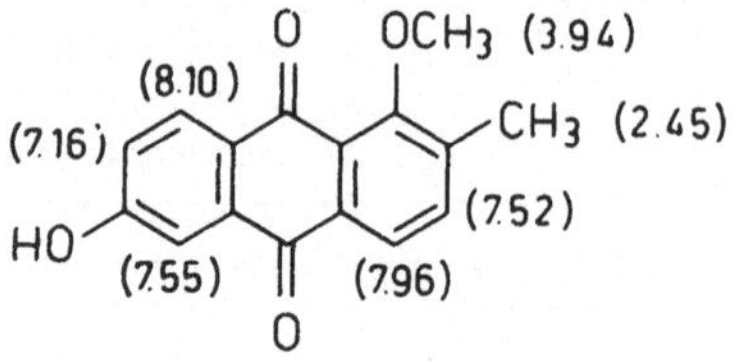

Soranjidiol 1-methyl ether

$C_{16}H_{12}O_4$
UV (EtOH): 245 (4.17), 268 (4.43), 286 (4.31), 355 (3.56). (EtOH/OH⁻): 246 (4.41), 286 (4.21), 307 (4.32), 470 (3.58) (*101*)
IR (KBr): 3265, 1670, 1645, 1600–1574, 1451, 1391, 1334, 1281, 1222, 1190, 1092, 1041, 988, 819, 751 (*101*)
MS (70 eV): 268 (M⁺, 100), 267 (8), 254 (6), 253 (21), 251 (14), 250 (9), 240 (5), 239 (27), 222 (12), 211 (5), 209 (3), 197 (4), 165 (5), 139 (6), 91 (5) (*101*)
^{1}H-NMR (60 MHz, CDCl$_3$/DMSO$_{d6}$): (*101*)

Sources: *Morinda rioic* L. (*101*)

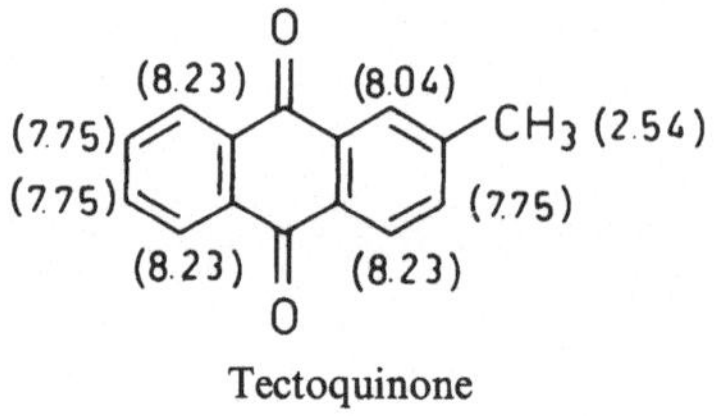

Tectoquinone

$C_{15}H_{10}O_2$
mp: 174–175° (EtOH) (*45*)
UV (EtOH): 255 (4.64), 264 (4.38), 274 (4.14), 327 (3.52). (EtOH/OH⁻): idem (*101*)
IR: 1676, 1594, 1322, 1291, 1170, 1152, 1134, 922, 836, 710 (*101*)
MS (70 eV): 222 (M⁺, 100), 207 (15), 194 (22), 162 (26) (*101*)
^{1}H-NMR (60 MHz, CDCl$_3$): (*101*)

Sources:
Morinda lucida Benth. (*2*)   *M. umbellata* L. (*39*)
M. rioic L. (*101*)   *Prismatomeris tetrandra* (Roxb.) K. Schum. (*122*)

References, pp. 143–149

Table 7 (*continued*)

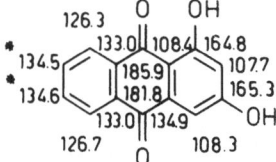

Tinctomorone

$C_{25}H_{28}O_7$
mp: 195° (*49*)
UV: 220, 245, 275, 320, 410 (*49*)
IR: 3600, 3400, 2950, 1725, 1670, 1638 (*49*)
MS: 440 (M^+), 412, 411, 383, 312, 297, 270, 269, 268, 254 (100), 240 (*49*)

Sources: *Morinda tinctoria* Roxb. (*49*)

Xanthopurpurin

$C_{14}H_8O_4$
1: yellow; 2, 3: yellow; 5: red (*55*)
mp: 266–268° (EtOH) (*38*)
^{13}C-NMR (22.63 MHz, DMSO$_{d6}$): (*17*)

Sources:

Asperula odorata L. (*37*)	*G. sterneri* Ehrend. (*37*)
Galium aparine L. (*37*)	*G. verum* L. (*37*)
G. fleuroti Jord. (*37*)	*Morinda umbellata* L. (*39*)
G. mollugo L. (*37*)	*Relbunium hypocarpium* Hemsl. (*9*)
G. pumilum L. (*37*)	*Rubia oliveri* L. (*55*)
G. purpureum L. (*55*)	*R. petiolaris* L. (*55*)
G. saxatile L. (*37*)	*R. tinctorum* L. (*31, 52, 55, 97*)
G. saxatile × *G. sterneri* (*37*)	

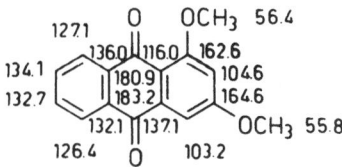

Xanthopurpurin dimethyl ether

Table 7 (*continued*)

$C_{16}H_{12}O_4$
1: bright yellow (*38*)
mp: 154–155° (EtOH) (*37*)
^{13}C-NMR (22.63 MHz, CDCl$_3$): (*17*)

Sources:
Galium aparine L. (*37*)   G. sterneri Ehrend. (*37*)
G. mollugo L. (*37*)   G. verum L. (*37*)
G. saxatile L. (*37*)   Rubia tinctorum L. (*38*)

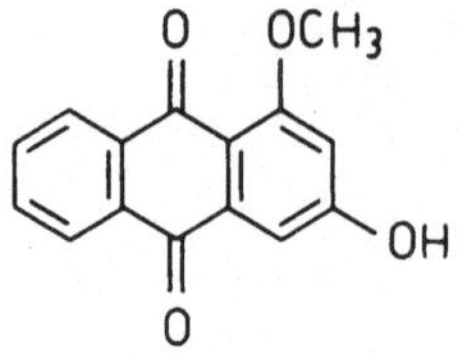

Xanthopurpurin 1-methyl ether

Sources: *Relbunium hypocarpium* Hemsl. (*113*)

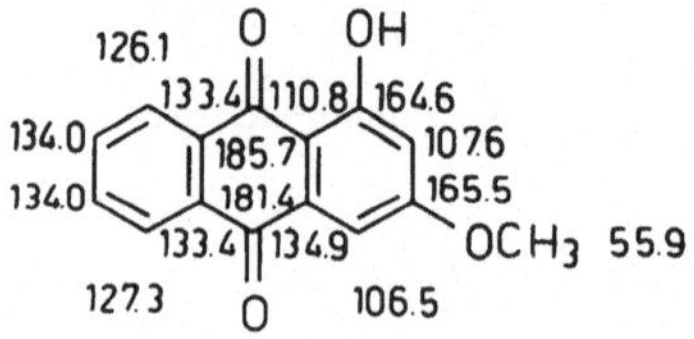

Xanthopurpurin 3-methyl ether (xanthopurpurin 2-methyl ether)

$C_{15}H_{10}O_4$
1: bright yellow (*38*)
mp: 193–194° (EtOH) (*38*)
^{13}C-NMR (22.63 MHz, CDCl$_3$): (*18*)

Sources:
Relbunium hypocarpium Hemsl. (*113*)   Rubia tinctorum L. (*38*)

Table 8. *List of Species of the Rubiaceae Reported to Contain Anthraquinones and the Anthraquinones Isolated from Them*[a]

Asperula azura L.	Root	alizarin (55) munjistin (55) pseudopurpurin (55) n.i. (56)
A. besseriana Klok.	Root	alizarin (21) lucidin (21) rubiadin (21) rubiadin 1-methyl ether (21) n.i. (22)
A. ciliata Rochel.	Root	alizarin (120)
A. cynanchica L.	Root	n.i. (22)
A. galioides M.B.	Root	alizarin (55) munjistin (55) pseudopurpurin (55) n.i. (22, 56)
A. glabra C. Koch (= *A. involucrata* Wahlenb.)	Root, PTC	n.i. (112)
A. octonaria Klok.	Root	n.i. (22)
A. odorata L.	Root	alizarin (37, 55) alizarin 1-methyl ether (37) alizarin 2-methyl ether (37) 2-hydroxyanthraquinone (37) 1-hydroxy-2-methylanthraquinone (37) lucidin (37) 2-methoxyanthraquinone (37) 1-methoxy-2-methylanthraquinone (37) pseudopurpurin (37, 55) purpurin (37) rubiadin (37) xanthopurpurin (37) n.i. (22)
A. tinctoria L.	Root	alizarin (55) munjistin (55) pseudopurpurin (55) n.i. (22, 56)
Cinchona sp.	Bark	n.i. (43)
C. ledgeriana Moens.	PTC	alizarin (94) alizarin 1-methyl ether (94) anthragallol 1,2-dimethyl ether (128) anthragallol 1,3-dimethyl ether (128) 1,8-dihydroxyanthraquinone (94) 1,3-dihydroxy-2,5-dimethoxyanthraquinone (128)

[a] n.i. not identified; PTC: plant tissue culture.

Table 8 *(continued)*

		1,3-dihydroxy-4-methoxyanthraquinone (*128*)
		2,5- (or 3,5-)dihydroxy-1,3,4- (or-1,2,4-)-trimethoxyanthraquinone (*128*)
		5,6-dimethoxy-1- (or -4-)hydroxy-2- (or -3-)hydroxymethylanthraquinone (*128*)
		1,4-dimethoxy-2,3-methylenedioxyanthraquinone (*128*)
		1-hydroxy-2-hydroxymethylanthraquinone (*94, 128*)
		1-hydroxy-2-methylanthraquinone (*128*)
		2-hydroxy-1,3,4-trimethoxyanthraquinone (*128*)
		5-methoxy-2- (or -3-)methyl-1,4,6-trihydroxyanthraquinone (*128*)
		4-methoxy-1,3,5-trihydroxyanthraquinone (*128*)
		morindone 5-methyl ether (*128*)
		purpurin (*128*)
		rubiadin (*94, 128*)
C. pubescens Vahl.	PTC	alizarin (*95*)
		alizarin 1-methyl ether (*95*)
		1,8-dihydroxyanthraquinone (*95*)
		1-hydroxy-2-hydroxymethylanthraquinone (*95*)
		rubiadin (*95*)
Coelospermum paniculatum F. Muell.	Root bark	damnacanthol (*24*)
		nordamnacanthal (*24*)
		rubiadin (*24*)
		rubiadin 1-methyl ether (*24*)
C. reticulatum Benth.	Root bark	coelulatin (*24*)
		lucidin (*24*)
		rubiadin (*24*)
Commitheca liebrechtsiana Brem.	Root bark, stem bark	lucidin (*66*)
		rubiadin (*66*)
Coprosma acerosa A. Cunn.	Stem bark	anthragallol 1,2-dimethyl ether (*34*)
		anthragallol 2-methyl ether (*34*)
		3-hydroxy-2-methylanthraquinone (*34*)
		lucidin (*34*)
		rubiadin (*34*)
		rubiadin 1-methyl ether (*34*)
		soranjidiol (*34*)
C. areolata Cheesem.	Stem bark	copareolatin (*28*)
		rubiadin 1-methyl ether (*28*)

Table 8 *(continued)*

C. australis Robinson	Stem bark	copareolatin *(32)* morindone *(29, 110)* rubiadin *(32)* rubiadin 1-methyl ether *(29, 110)* soranjidiol *(32)*
C. linariifolia Hook. f.	Stem bark	anthragallol 1,3-dimethyl ether *(120)* anthragallol 2-methyl ether *(27, 120)* 3-hydroxy-2-methylanthraquinone *(27)* nordamnacanthal *(27)* rubiadin *(27)* rubiadin 1-methyl ether *(27)*
C. lucida Forst.	Stem bark	anthragallol *(31)* anthragallol 1,2-dimethyl ether *(31)* anthragallol 2-methyl ether *(31)* 3-hydroxy-2-methylanthraquinone *(31)* lucidin *(31)* rubiadin *(31)* rubiadin 1-methyl ether *(31)*
C. parviflora Hook.	Bark	6-methylalizarin *(120)* rubiadin 1-methyl ether *(120)*
C. propinqua A. Cunn.	Bark	n.i. *(27)*
C. rhamnoides A. Cunn.	Root bark	anthragallol 1,2-dimethyl ether *(33)* rubiadin 1-methyl ether *(33)*
C. robusta Raoul.	Heartwood	rubiadin *(27)* rubiadin 1-methyl ether *(27)*
C. rotundifolia A. Cunn.	Bark	damnacanthal *(27)* lucidin *(27)* rubiadin *(27)* rubiadin 1-methyl ether *(27)*
C. tenuicaulis Hook. f.	Bark	3-hydroxy-2-methylanthraquinone *(27)* 6-methylalizarin *(27)* rubiadin *(27)* rubiadin 1-methyl ether *(27)*
Crucianella maritima L.	Root	alizarin *(120)*
Damnacanthus indicus Gaertl. fil. var. *microphyllus* Makino	Root	damnacanthal *(120)* juzunal *(120)*
D. major Sieb. & Zucc.	Root	alizarin 1-methyl ether *(26)* 2-benzylxanthopurpurin *(26)* damnacanthal *(64, 120)* damnacanthol *(120)* 5- (or 8-)hydroxyalizarin 1-methyl ether *(26)* juzunal *(26, 64, 120)* juzunol *(64)* majoranal *(26)* nordamnacanthal *(26)* norjuzunal *(26)*

Table 8 *(continued)*

D. major Sieb. & Zucc. var. *parviflorius* Koidz.	Root	damnacanthal (*64, 120*) juzunal (*64, 120*) juzunol (*64*)
D. subspinosus Hand-Mazz.	Root	3,5-dihydroxy-2-ethoxymethyl-4-methoxyanthraquinone (*89*) 2-ethoxymethyl-3-hydroxy-4-methoxyanthraquinone (*89*) 8-hydroxydamnacanthol ω-ethyl ether (*88*) rubiadin 1-methyl ether (*89*)
Danais fragrams Gaertn.	Root bark	rubiadin (*4*)
Galium album Mill.	Root	alizarin (*75*) alizarin 1-methyl ether (*75*) 1,6- (or 1,7-)dihydroxy-4-methoxyanthraquinone (*75*) 1,3-dihydroxy-2-methoxymethyl-anthraquinone (*75*) galiprenylin (*76*) 1-hydroxy-2-hydroxyethyl-3-methoxyanthraquinone (*75*) 1-hydroxy-2-hydroxymethylanthra-quinone (*75*) purpurin (*75*) rubiadin (*75*) rubiadin 3-methyl ether (*75*)
G. aparine L.	Root (*37, 55, 112*), PTC (*112*)	alizarin (*37, 55*) alizarin 1-methyl ether (*37*) alizarin 2-methyl ether (*37*) 1-hydroxy-2-methylanthraquinone (*37*) lucidin (*37*) 2-methoxyanthraquinone (*37*) 1-methoxy-2-methylanthraquinone (*37*) munjistin (*55*) pseudopurpurin (*37, 55*) purpurin (*37*) rubiadin (*37, 55*) xanthopurpurin (*37*) xanthopurpurin dimethyl ether (*37*) n.i. (*56, 112*)
G. atherodes Spreng.	Root	alizarin 1-methyl ether (*120*)
G. cruciata L.	Root	alizarin (*55*) munjistin (*55*) pseudopurpurin (*55*)
G. dasypodum Klok.	Root	alizarin (*131, 133*) 1,3-dihydroxy-2-ethoxymethyl-anthraquinone (*131, 133*) lucidin (*131, 133*)

References, pp. 143–149

Table 8 *(continued)*

		rubiadin (*131, 133*)
		rubiadin 1-methyl ether (*131, 133*)
		n.i. (*22*)
G. divaricatum Pourret *ex* Lam.	Root, PTC	n.i. (*112*)
G. erectum Hudson (= *G. album* Mill.)	Root, PTC	n.i. (*112*)
G. fleuroti Jord.	Root	alizarin (*37*)
		pseudopurpurin (*37*)
		purpurin (*37*)
		xanthopurpurin (*37*)
G. glabratum Klok.	Root	n.i. (*22*)
G. glaucum L.	Root, PTC	n.i. (*112*)
G. mollugo L.	Root (*37, 55, 112*) PTC (*3, 11, 12, 62, 71, 72, 112, 126*)	alizarin (*11, 12, 37, 55, 126*)
		alizarin 1-methyl ether (*37*)
		alizarin 2-methyl ether (*37*)
		2-hydroxyanthraquinone (*37*)
		1-hydroxy-2-methylanthraquinone (*37*)
		lucidin (*11, 12, 37, 71, 72, 126*)
		lucidin ω-ethyl ether (*11*)
		2-methoxyanthraquinone (*37*)
		munjistin (*55*)
		pseudopurpurin (*11, 12, 37, 55, 126*)
		purpurin (*11, 12, 37, 72*)
		rubiadin (*55*)
		xanthopurpurin (*37*)
		xanthopurpurin dimethyl ether (*37*)
		n.i. (*3, 61, 62, 112*)
G. normani Dahl.	Root	alizarin (*37*)
		pseudopurpurin (*37*)
		purpurin (*37*)
		rubiadin (*37*)
G. odoratum (L.) Scop.	Root, PTC	n.i. (*112*)
G. parisiense L.	Root, PTC	n.i. (*112*)
G. polonicum Br.	Root	alizarin (*55*)
		munjistin (*55*)
		pseudopurpurin (*55*)
		rubiadin (*55*)
		n.i. (*56*)
G. praeboreale Klok.	Root	n.i. (*22*)
G. pseudomollugo Klok.	Root	n.i. (*22*)
G. pumilum L.	Root	alizarin (*37, 55*)
		alizarin 1-methyl ether (*37*)
		alizarin 2-methyl ether (*37*)

Table 8 *(continued)*

		2-hydroxyanthraquinone *(37)* 1-hydroxy-2-methylanthraquinone *(37)* lucidin *(37)* munjistin *(55)* pseudopurpurin *(37, 55)* purpurin *(37)* rubiadin *(37, 55)* xanthopurpurin *(37)*
G. purpureum L.	Root	alizarin *(55)* munjistin *(55)* pseudopurpurin *(55)* rubiadin *(55)* xanthopurpurin *(55)* n.i. *(56)*
G. rubioides L.	Root *(55, 56, 112)* PTC *(112)*	alizarin *(55)* munjistin *(55)* pseudopurpurin *(55)* rubiadin *(55)* xanthopurpurin *(55)* n.i. *(56, 112)*
G. ruthenicum Willd.	Root	alizarin *(21)* lucidin *(21)* rubiadin *(21)* n.i. *(22)*
G. saccharatum All. (= *G. verrucosum* Hudson)	Root, PTC	n.i. *(112)*
G. salicifolium Klok.	Root	n.i. *(22)*
G. saxatile L.	Root	alizarin *(37)* alizarin 1-methyl ether *(37)* alizarin 2-methyl ether *(37)* 2-hydroxyanthraquinone *(37)* 1-hydroxy-2-methylanthraquinone *(37)* lucidin *(37)* 2-methoxyanthraquinone *(37)* 1-methoxy-2-methylanthraquinone *(37)* pseudopurpurin *(37)* purpurin *(37)* rubiadin *(37)* xanthopurpurin *(37)* xanthopurpurin dimethyl ether *(37)*
G. saxatile × *G. sterneri*	Root	alizarin *(37)* alizarin 1-methyl ether *(37)* lucidin *(37)* 2-methoxyanthraquinone *(37)* pseudopurpurin *(37)* purpurin *(37)* rubiadin *(37)* xanthopurpurin *(37)*

Table 8 *(continued)*

G. schultesii Vest.	Root	alizarin *(55)*
		munjistin *(55)*
		pseudopurpurin *(55)*
		rubiadin *(55)*
G. semiamictum Klok.		lucidin *(130)*
		rubiadin *(130)*
		n.i. *(22)*
G. spurium L.	Root, PTC	n.i. *(112)*
G. sterneri Ehrend.	Root *(37, 112)*, PTC *(112)*	alizarin *(37)*
		alizarin 1-methyl ether *(37)*
		alizarin 2-methyl ether *(37)*
		lucidin *(37)*
		pseudopupurin *(37)*
		purpurin *(37)*
		rubiadin *(37)*
		xanthopurpurin *(37)*
		xanthopurpurin dimethyl ether *(37)*
		n.i. *(112)*
G. tauricum (Willd.) Roem. et Schult.	Root	n.i. *(22)*
G. tenuissimum Bieb.	Root, PTC	n.i. *(112)*
G. triflorum Michx.	Root, PTC	n.i. *(112)*
G. verrucosum Hudson	Root, PTC	n.i. *(112)*
G. verum L.	Root *(37, 22, 112)*, PTC *(112)*	alizarin *(37)*
		alizarin 1-methyl ether *(37)*
		alizarin 2-methyl ether *(37)*
		2-hydroxyanthraquinone *(37)*
		1-hydroxy-2-methylanthraquinone *(37)*
		lucidin *(37)*
		2-methoxyanthraquinone *(37)*
		1-methoxy-2-methylanthraquinone *(37)*
		pseudopurpurin *(37)*
		purpurin *(37)*
		rubiadin *(37)*
		xanthopurpurin *(37)*
		xanthopurpurin dimethyl ether *(37)*
		n.i. *(22, 112)*
G. volgense Pobed.	Root	n.i. *(56)*
G. vulgare S.F. Gray (= *G. mollugo* L.)	Root	alizarin *(55)*
		munjistin *(55)*
		pseudopurpurin *(55)*
		rubiadin *(55)*
Hedyotis auricularia L.	Stem, root	alizarin *(107)*
Hymenodictyon excelsum Wall.	Root	anthragallol *(26)*
		2-benzylxanthopurpurin *(26)*

Table 8 *(continued)*

		damnacanthal (*26*)
		lucidin (*26*)
		6-methylalizarin (*26*)
		morindone (*26*)
		nordamnacanthal (*26*)
		rubiadin (*26*)
		rubiadin 1-methyl ether (*26*)
		soranjidiol (*26*)
Lasianthus chinensis Benth.	Stem	damnacanthal (*68*)
Morinda angustifolia Roxb.	Heartwood, leaf	morindone (*105*), and according to the author: rhein (*105*) aloe-emodin (*105*)
M. citrifolia L.	Flower (*121*), heartwood (*120*) leaf (*93*) root (*93*) root bark (*23, 115, 120*) PTC (*70, 79, 80, 129*)	alizarin (*79, 80, 120*) alizarin 1-methyl ether (*115*) anthragallol 1,2-dimethyl ether (*120*) anthragallol 2,3-dimethyl ether (*120*) damnacanthal (*120*) damnacanthol (*23*) 5,6-dihydroxylucidin (*70*) 5,7-dimethoxy-4-hydroxy-2-methyl-anthraquinone (*121*) 2-ethoxymethyl-3-methoxy-1,5,6-trihydroxyanthraquinone (*80*) 3-hydroxymorindone (*70*) lucidin (*70, 80*) lucidin ω-ethyl ether (*80*) lucidin ω-methyl ether (*80*) 2-methyl-3,5,6-trihydroxyanthraquinone (*70*) morindone (*23, 70, 79, 80, 115, 120*) nordamnacanthal (*23, 80, 120*) rubiadin (*23, 70, 80, 120*) rubiadin 1-methyl ether (*23, 120*) soranjidiol (*23, 120*) n.i. (*93, 129*)
M. jasminoides A. Cunn.	Root bark	rubiadin (*23*) rubiadin 1-methyl ether (*23*)
M. longiflora G. Don.	Root, leaf	alizarin 1-methyl ether (*10*) rubiadin 1-methyl ether (*10*)
M. lucida Benth.	Stem (*1, 2*) heartwood (*44*)	alizarin 1-methyl ether (*2*) anthraquinone-2-carbaldehyde (*2*) damnacanthal (*2*) damnacanthol ω-methyl ether (*44*) 3-hydroxyanthraquinone-2-carbaldehyde (*44*)

Table 8 *(continued)*

		1-hydroxy-2-methylanthraquinone (*2*) lucidin (*44*) 1-methoxy-2-methylanthraquinone (*2*) 2-methylanthraquinone (*2*) moridone 5-methyl ether (*44*) munjistin methyl ester (*44*) nordamnacanthal (*2*) oruwal (*1, 2*) oruwalol (*1, 2*) rubiadin (*2*) rubiadin 1-methyl ether (*2*) soranjidiol (*2, 44*)
M. parvifolia Bartl.	Root	alizarin 1-methyl ether (*41*) 1-hydroxy-2-hydroxymethylanthra- quinone (*42*) 1-hydroxy-6-(or -7-)hydroxymethyl- anthraquinone (*42*) 2-hydroxymethylanthraquinone (*42*) lucidin ω-ethyl ether (*42*) lucidin ω-methyl ether (*42*) morindaparvin-A (*41*) morindaparvin-B (*42*)
M. persicaefolia Buch.-Ham.	Root, stem	morindone (*120*)
M. roioc L.	Root	copareolatin 1-(or -5),6-dimethyl ether (*101*) copareolatin 6-methyl ether (*101*) damnacanthal (*101*) 3-hydroxyanthraquinone- 2-carbaldehyde (*101*) lucidin (*101*) morindone (*101*) morindone 5-methyl ether (*101*) nordamnacanthal (*101*) rubiadin 1-methyl ether (*101*) soranjidiol (*101*) soranjidiol 1-methyl ether (*101*) tectoquinone (*101*)
M. tinctoria Roxb.	Root (*120*), root bark (*106*), heartwood (*49, 120*)	alizarin-1-methyl ether (*120*) damnacanthal (*49, 120*) morindone (*49, 92, 106, 120*) nordamnacanthal (*49, 92, 120*) tinctomorone (*49*)
M. umbellata L.	Root (*39, 120*) stem (*39*)	alizarin (*39*) alizarin 1-methyl ether (*39*) alizarin 2-methyl ether (*39*) damnacanthal (*120*) 2-hydroxyanthraquinone (*39*)

Table 8 *(continued)*

		1-hydroxy-2-methylanthraquinone *(39)* lucidin *(39)* 2-methoxyanthraquinone *(39)* 1-methoxy-2-methylanthraquinone *(39)* 2-methylanthraquinone *(39)* 6-methylxanthopurpurin *(120)* morindone *(120)* munjistin *(39)* rubiadin *(39)* rubiadin 1-methyl ether *(39)* soranjidiol *(120)* xanthopurpurin *(39)*
Oldenlandia umbellata L.	Root *(120)* whole plant *(102)*	alizarin *(120)* alizarin 1-methyl ether *(102, 120)* anthragallol 1,2-dimethyl ether *(102, 120)* anthragallol 1,3-dimethyl ether *(102, 120)* anthragallol 1,2,3-trimethyl ether *(102)* 2-hydroxyanthraquinone *(120)* hystazarin monomethyl ether *(120)*
Ploclama pendula Ait.	Wood	alizarin 1-methyl ether *(58)* damnacanthol ω-ethyl ether *(58)* 1,5-dihydroxy-2-methylanthra- quinone *(58)* 1-hydroxy-2-methylanthraquinone *(58)* 1-hydroxy-2-methyl-5-methoxy- anthraquinone *(58)* morindone 5-methyl ether *(58)* rubiadin *(58)*
Prismatomeris malayana Ridley	Root	rubiadin *(77)* rubiadin 1-methyl ether *(77)*
Prismatomeris tetrandra (Roxb.) K. Schum.		damnacanthal *(122)* 2-methylanthraquinone *(122)* rubiadin *(122)* rubiadin 1-methyl ether *(122)*
Putoria calabrica Perss.	Aerial parts	damnacanthol ω-ethyl ether *(57)* 2-ethoxymethyl-1-methoxy- 3,5,6-trihydroxyanthraquinone *(57)* 1-hydroxy-2-methylanthraquinone *(57)* 1-methoxy-2-methyl-3,5,6-tri- hydroxyanthraquinone *(57)* rubiadin *(57)*
Relbunium hypocarpium Hemsl.	Root *(9, 113, 120)*	alizarin 1-methyl ether *(113)* anthragallol 1,2-dimethyl ether *(113)* anthragallol 1,3-dimethyl ether *(113)*

Table 8 *(continued)*

	stem *(9)*	munjistin *(9, 120)*
		pseudopurpurin *(9, 120)*
	leaf *(9)*	purpurin *(9, 113, 120)*
		xanthopurpurin *(9)*
		xanthopurpurin 1-methyl ether *(113, 120)*
		xanthopurpurin 3-methyl ether *(113, 120)*
Rubia akane (= *R. cordifolia* L. var. *mungista* Miq.)	Root *(74, 120)* PTC *(117, 118)*	alizarin *(118)* lucidin 1-ethyl ether *(118)* 2-methyl-1,3,6-trihydroxyanthraquinone *(74)* pseudopurpurin *(118)* purpurin *(118, 120)* munjistin (120) n.i. *(117)*
R. chinensis Regel & Maack. var. *glabrescens* Kitagawa	Root	munjistin *(120)*
R. cordifolia L.	Root	alizarin *(74, 96)* alizarin 2-methyl ether *(47)* 1,4-dihydroxy-5-(or -8-)methoxyanthraquinone *(47)* 4,5-dihydroxy-7-methoxy-2-methylanthraquinone *(119)* 1,4-dihydroxy-2-methylanthraquinone *(46)* 1,5-dihydroxy-2-methylanthraquinone *(46)* 1,4-dihydroxy-6-methylanthraquinone *(119)* 1,3-dimethoxy-2-carboxyanthraquinone *(47)* 1-hydroxy-2-methylanthraquinone (74, 119) lucidin *(74)* lucidin ω-ethyl ether *(74)* 2-methyl-1,3,6-trihydroxyanthraquinone *(74)* munjistin *(120)* nordamnacanthal *(119)* pseudopurpurin *(120)* purpurin *(96, 120)* rubiadin *(47)* xanthopurpurin *(96, 120)*
R. fruticosa Ait.	Root PTC	n.i. *(112)*

Table 8 *(continued)*

R. iberica C. Koch.	Root	alizarin *(98)* lucidin *(98)* nordamnacanthal *(98)*
R. oliveri L.	Root	alizarin *(55)* munjistin *(55)* pseudopurpurin *(55)* rubiadin *(55)* xanthopurpurin *(55)* n.i. *(56)*
R. peregrina L.	Root	pseudopurpurin *(120)*
R. petiolaris L.	Root	alizarin *(55)* munjistin *(55)* pseudopurpurin *(55)* rubiadin *(55)* xanthopurpurin *(55)* n.i. *(56)*
R. sikkimensis Kurz.	Root	munjistin *(120)* purpurin *(120)* xanthopurpurin *(120)*
R. tetragona Schum.	Root	alizarin 1-methyl ether *(120)* purpurin *(120)* xanthopurpurin 1-methyl ether *(120)* xanthopurpurin 3-methyl ether *(120)*
R. tinctorum L.	Root *(15, 16, 36, 38, 52, 53, 54, 55, 56, 78, 82, 83, 84, 85, 87, 90, 97, 112, 120)* PTC *(112)*	alizarin *(15, 36, 38, 52, 55, 78, 82, 83, 84, 87, 97, 120)* alizarin dimethyl ether *(38)* alizarin 1-methyl ether *(38)* alizarin 2-methyl ether *(38)* anthragallol *(38)* anthragallol 2,3-dimethyl ether *(38)* anthragallol 3-methyl ether *(38)* 1,4-dihydroxy-2-ethoxymethyl-anthraquinone *(16)* 1,4-dihydroxy-2-hydroxymethyl-anthraquinone *(15)* 2-hydroxyanthraquinone *(38)* 1-hydroxy-2-hydroxymethylanthra-quinone *(15)* 1-hydroxy-2-methylanthraquinone *(38)* lucidin *(38, 97, 120)* 2-methoxyanthraquinone *(38)* 1-methoxy-2-methylanthraquinone *(38)* munjistin *(38, 52, 55, 120)* pseudopurpurin *(15, 36, 38, 52, 55, 120)* purpurin *(15, 36, 38, 52, 83, 97, 120)* quinizarin *(15)* rubiadin *(36, 38, 52, 55, 120)*

References, pp. 143–149

Table 8 (continued)

		xanthopurpurin (*38, 52, 55, 97, 120*)
		xanthopurpurin dimethyl ether (*38*)
		xanthopurpurin 3-methyl ether (*38*)
		n.i. (*53, 54, 56, 85, 90, 112*)
Sherardia arvensis L.	Root (*112, 120*) PTC (*112*)	pseudopurpurin (*120*) n.i. (*112*)

References

1. ADESIDA, G.A., and E.K. ADESOGAN: Oruwal, a Novel Dihydroanthraquinone Pigment from *Morinda lucida* Benth. J. Chem. Soc. Chem. Commun. 1972, 405.
2. ADESOGAN, E.K.: Anthraquinones and Anthraquinols from *Morinda lucida*. Tetrahedron **29**, 4099 (1973).
3. AMRHEIN, N., B. DEUS, P. GEHRKE, and H.C. STEINRÜCKEN: The Site of Inhibition of the Shikimate Pathway by Glyphosate. Plant Physiol. **66**, 830 (1980).
4. ANDRÉ, R., F. BAILLEUL, P. DELAVEAU, R.R. PARIS, and A. JACQUEMIN: Etude chimique du *Danais fragrans* Gaertn. (Rubiacées). Plant. Med. Phytother. **10** (2), 110 (1976).
5. ARNONE, A., G. FRONZA, R. MONDELLI, and J.ST. PYREK: ^{13}C-NMR Analysis of Anthraquinones as Models for Anthracycline Antibiotics. J. Magn. Reson. **28**, 69 (1977).
6. ASZALOS, A.: Analysis of Antitumour Antibiotics by HPLC. J. Liquid Chromatogr. **7** (s–1), 69 (1984).
7. BANKS, S., S.J. SAUNDERS, I.N. MARKS, B.H. NOVIS, and B.O. BARBEZAT: In: Drug Treatment, 2nd ed. (AVERY, G.S., ed.), p. 712. Sydney-New York: ADIS Press. 1980.
8. BANVILLE, J., J.-L. GRANDMAISON, G. LANG, and P. BRASSARD: Reactions of ketene acetals. Part I. A Simple Synthesis of Some Naturally Occurring Anthraquinones. Can. J. Chem. **52**, 80 (1974).
9. BARRE, F.P.: Anthraquinone Substances in *Relbunium hypocarpium*. Soc. Venez. Cienc. Natur. Bol. **27** (112), 314 (1967).
10. BARROWCLIFF, M., and F. TUTIN: Chemical Examination of the Root and Leaves of *Morinda longiflora*. J. Chem. Soc. **91**, 1907 (1907).
11. BAUCH, H.-J., and E. LEISTNER: Aromatic Metabolites in Cell Suspension Cultures of *Galium mollugo* L. Planta Med. **33**, 105 (1978).
12. – – Attempts to Demonstrate Incorporation of Labelled Precursors into Aromatic Metabolites in Cell Suspension Cultures of *Galium mollugo* L. Planta Med. **33**, 124 (1978).
13. BENDER, M.: Colours for Textiles (Ancient and Modern). J. Chem. Educ. **24**, 2 (1947)
14. BERG, A. VAN DEN: Personal Communication.
15. BERG, W., A. HESSE, M. HERRMANN, and R. KRAFT: Zur Strukturaufklärung von neuen Anthrachinonderivaten aus *Rubia tinctorum* L. Pharmazie **30**, 330 (1975).
16. BERG, W., A. HESSE, R. KRAFT, and M. HERRMANN: Zur Strukturaufklärung von neuen Anthrachinonderivaten aus *Rubia tinctorum* L. Pharmazie **29**, 478 (1974).

17. BERGER, Y., and A. CASTONGUAY: The ^{13}C-NMR spectra of Anthraquinone, Eight Polyhydroxyanthraquinones and Eight Polymethoxyanthraquinones. Org. Magn. Reson. 11 (8), 375 (1978).
18. BERGER, Y., A. CASTONGUAY, and P. BRASSARD: Carbon-13 Nuclear Magnetic Resonance Studies of Anthraquinones. Part II. Hydroxymethoxyanthraquinones, Acetoxymethoxyanthraquinones and Naturally Occurring Anthraquinone Analogues. Org. Magn. Reson. 14 (2), 103 (1980).
19. BERGER, S., and A. RIEKER: Identification and Determination of Quinones. In: The Chemistry of the Quinonoid Compounds, Part I (PATAI, S., ed.), p. 215. London-New York-Sydney-Toronto: J. Wiley and Sons. 1974.
20. BEYNON, J.H., and A.E. WILLIAMS: Mass Spectra of Various Quinones and Polycylic Ketones. Appl. Spectrosc. 14 (6), 156 (1960).
21. BORISOV, M.I.: Anthraquinone Glycosides of *Asperula besseriana* and *Galium ruthenicum*. Rastit. Resur. 11 (3), 362 (1975).
22. BORISOV, M.I., N.S. ZHURAVLEV, and T.I. ISAKOVA: Quantitative Content of Anthraquinones in Some Woodruff and Bedstraw Species. Rastit. Resur. 12 (4), 536 (1976).
23. BOWIE, J.H., and R.G. COOKE: Colouring Matters of Australian Plants. IX Anthraquinones from *Morinda* Species. Aust. J. Chem. 15, 332 (1962).
24. BOWIE, J.H., R.G. COOKE, and P.E. WILKIN: Colouring Matters of Australian Plants. X. Anthraquinones from *Coelospermum* species. Aust. J. Chem. 15, 336 (1962).
25. BOWIE, J.H., and P.Y. WHITE: Electron Impact Studies. Part XXXIX. Proximity Effects in the Mass Spectra of Aromatic Carbonyl Compounds Containing Adjacent Methoxy-substituents. J. Chem. Soc. (b) 1969, 89.
26. BREW, E.J.C., and R.H. THOMSON: Naturally occurring Quinones. Part XIX. Anthraquinones in *Hymenodictyon excelsum* and *Damnacanthus major* J. Chem. Soc. (c) 1971, 2001.
27. BRIGGS, L.H., J.F. BEACHEN, R.C. CAMBIE, N.P.B. DUDMAN, A.W. STEGGLES, and P.S. RUTLEDGE: Chemistry of the *Coprosma* Genus. Part XIV. Constituents of Five New Zealand Species. J. Chem. Soc., Perk. Trans. I 1976, 1789.
28. BRIGGS, L.H., M.R. CRAW, and J.C. DACRE: Chemistry of the *Coprosma* Genus. Part II. The Colouring Matters from *Coprosma areolata*. J. Chem. Soc. 1948, 568.
29. BRIGGS, L.H., and J.C.DACRE: Chemistry of the *Coprosma* Genus. Part I. The Colouring Matters from *Coprosma australis*. J. Chem. Soc. 1948, 564.
30. BRIGGS, L.H., and P.W. LE QUESNE: Chemistry of the *Coprosma* Genus. Part XII. The Glycoside of Morindone from *Coprosma australis*. J. Chem. Soc. 1963, 3471.
31. BRIGGS, L.H., and G.A. NICHOLLS: Chemistry of the *Coprosma* Genus. Part IV. The Monoglycosidic Anthraquinone Compounds from *Coprosma lucida*. J. Chem. Soc. 1949, 1241.
32. BRIGGS, L.H., G.A. NICHOLLS, and R.M.L. PATERSON: Chemistry of the *Coprosma* Genus. Part VI. Minor Anthraquinone Colouring Matters from *Coprosma australis*. J. Chem. Soc. 1952, 1718.
33. BRIGGS, L.H., and A.R. Taylor: Chemistry of the *Coprosma* Genus. Part X. The Colouring Matters from *Coprosma rhamnoides*. J. Chem. Soc. 1955, 3298.
34. BRIGGS, L.H., and B.R. THOMAS: Chemistry of the *Coprosma* Genus. Part V. The Anthraquinone Colouring Matters from *Coprosma acerosa*. J. Chem. Soc. 1949, 1246.
35. BRISSON, C., and P. BRASSARD: Regio-specific Reactions of some Vinylogous Ketene Acetals with Haloquinones and Their Regio-selective Formation by Dienolization. J. Org. Chem. 46, 1810 (1981).
36. BURNETT, A.R., and R.H. THOMSON: Biogenesis of Anthraquinones in Rubiaceae. J. Chem. Soc. Chem. Commun. 1967, 1125.
37. - - Naturally Occurring Quinones. Part XIII. Anthraquinones and Related Naphta-

lenic Compounds in *Galium* spp. and in *Asperula odorata.* J. Chem. Soc. (c) **1968**, 854.
38. – – Naturally Occurring Quinones. Part XV. Biogenesis of the Anthraquinones in *Rubia tinctorum* L. (Madder). J. Chem. Soc. (c) **1968**, 2437.
39. – – Anthraquinones in *Morinda umbellata* L. Phytochemistry **7**, 1421 (1968)
40. CAMERON D.W., and M.J. CROSSLEY: Synthesis of Emodin Methyl Ethers. Aust. J. Chem. **30**, 1161 (1977).
41. CHANG, P., K.-H. LEE, T. SHINGU, T. HIRAYAMA, and I.H. HALL: Antitumor Agents 50. Morindaparvin-A, a New Antileukemic Anthraquinone, and Alizarin 1-Methyl Ether from *Morinda parvifolia*, and the Antileukemic Activity of the Related Derivatives. J. Nat. Prod. **45** (2), 206 (1982).
42. CHANG, P., and K.-H. LEE: Cytotoxic Antileukemic Anthraquinones from *Morinda parvifolia.* Phytochemistry **23** (8), 1733 (1984).
43. COVELLO, M., O. SCHETTINO, M.I. LA ROTONDA, and P. FORGIONE: Riconoscimento e determinazione quantitativa dei derivati anthrachinonici di origine vegetale per via cromatografica. Boll. Soc. Ital. Biol. Sper. **46**, 500 (1970).
44. DEMAGOS, G.P., W. BALTUS, and G. HÖFLE: New Anthraquinones and Anthraquinone Glycosides from *Morinda lucida.* Z. Naturforsch. **36b**, 1180 (1981).
45. DODSWORTH, D.J., M.-P. CALCAGNO, E.U. EHRMANN, B. DEVADAS, and P.G. SOMMES: A New Route to Anthraquinones. J. Chem. Soc., Perkin Trans. I **1981**, 2120.
46. DOSSEH, CH., A.M. TESSIER, and P. DELAVEAU: Racines de *Rubia cordifolia.* II: Nouvelles quinones. Planta Med. **43**, 141 (1981).
47. – – – Nouvelles quinones des racines de *Rubia cordifolia* L. III. Planta Med. **43**, 360 (1981).
48. EL EZABY, M.S., T.M. SALEM, A.H. ZEWAIL, and R. ISSA: Spectral Studies of some Hydroxy Derivatives of Anthraquinones. J. Chem. Soc. (b) **1970**, 1293.
49. ESWARAN, V., V. NARAYANAN, S. NEELAKANTAN, and P.V. RAMAN: Tinctomorone – A New Anthraquinone Ester from the Heart Wood of *Morinda tinctoria.* Indian J. Chem. **17B**, 650 (1979).
50. FIESER, L.F.: The Discovery of Synthetic Alizarin. J. Chem. Educ. **7** (11), 2609 (1930).
51. FINGL, E.: In: The Pharmacological Basis of Therapeutics, 6th ed. (GOODMAN GILMAN, A., L.S. GOODMAN, and A. GILMAN, eds.), p. 1007. New York: MacMillan Publishing Co., Inc. 1980.
52. FORMANEK, I.: Studiul chromatographic al principilor anthrachinonice din Roiba (*Rubia tinctorum* L.). Rev. Med. (Tirgu-Mures, Rom.) **15** (3), 337 (1969).
53. – Studiul metodelor de dozare a derivatilor anthrachinonici din Roiba (*Rubia tinctorum* L.). Rev. Med. (Tirgu-Mures, Rom.) **16** (2), 206 (1970).
54. – Continutul in derivati anthrachinonici al radacinilor de Roiba (*Rubia tinctorum* L.) in functie de diferiti factori. Rev. Med. (Tirgu-Mures, Rom.) **16** (3–4) 380 (1970)
55. FORMANEK, I., and G. RÁCZ: Prezenta principiilor anthracenice din *Rubia tinctorum* L. in alti reprezentanti ai familiei Rubiaceae. Farmacia (Bucharest) **21** (4), 201 (1973).
56. – – Date comparative privind continutul de derivati anthrachinonic al unor specii din Familia Rubiaceae. Rev. Med. (Tirgu-Mures, Rom.) **25** (1–2), 138 (1979).
57. GONZALEZ, A.C., J.T. BARROSO, R.J. CARDONA, J.M. MEDINA, and L.F. RODRIGUEZ: Quimica de las Rubiaceas. II. Componentes de la *Putoria calabrica* Perss. An. Quim. **73**, 538 (1977).
58. GONZALEZ, A.G., R. J. CARDONA, H. LOPEZ DORTA, J.M.MEDINA, and L.F. RODRIGUEZ: Quimica de las Rubiaceas. III. antraquinonas de la *Ploclama pendula* Ait. An. Quim. **73**, 869 (1977).

59. GONZALEZ, A.G., R. FREIRE, J. SALAZAR, and E. SUAREZ: Quinonas naturales I. Antraquinonas de la *Isoplexis sceptrum*. An. Quim. **68**, 53 (1971)
60. HASSAL, C.H., and B.A. MORGAN: Tetracycline Studies. Part IV. Some Novel Cyclisations through Benzophenone Carbanions, Including a New Synthesis of Anthraquinones. J. Chem. Soc., Perkin Trans. I **1973**, 2853.
61. HEIDE L., and E. LEISTNER: 2-Methoxycarbonyl-3-prenyl-1,4-naphtoquinone, a Metabolite related to the Biosynthesis of Mollugin and Anthraquinones in *Galium mollugo* L. J. Chem. Soc. Chem. Commun. **1981**, 334.
62. – – Enzyme Activities in Extracts of Anthraquinone-containing Cells of *Galium mollugo*. Phytochemistry **22**(3), 659 (1983).
63. HENRIKSEN, L.M., and H. KJØSEN: Derivatization of Natural Anthraquinones by Reductive Silylation for Gas Chromatographic and Gas Chromatographic-Mass Spectrometric Analysis. J. Chromatogr. **258**, 252 (1983)
64. HIROSE, Y.: Synthesis of Damnacanthal, Damnacanthol, Norjuzunal and Norjuzunol, the Coloring Matters of *Damnacanthus* Spp. Chem. Pharm. Bull. (Tokyo) **8**, 417 (1960).
65. HIROSE, Y., J. KUSUDA, S. NONOMURA, AND H. FUKUI: Studies on the Synthesis of Munjistin. IV. Synthesis of Munjistin through 1,4-Dihydroxy-2-methyl-anthraquinone. Chem. Parm. Bull. (Tokyo) **16** (7), 1377 (1968).
66. HOCQUEMILLER, R., A. FOURNET, A. BOUQUET, J. BRUNETON, and A. CAVÉ: Note sur le *Commitheca liebrechtsiana* (Rubiacées). Plant. Med. Phytothér. **10** (2), 110 (1976).
67. HÖFLE, G.: ^{13}C-NMR-Spektroscopie chinoider Verbindungen – II. Substituierte 1,4-Naphthochinone und Anthrachinone. Tetrahedron **33**, 1693 (1977).
68. HUI, W.H., S.K. SZETO, and C.W. YEE: an Examination of the Rubiaceae of Hong Kong. Phytochemistry **6**, 1299 (1967).
69. IMRE, S., and L. ERSOY: Die Struktur Digiferrol und Digiferruginol. Z. Naturforsch. **28c**, 471 (1973).
70. INOUE, K., H. NAYESHIRO, H. INOUYE, and M.H. ZENK: Anthraquinones in Cell Suspension Cultures of *Morinda citrifolia*. Phytochemistry **20** (7), 1693 (1981).
71. INOUE, K., Y. SHIOBARA, H. NAYESHIRO, H. INOUYE, G. WILSON, and M.H. ZENK: Site of Prenylation in Anthraquinone Biosynthesis in Cell Cultures of *Galium mollugo*. J. Chem. Soc. Chem. Commun. **1979**, 957.
72. – – – – – – Biosynthesis of Anthraquinones and Related Compounds in *Galium mollugo* Cell Suspension Cultures. Phytochemistry **23** (2), 307 (1984).
73. INOUE, K., S. UEDA, H. NAYESHIRO, and H. INOUYE: Quinones of *Streptocarpus dunnii*. Phytochemistry **22** (3), 737 (1984).
74. ITOKAWA, H., K. MIHARA, and K. TAKEYA: Studies on a Novel Anthraquinone and its Glycosides Isolated from *Rubia cordifolia* and *R. akane*. Chem. Pharm. Bull. (Tokyo) **31** (7), 2353 (1983).
75. KUIPER, J., and R.P. LABADIE: Polyploid Complexes Within the Genus *Galium*. Part I: Anthraquinones of *Galium album*. Planta Med. **42**, 390 (1981).
76. – – Polyploid Complexes Within the Genus *Galium*. Part 2: Galiprenylin a New A-Ring Prenylated Anthraquinone of *Galium album*. Planta Med. **48**, 24 (1983).
77. LEE, H.H.: Colouring Matters from *Prismatomeris malayana*. Phytochemistry **8**, 501 (1969).
78. LEISTNER, E.: Mode of Incorporation of Precursors Into Alizarin (1,2-dihydroxy-9,10-anthraquinone). Phytochemistry **12**, 337 (1973).
79. – Biosynthesis of Morindone and Alizarin in Intact Plants and Cell Suspension Cultures of *Morinda citrifolia*. Phytochemistry **12**, 1669 (1973).
80. – Isolation, Identification and Biosynthesis of Anthraquinones in Cell Suspension Cultures of *Morinda citrifolia*. Planta Med. Suppl. **1975**, 214.

81. – Biosynthesis of Plant Quinones. In: The Biochemistry of Plants, Vol. 7 (P.K. STUMPF, and E.E. CONN, eds.), p. 403. New York-London-Toronto-Sydney-San Francisco: Academic Press 1981.
82. LEISTNER, E., and M.H. ZENK: Incorporation of Shikimic Acid into 1,2-Dihydroxyanthraquinone (Alizarin) by *Rubia tinctorum* L. Tetrahedron Lett. **1967**, 475.
83. – – Ein neuer Biosyntheseweg für Anthrachinone: Der Einbau von Shikimisäure in 1,2-dihydroxyanthrachinon (Alizarin) und 1,2,4-trihydroxyanthrachinon (Purpurin) in *Rubia tinctorum* L. Z. Naturforsch. **22b**, 865 (1967).
84. – – Incorporation of 1,4-Naphtoquinone into 1,2-Dihydroxyanthraquinone (Alizarin) in *Rubia tinctorum* L. Tetrahedron Lett. **1968**, 861.
85. – – Mevalonic Acid, a Precursor of the Substituted Benzenoid Ring of Rubiaceae Anthraquinones. Tetrahedron Lett. **1968** (11), 1395.
86. – – Chrysophanol (1,8-Dihydroxy-3-methylanthraquinone) Biosynthesis in Higher Plants. J. Chem. Soc. Chem. Commun. **1969**, 210.
87. – – Nonsymmetric Incorporation of Carboxyl-^{14}C-shikimic Acid into Alizarin (1,2-Dihydroxyanthraquinone) in *Rubia tinctorum* L. Tetrahedron Lett. **20**, 1677 (1971).
88. LI, G.-W., Q.-C. PAN, X.-P. YANG, and B.-P. YING: Isolation and Structural Determination of 8-Hydroxydamnacanthol ω-Ethyl Ether From the Root of *Damnacanthus subspinosus* Hand-Mazz. Acta Pharm. Sinica **19** (9), 681 (1984).
89. LI, G., Z. ZHAO, R. XU, B. YING, and Q. PAN: Studies on the Chemical Constituents of the Root of *Damnacanthus subspinosus* Hand-Mazz. I. The Isolation and Structural Determination of Subspinosin and 8-Hydroxysubspinosin. Acta Pharm. Sinica **16** (8), 576 (1981).
90. LUTOMSKI, J., and W. RASZEJA: Jakósciowe i ilosciove wahania glikozydow antrachinonowych w marzannie barwierskiej (*Rubia tinctorum* L.) w zaleznosci od organurosliny i okresu zbioru. Qualitative and Quantitative Variations of Anthraquinone Glucosides in Madder Plant Parts in Relation to the Harvesting Time. Farm. Pol. **23**, 613 (1967).
91. MIHAI, G.G., P.G. TARASSOFF, and N.J. FILIPESCU: Photohydroxylation of Anthraquinone in Concentrated Sulphuric Acid. J. Chem. Soc., Perkin Trans. I **1975**, 1374.
92. MISHRA, G., and N. GUPTA: Chemical Investigation of Roots of *Morinda tinctoria* Roxb. J. Inst. Chem., Calcutta **54** (1), 22 (1982).
93. MOORTHY, N.K., and G.S. REDDY: Preliminary Phytochemical and Pharmacological Study of *Morinda citrifolia* L. Antiseptic **67** (3), 167 (1970).
94. MULDER-KRIEGER, TH., R. VERPOORTE, A. DE WATER, M. VAN GESSEL, B.C.J.A. VAN OEVEREN, and A. BAERHEIM SVENDSEN: Identification of the Alkaloids and Anthraquinones in *Cinchona ledgeriana* Callus Cultures. Planta Med. **46**, 19 (1982).
95. MULDER-KRIEGER, TH., R. VERPOORTE, M. VAN DER KREEK, and A. BAERHEIM SVENDSEN: Identification of Alkaloids and Anthraquinones in *Cinchona pubescens* Callus Cultures; the Effect of Plant Growth Regulators and Light on the Alkaloid Content. Planta Med. **50** (1), 17 (1984)
96. MURTI, V.V.S., T.R. SESHADRI, and S. SIVAKUMARAN: Anthraquinones of *Rubia cordifolia* L. Phytochemistry **11**, 1524 (1972).
97. – – – A Study of Madder, the Roots of *Rubia tinctorum* L. Indian J. Chem. **8**, 779 (1970).
98. – – – Chemical Components of *Rubia iberica* C. Koch. Indian J. Chem. **10**, 246 (1972).
99. OSHIMA, Y., and K. TAKAHASHI: Separation Methods for Sennosides. J. Chromatogr. **258**, 292 (1983).
100. PARHAM, W.E., C.K. BRADSHER, and K.J. EDGAR: *o*-Benzoylbenzoic Acids by the Reaction of Lithium 2-Lithiobenzoates With Acid Chlorides. A Contribution to the Chemistry of Alizarin and Podophyllotoxin. J. Org. Chem. **46**, 1057 (1981).

101. PARK, Y.H. : Part I. A Phytochemical Study of *Morinda roioc* L. (Family Rubiaceae). Part II. Alkaloids in Aged Potatoes, *Solanum tuberosum* L. (Family Solanaceae). Dissertation, Univ. of Mississippi **1977**.
102. PURUSHOTHAMAN, K.K., S. SARADAMBAL, and V. NARAYANASWAMI: Isolation and Identification of Some Anthraquinone Derivatives from *Oldenlandia umbellata* Linn. Leather Sci. (Madras) **15**, 49 (1968).
103. QUERCIA, V.: HPLC in the Determination of Some Anthraquinone Aglucones. Pharmacology **20** (Suppl. 1), 76 (1980).
104. RAI, P.P., T.D. TURNER, and S.A. MATLIN: HPLC of Naturally Occurring Anthraquinones. J. Chromatogr. **110** (2), 401 (1975).
105. RAO, R.V.K., J.V.L.N.S. RAO, and C.V. SUDHAKAR: Chemical Examination of *Morinda angustifolia* (Heartwood and Leaves). Indian J. Pharm. Sci. **40** (5), 169 (1978).
106. RAO, P.S., and G.C. VEERA-REDDY: Isolation and Characterization of the Glycoside of Morindone From the Root Bark of *Morinda tinctoria* var. *tomentosa*. Indian J. Chem. **15 B**, 497 (1977).
107. RATNAGIRISWARAN, A.N., and K. VENKATACHALAM: Auricularine – a New Alkaloid from the Roots and Stems of *Hedyotis auricularia*. J. Indian Chem. Soc. **19**, 389 (1942).
108. ROBERGE, G., and P. BRASSARD: Reactions of Ketene Acetals: 12. A Regiospecific Synthesis of Anthragallols. Synthesis, **1981**, 381.
109. ROBERGE, G., and P. BRASSARD: Reactions of Ketene Acetals 13. Synthesis of Contiguously Trihydroxylated Naphto- and Anthraquinones. J. Org. Chem. **46**, 1461 (1981).
110. ROBERTS, J.L., P.S. RUTLEDGE, and M.J. TREBILCOCK: Experiments Directed Toward the Synthesis of Anthracyclinones. I. Synthesis of 2-Formylmethoxyanthraquinones. Aust. J. Chem. **30**, 1553 (1977).
111. SCHILCHER, H.: Pflanzliche Urologika. Dtsch. Apoth. Ztg. **124**, 47, 2431 (1984).
112. SCHULTE, U., H. EL-SHAGI, and M.H. ZENK: Optimization of 19 Rubiaceae Species in Cell Culture for the Production of Anthraquinones. Plant Cell Rep. **3** (2), 51 (1984).
113. SEELKOPF, C.: Anthraquinones of *Relbunium hypocarpium*. Praep. Pharm. **6** (1), 13 (1970); Chemical Abstracts **73**: 22149 q (1970).
114. SIMONSEN, J.L.: LXVI – Morindone. J. Chem. Soc. **113**, 766 (1918).
115. – LIX – Note on the Constituents of *Morinda citrifolia*. J. Chem. Soc. **117**, 561 (1920).
116. STÖCKIGT, J., U. SROCKA, and M.H. ZENK: Structure and Biosynthesis of New Antraquinone from *Streptocarpus dunnii*. Phytochemistry **12**, 2389 (1973).
117. SUZUKI, H., T. MATSUMOTO, and Y. MIKAMI: Effects of Nutritional Factors on the Formation of Anthraquinones by *Rubia cordifolia* Plant Cells in Suspension Culture. Agric. Biol. Chem. **48** (3), 603 (1984).
118. SUZUKI, H., T. MATSUMOTO, and Y. OBI: Anthraquinones in Cell Suspension Cultures of *Rubia cordifolia* var. *mungista* MIQ. Proc. 5th Int. Cong. Plant Tissue and Cell Culture 1982, ed. by A. FUJIWARA, Tokyo, **1982**, 285.
119. TESSIER, A.M., P. DELAVEAU, and B. CHAMPION: Nouvelles anthraquinones des racines de *Rubia cordifolia*. Planta Med. **41**, 337 (1981).
120. THOMSON, R.H.: The Naturally Occurring Quinones, 2nd ed. London and New York: Academic Press 1971.
121. TIWARI, R.D., and J. SINGH: Structural Study of the Anthraquinone Glycoside from the Flowers of *Morinda citrifolia*. J. Indian Chem. Soc. **54**, 429 (1977).
122. TU, D., Z. PANG, and N. BI: Studies on Chemical Constituents of *Prismatomeris tetrandra* (Roxb.) K. Schum. Yaoxue Xuebao **16** (8), 631 (1981); Chemical Abstracts **96**:65676 (1982).

123. VAN EIJK, G.W., and H.J. ROEYMANS: Gas-Liquid Chromatography of Trimethylsilyl Ethers of Naturally Occurring Anthraquinones. J. Chromatogr. **124**, 66 (1976).
124. VAN EIJK, G.W., and H.J. ROEYMANS: Separation and Identification of Naturally Occurring Anthraquinones by Capillary Gas Chromatography and Gas Chromatography-Mass Spectrometry. J. Chromatogr. **295**, 497 (1984).
125. VERMES, B., L. FARKAS, and H. WAGNER: Synthesis and Structure Proof of Morindone 6-0-Primeveroside and 6-0-Rutinoside. Phytochemistry **19**, 119 (1980).
126. WILSON, G., and P. MARRON: Growth and Anthraquinone Biosynthesis by *Galium mollugo* L. Cells in Batch and Chemostat Culture. J. Exp. Botany **29**, 837 (1978).
127. WIJNSMA, R., and R. VERPOORTE: unpublished results.
128. WIJNSMA, R., R. VERPOORTE, TH. MULDER-KRIEGER, and A. BAERHEIM SVENDSEN: Anthraquinones in Callus Cultures of *Cinchona ledgeriana*. Phytochemistry **23** (10), 2307 (1984).
129. ZENK, M.H., H. EL-SHAGI, and U. SCHULTE: Anthraquinone Production by Cell Suspension Cultures of *Morinda citrifolia*. Planta Med. Suppl. **1975**, 79.
130. ZHURAVLEV, N.S.: Anthraquinones of *Galium semiamictum*. Khim. Prir. Soedin. **10** (5), 656 (1974).
131. ZHURAVLEV, N.S., and M.I. BORISOV: Anthraquinones of *Galium dasypodum*. Khim. Prir. Soedin. **5** (2), 118 (1969).
132. – – Anthraquinones of *Galium dasypodum*. Khim. Prir. Soedin. **5** (3), 176 (1969).
133. – – Anthraquinones of *Galium dasypodum*. Farm. Zh. (Kiev) **25** (1), 76 (1970)
134. WEISCHE, A., and E. LEISTNER: Cell Free Synthesis of o-Succinylbenzoic Acid from Iso-chorismic acid, the Key Reaction in Vitamin K_2 (menaquinone) Biosynthesis. Tetrahedron Lett. **26** (12), 1487 (1985).
135. LEISTNER, E.: Occurrence and Biosynthesis of Quinones in Woody Plants. In: Biosynthesis and Biodegradation of Wood Components, p. 273. New York: Academic Press. 1985.
136. KOLKMANN, R., and E. LEISTNER: Synthesis and Revised Structure of the o-Succinyl benzoic Acid Coenzyme A Ester, an Intermediate in Menaquinone Biosynthesis. Tetrahedron Lett. **26** (14), 1703 (1985)

(*Received July 18, 1985*)

Recent Developments in the Field of Marine Natural Products with Emphasis on Biologically Active Compounds

By H.CHR. KREBS, Chemisches Institut, Tierärztliche Hochschule, Hannover, Federal Republic of Germany

Contents

I. Introduction 152

II. Porifera 153
    II.1. Steroids from Porifera 154
    II.2. Terpenoid Constituents from Porifera 157
    II.3. Amino Acid Derived Metabolites from Porifera 183
    II.4. Peptide Alkaloids, Peptides, and Proteins from Porifera 190
    II.5. Nucleosides from Porifera 193
    II.6. Alkaloids and Other Heterocyclic Compounds from Porifera 194
    II.7. Macrolides from Porifera 199
    II.8. Phenols and Aromatic Ethers from Porifera 202
    II.9. Carboxylic Acids from Porifera 203
    II.10. Miscellaneous Other Compounds from Porifera 205

III. Coelenterata (Cnidaria) 208
    III.1. Hydrozoa, Cubozoa, and Scyphozoa 208
    III.2. Hexacorallia: Sea Anemones 211
    III.3. Hexacorallia: Other Organisms 213
    III.4. Octocorallia 216
        III.4.1. Steroids from Octocorallia 216
        III.4.2. Terpenes from Octocorallia 222
        III.4.3. Miscellaneous Compounds from Octocorallia 248

IV. Bryozoa 253

V. Mollusca 257
    V.1. Gastropoda 257
        V.1.1. Nudibranchia 257
            V.1.1.1. Steroids from Nudibranchia 261
            V.1.1.2. Terpenes from Nudibranchia 261
            V.1.1.3. Miscellaneous Compounds from Nudibranchia . . . 267
        V.1.2. Aplysiidae 269
            V.1.2.1. Terpenes from Aplysiidae 269
            V.1.2.2. Miscellaneous Compounds from Aplysiidae 271

            V.1.3. Conidae 273
            V.1.4. Miscellaneous Compounds from Other Marine Snails 275
    V.2. Bivalvia 282
    V.3. Cephalopoda 282

VI. Echinodermata 282
    VI.1. Saponins from Starfish 283
    VI.2. Steroids from Starfish 295
    VI.3. Miscellaneous Compounds from Starfish 298
    VI.4. Saponins from Sea Cucumbers 299
    VI.5. Steroids from Sea Cucumbers 307
    VI.6. Crinoids and Ophiuroids 307
    VI.7. Sea Urchins 308

VII. Tunicata 309

VIII. Miscellaneous Other Sources 317

References 320

I. Introduction

The marine flora and animal world are a rich source of biologically active compounds. A large number of sometimes highly toxic metabolites has been isolated *inter alia* from protozoans, sponges, coelenterates, echinoderms, molluscs, nemertines, sea snakes, and fishes (*1–4*) and the great interest in these chemically varied compounds has resulted in extensive publications (*5–13*). Among the best known of the toxic substances are tetrodotoxin, saxitoxin, and the polypeptides from sea anemones, but they will not be the main subject of this report.

Marine species studied in recent years have yielded a variety of compounds which possess known or novel pharmacological activities in mammalian species or have exhibited antimicrobial, antibacterial, antiviral or antineoplastic properties (*14*). JACOBS *et al.* (*15*) used the fertilized sea urchin egg as a model for the study of the site and mode of action of drugs that inhibit cell division. The authors tested 130 purified marine natural products as well as 14 known antineoplastic agents. Many new metabolites may be useful as drugs (*16*) or as biochemical, physiological, and pharmacological tools in biomedical research (*17–20*). Just as many potent drugs have been found in terrestrial plants it seems likely that a search for new natural products from marine organisms might uncover unknown entities which could be added to the already existing drug armamentarium.

Isolation, biosynthesis, and metabolism of sterols in marine invertebrates constitutes an active field of reseach (*21–26*); the discovery of

References, pp. 320–363

unusual sterols may contribute to the solution of a variety of problems dealing with food chain and symbiotic relationships. Furthermore, marine wax esters (*27*) and terpenes (*28, 29*) have been studied.

As it is uneconomical to extract and purify material from species that would have to be collected in large quantities from remote corners of the world the development of useful drugs from these sources has been limited. COLWELL (*30*) reported that "genetic engineering can change this situation dramatically, by revealing the vast and diverse genetic composition of marine life for pharmacological application". The potential use of microorganisms, isolated from marine environment and cultivated under conditions pertinent to sea water, for antibiotics and enzyme production has been discussed by OKAMI (*31*).

The present report is devoted to recent developments in many natural products derived from marine invertebrates, with emphasis given to those metabolites which show biological activity. As it is impossible to decide whether a newly described substance for which no activity is mentioned is inactive or whether such a substance has not been tested, several such metabolites are included as well as chemically related compounds with various biological activities. The literature since 1980 dealing with isolation of organic compounds will be reviewed. Papers dealing with syntheses are cited only when they provide confirmation for the proposed structure of a natural product. Metabolites from dinoflagellates and the "paralytic shellfish poisons" will not be discussed because of an extensive review by SHIMIZU in a recent volume of this series (*32*).

II. Porifera

The phylum Porifera or sponges which comprises the most primitive multicellular animals represents a rich source for discovery of natural compounds, most of them with considerable biological activity. This is demonstrated by numerous publications. Most investigations have dealt with terpenoids, but the steroidal constituents have also been examined thoroughly. The literature until about 1980 is reviewed by BERGQUIST and WELLS (*33*) in a publication dealing with the chemotaxonomy of the Porifera.

Sponges have world-wide distribution and are well represented in coral reefs where they often make an important contribution to the biomass and could potentially constitute an interesting food resource. Some species have physical means of defence; others are protected by compounds which exhibit toxic or antifeeding activity. MCCAFFREY

(34) prepared a crude extract of *Haliclona* sp. which was lethal to goldfish, but not to mice although it induced ptosis in the latter. The extract exerted a positive inotropic effect on the isolated guinea pig atrium and induced contraction of the rat uterine smooth muscle. SEVCIK and BARBOZA *(35)* isolated two fractions from the alcoholic extract of *Tedania ignis* which exhibited reversible presynaptic effects on the frog neuromuscular junction blocking the release of acetylcholine, although animals of the same species collected in a different area were found to be essentially free of those neurotoxic compounds *(36)*. Predators of the sponges are the nudibranch mollusks which accumulate the bioactive metabolites from their food. Many sponges living in tropical waters contain antibiotics. It has been shown that several sponges are associated with various kinds of microorganisms, which might be the source of some of the constituents from Porifera.

II.1. Steroids from Porifera

Most publications have dealt with 3β-hydroxy- and 3β-(hydroxymethyl)-A-*nor*-sterols containing unique or partially novel side chains *(37–65)*. Presence of unusual sterols may be of phylogenetic significance and can also shed light on the complex structure-function role of the sterol in membranes. New sterols present in minor and trace amounts may offer important clues to biosynthetic or dietary pathways of the major compounds. Thus the discovery of nuclearly modified sterols, the 3β-(hydroxymethyl)-A-*nor*-sterans, has been of considerable value in studying the food chain, biosynthesis and chemotaxonomy of certain sponges. On the basis of direct incorporation experiments DE STEFANO and SODANO *(66)* suggested that cholest-4-en-3-one is an intermediate in the conversion of cholesterol to 3β-(hydroxymethyl)-A-*nor*-cholestane in the marine sponge *Axinella verrucosa*. Two sterols with a 5α-methoxy-$\Delta^{6,8(14)}$-nucleus **1** and **2** were isolated from *Axinella cannabina*, but the authors presumed that these compounds may be artifacts, produced during the isolation procedure by reaction of methanol with the corresponding 5α,8α-epidioxysterols *(60)*. Several 5α,8α-epidioxides (**3–12**) were isolated from a *Hyrtios* species *(62)*, while *Thalysias junipertina* contained the same steroids (**3–12**) in addition to four related compounds (**13–16**) *(67)*. Furthermore, KOCH *et al.* *(62)* described some sterols with functionalized side chains (**17–24**) from a *Hyrtios* species. Compound **20** as well as the C-24 aldehyde (**25**) and cholesta-5,25-diene-3β-24ξ-diol (**26**) were found in the extract of the Far Eastern sponge *Esperiopsis digitata* *(61)*. The biogenesis of the last named substances is not clear; they are possibly formed through oxidation, *in*

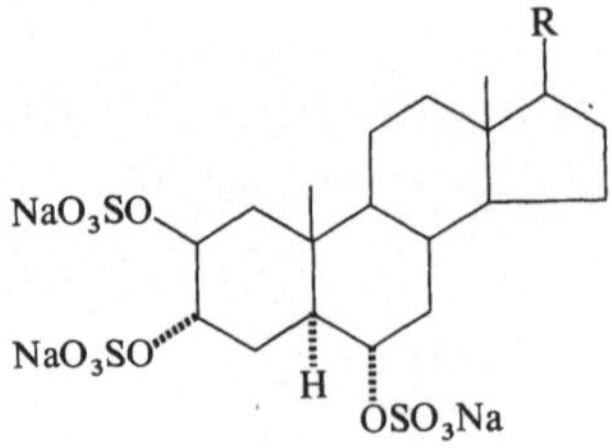

(27) R = [side chain]

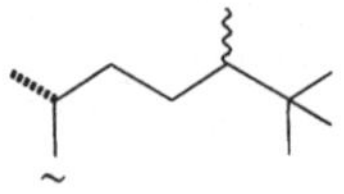

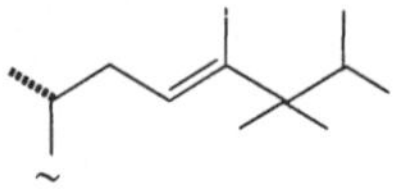

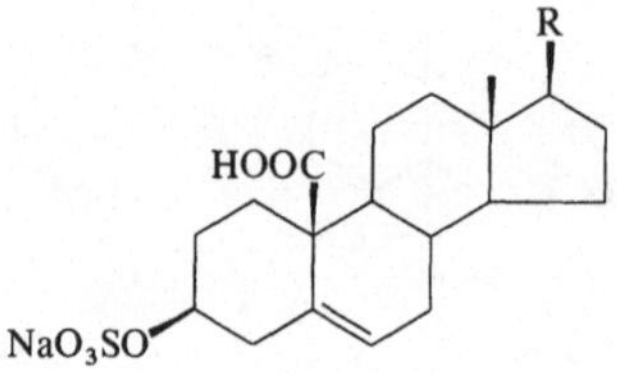

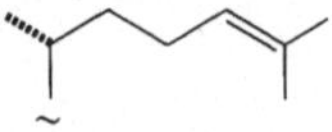

(28) R = [side chain]

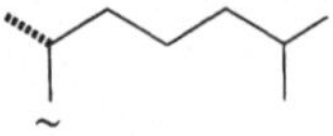

(29) R = [side chain]

(30) R = [side chain]

(31) R = [side chain]

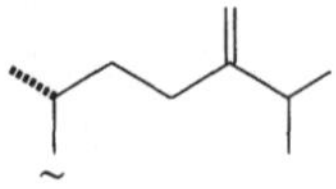

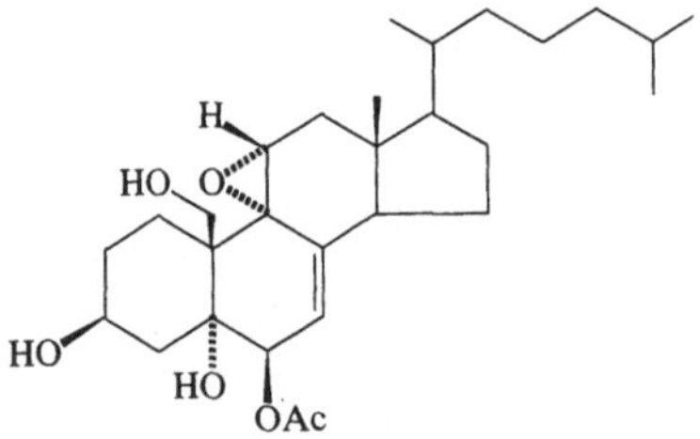

(32)

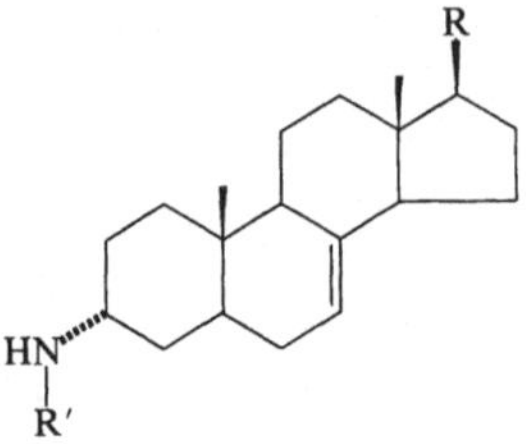

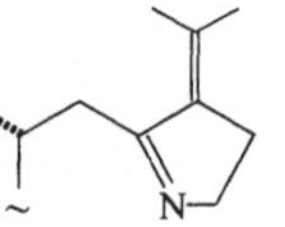

(33) R = [side chain]
R' = H

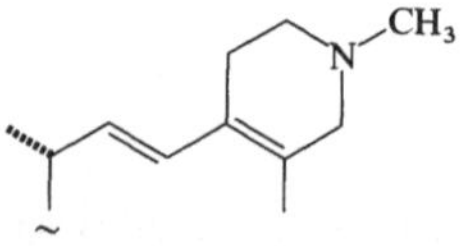

(34) R = [side chain]
R' = CH₃

vivo, of sterol precursors and as the result of oxidation during extraction and chromatography. Although there was no reference to biological activity of these sterols, there are reports of cytotoxic sulfated steroids from sponges. Halistanol sulfate (**27**), a trisodium sulfate of 24ξ,25-dimethylcholestane-2β,3α,6α-triol, has been isolated from the Okinawan species *Halichondria* cf. *moorei* as an antimicrobial constituent (*68*). MAKARIEVA et al. (*64*) obtained a similar compound, sokotrasterol sulfate (**28**), differing only in the side chain, from *Halichondria* sp.

The authors presumed, that biosynthesis of free sterols and trisulfated derivatives in these animals involves different precursors. Another group of sulfated sterols (**29–31**) having a wide variety of biological activities has been isolated from *Toxadocia zumi* (*69*). These compounds were highly active against *Staphylococcus aureus* and *Bacillus subtilis*, inhibiting the growth at 100 µg/disc and 50 µg/disk, respectively. The sterol sulfate mixture inhibited cell division in the fertilized sea urchin egg assay (*15*) at 5 µg/ml but did not cause cell lysis. The sulfates were toxic to shrimp and fish. A new hydroxylated sterol with the unusual feature of a 9,11-epoxide and a 19-hydroxyl group was isolated by GUNASEKERA and SCHMITZ (*70*) from a *Dysidea* species. Its structure was established as 9α,11α-epoxycholest-7-ene-3β,5α,6β,19-tetrol 6-acetate (**32**) by spectroscopic means, by acetylation and by hydrolysis. Recently ROSSER and FAULKNER (*71*) found the steroidal alkaloids plakinamine A (**33**) and plakinamine B (**34**) as antimicrobial metabolites of *Plakina* sp. This type of compounds has not been reported previously from marine organisms. **33** and **34** inhibited the growth of *Staphylococcus aureus* and *Candida albicans*.

II.2. Terpenoid Constituents from Porifera

The occurrence of terpenoids in sponges is widespread but certain types, such as linear furanoterpenes, isoprenyl quinols, linear sesqui- and sesterterpenes and diterpenes are found most often. Most of these compounds have shown biological activity of some sort. Tests to evaluate the ability of terpenoids from sponges to inhibit cleavage and development of fertilized sea urchin eggs (*72*) and to determine their antimicrobial activity (*73*) are described for avarol, a sesquiterpenoid hydroquinone from *Dysidea avara*. Furanoid sesquiterpenes have been reported as constituents of *Dysidea*, *Euryspongia* and *Siphonodictyon* species. SCHULTE *et al.* (*74*) isolated nakafuran-8 (**35**) and nakafuran-9 (**36**) from *Dysidea fragilis* from Hawaiian waters as well as from its prey, two species of nudibranchs. Both compounds possessed antifeedant properties against fish. Penlanfuran (**37**) from the same species, but from the north coast of Brittany, had an unprecedented skeleton among sponge products (*75*), and eight more new sesquiterpenoids (**38–45**), seven of them related to penlanfuran, were subsequently reported from the same source (*76*). Similarity of structures and the same absolute configuration at the isopropyl-bearing carbon atom led the authors to propose a common biogenesis for these *p*-menthene-type sesquiterpenoids. GUELLA *et al.* (*76a*) used a mixture of different species of Mediterranean sponges as a source of tavacfuran (**46**), tavacbutenolide-1 (**47**), tavacbutenolide-2 (**48**), and tavacpallescensin (**49**) as well as some known furanosesquiterpenes.

Dysidea amblia provided different metabolites, depending on the site of collection. In addition to some diterpenes, two sesquiterpenes, pallescensin A (**50**) and the new compound pallescensolide (**51**), were found (*77*). The authors suspected that oxidation of furans can occur in the sponge but that the methoxy group, present for example in **51**, might be introduced during extraction with methanol. While investigating metabolites of nudibranchs, HOCHLOWSKI et al. (*78*) found a sponge *Euryspongia* sp. to be the source of euryfuran (**52**). DUNLOP et al. (*79*) found **52**, too, together with seven new sesquiterpenes (**53**–**59**) while investigating four separate collections of *Dysidea herbacea*. GRODE and CARDELLINA (*80*) examined the chemical constituents of the sponge *Dysidea etheria* and its predator, the nudibranch *Hypselodoris zebra*, and found furodysinin (**60**) in both kinds of animals. In addition, they obtained a new sesquiterpene lactone (**61**) from the sponge.

Siphonodictidine (**62**), isolated from an Indo-Pacific *Siphonodictyon* species, may be responsible for inhibiting coral growth (*81*). The methanol extract of this sponge showed antimicrobial activity against *Staphylococcus aureus* and *Bacillus subtilis*. Upial (**63**) was a nonisoprenoid sesquiterpene aldehyde lactone with a rare bicyclo[3.3.1]nonane skeleton, isolated from *Dysidea fragilis* (*82*). Its structure has been deduced from spectral correlations and chemical transformations. A possible biological pathway to **63**, starting from a furanoid sesquiterpene, was formulated in the paper. Recently, TASCHNER and SHAHRIPOUR (*82a*) described the synthesis of its enantiomer (−)-upial.

Sesquiterpenes bearing a phenolic or chinoide moiety are also common in Porifera. SULLIVAN et al. (*81, 83*) examined *Siphonodictyon coralliphagum*, a sponge that burrows into coral heads, and *S. mucosa*, a species that grows partially buried in coral debris, and found two antimicrobial compounds, siphonodictyal-A (**64**) and siphonodictyal-B (**65**) which were suspected to be toxic to corals. Mildly cytotoxic compounds with similar structures established by X-ray analysis were isolated from *Dysidea arenaria* (*84*). Arenarol (**66**) and arenarone (**67**) had the same rearranged sesquiterpene skeleton with a hydroquinone and quinone moiety, respectively. Crude extracts of a sponge tentatively identified as a species of *Fenestraspongia* exhibited *in vitro* antimicrobial activity and inhibited cell division in the fertilized sea urchin egg assay (*84a*). The active compounds have been identified as ilimaquinone (**68**) and 5-*epi*-ilimaquinone (**69**). A colorless dimer, bispuupehenone (**70**), has been isolated from a Tahitian sponge, *Hyrtios* (= *Inodes*) *eubamma*, in addition to the previously known puupehenone (**71**) (*85*). DJURA

References, pp. 320–363

et al. (*86*) in examining two samples of *Smenospongia echina* and *S. aurea* found some brominated tryptamines and two related sesquiterpene phenols (**72, 73**) in the first species as well as another sesquiterpene phenol (**74**) from one sample of *S. echina*. Three new aromatic sesquiterpenes (**75–77**) have been isolated from a *Halichondria* species found on the coast of West Australia (*87*).

A group of nitrogenous sesquiterpenes consists of isonitriles, isothiocynates and formamides. The sponge *Axinella cannabina* provided one previously known (**78**) and eight new compounds (**79–86**) belonging to this group (*88, 89*). The isolation of three formamide-isothiocyanate-isocyanide series from *A. cannabina* supported the argument that these functionalities are biogenetically related. NAKAMURA *et al.* (*90*) isolated a novel sesquiterpene named theonellin (**87**) with a conjugated diene unit and its corresponding isothiocyanate (**88**) and formamide (**89**) from the sponge *Theonella* cf. *swinhoei* collected near Okinawa.

Methanolic extraction of the Okinawa sea sponge *Agelas* sp. and fractionation by monitoring antispasmodic activity using isolated guinea pig ileum furnished agelasidine-A (**90**) (*91*) which was the first marine natural product containing guanine and sulfone units. CAPON and FAULKNER (*92*) described the same sesquiterpene (**90**) as a constituent of an unidentified *Agelas* species collected at Palau, Western Caroline Islands.

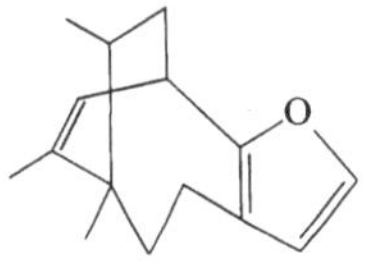

(**35**)

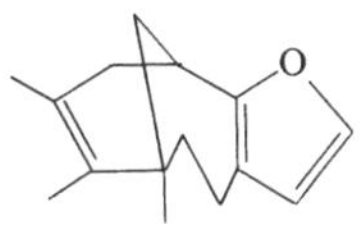

(**36**)

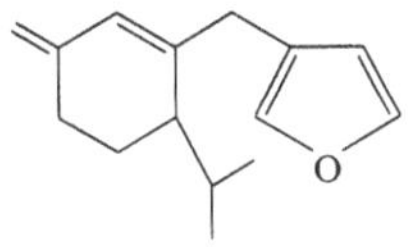

(**37**)

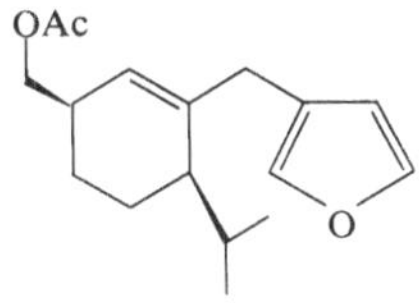

(**38**)

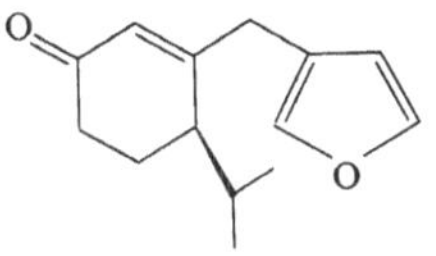

(**39**)

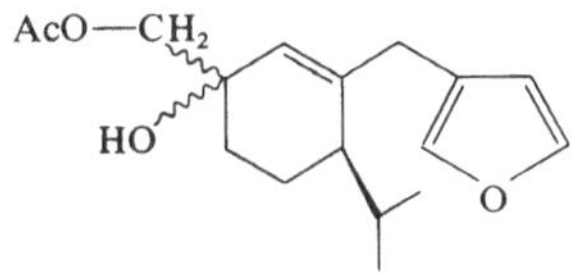

(**40**) α-CH_2—OAc
(**41**) β-CH_2—OAc

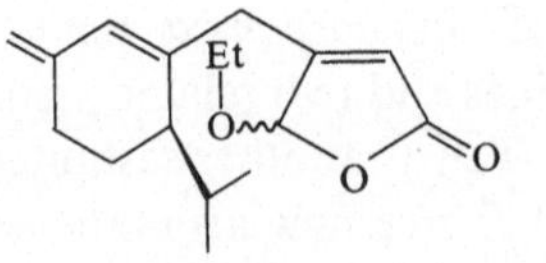

(42) (+)–
(43) (−)–

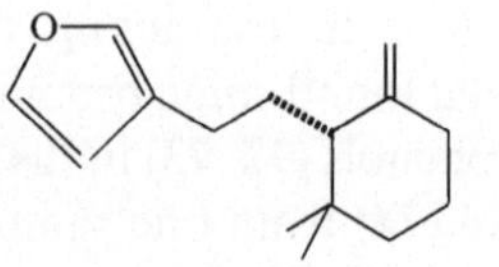

(44)

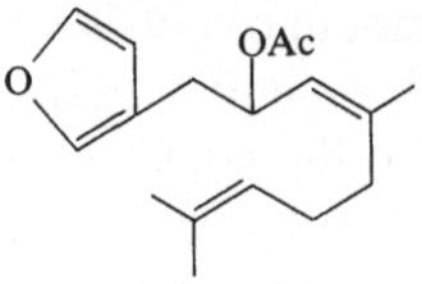

(45)

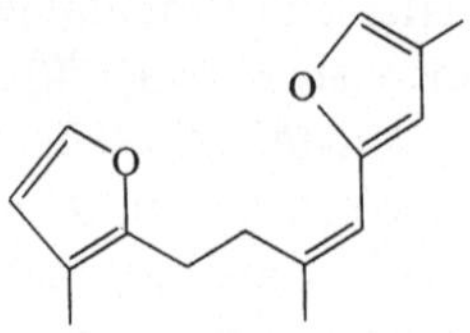

(46)

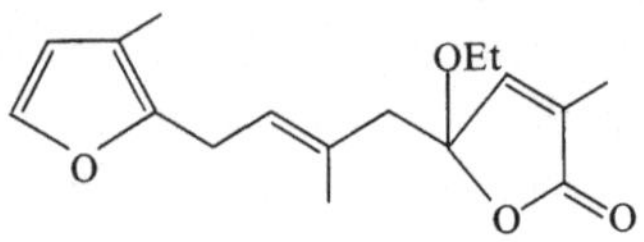

(47)

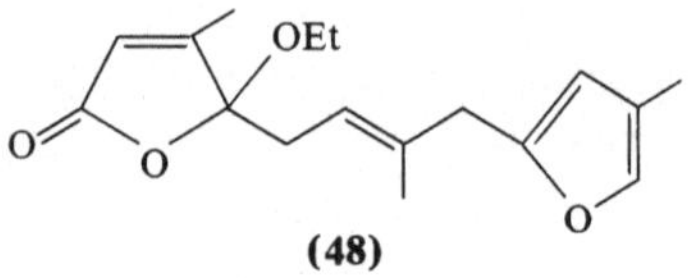

(48)

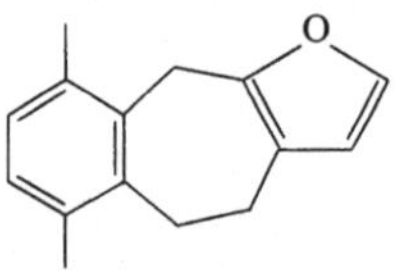

(49)

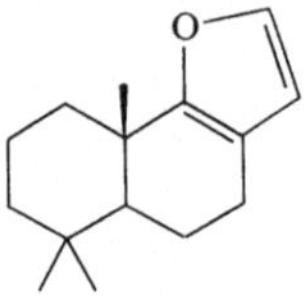

(50)

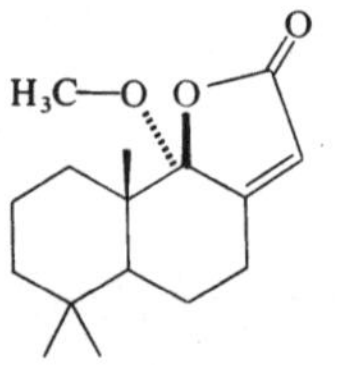

(51)

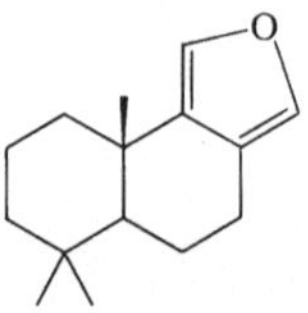

(52)

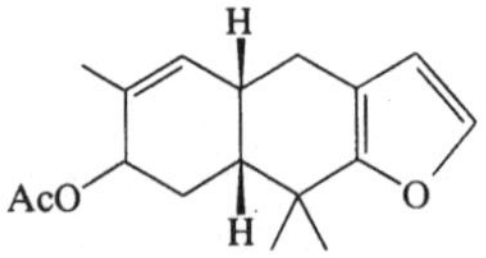

(53)

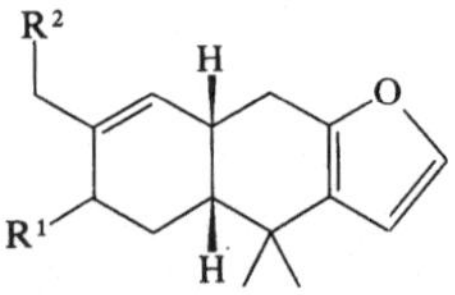

(54) R^1=β-OH   R^2=H
(55) R^1=β-OAc R^2=H
(56) R^1=α-OH   R^2=H
(57) R^1=H   R^2=OH
(60) R^1=H   R^2=H

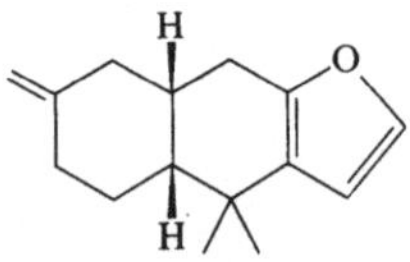

(58)

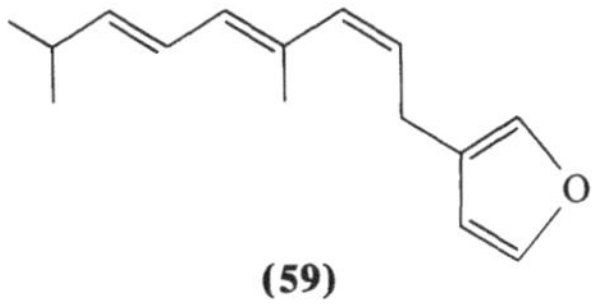

(59)

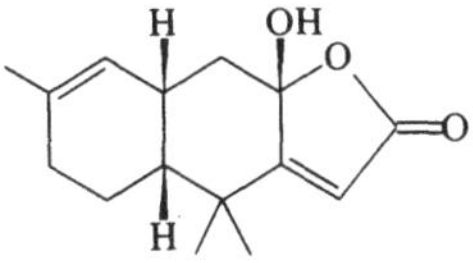

(61)

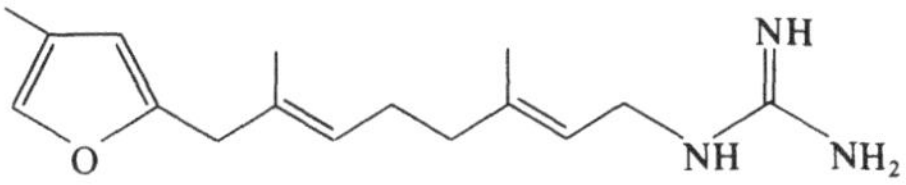

(62)

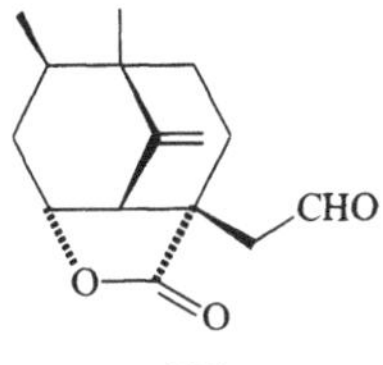

(63)

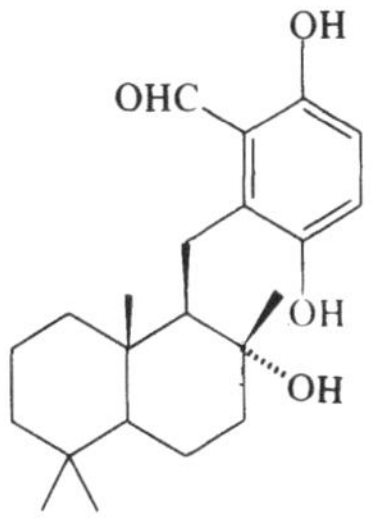

(64)

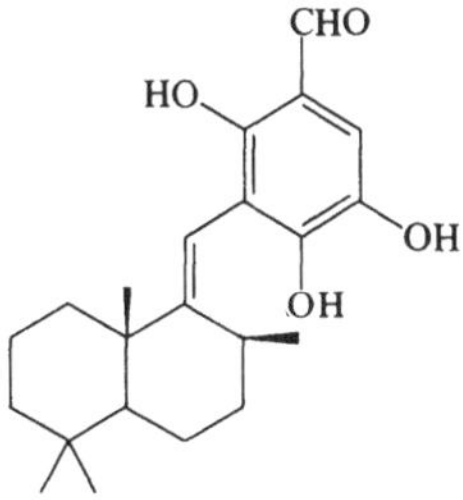

(65)

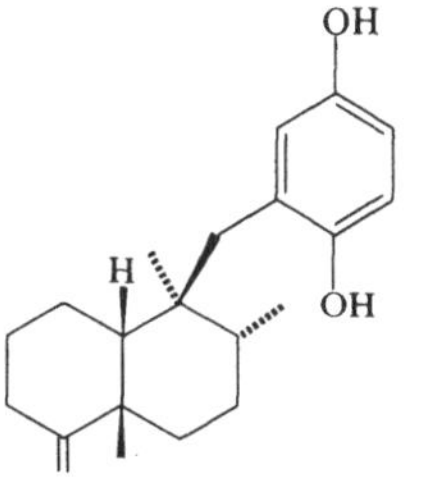

(66)

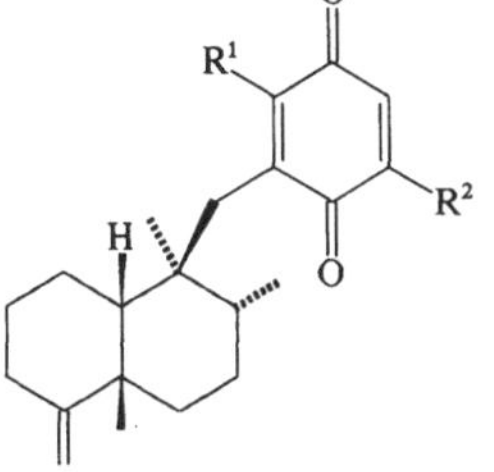

(67) R¹=H R²=H
(69) R¹=OH R²=OCH₃

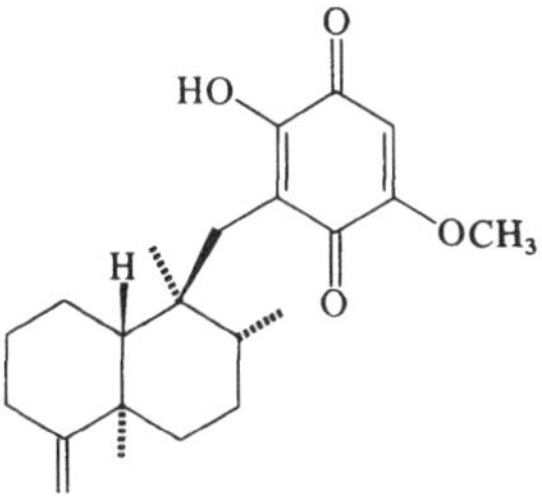

(68)

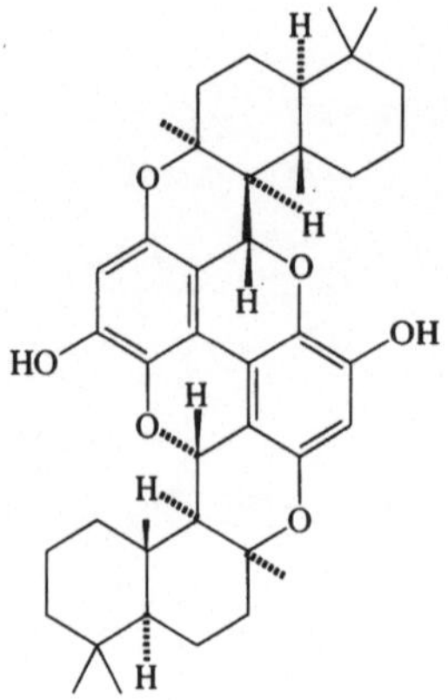

(70)

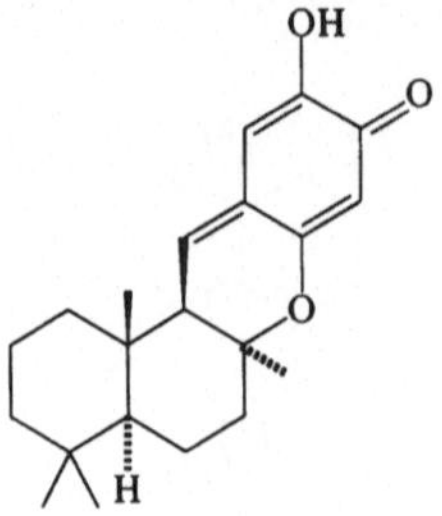

(71)

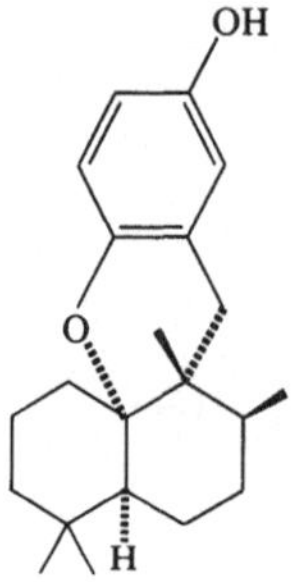

(72)

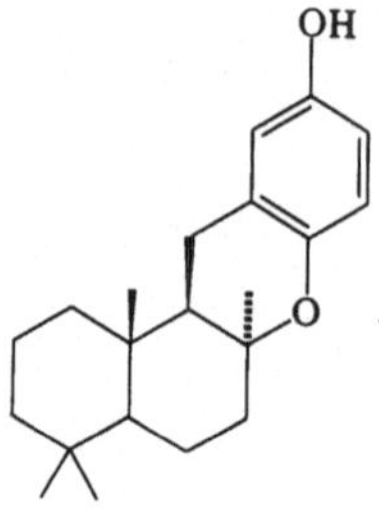

(73)

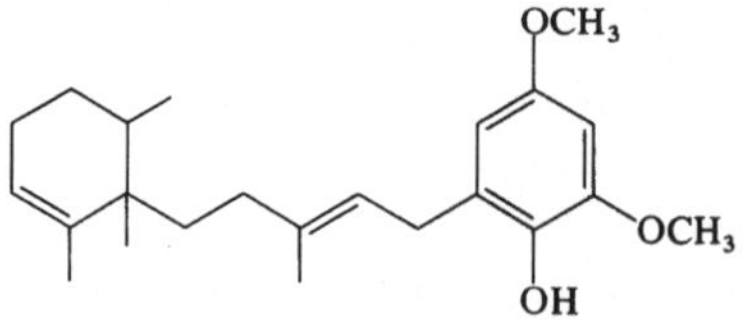

(74)

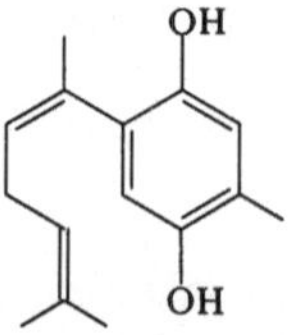

(75)

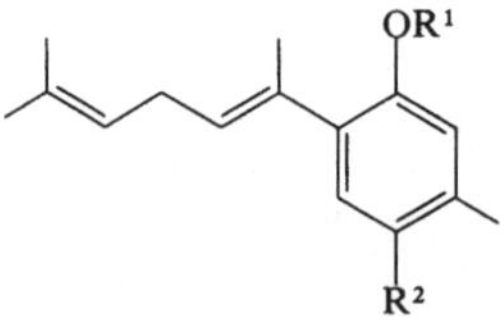

(76) $R^1 = H$ $R^2 = OH$
(77) $R^1 = Ac$ $R^2 = H$

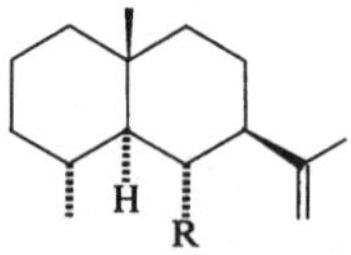

(78) $R = \overset{\oplus}{N} \equiv \overset{\ominus}{C}$
(79) $R = N{=}C{=}S$
(80) R = NH—CO—H

References, pp. 320–363

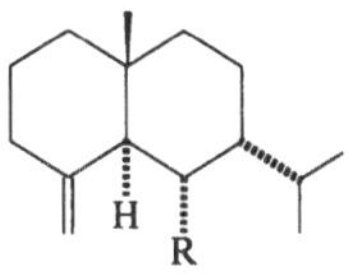

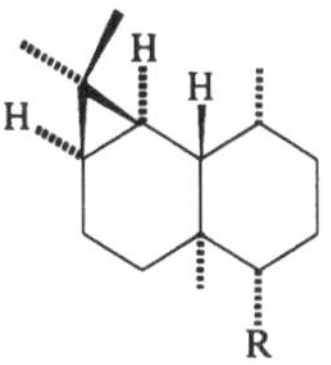

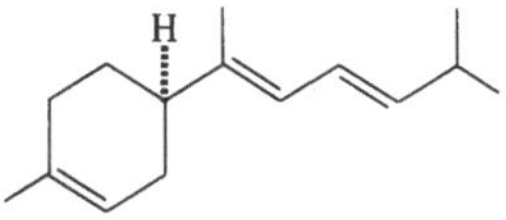

(81) R=N≡C
(82) R=N=C=S
(83) R=NH—CO—H

(84) R=N≡C
(85) R=N=C=S
(86) R=NH—CO—H

(87)

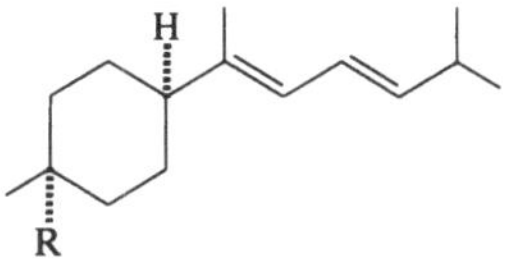

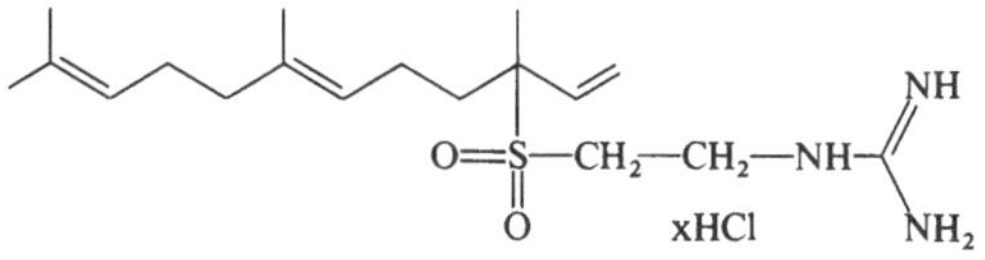

(88) R=N=C=S
(89) R=NH—CO—H

(90)

Most diterpenes isolated so far from Porifera contain furan rings. Ambliofuran (**91**) has been isolated from *Dysidea amblia* (*77, 93*) and is the diterpene analogue of previously known sesqui- and sesterterpenes from sponges. Ambliol-A (**92**) is a major furanoid diterpene alcohol of this species and contains one additional carbocyclic ring. Two of the minor metabolites from *D. amblia*, dehydroambliol-A (**93**) and ambliolide (**94**), had the same basic structure as **92**, but **94** contains a lactone ring in place of the furan ring. A *Dendrilla* species also furnished **93** together with a new metabolite of similar structure, 1-bromo-8-ketoambliol-A acetate (**95**) (*94*), which was obtained as an unstable oil which decomposed autocatalytically in chloroform solution. Ambliol-B (**96**), a further major constituent from *D. amblia*, was first reported to contain a *cis*-fused bicyclic ring system (*93*), but X-ray analysis after resolation from a different locality showed it to be a *trans*-fused decalin (*77*). A further compound from the same species was ambliol-C (**97**), which was stereoisomeric with ambliol-B (**96**).

The common and widely distributed bath sponge *Spongia officinalis* is a rich source of terpenoids. Several publications report the presence of antifungal and antimicrobial tetracyclic furanoditerpenes whose structure is based on that of the spongiane skeleton. Spongia-13(16),14-dien-19-oic acid (**98**), spongia-13(16),14-dien-19-al (**99**), and spongia-13(16),14-diene (**100**) itself were isolated by CAPELLE *et al.* (*95*). Another five diterpenes (**101–105**) from the same species were isolated by CIMINO *et al.* (*96*). Repeated chromatography of the diethyl ether solubles from an acetone extract of the fresh sponges gave the previously known isoagatholactone (**101**) in addition to 15α,16α-diacetoxyspongiane (**102**) and three new tricyclic diterpenes (**103–105**) which could be considered precursors of the spongiane diterpenoids. More recently a Spanish group (*97*) investigated the same sponge and found four additional new diterpenes (**106–109**), isoagatholactone (**101**) and another known compound, aplysillin (**110**). Isolation of these tetracyclic diterpenes has also prompted some synthetic studies (*98*). Recently, three new diterpene lactones (**111–113**), **111** with mildly cytotoxic activity, have been reported from *Igernella notabilis* (*98a*).

SCHMITZ *et al.* (*99*) isolated a new hydroxylated atisane, atisane-3β,16α-diol (**114**), from the Caribbean sponge *Tedania ignis* which showed some cytotoxic activity. In the course of a search for constituents of Mediterranean marine organisms, MAYOL *et al.* (*100*) investigated the sponge *Spongionella gracilis* and reported isolation and structure elucidation of gracilin A (**115**), a tricyclic *nor*-diterpene acetate, which can be related to the spongiane skeleton. Gracilin B (**116**) was a second unusual compound from *S. gracilis*, which has a *bis-nor*-diterpene skeleton (*101*). Its structure was established mainly by means of 2D-^{13}C-^{1}H shift correlation spectroscopy. A series of compounds (**117–120**) having a similar fused γ-lactone-tetrahydrofuran ring system has been isolated from *Dendrilla* sp. (*94*). Dendrillolide A (**117**), dendrillolide B (**118**) and norrisolide (**119**) – the latter having been previously isolated from the dorid nudibranch *Chromodoris norrisi* (*102*) – had the same molecular formula, $C_{22}H_{32}O_5$, while dendrillolide C (**120**) had lost the elements of acetic acid. Compounds **117**, **118**, and **120** have the same perhydroazulene skeleton. A biosynthetic pathway to these metabolites starting from a spongiane precursor was discussed. Extraction of the sponge *Aplysilla sulphurea* with light petroleum yielded aplysulphurin (**121**), an aromatic furanoid diterpene (*103*). The structure has been confirmed by X-ray analysis.

Different *Agelas* species from the Pacific (Palau, Okinawa) and the Caribbean have provided diterpenoids containing a purine or a 9-methyladenine unit. These compounds showed antimicrobial and Na,K-ATPase inhibitory effects. Ageline A (**122**) and Ageline B (**123**) from an unidentified *Agelas* species were active against *Bacillus subtilis*, *Staphylococcus aureus*, *Candida albicans* and the marine bacterium B-392 (*92*). They also showed mild ichthyotoxic properties. Four novel bicyclic (*104*) and two monocyclic (*105*) diterpenoids with a 9-methyl adenine unit possessing inhibitory effects on Na,K-ATPase have been isolated from the Okinawan sea sponge *Agelas nakamurai*. The structures of agelasines-A, -B, -C, and -D (**124–127**) (*104*) as well as those of agelasine-E (**128**), and agelasine-F (**122**) (*105*) have been elucidated spectroscopically and through various chemical conversions. Agelasine-

References, pp. 320–363

F (**122**) was found to be identical with ageline A (**122**) previously reported by CAPON and FAULKNER (*92*). NAKATSU *et al.* (*106*) determined the structure of an unusual purine-containing diterpene (**129**), an artifact derived from an unstable base that was the major antimicrobial metabolite of *Agelas mauritiana*.

The isocyano function occurs naturally in compounds isolated from terrestrial microorganisms and from marine sponges as has already been mentioned in the section on sesquiterpenes. A series of tricyclic diterpenes which bear isocyano, hydroxyl, tetrahydropyranyl, and chlorine functions have been reported as antibiotics from *Acanthella* species (*107, 108*). Kalihinol-A (**130**), which exhibited *in vitro* activity against *Bacillus subtilis*, *Staphylococcus aureus*, and *Candida albicans*, but was inactive against *Escherichia coli*, was a richly functionalized tricyclic diterpene. The same group (*108*) also reported the structures of kalihinols-B (**131**), -C (**132**), -E (**133**), and -F (**134**). While **133** was a C-14 epimer of **130**, the tetrahydropyranyl moiety of **130** or **133** was replaced by a tetrahydrofuran moiety in the other metabolites. A sponge, *Adocia* sp., furnished six new diterpene isocyanides (**135–140**) in addition to one previously reported compound (**141**) belonging to this group (*109*). Compound **141** and the isocyanide mixture **135–140** were said to display marked *in vitro* antimicrobial activity, particularly against gram positive bacteria, but did not show *in vivo* activity other than marked toxicity.

MANES *et al.* (*110*) isolated two substances from an undescribed *Prianos* species. While the major compound was a *nor*-sesterterpene which will be discussed in the next section together with other sesterterpenes, the minor one was a *nor*-diterpene peroxide, methyl nuapapuanoate (**142**). CAPON and MACLEOD (*110a*) obtained another *nor*-diterpene peroxide (**143**) from an unidentified sponge.

Recently NAKAMURA *et al.* (*110b*) isolated two new diterpene derivatives of hypotaurocyamine, agelasidine B (**144**) and agelasidine C (**145**), from the Okinawan sponge *Agelas nakamurai*. The agelasidines inhibited the growth of microorganisms such as *Staphylococcus aureus* and showed inhibitory effects on contractile response of smooth muscles and enzymatic reactions on Na,K-ATPase isolated from pig brain.

(**91**)   (**92**)

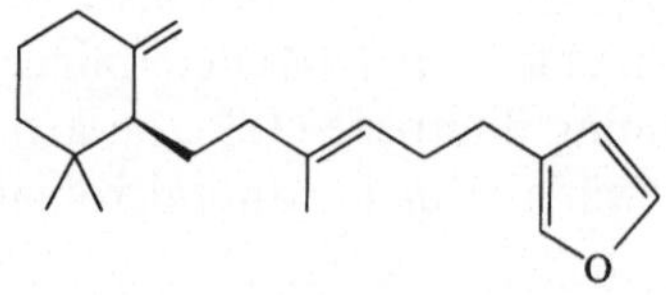

(93)

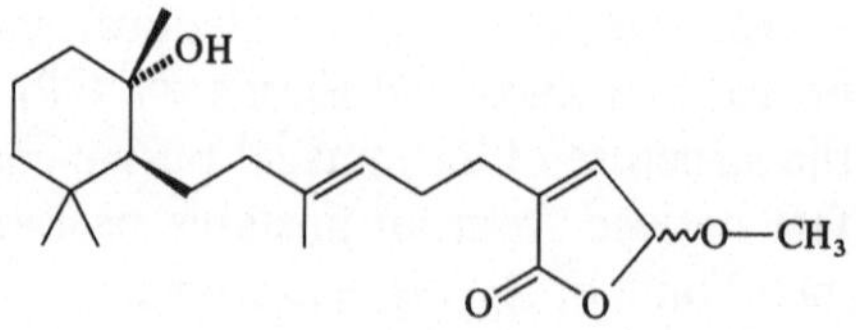

(94)

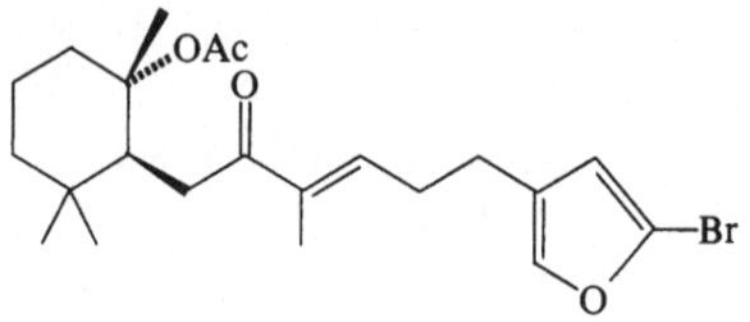

(95)

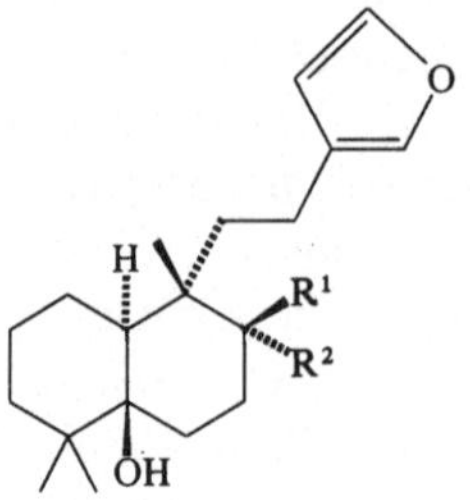

(96) R¹=CH₃ R²=H
(97) R¹=H   R²=CH₃

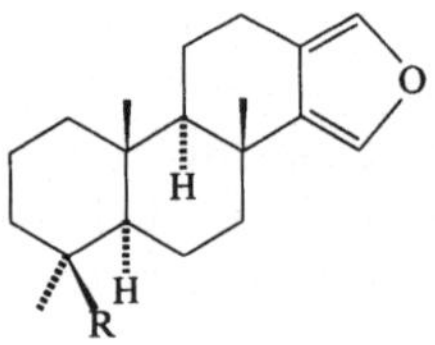

(98) R=COOH
(99) R=CHO
(100) R=CH₃

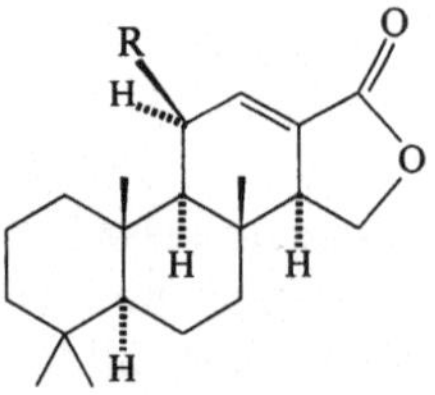

(101) R=H
(106) R=OH
(107) R=OAc

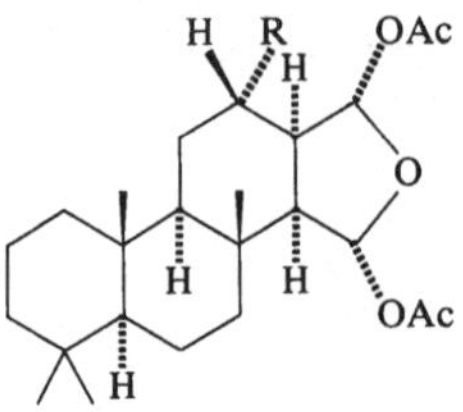

(102) R=H
(110) R=OAc

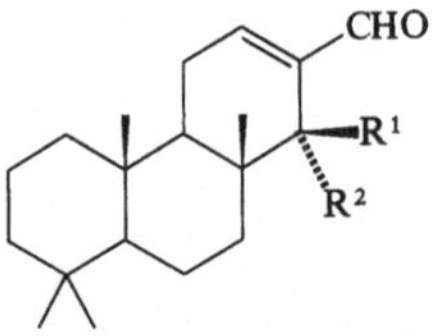

(103) R¹=CHO   R²=H
(104) R¹=CH₂OAc R²=H
(105) R¹=H   R²=CHO

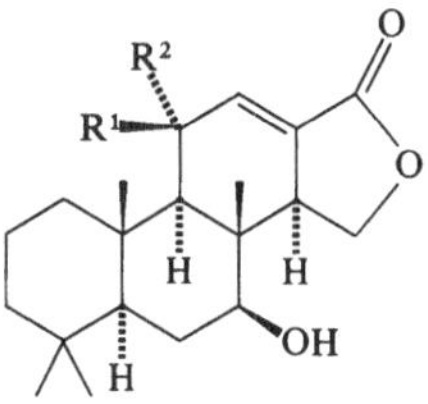

(108) R¹=OH, R²=H
(109) R¹=H, R²=OH

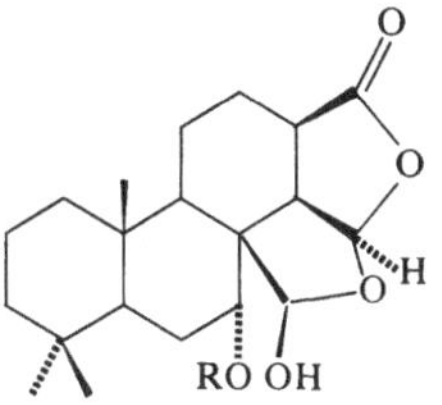

(111) R=COCH₂CH₂CH₃
(112) R=COCH₃
(113) R=H

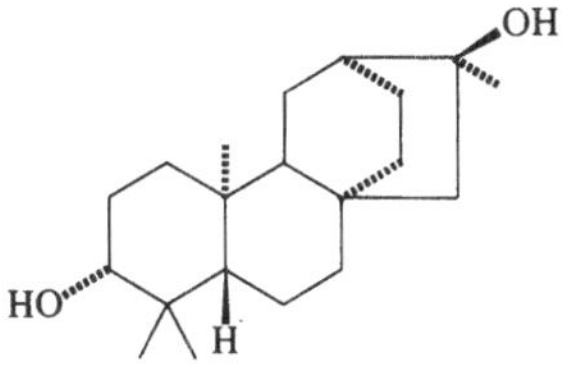

(114)

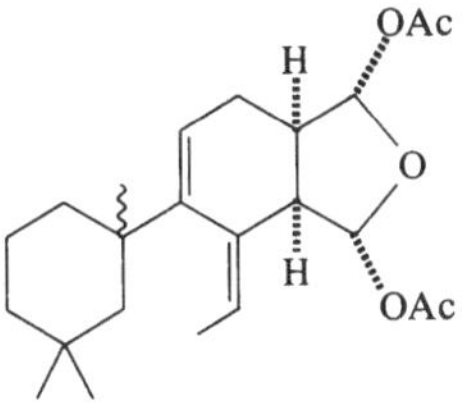

(115)

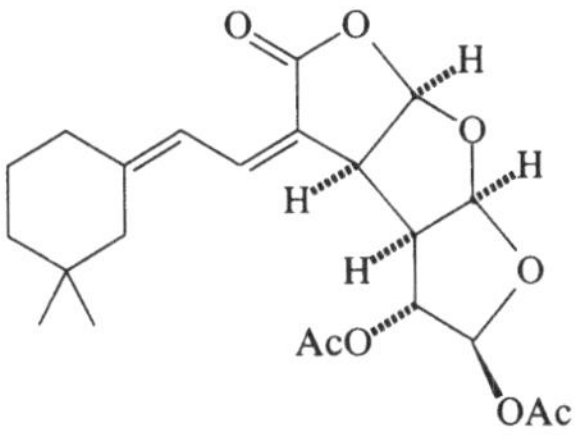

(116)

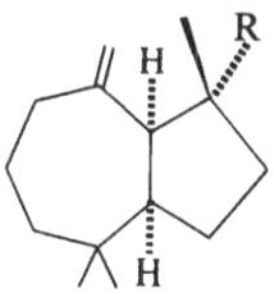

(117) R=

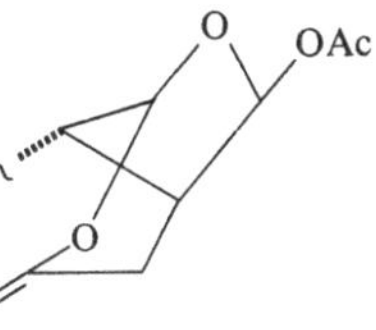

(118) R=

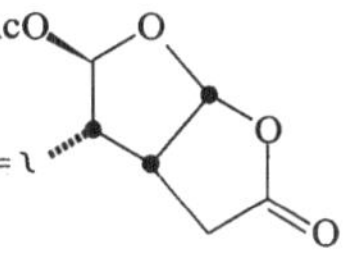

(120) R=

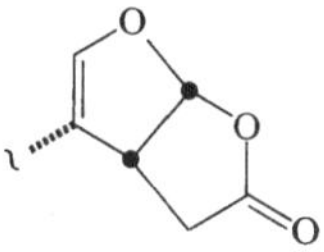

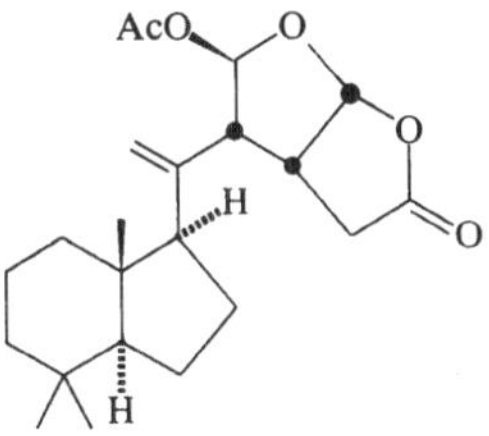

(119)

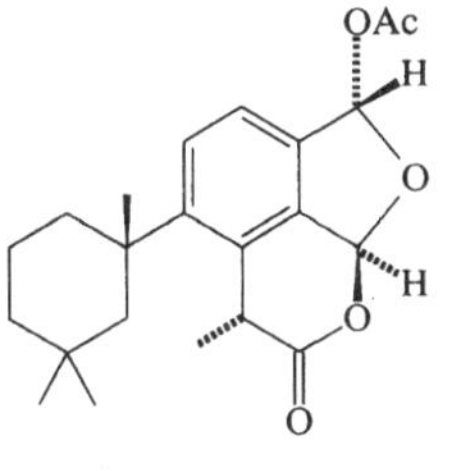

(121)

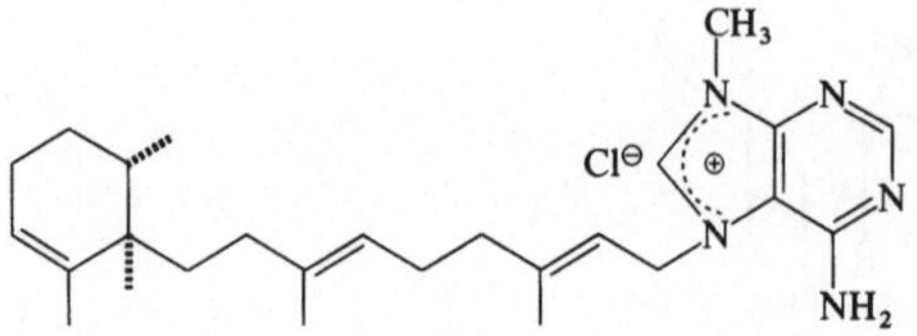

(122)

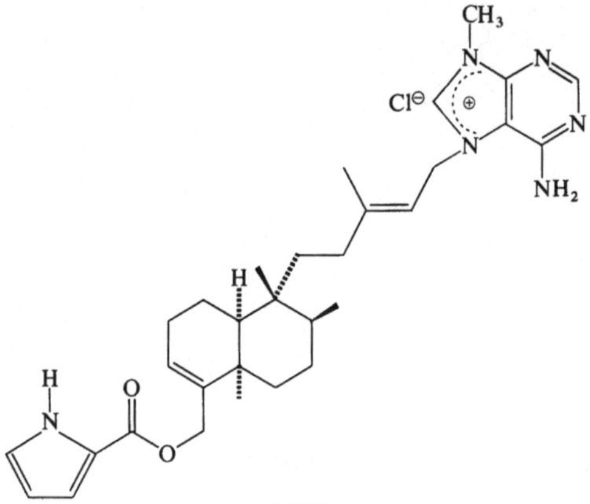

(123)

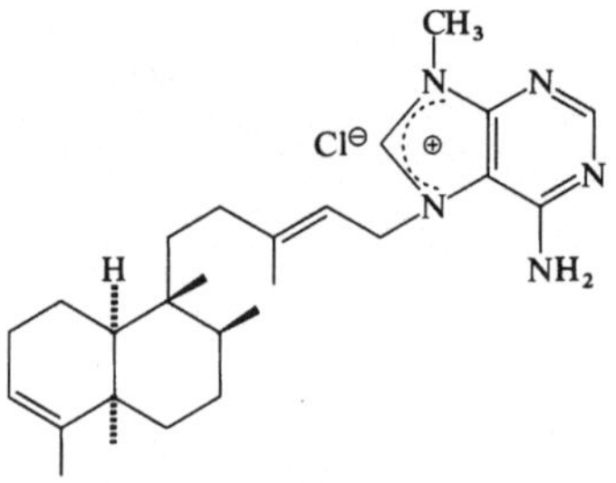

(124)

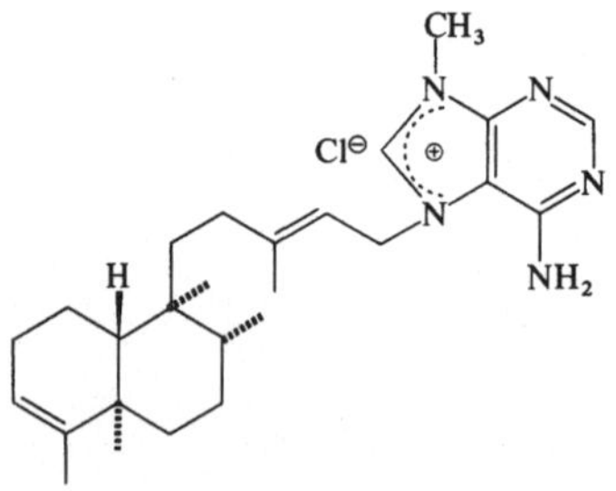

(125) 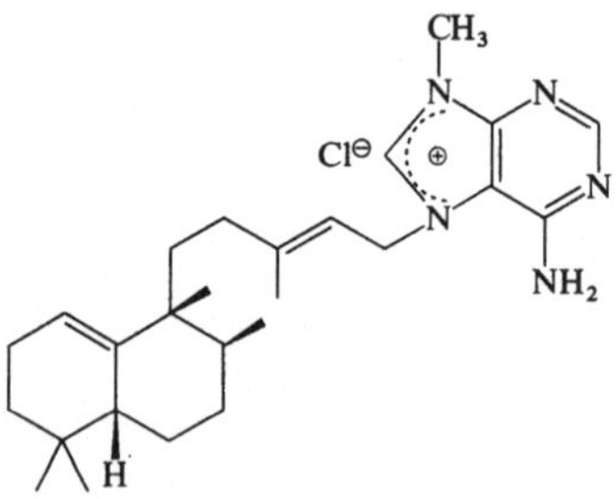 (126)

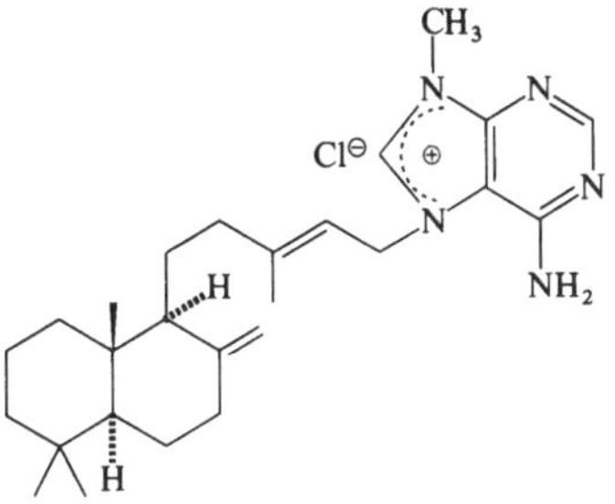

(127)

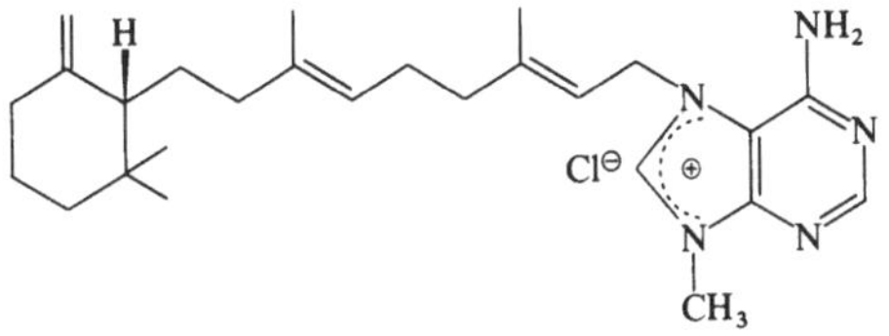

(128)

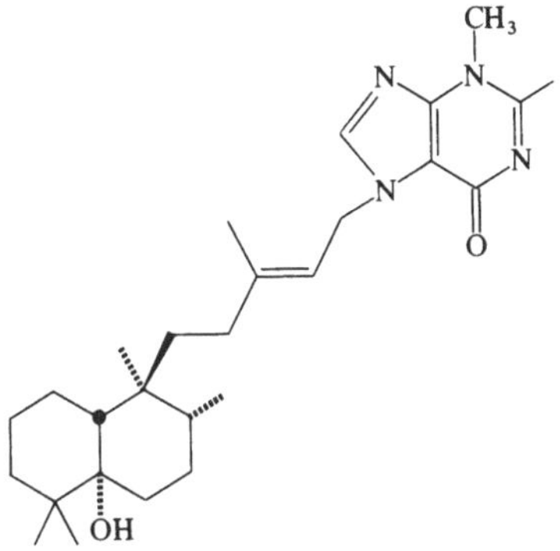

(129)

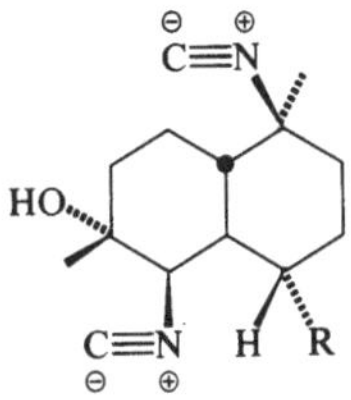

(130) R =

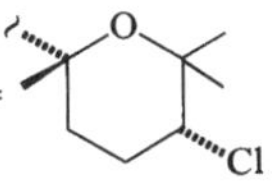

(131) R =

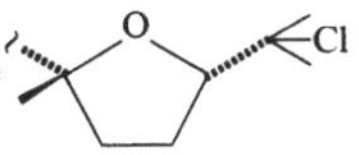

(132) R =

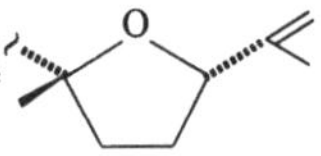

(133) R =

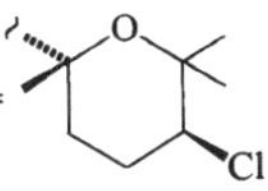

(134) R =

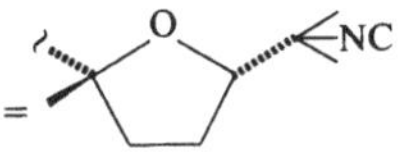

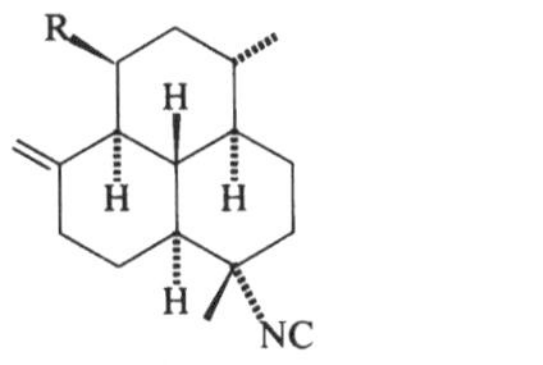

(135) R = CH=C(CH₃)₂
(138) R = CH₂—C(CH₃)=CH₂
(139) R = CH₂—C(NC)(CH₃)₂

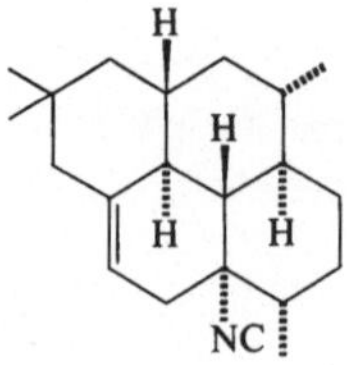

(136) 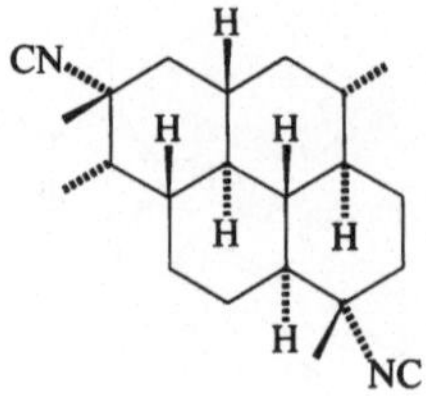 (137) 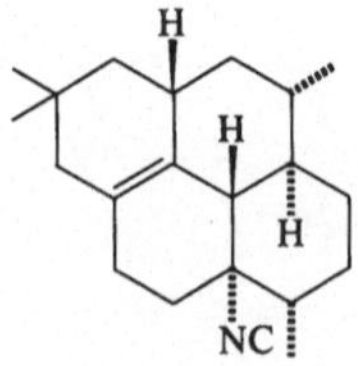 (140)

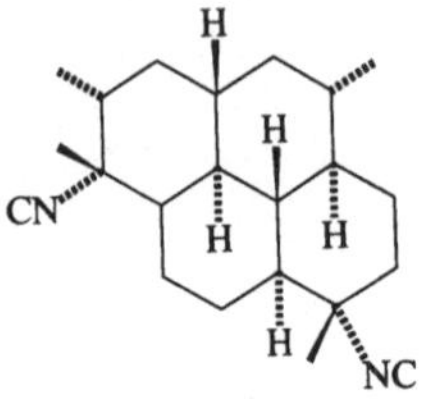

(141) 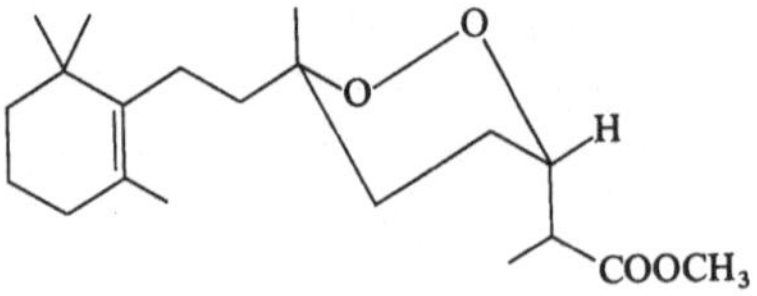 (142)

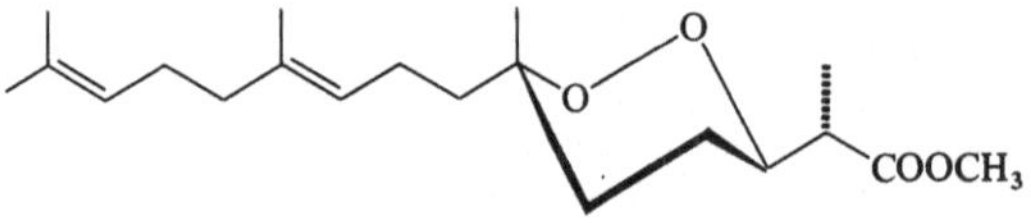

(143)

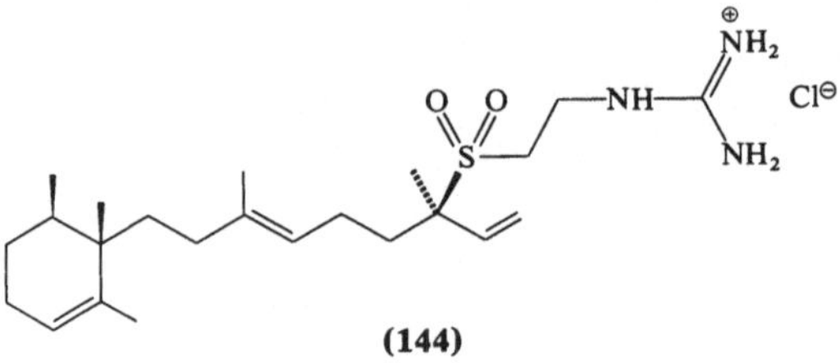

(144)

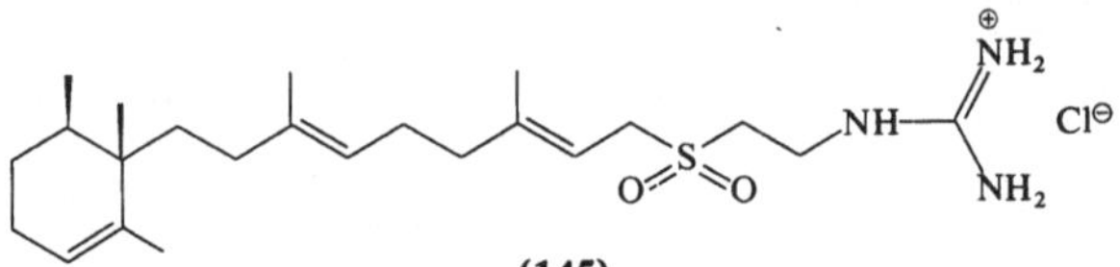

(145)

Many species of the sponge genus *Spongia* contain biosynthetically intriguing C_{21} difuranoterpenes probably derived from linear sesterterpene tetronic acids such as variabilin (**146**) which has long been known as a constitutent of *Ircinia variabilis*. A detailed discussion of sesterterpenes from terrestrial as well as marine sources has been given in the previous volume of this series (*110c*). Sponges of the genus *Ircinia* are a source of linear furanoterpenes such as **146** and isomeric compounds together with C_{21} furanoterpenes. GONZÁLEZ et al. (*111*) isolated **147** and **148**, two sesterterpenes, together with their related C_{21} compounds **149** and **150** from *I. dendroides* and deduced the stereochemistry of the trisubstituted isoprenoid double bonds shown in the formulas. The co-occurrence of furanosesterterpenes and C_{21} furanoterpenes with related structures gave rise to the presumption that the C_{21} compounds are derived by degradation of the furanosesterterpenes. The authors converted the acetates of **147** and **148** into the acetates of **149** and **150** by oxidative degradation with H_2O_2 in the presence of basic Al_2O_3, thus providing a chemical analogy for the proposed biogenetic scheme.

Three more new C_{21} compounds, the difuranoterpene didehydrofuranospongin-1 (**151**) from *Phyllospongia foliascens* (*112*), furospongolide (**152**), containing an α,β-unsaturated butenolide terminus, from *Dysidea herbacea* (*113*), and compound **153** from a *Spongia* species (*114*) have been reported in the last few years. A C_{22} furanoterpene, furodendin (**154**), has been isolated together with tetracyclic sesterterpenes from the dichloromethane extract of the freeze-dried sponge *Phyllospongia dendyi* (*112*). Furodendin (**154**) which possesses, a β,γ-unsaturated lactone ring was the first example of a compound, presumably derived from a geranylfarnesol precursor, in which a C_3 unit has been lost to give a C_{22} degraded terpenoid.

Two linear sesterterpenes from *Spongia idia* were identified as furospinulosin-1 (**155**) and idiadione (**156**) (*115*). Substance **156** was toxic to the starfish *Pisaster giganteus*, the brine shrimp *Artemia* sp., and to the ectoproct *Membranipora membranacea* and immobilized the larvae of the red abalone *Haliotis rufescens*. **156** has also been isolated from the dorid nudibranch *Cadlina marginata*, which was the only predator observed to feed on *S. idia*. FUSETANI et al. (*116*) found that the lipophilic extract of a marine sponge, *Cacospongia scalaris*, showed significant activity in the starfish egg assay by inhibiting cell division of the fertilized eggs. After purification by chromatography they obtained the active material, two furanosesterterpenes **157** and **158**, both possessing a conjugated tetronic acid moiety. **157** and **158** were also active against microorganisms.

Luffariella variabilis provided four sesterterpenoid antibiotics (**159**–**162**) (*117, 118*). Manoalide (**159**) which contains an α,β-unsaturated δ-lactol function was without close structural analogy. **160** was an open chain isomer of **159**, with *E*-configuration of the C-6,7 double bond which prevents ring closure between the C-24 aldehyde and the δ-OH at C-4, and was therefore named *seco*-manoalide. **161** and **162** were *cis-trans* isomers around the C-6,7 double bond. (*E*)- (**161**) and (*Z*)-neomanoalide (**162**) are formally derived from **160** by ring opening of the γ-hydroxybutenolide to the aldehyde acid, reclosure to form a δ-lactone with the C-4 hydroxy group and reduction of the free C-24 and C-25 aldehyde functions. Manoalide (**159**) had antiinflammatory properties and was an inhibitor of phospholipase A_2 (*119, 120*). Cavernosine (**163**), an ichthyotoxic terpenoid lactone from *Fasciospongia cavernosa*, appears to be a degraded manoalide (**159**) (*121*). An unidentified sponge from Palau contained an antimicrobial sesterterpene, palauolide (**164**) which showed activity against *Bacillus subtilis* and *Staphylococcus aureus* (*122*).

Several tetracarbocyclic sesterterpenes of marine origin are related to scalarin (**165**), the first compound of this series, which has been isolated from *Cacospongia scalaris*. Two more compounds of the same type from this source are scalaradial (**166**) and desacetylscalaradial (**167**) (*123*). **167** showed potent cytotoxic activity *in vitro*. The basic hydrocarbon skeleton of these compounds was named scalarane (**168**) (*124*). Sponges of the genus *Phyllospongia* are good sources of such compounds. *P. dendyi* provided 24-methyl-24,25-dioxoscalar-16-en-12β-yl 3-hydroxybutanoate (**169**) and 12α-acetoxy-24-methyl-24-oxoscalar-16-en-25-al (**170**) (*112, 124*). **169** was responsible for the biological activity of the crude extract which showed antifungal and antiinflammatory properties. In the same paper (*124*) KAZLAUSKAS *et al.* described isolation from a second unclassified *Phyllospongia* sp. of four new methylscalaranes (**171–174**) together with a compound (**175**) which had been isolated previously from *P. radiata*. The dichloromethane extract of this sponge showed biological activity similar to that of *P. dendyi*. Foliaspongin (**176**) is an antiinflammatory bishomosesterterpene from *P. foliascens* (*125, 126*), possessing a 4α-ethyl group. Similarly alkylated scalarins (**177–180**), sometimes also with a 4α-ethyl residue, were previously isolated from *Dysidea herbacea* (*127*). The extra methyl groups in **169–180** presumably derive from methionine.

A compound with the unstable configuration at C-18, 12-deacetyl-12,18-diepiscalaradial (**181**) (*115*) from *Spongia idia*, was toxic to starfish, brine shrimp, and to the hydroid *Bougainvilla* sp. and also immobilized the larvae of the red abalone *Haliotis rufescens*. KAZLAUSKAS *et al.* (*128*) reported isolation of five new 24-methylscalarane derivatives (**182–186**), the last three pentacyclic, from the dichloromethane extract of a *Lendenfeldia* sp. which is closely related to the genus *Phyllospongia*. *S. idia* furnished also four pentacyclic sesterterpenes, heteronemin (**187**), 12-epideoxoscalarin (**188**), scalarafuran (**189**), and scalarolide (**190**) (*115*). **187** and **188** showed various biological activities similar

References, pp. 320–363

to those exhibited by **181**. CROFT *et al.* (*129*) isolated from a *Carteriospongia* sp. (previously *Phyllospongia*) the dimethylated scalarane derivative (**191**) possessing a C-20 acetoxy function whose structure and relative stereochemistry were established by X-ray analysis. Pentacyclic alkylated scalaranes have also been found in the same *Dysidea herbacea* collection which gave **177–180** and were named scalardysin-A (**192**) and scalardysin-B (**193**) (*127*). Scalarobutenolide (**194**) from *Spongia nitens* also has a scalarane skeleton but differs in the arrangement of the butenolide ring (*130*).

A unique tetracyclic furanosesterterpene, wistarin (**195**), has been discovered in *Ircinia wistarii* in addition to the previously known isomer ircinianin (**196**) (*131*). **196** decomposed readily, whereas **195** was relatively stable. As wistarin (**195**) was not detected among the decomposition products of **196** and **196** did not undergo cyclization to **195** on treatment with dilute acid, base or Lewis acids, it was proposed that **195** was a natural product and not an artifact. The biosynthesis of these sesterterpenes must involve a different folding of the acyclic C_{25}-precursor.

Some *nor*-sesterterpene peroxides have also been isolated. Sigmosceptrellin-A, -B, and -C (**197–199**), which are related structurally to **164**, constituted the ichthyotoxic fraction of the sponge *Sigmosceptrella laevis* (*132*). Structure and relative configuration of the methyl ester of **197** were deduced by X-ray diffraction; **198** is the C-3 epimer and **199** is the C-2 epimer of **197**. Recently CAPON and MACLEOD (*110a*) revised the previously assigned absolute stereochemistry of the sigmosceptrellins as shown in structures **197–199**. SOKOLOFF *et al.* (*133*) have investigated *nor*-sesterterpenoid peroxide antibiotics from the Red Sea sponge *Prianos* sp. The name "prianicin" was initially given to the active principle which was isolated from the methanolic extract of the whole sponge. The major component, prianicin-A to which formula (**200**) was assigned, was identical with muqubilin previously isolated from *Prianos* sp. by KASHMAN and ROTEM (*134*) whereas prianicin-B had the same structure as sigmosceptrillin-A (**197**). The prianicins strongly inhibited the growth of gram positive bacteria, but were noneffective against gram negative bacteria. Of particular interest was the activity of the prianicins against *Saccharomyces cerevisiae* with the authors claiming that this demonstrates the ability of the prianicins to prevent secondary infections due to pathogenic fungi. MANES *et al.* (*110*) also found muqubilin in an undescribed *Prianos* species from the Tongan coral reefs and provided evidence for revision of its stereochemistry to **201**, with the configuration at C-2 remaining undefined. The C-2 configuration, as shown in structure **201**, has recently been clarified in an article (*110a*) dealing also with some other new peroxides

(**143**, **202–204**). Muqubilin (**201**) was responsible for the biological activity of the crude extract, because at 16 µg/mL cleavage of fertilized sea urchin eggs was 100% inhibited (*110*). A mixture of compounds **202** and **203** which has been isolated from an unidentified sponge showed antimicrobial activity (*110a*).

Suvanine (**205**) is a novel sesterterpene which contains a guanidinium bisulfate and a furan functionality (*135*). This ichthyotoxic metabolite (**205**) was obtained from an undescribed *Ircinia* species and exhibited more than 90% inhibition of sea urchin cell division at 16 µg/mL.

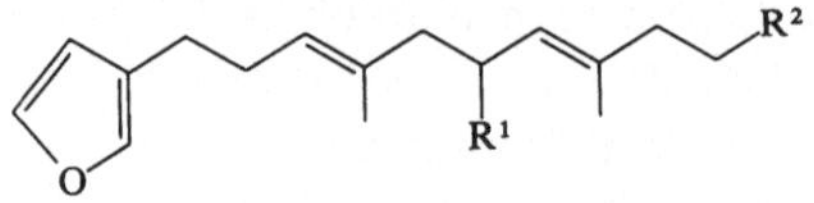

(**146**) $R^1 = H$ $R^2 = H_2C$–

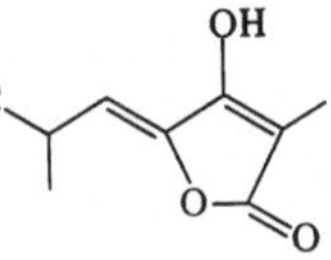

(**151**) $R^1 = OH$ $R^2 = H_2C$–

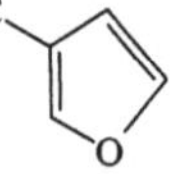

(**152**) $R^1 = H$ $R^2 = H_2C$–

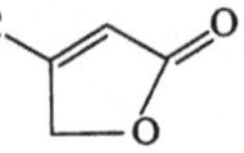

(**154**) $R^1 = H$ $R^2 = H_2C$–

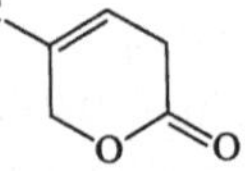

(**155**) $R^1 = H$ $R^2 = HC$=

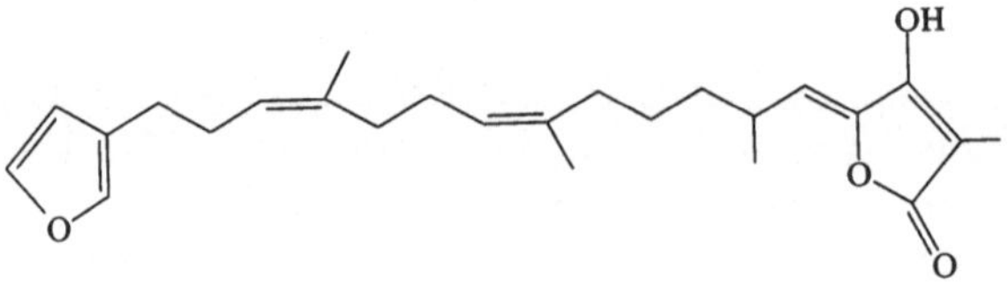

(**147**)

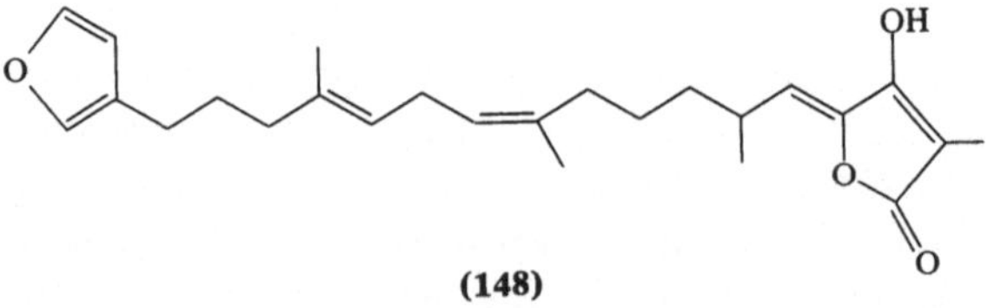

(**148**)

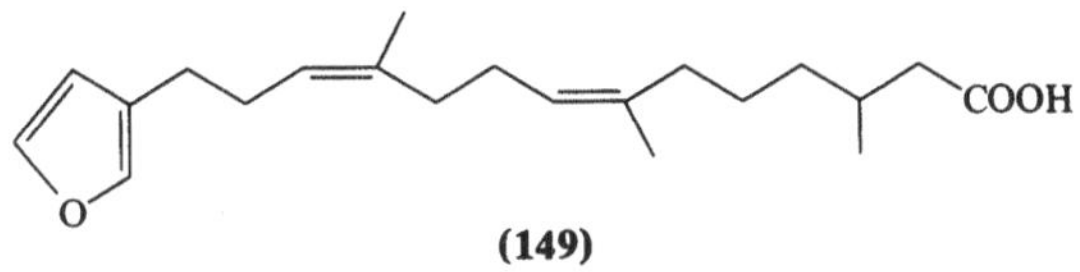

(149)

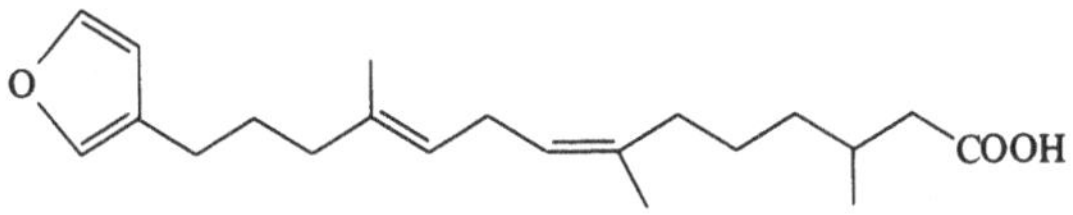

(150)

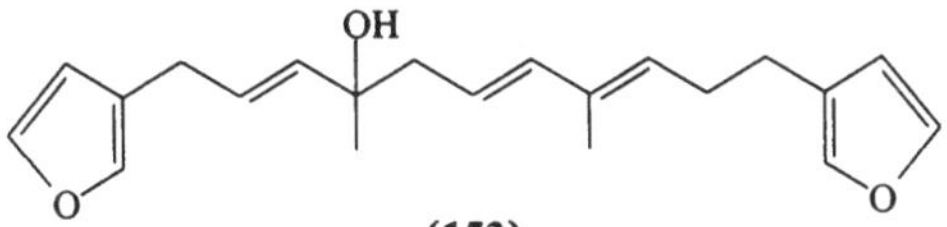

(153)

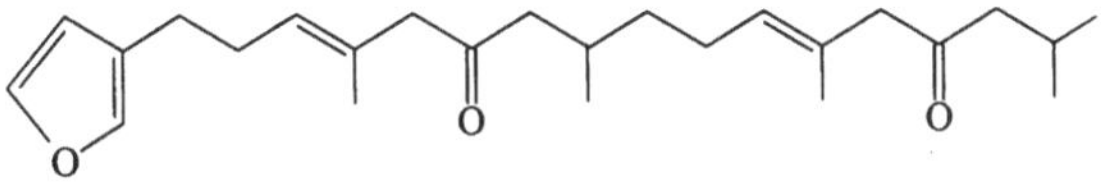

(156)

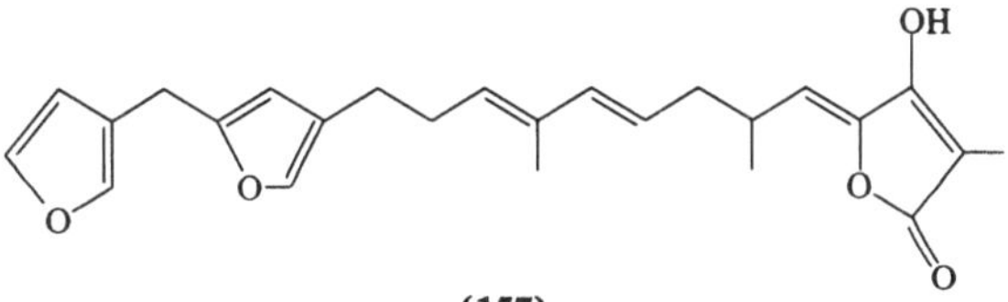

(157)

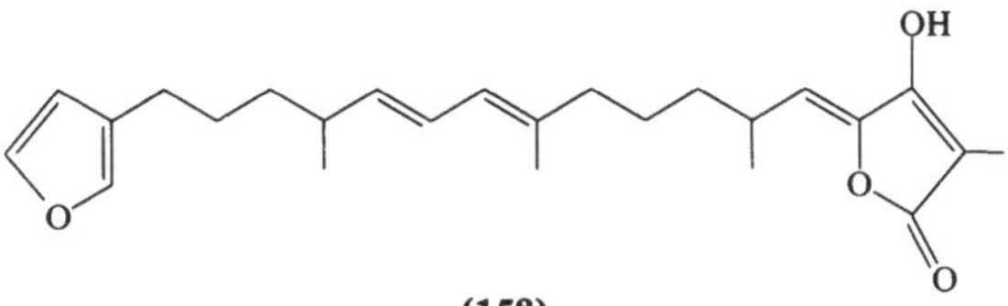

(158)

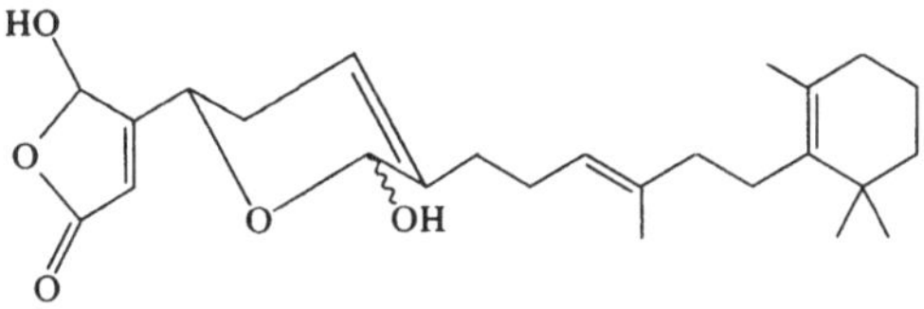

(159)

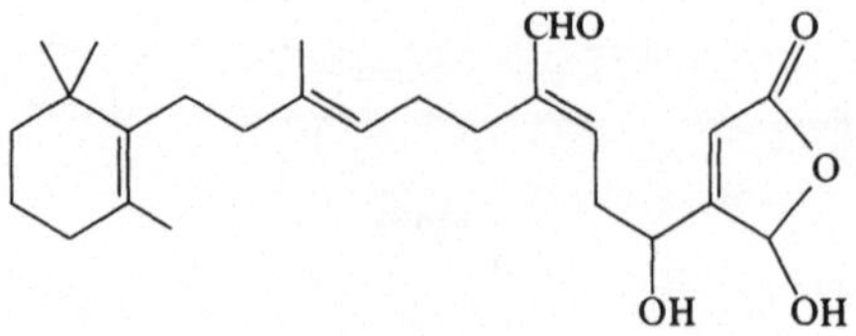

(160)

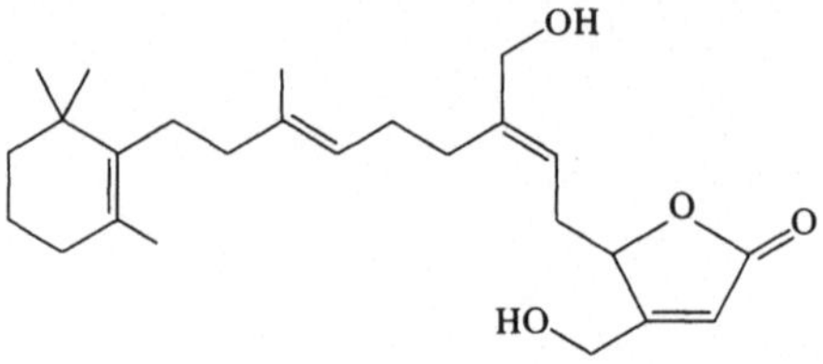

(161)

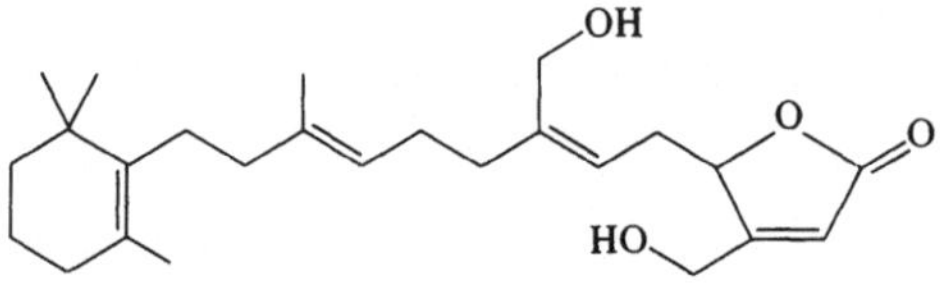

(162)

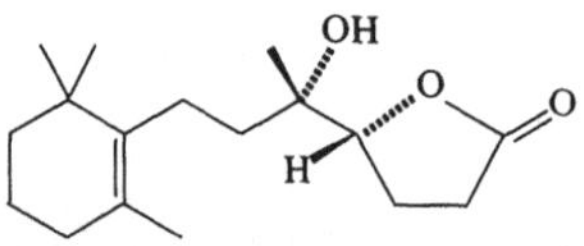

(163)

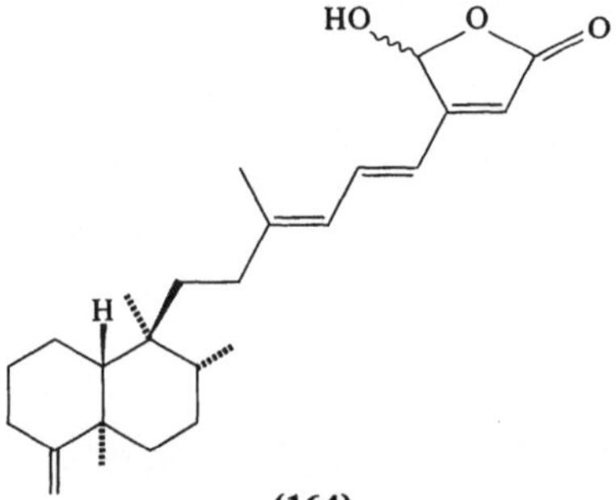

(164)

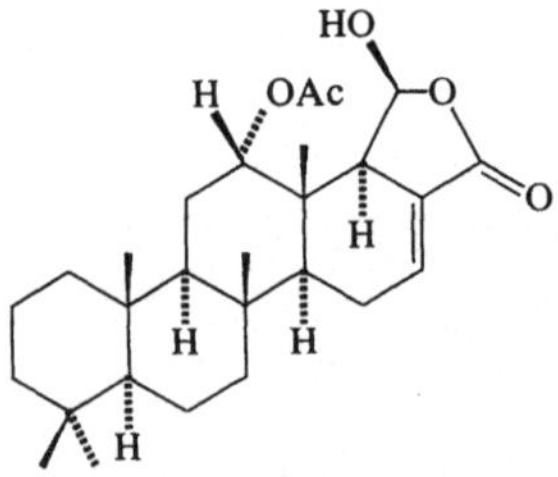

(165)

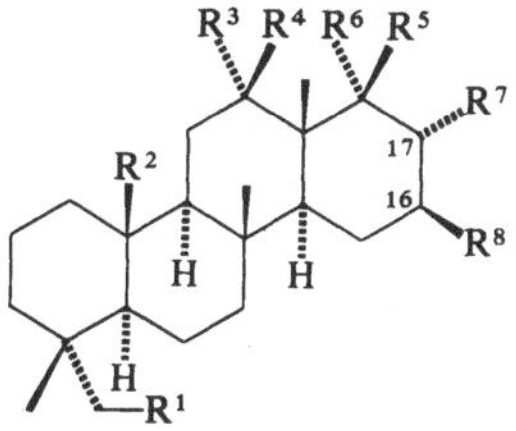

	R^1	R^2	R^3	R^4	R^5	R^6	R^7	R^8	
(166)	H	CH_3	OAc	H	CHO	H	CHO	H	$\Delta^{16(17)}$
(167)	H	CH_3	OH	H	CHO	H	CHO	H	$\Delta^{16(17)}$
(168)	H	CH_3	H	H	CH_3	H	CH_3	H	
(169)	H	CH_3	H	X	CHO	H	CO—CH_3	H	$\Delta^{16(17)}$
(170)	H	CH_3	OAc	H	CHO	H	CO—CH_3	H	$\Delta^{16(17)}$
(171)	H	CH_2OH	=O		CHO	H	CO—CH_3	H	$\Delta^{16(17)}$
(172)	H	CH_2OAc	=O		CHO	H	CO—CH_3	H	$\Delta^{16(17)}$
(173)	H	CH_3	H	OH	CHO	H	CO—CH_3	OH	
(174)	H	CH_2OAc	=O		CHO	H	CO—CH_3	OAc	
(175)	H	CH_3	H	OH	CHO	H	CO—CH_3	OAc	
(176)	CH_3	CH_3	H	OH	CHO	H	CO—CH_3	Y [b]	
(177)	H	CH_3	OH	H	CHO	H	CO—CH_3	OAc [b]	
(178)	CH_3 [a]	CH_3	OH	H	CHO	H	CO—CH_3	OAc [b]	
(179)	H	CH_3	OAc	H	CHO	H	CO—CH_3	OAc [b]	
(180)	CH_3 [a]	CH_3	OAc	H	CHO	H	CO—CH_3	OAc [b]	
(181)	H	CH_3	H	OH	H	CHO	CHO	H	$\Delta^{16(17)}$
(182)	H	COOH	=O		CHO	H	CO—CH_3	H	$\Delta^{16(17)}$
(183)	H	CH_2OH	OAc	H	CHO	H	CO—CH_3	H	$\Delta^{16(17)}$

X = —O—CO—CH_2—CH(OH)—CH_3
Y = —O—CO—CH_2—CH(OH)—CH_2—CH_3
[a] Stereochemistry at C-4 is not defined
[b] Group α-orientated

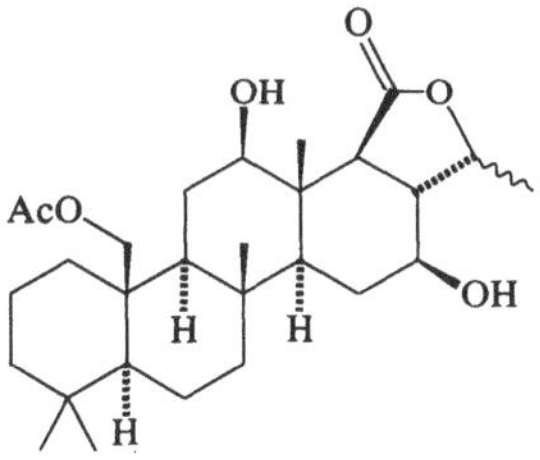

(184)

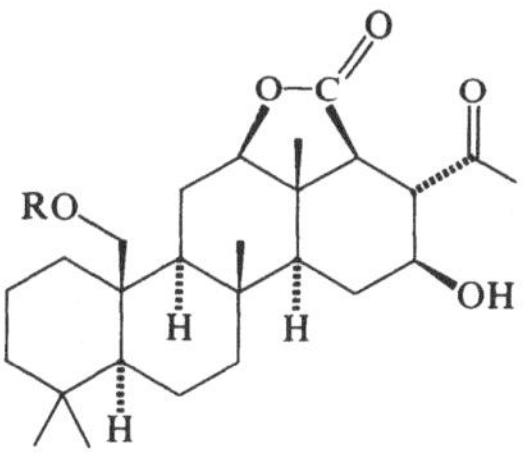

(185) R = Ac
(186) R = H

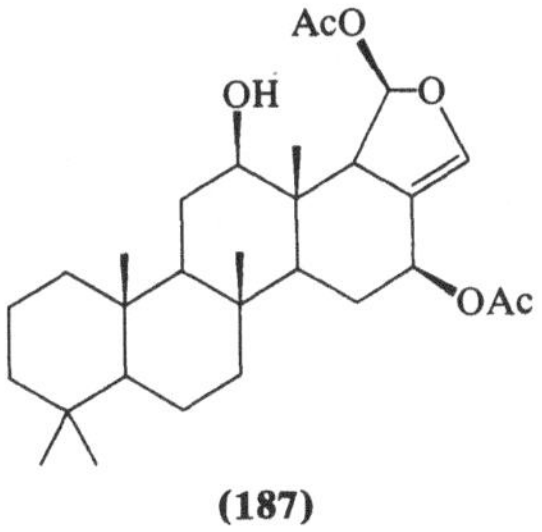

(187) 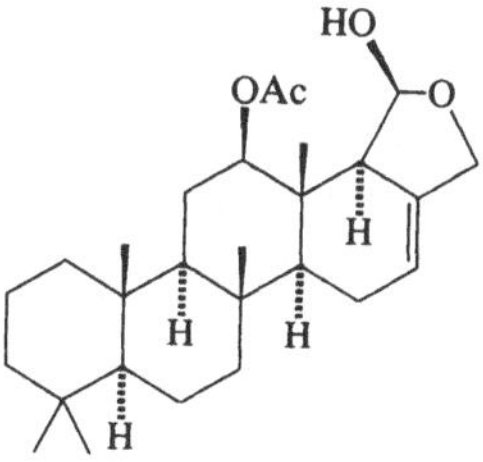   (188) 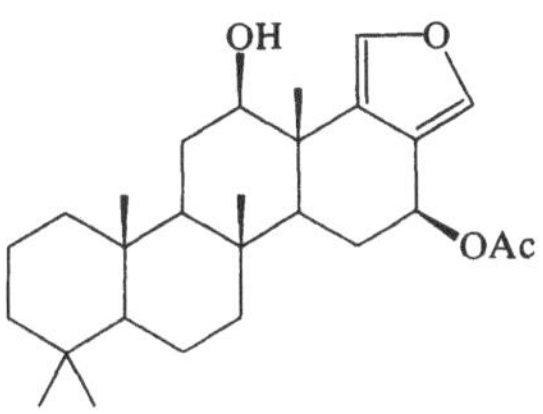   (189)

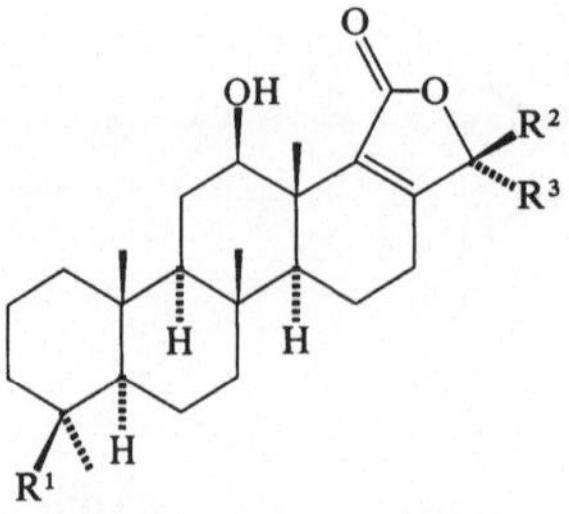

(190) R¹=CH₃   R²=R³=H
(191) R¹= [AcO, H]   R²=CH₃, R³=H

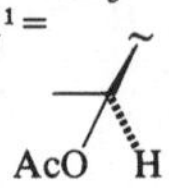

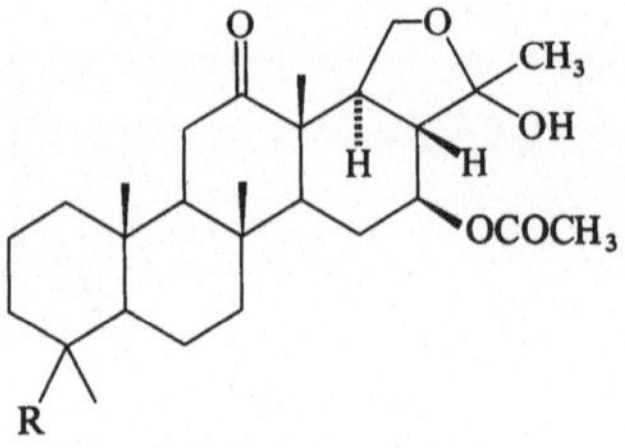

(192) R=CH₃
(193) R=CH₂—CH₃

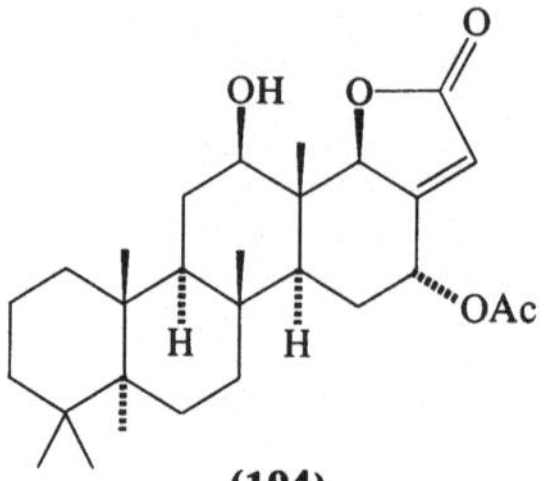

(194)

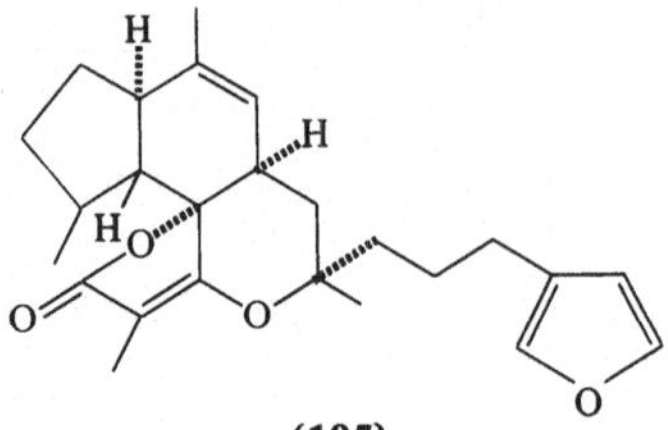

(195)

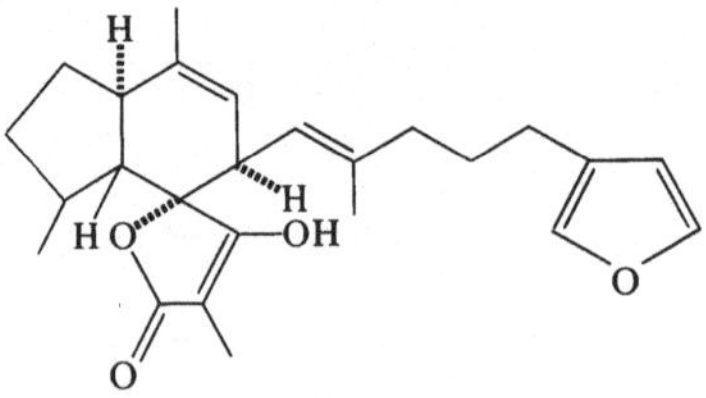

(196)

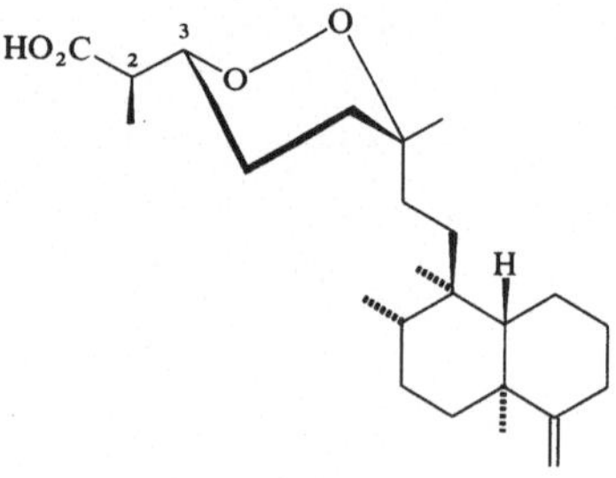

(197)

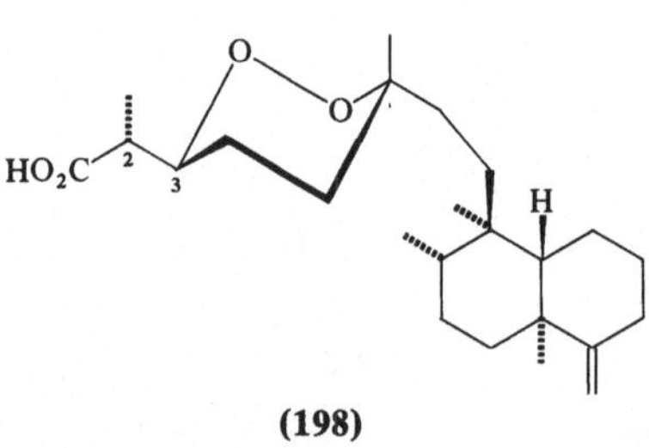

(198)

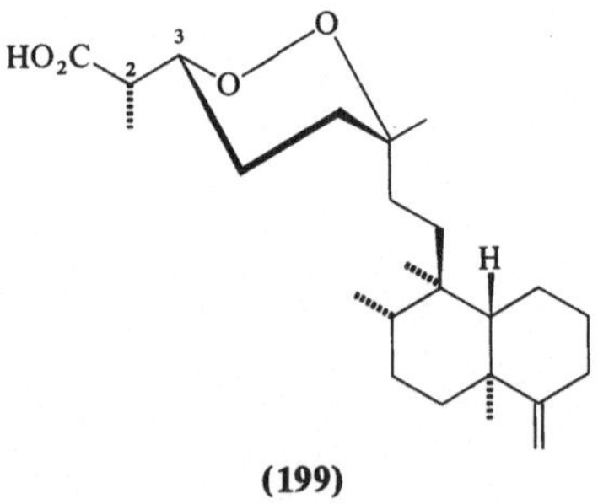

(199)

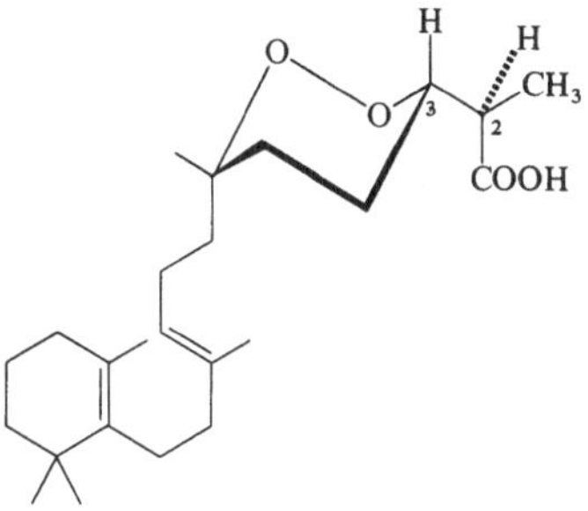

(200)

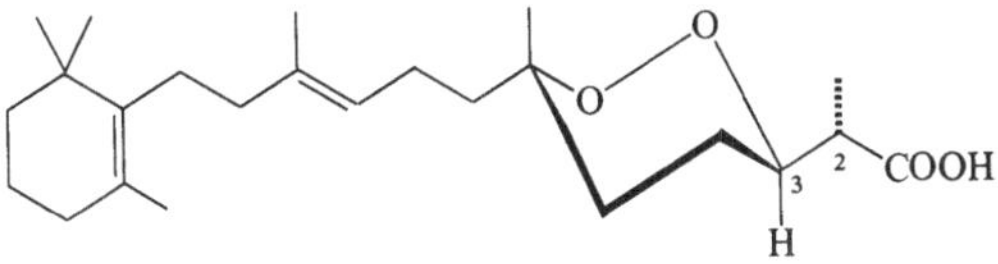

(201)

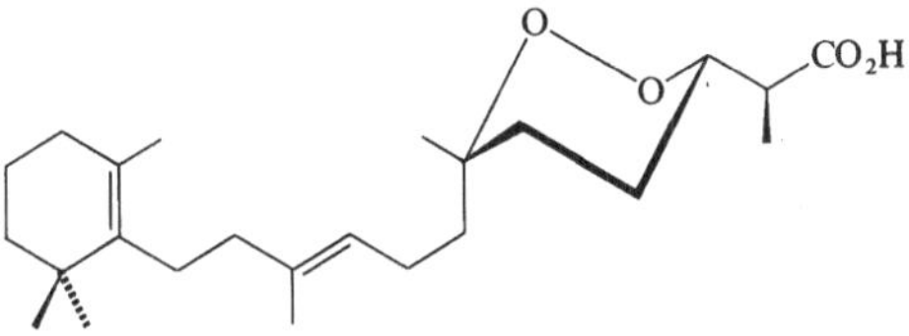

(202)

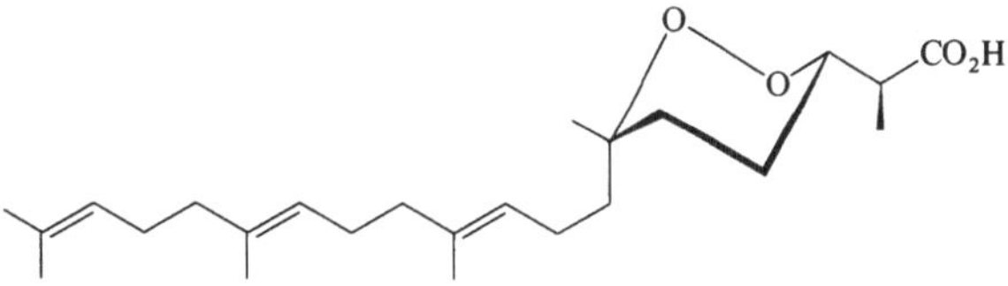

(203)

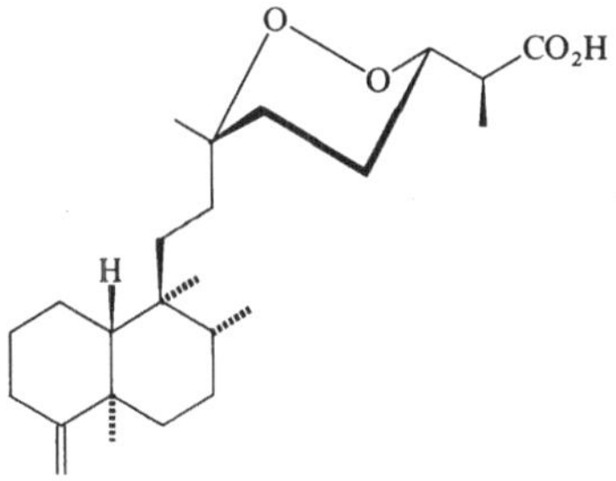

(204) 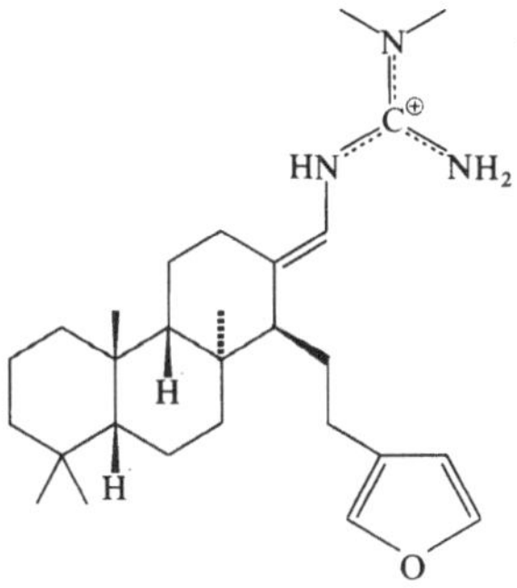   (205)

Although sponges have proved to be a particularly fruitful source of new sesqui-, di-, and sesterterpenes, triterpenes are very rare in these animals. Eight squalene-derived triterpenes (**206–213**) possessing a hitherto-unknown ring system have been isolated from the Red Sea sponge *Siphonochalina siphonella* (*136–138*); the so-called sipholane skeleton consists of a *cis*-fused octahydroazulene linked by a two-carbon bridge to a *trans*-decahydrobenzoxepine. The structure of the major triterpene, sipholenol-A (**206**), was established by X-ray analysis of its monoacetate. Spectroscopic data revealed that sipholenol-B (**207**) and sipholenol-C (**208**) were the C-10 epimer and the Δ^{13}-isomer of **206**, respectively, Sipholenol-D (**209**) and sipholenol-E (**210**), which are closely related, have been purified as the 4-mono-acetates and were interrelated by oxidation of **210** to **209** with Jones reagent. Another major constituent is sipholenone-A (**211**), the 4-keto derivative of **206**. Other sipholenones included the epoxide (**212**) and the 15(28)-en-16-one (**213**). Siphonellinol (**214**) was another unusual tricyclic triterpene from the same sponge which also contains the *trans*-decahydrobenzoxepine ring system but lacks the octahydroazulene moiety (*138*). The proposed biogenesis of these compounds involves two independent cyclisations of the two farnesyl halves of 2,3:6,7:18,19-triepoxysqualene.

Seven new triterpenes belonging to the rare malabaricane class from the sponge *Jaspis stellifera* were identified by RAVI *et al.* (*139*) by a combination of spectral evidence and chemical degradation as (13*Z*,15*E*,17*E*,22*E*,24*Z*)-3,12-dioxomalabarica-13,15,17,22,24-pentaen-26-oic acid (**215**), (13*Z*,15*E*,17*E*,22*Z*,24*Z*)-3,12-dioxomalabarica-13,15,17,22,24-pentaen-22,26-olide (**216**), (13*Z*,15*E*,17*E*,22*E*,24*Z*)-3β-acetoxy-12-oxomalabarica-13,15,17,22,24-pentaen-26-oic acid (**217**), (13*Z*,15*E*,17*E*,22*E*)-3β-hydroxy-12-oxomalabarica-13,15,17,22,24-pentaen-28-oic acid (**218**), (13*Z*,15*E*,17*E*,22*E*)-3β-acetoxy-12-oxomalabarica-13,15,17,22,24-pentaen-28-oic acid (**219**), (13*Z*,15*E*,17*E*)-3β,28-diacetoxy-22-hydroxymalabarica-13,15,17,24-tetraen-12-one (**220**), and (13*Z*,15*E*,17*E*,22*E*)-3β-hydroxymalabarica-13,15,17,22,24-pentaen-12-one (**221**) (*139, 140*). Previously, malabaricane triterpenes had been obtained only from a single natural source, the wood of the tree *Ailanthus malacarica*. Subsequently, Mc CABE *et al.* (*141*) reported the constitution and stereochemistry of a yellow triterpenoid pigment (**222**) from a Somalian collection of the sponge *Stelletta* sp. The *trans-syn-trans* stereochemistry of the tricyclic nucleus was deduced by spectral and X-ray crystallographic methods. Their results contrasted with the earlier conclusions of RAVI *et al.* (*139*), who had postulated *trans-anti-trans* stereochemistry for the carbon skeleton of **215–217**. The spectroscopic data given for **216** were in close agreement with those of **222**, so that the two compounds may be identical.

References, pp. 320–363

Several carotenoids have been isolated from marine sponges (*142*), for example the main carotenoids of *Microciona prolifera* (*143*) and a carotenoid sulfate, bastaxanthin (**222a**), from *Ianthella basta* (*144*). Bastaxanthin (**222a**) was the first known naturally occurring carotenoid sulfate.

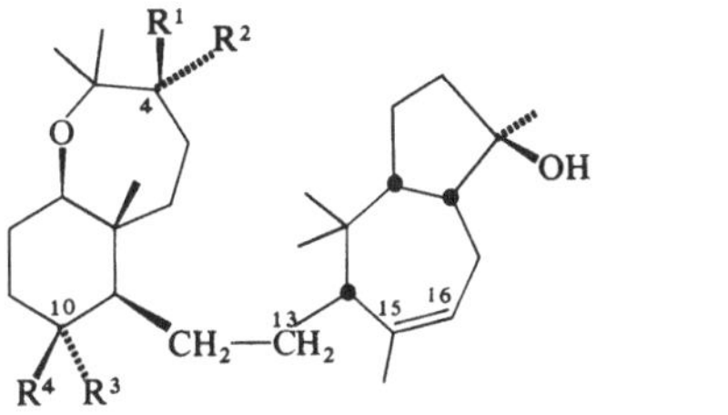

(**206**) R^1=H R^2=OH R^3=CH_3 R^4=OH
(**207**) R^1=H R^2=OH R^3=OH R^4=CH_3
(**211**) R^1/R^2=O R^3=CH_3 R^4=OH

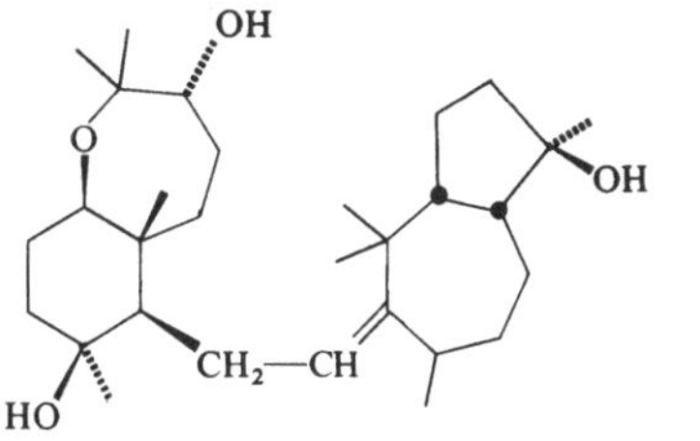

(**208**)

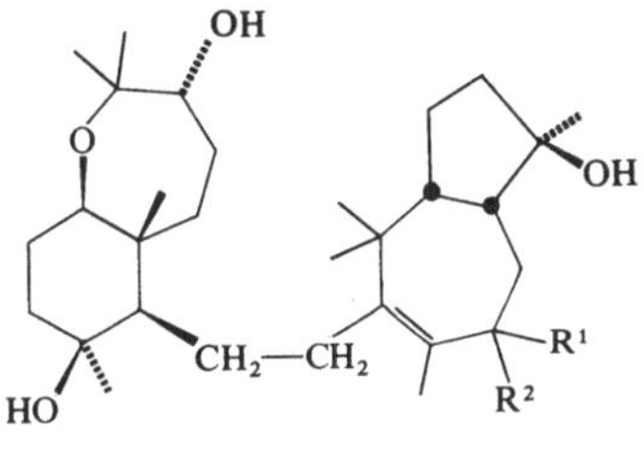

(**209**) R^1/R^2=O
(**210**) R^1=H R^2=OH

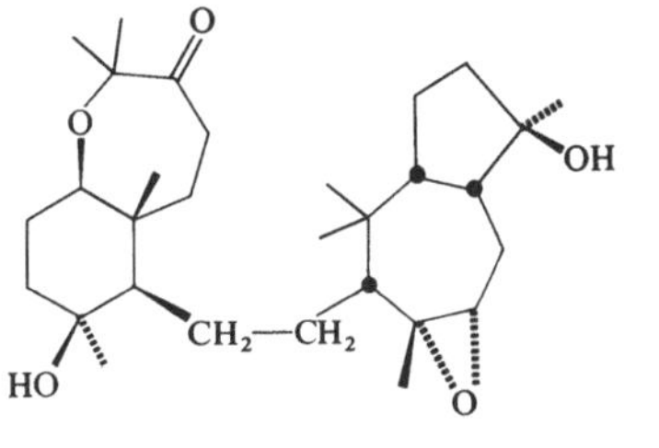

(**212**)

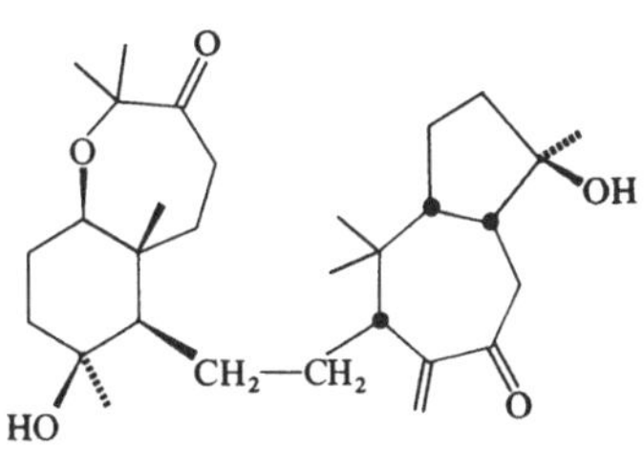

(**213**)

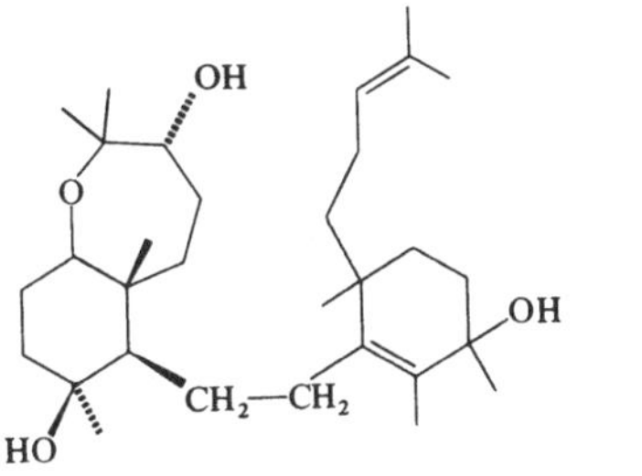

(**214**)

(215) $R^1/R^2=O$ $R^3=CH_3$ $R^4=$ [pentadienoic acid group]

(216) $R^1/R^2=O$ $R^3=CH_3$ $R^4=$ [methylpyranone group]

(217) $R^1=H$ $R^2=OAc$ $R^3=CH_3$ $R^4=$ [pentadienoic acid group]

(218) $R^1=H$ $R^2=OH$ $R^3=COOH$ $R^4=$ [dienyl group]

(219) $R^1=H$ $R^2=OAc$ $R^3=COOH$ $R^4=$ [dienyl group]

(220) $R^1=H$ $R^2=OAc$ $R^3=CH_2OAc$ $R^4=$ [hydroxy dienyl group]

(221) $R^1=H$ $R^2=OH$ $R^3=CH_3$ $R^4=$ [dienyl group]

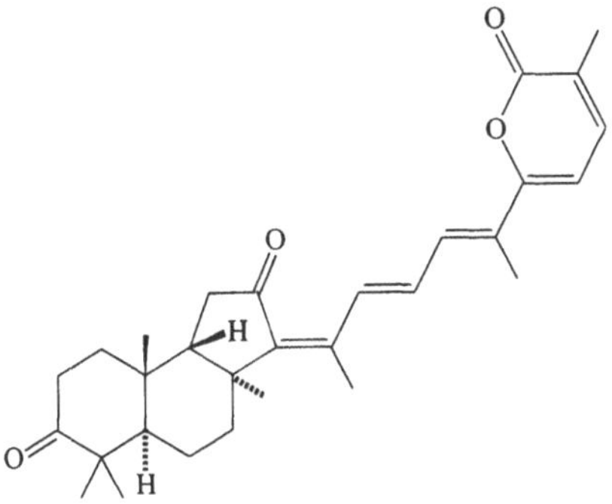

(222)

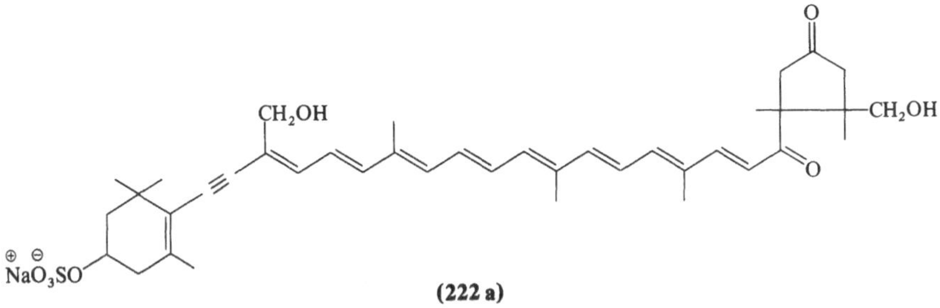

(222 a)

II.3. Amino Acid Derived Metabolites from Porifera

Metabolites in Porifera derived from amino acids may be subdivided into those which are biogenetically related to tryptophane and tyrosine, respectively. Brominated tryptamines were antimicrobial metabolites reported as constituents from *Smenospongia* sp. (*86*). While 5-bromo-N,N-dimethyltryptamine (**223**) and a brominated indole (**224**) were found in *S. aurea*, *S. echina* contained 5,6-dibromo-N,N-dimethyltryptamine (**225**). Two compounds related to **224** have been described by the same group (*145*) as constituents from *Dercitus* species. Spectroscopic data and confirmation by total synthesis led to their formulation as 2'-de-N-methylaplysinopsin (**226**), and 6-bromo-2'-de-N-methylaplysinopsin (**227**). The previously known aplysinopsin (**228**) was obtained as the major metabolite. Methylaplysinopsin (**229**), which has been isolated as a yellow crystalline material from a methanolic extract of the dictyoceratid sponge *Aplysinopsis reticulata*, was found to be active on oral administration in preventing ptosis induced in mice by tetrabenacine (*146*). **229** is also a short-acting inhibitor of monoamine oxidase activity with the greatest potency when serotonine was the substrate. The reversal of tetrabenacine ptosis could be explained by inhibition of MAO activity. Aplysinopsin (**228**) was also active but much less potent. Recently TYMIAK *et al.* (*147*) reported isolation and characterization of 6-bromoaplysinopsin (**230**) and 6-bromo-4′-N-demethylaplysinopsin (**231**) in addition to **223** and **225** from *S. aurea*. Methyl-(*E*)-3-(6-bromoindol-3-yl)prop-2-enoate (**232**) has been found in the extract of a sponge of the genus *Iotrochota* (*148*).

(**223**) $R^1 = Br$ $R^2 = H$
(**225**) $R^1 = Br$ $R^2 = Br$

(**232**)

(**224**) $R^1 = H$ $R^2 = Br$ $R^3 = H$ $R^4 = O$ $R^5 = H$
(**226**) $R^1 = H$ $R^2 = H$ $R^3 = H$ $R^4 = NH$ $R^5 = CH_3$
(**227**) $R^1 = H$ $R^2 = Br$ $R^3 = H$ $R^4 = NH$ $R^5 = CH_3$
(**228**) $R^1 = H$ $R^2 = H$ $R^3 = CH_3$ $R^4 = NH$ $R^5 = CH_3$
(**229**) $R^1 = H$ $R^2 = H$ $R^3 = CH_3$ $R^4 = N-CH_3$ $R^5 = CH_3$
(**230**) $R^1 = H$ $R^2 = Br$ $R^3 = CH_3$ $R^4 = NH$ $R^5 = CH_3$
(**231**) $R^1 = H$ $R^2 = Br$ $R^3 = CH_3$ $R^4 = NH$ $R^5 = H$

While sponges of the orders Spongidae and Dysideidae usually contain terpenoids without halogen substituents as secondary metabolites, genera of the order Aplysinidae (Verongidae) have proved to be a rich source of brominated metabolites derived from tyrosine. The tyrosine side chain has been either changed into (oximino)(N-vinyl)amide, nitrile, lactone, 2-oxazolidone and isoxazole groupings or has been oxidatively removed, while the aromatic nucleus has been either retained, rearranged or partially reduced.

In an effort to compare the halogenated tyrosine derivatives from Desmospongiae MAKARIEVA *et al.* (*149*) investigated 22 species of eleven Demospongiae families and identified 10 substances of this type in *Aplysina* (*Verongia*), *Verongula*, and *Aiolochroia* species. It has been shown that animals of the Aplysinidae family biosynthesize all three forms [(+), (−), (±) = racemate] of aeroplysinin-1 (**233**) and of an oxazolidone derivative (**234**), as well as ($\pm$)-aeroplysinin-2 (**235**), two ketals (**236, 237**), and a dienone (**238**). The action of the compounds which possessed cytotoxic activities on KB and HL-5 cells on $Na^+ - K^+$-ATPase activity was investigated (*150*). Inhibition of $Na^+ - K^+$-ATPase was found for several compounds, even at 6×10^{-6} M concentrations. Later, GORSHKOV et al. (*151*) prepared 3,5-dibromo-1-acetoxy-4-oxo-2,5-cyclohexadien-1-acetonitrile from aeroplysinin (**233**) and investigated the inhibitory characteristics of this compound.

D'AMBROSIO et al. (*152, 153*) isolated three antibacterial compounds similar to **238** from the butanolic extract of the Mediterranean sponge *Aplysina* (= *Verongia*) *cavernicola*, *i.e.* ($\pm$)-3-bromoverongiaquinol (**239**), ($\pm$)-3-bromo-5-chloroverongiaquinol (**240**), and dichloroverongiaquinol (**241**). Products of the formal cyclization of **239** or **240** were also found, namely, 5-chlorocavernicolin (**242**), the C(7)-epimer 7β-bromo-5-chlorocavernicolin (**243**) and 7α-bromo-5-chlorocavernicolin (**244**), as well as 5-bromo-7β-chlorocavernicolin (**245**) and 5-bromo-7α-chlorocavernicolin (**246**). Structures **239** and **240** were confirmed by synthesis of the racemic compounds. In two earlier papers the same group had already reported the structures of cavernicolin-1 (**247**) and cavernicolin-2 (**248**) (*154*) as racemates and monobromocavernicolin (**249**) (*155*) from *A. cavernicola*, which contain only bromine but no chlorine. Compound **249** inhibited the growth of *Sarcina lutea*, *Bacillus subtilis*, *Alcaligenes faecalis*, and *Proteus vulgaris*. In studying the biosynthesis of dibromotyrosine derived antimicrobial compounds in the marine sponge *Aplysina fistularis* (*Verongia aurea*), TYMIAK and RINEHART (*156*) demonstrated the conversion of phenylalanine and tyrosine to the dienone **238** as well as to the rearranged product dibromohomogentisamide (**250**) by using radiolabeled precursors.

Other bromine-containing metabolites from sponges, each originating from more than one tyrosine unit, were psammaplysin-A and psammaplysin-B, presumably **251** and **252**, from *Psammaplysilla purpurea* (*157*), aerothionin (**253**) from *Aplysina fistularis* (*158*), isofistularin-3 (**254**) from *Verongia aerophoba* (*159*) and bastadins-1 to -7 (**255–261**) from *Ianthella basta* (*160, 161*). While **251** and **252** which possess a previously unknown spirocyclohexadienyl-oxazoline moiety were antibiotics, extensive testing revealed no significant pharmacological properties for aerothionin (**253**). In a more recent study of two tetrabromo compounds from *Psammaplysilla purpurea*, ROLL et al. (*162*) found that the major metabolite was identical with an authentic sample of diacetylpsammaplysin-A and revised the previously published structures **251** and **252** of psammaplysin-A and psammaplysin-B (*157*) to **262** and **263**, on the basis of ^{13}C-^{13}C connectivity studies and X-ray diffraction. Compound **254** was cytotoxic *in vitro* (KB cells), the effective dose being 4 µg/ml. The methanolic extract of *I. basta* as well as the bastadins (**255–261**) themselves showed potent *in vitro* and some *in vivo* activity against gram positive bacteria. In addition to isofistularin-3 (**254**) CIMINO *et al.* (*159*) isolated the dienone **238**, aeroplysinin-1 (**233**) and two new bromo-compounds, aerophobin-1 (**264**), and aerophobin-2 (**265**) from *V. aerophoba*. *Psammaplysilla purea* furnished a novel secondary metabolite, purealin (**266**), which inhibited the activity of myosin Ca-ATPase and Na,K-ATPase (*162a*). However, the activity of myosin K,EDTA-ATPase was enhanced by **266**. The structure of **266** has been determined by ^{1}H-^{1}H homonuclear and ^{1}H-^{13}C heteronuclear NMR chemical shift correlations and CD spectra.

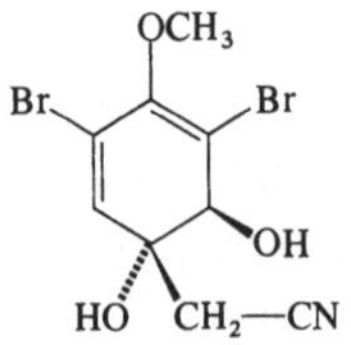

(**233**)

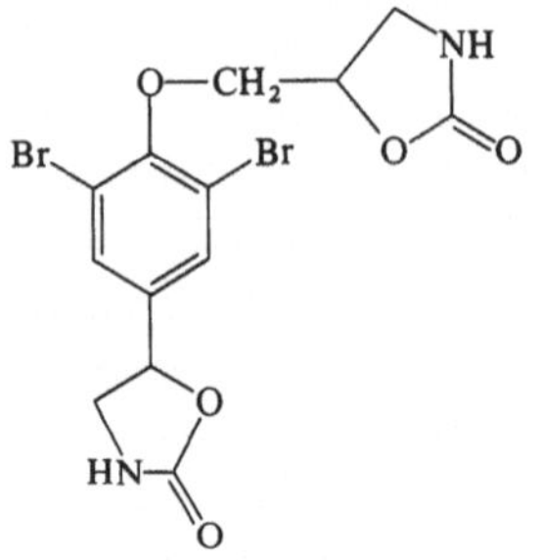

(**234**)

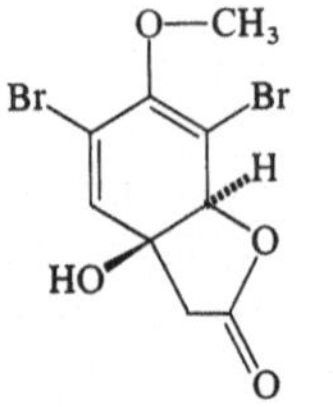

(**235**)

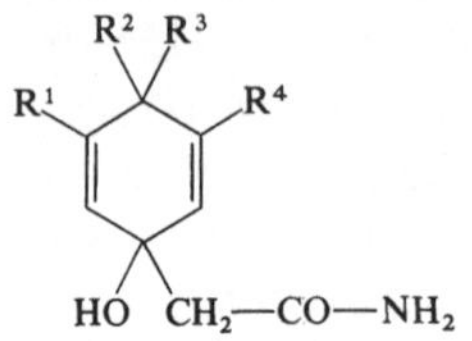

(**236**) R^1=Br R^2=O—CH$_3$ R^3=O—CH$_3$ R^4=Br
(**237**) R^1=Br R^2=O—CH$_3$ R^3=O—C$_2$H$_5$ R^4=Br
(**238**) R^1=Br R^2/R^3=O R^4=Br
(**239**) R^1=Br R^2/R^3=O R^4=H
(**240**) R^1=Br R^2/R^3=O R^4=Cl
(**241**) R^1=Cl R^2/R^3=O R^4=Cl

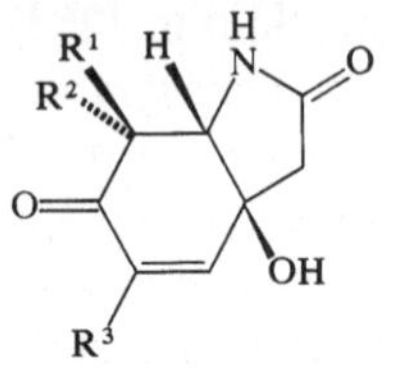

(**242**) R^1=H R^2=H R^3=Cl
(**243**) R^1=Br R^2=H R^3=Cl
(**244**) R^1=H R^2=Br R^3=Cl
(**245**) R^1=Cl R^2=H R^3=Br
(**246**) R^1=H R^2=Cl R^3=Br
(**247**) R^1=Br R^2=H R^3=Br
(**248**) R^1=H R^2=Br R^3=Br
(**249**) R^1=H R^2=H R^3=Br

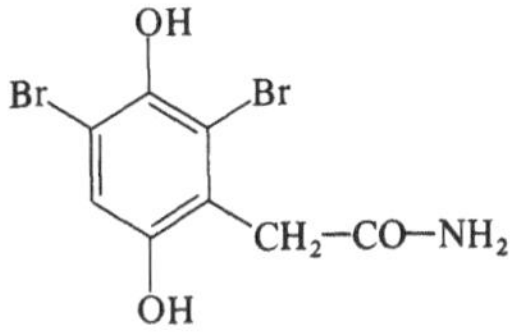

(250)

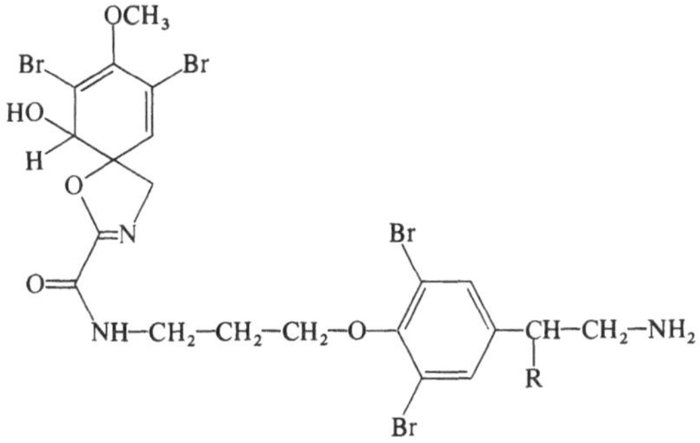

(251) R = H
(252) R = OH

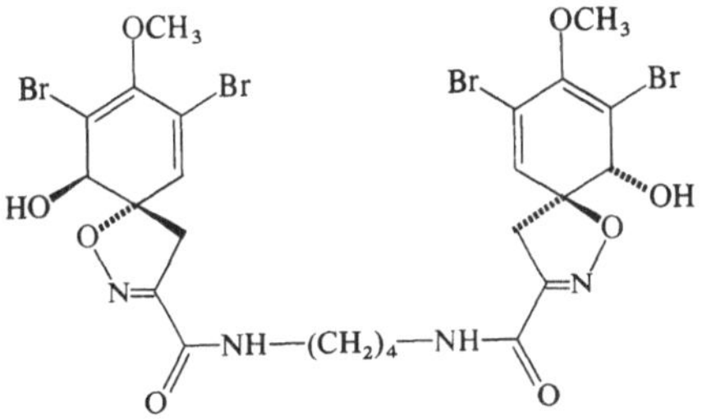

(253)

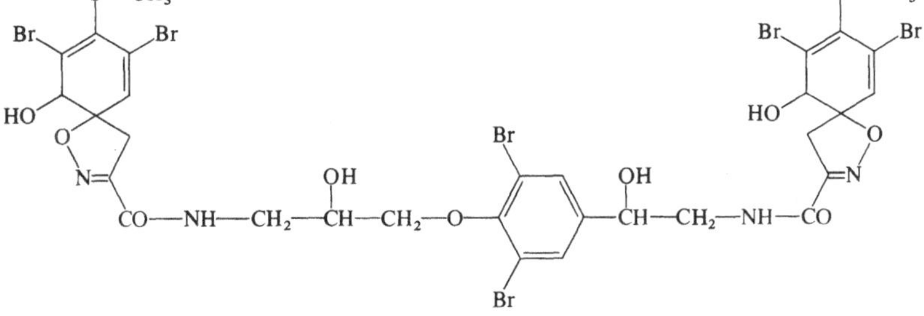

(254)

(255) R = H
(256) R = Br

(257)

(258) R = Br
(261) R = H

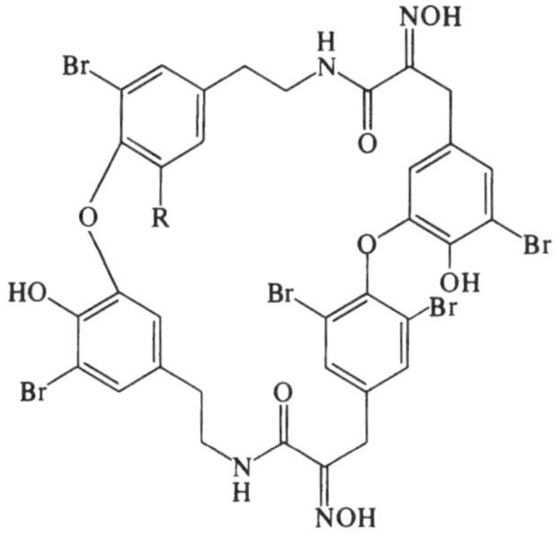

(259) R=H
(260) R=Br

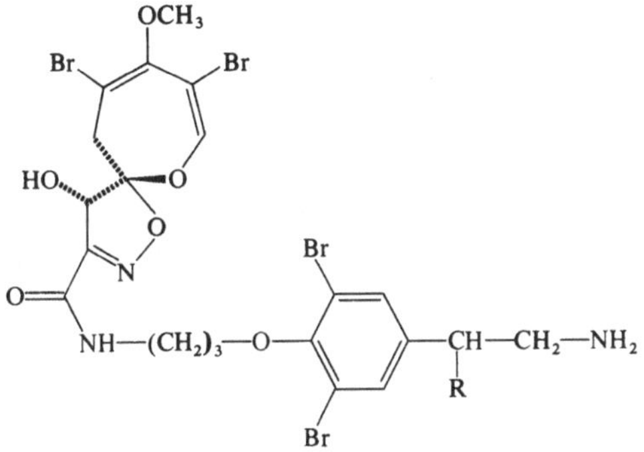

(262) R=H
(263) R=OH

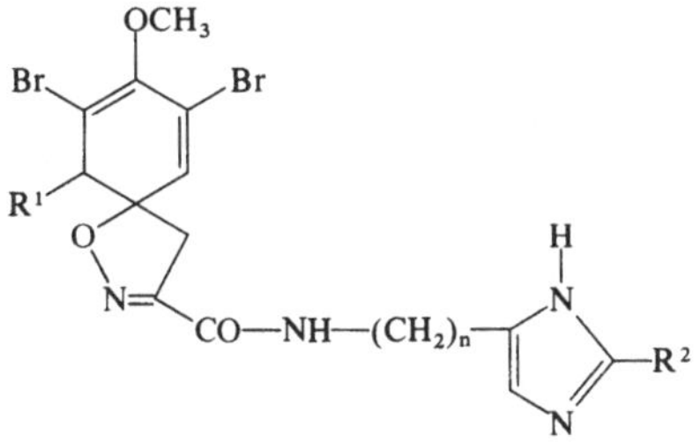

(264) R¹=OH R²=H   n=2
(265) R¹=H   R²=NH₂   n=3

(266)

II.4. Peptide Alkaloids, Peptides, and Proteins from Porifera

Dysidenin (**267**) and the toxic isodysidenin (**268**) were two highly modified peptides in collections of the sponge *Dysidea herbacea* from various locations in the region of the Great Barrier Reef. Their structures, that of **267** without stereochemistry, had already been established prior to 1980. CHARLES *et al.* (*163*) correlated **267** and **268**, thus confirming the hypothesis that **267** was the C-5 epimer of **268** and deduced the relative configuration of isodysidenin (**268**) on the basis of an X-ray analysis of a derivative of the latter (*163a*). The absolute stereochemistry of the remaining three asymmetric centers in **267** and **268** was assigned as 2R, 7R, and 13R. More recently BISKUPIAK and IRELAND (*164*) have proved, by using a new method for determining the chirality of 2-(1'-aminoalkyl)thiazoles, that C-13 is S in both dysidenin and isodysidenin and have proposed that dysidenin and isodysidenin have the opposite absolute configuration shown in formulas **269** and **270**.

Extraction of a sample of *D. herbacea* from a different location along the Great Barrier Reef by ERICKSON and WELLS (*165*) produced four additional polychloro amino acid-derived metabolites which were assigned formulas **271**–**274**. Two of these compounds, demethyldysidenin (**271**) and demethylisodysidenin (**272**) were C-5 epimers and lower homologues of **267** and **268** (or **269** and **270**) which lack the chiral center at C-13. **272** was reported to exhibit antihypertensive activity when administered intravenously. DE LASZLO and WILLIARD (*166*) synthesized (+)-demethyldysidenin (**271**) and (−)-demethylisodysidenin (**272**). This asymmetric synthesis led to a revision of the absolute stereochemistry assigned to the natural products and showed them to be **275** and **276**. The same group synthesized another unique halogenated

References, pp. 320–363

marine natural product, dysidin (**277**) (*167*), which had already been isolated from *D. herbacea* prior to 1980.

Cliona celata has furnished a series of linear peptide alkaloids, the celenamides A to D (**278–281**), in addition to the previously known clionamide (**282**) (*168, 169*). These compounds were isolated in the form of their polyacetates after acetylation of the methanolic extract of the homogenized sponge. The celenamides contain a unique amino acid α,β-didehydro-3,4,5-trihydroxyphenylalanine, as a subunit. The configuration of the double bond in the dehydroamino acid residue has been determined by synthesis of the hexa-acetate of **278** (*170*).

Three diketopiperazines cyclo-(L-Pro-L-Leu) (**283**), cyclo-(L-Pro-L-Val) (**284**), and cyclo-(Pro-Ala) (**285**) have been identified from an extract of *Tedania ignis* (*99*).

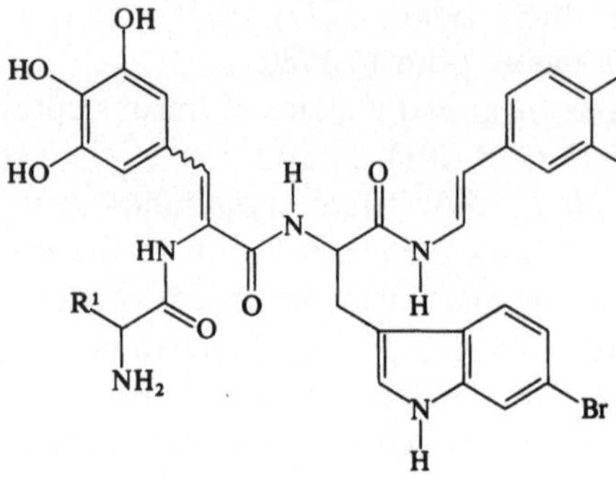

(278) $R^1 = CH_2-CH(CH_3)_2$ $R^2 = OH$
(279) $R^1 = CH(CH_3)_2$ $R^2 = OH$
(280) $R^1 = CH_2-CH(CH_3)_2$ $R^2 = H$

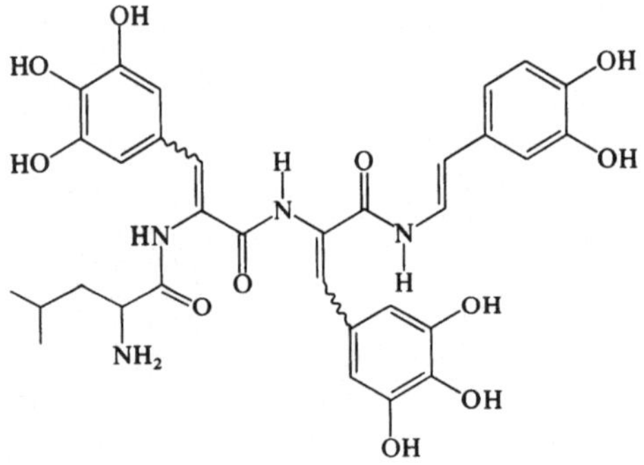

(281)

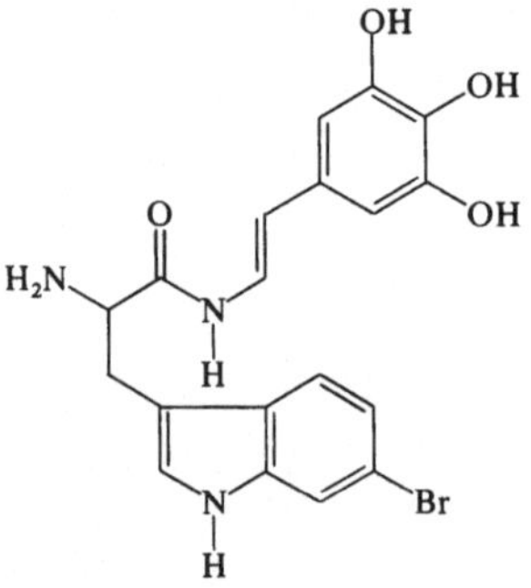

(282)

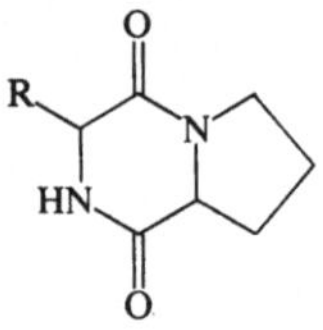

(283) $R = CH_2CH(CH_3)_2$
(284) $R = CH(CH_3)_2$
(285) $R = CH_3$

MATSUNAGA et al. (*171–173*) have described the first bioactive polypeptides derived from a sponge, the antimicrobial discodermins A, B, C, and D (**286–289**) from *Discodermia kiiensis*. Among other methods, fast atom bombardment (FAB) mass spectrometry was applied for determination of their structures. These were unusual since **286–289** contain several D-amino acids and **286–288** t-Leu residues. Discodermin A (**286**), C (**288**), and D (**289**) also inhibited the development of starfish embryo.

HCO—D—Ala—L—Phe—L—Pro—X—D—Trp—L—Arg—D—Cys(O_3H)—L—Thr—L—MeGln—L—Leu
$\qquad\qquad\qquad\qquad\qquad\qquad\qquad\qquad\qquad\qquad\qquad\qquad\qquad$ \ $\qquad\qquad$ /
$\qquad\qquad\qquad\qquad\qquad\qquad\qquad\qquad\qquad\qquad\qquad\qquad\qquad\quad$ Sar—L—Thr—L—Asn

(**286**) X = D—t—Leu—L—t—Leu
(**287**) X = D—Val—L—t—Leu
(**288**) X = D—t—Leu—L—Val
(**289**) X = D—Val—L—Val

An isopropyl alcohol extract of *Geodia mesotriaena*, a sponge collected in the Gulf of California, exhibited confirmed activity in the National Cancer Institute's *in vitro* 9KB test system and *in vivo* toxicity in the murine P 388 lymphocytic leukemia (PS) screen (*174*). Recollections of the same species gave different activities. Various extraction and separation steps furnished three fractions, geodiastatins 1 and 2 as well as geodiatoxin 1. The geodiastatins, which seem to be novel silicon-containing chromoproteins, caused 26% (at 25 and 5 mg/kg, respectively) life extension in the *in vivo* system. Geodiatoxin, which contained only a minor amount of silicon, showed an LD_{100} of 6 mg/kg mouse. MÜLLER et al. (*175*) described the isolation and purification of an inhibitory aggregation factor from *Geodia cydonium*. This factor which has a molecular weight of 27,000 and has been characterized as a glycoprotein is involved in a graft rejection mechanism of the sponge.

II.5. Nucleosides from Porifera

The first isolation of nucleosides from sponges dates from the 1950's. 1-Methylisoguanosine (**290**), which has been isolated from a nudibranch, was also found in the sponge *Tedania digitata* (*176, 177*). Synthesis by two routes confirmed the structure of this product which had a number of pharmacological properties. Skeletal muscle relaxant, hypothermic and cardiovascular (hypotensin associated with bradycardia) effects could be observed after oral administration in mice and rats (*178*). Compound **290** interacted directly with adenosine receptors in guinea pig brain slices to stimulate adenylate cyclase (*179*) and in contrast to adenosine it was resistant to deamination. The pharmacological effects of **290** are apparently due to its action as a long lasting adenosine analogue. Furthermore, **290** was effective in displacing [^{3}H]-diazepam bound to rat brain membranes (*180*). Two novel nucleosides, mycalisine A (**291**) and mycalisine B (**292**), have been isolated from a *Mycale* species collected in the Golf of Sagami, Japan (*180a*). Both compounds inhibited cell division of fertilized starfish eggs.

Isopropyl alcohol extracts of the sponge *Tethya aurantia* were fractionated for a component with negative chronotropic and inotropic activity on isolated guinea pig atria (*181*). The bioactive substance was found to be adenosine. The extracts also contained allantoin (**293**), which is an intermediate in the purinolytic sequence. Dichloromethane and methanol extracts of a new species of the genus *Echinodictyum* showed strong activities on the isolated guinea-pig trachea (*182*). The active constituent was identified as 4-amino-5-bromopyrrolo[2,3-d]pyrimidine (**294**).

II.6. Alkaloids and Other Heterocyclic Compounds from Porifera

A marine sponge belonging to the genus *Laxosuberites* contained a mixture of four 5-alkylpyrrole-2-carboxaldehydes (**295**), (6′Z)-5-(12′-

References, pp. 320–363

cyano-6'-dodecenyl)pyrrole-2-carboxaldehyde (**296**), and (6'Z)-5-(23'-cyano-23'-hydroxy-6'-tricosenyl)pyrrole-2-carboxaldehyde (**297**) (*183*). Such alkylated pyrrole-2-carboxaldehydes are well known from marine sources but whether they are biologically active is not known. Another pyrrole derivative, keramadine (**298**), a bromine-containing alkaloid isolated from an *Agelas* species, was a novel antagonist of serotonergic receptors (*184*); it appears to be closely related biogenetically to sceptrin (**299**) which has been isolated from *Agelas sceptrum* as an antimicrobial substance (*185*).

Several compounds containing a pyrrole ring in combination with a guanidine moiety have been isolated from marine sponges. One of these was a "yellow compound" from *Phakellia flabellata* (*186*), whose structure was shown to be **300**. A similar but bromine-containing compound (**301**) was a constituent of *Axinella verrucosa* and *Acanthella aurantiaca* (*187*). **301** was moderately cytotoxic *in vitro* (KB cells), but was inactive in the P 388 leukemia screen. Later, KITAGAWA *et al.* (*65*) isolated **300** and **301** together with the free base of **300** (**302**) from the Okinawan sponge *Hymeniacidon aldis*. Two pyrrololactams (**303**, **304**), likely degradation products of the guanidines **300–302**, have been obtained from the chloroform-methanol extract of *H. aldis* (*188*).

Ptilocaulin (**305**) and isoptilocaulin (**306**) were antimicrobial and cytotoxic cyclic guanidines from the Caribbean sponge *Ptilocaulis* aff. *P. spiculifer* (*189*). They were highly active against gram positive and gram negative bacteria, yeasts, and filamentous fungi. Structure and absolute stereochemistry of **305** have been confirmed by total syntheses of the racemate (*189a*, *198b*) and the $(-)$-enantiomer (*189b–d*).

The yellow zoochrome of the sponge *Verongia aerophoba* has been isolated by CIMINO *et al.* (*190*) and its structure established as 3,5,8-trihydroxy-4-quinolone (**307**). The pigment was easily oxidized to an unstable blue quinone. The dichloromethane and acetone extracts of the calcareous sponge *Leucetta microraphis* furnished a new pteridine, leucettidine (**308**) (*191*). Recently CIMINO *et al.* (*191a*) described the structure of a novel 8-oxopurine compound (**309**), isolated from *Hymeniacidon sanguinea*. Another heteroaromatic substance, with the new skeleton 1*H*-benzo[*de*]-1,6-naphthyridine, was aaptamine (**310**) from *Aaptos aaptos* which possessed α-adrenoceptor blocking activity in the isolated rabbit aorta (*192*). Amphimedine (**311**) is a new, cytotoxic fused pentacyclic aromatic alkaloid from a Pacific sponge, *Amphimedon* sp. (*193*).

From the methanolic extract of sun-dried specimens of the sponge *Petrosia seriata* BRAEKMAN *et al.* (*194*) isolated a novel type of bisquinolizidine alkaloid, the ichthyotoxic petrosin (**312**), and determined its structure by X-ray diffraction. **312**, which contains a C-16 macrocycle, was toxic to the fish *Lebistes reticulatus* (LD = 50 mg/L). Later, the authors described two additional ichthyotoxic compounds from the same species which were stereoisomers of **312**, petrosin-A (**313**) and petrosin-B (**314**), (*195*). Xestospongin A, B, C, and D (**315–318**) from *Xestospongia exigua* are also a new class of macrocycles incorporating two 1-oxaquinolizidine rings (*196*). They were described as vasodilative compounds inducing relaxation of blood vessel *in vivo*.

A sponge belonging to the genus *Reniera* furnished a series of antimicrobial isoquinolines and an isoindole derivative (*197*), renierone (**319**), mimosamycin (**320**), N-formyl-1,2-dihydrorenierone (**321**), 0-demethylrenierone (**322**), 1,6-dimethyl-7-methoxy-5,8-dihydroisoquinoline-5,8-dione (**323**), 2,5-dimethyl-6-methoxy-4,7-dihydroisoindole-4,7-dione (**324**), and renieramycins A–D (**325–328**). Mimosamycin (**320**) was originally isolated from *Streptomyces lavendulae*. Compounds **319** and **321**, both inhibited cell division in the fertilized sea urchin egg assay.

(**295**) $R = (CH_2)_n-CH_3$, $n = 14, 15, 16, 18$
(**296**) $R = (CH_2)_5-CH=CH-(CH_2)_5-CN$
(**297**) $R = (CH_2)_5-CH=CH-(CH_2)_{15}-CH(OH)-CN$

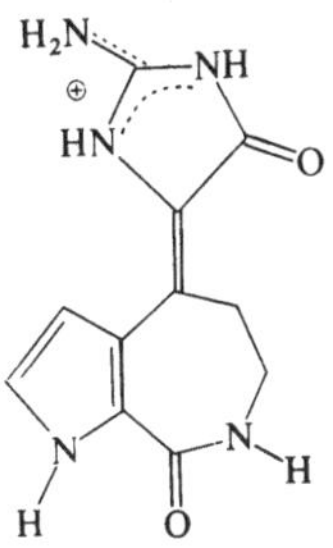

(300)

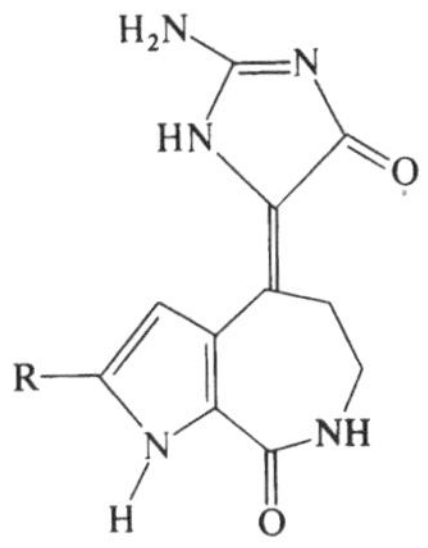

(301) R = Br
(302) R = H

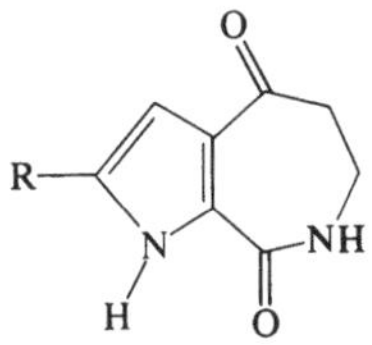

(303) R = H
(304) R = Br

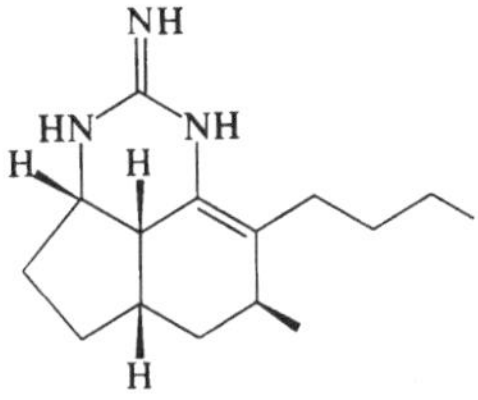

(305)

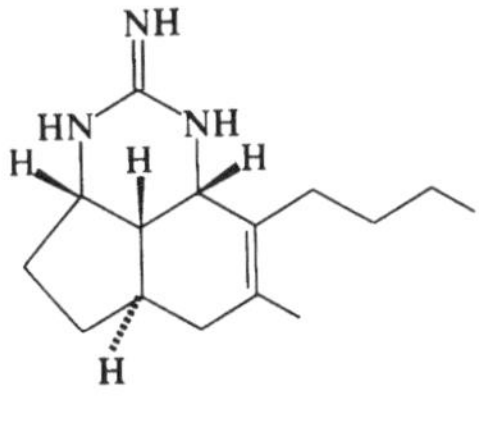

(306)

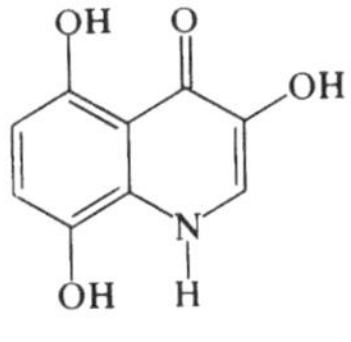

(307)

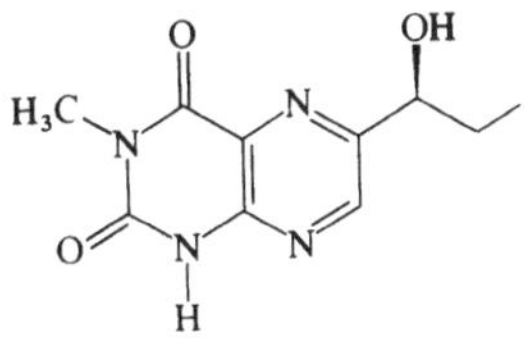

(308)

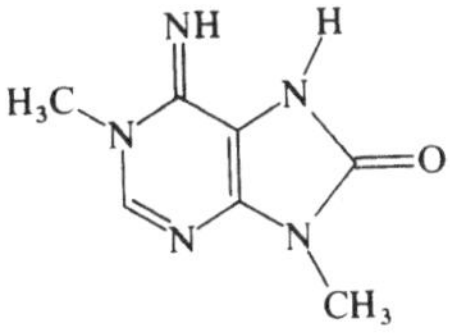

(309)

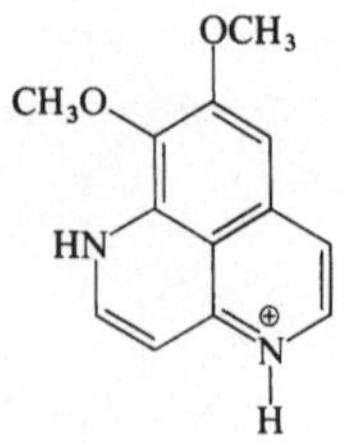

(310)

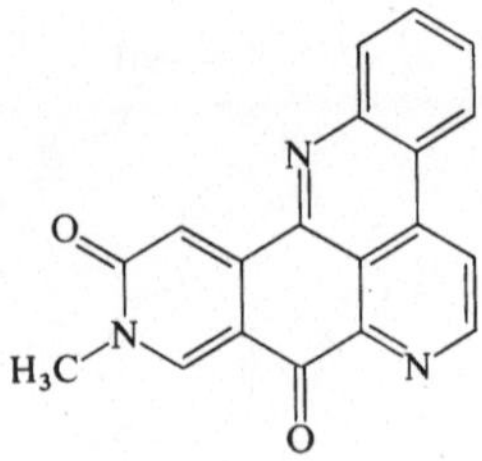

(311)

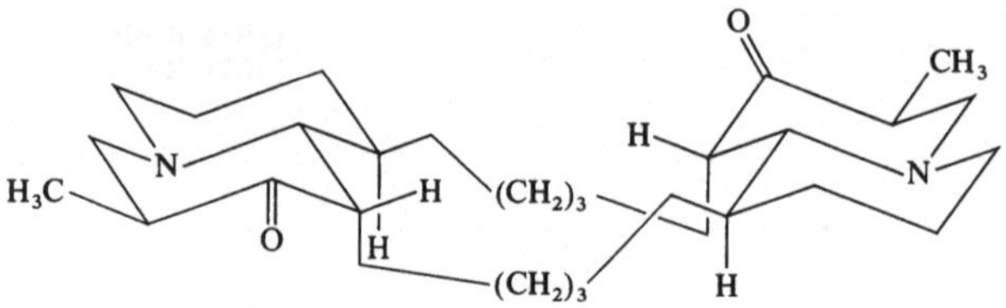

(312)

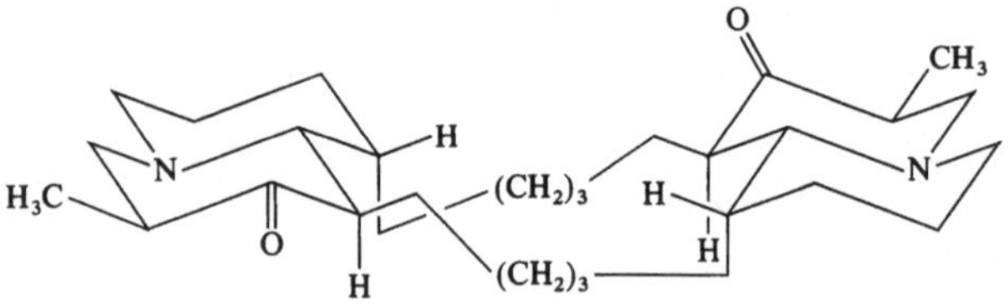

(313)

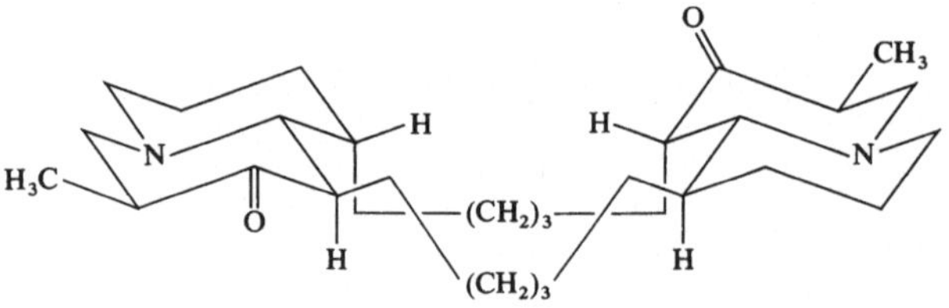

(314)

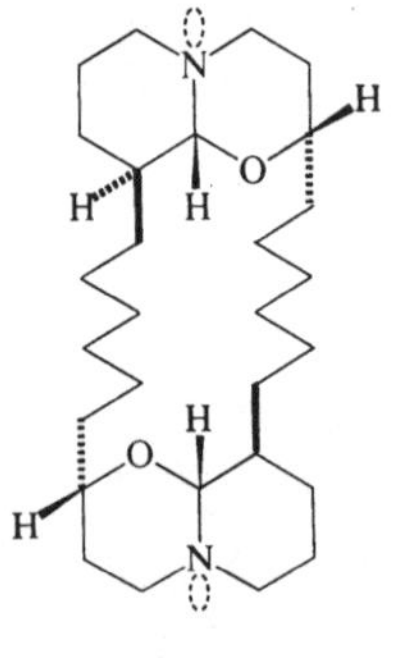

(315)

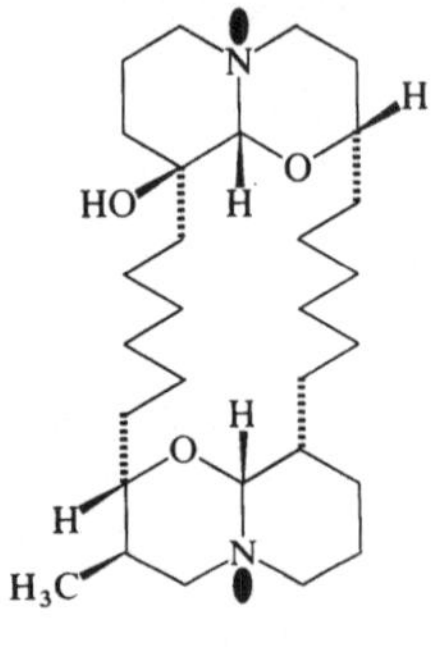

(316)

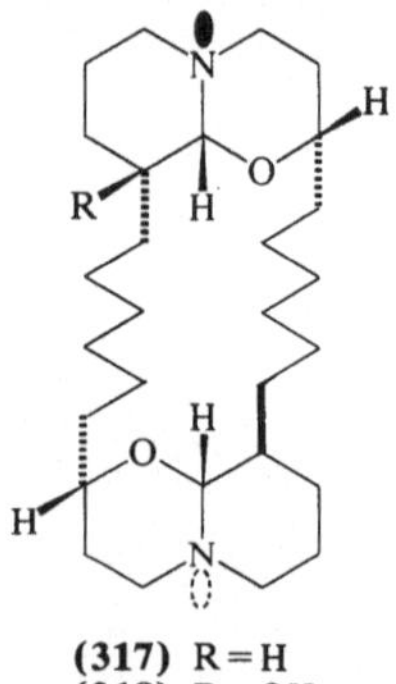

(317) R=H
(318) R=OH

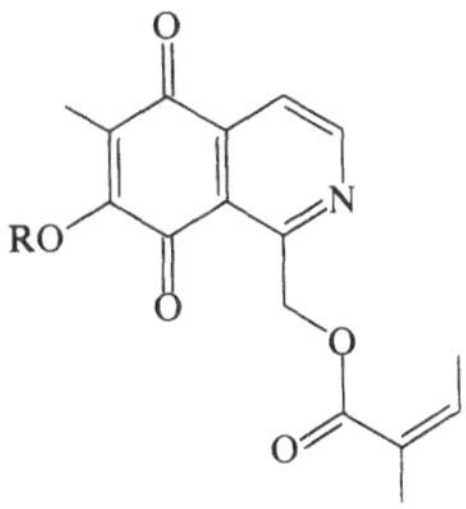

(319) R=CH₃
(322) R=H

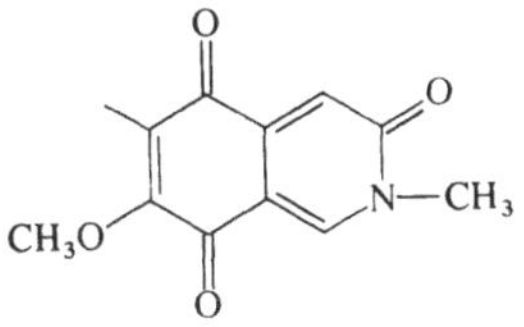

(320)

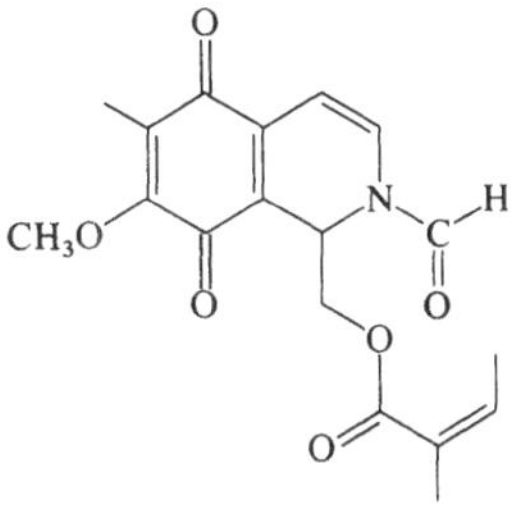

(321)

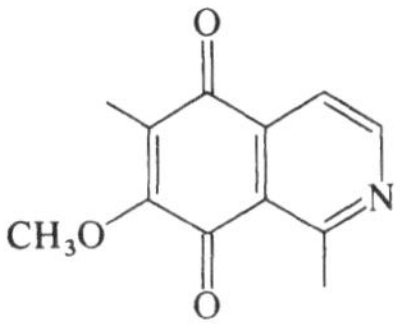

(323)

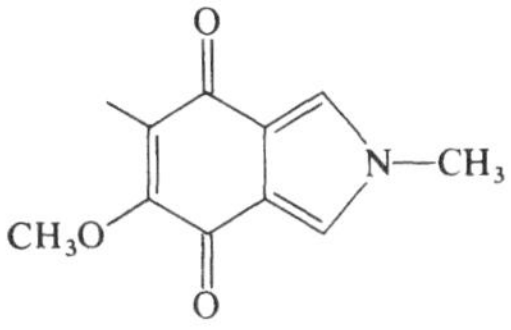

(324)

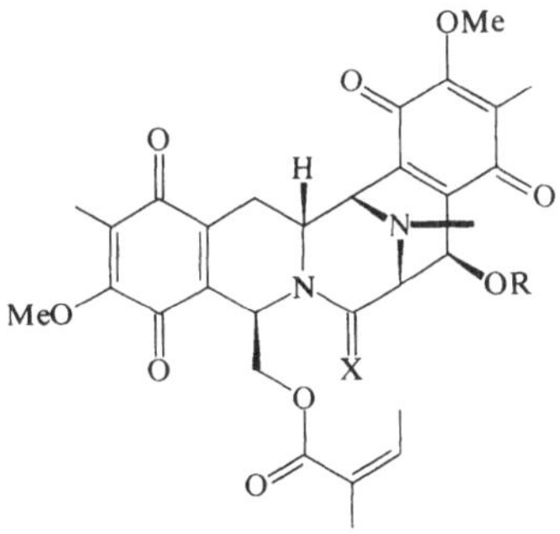

(325) R=H   X=H₂
(326) R=Et X=H₂
(327) R=H   X=O
(328) R=Et X=O

II.7. Macrolides from Porifera

Seven macrocyclic alkaloids have already been mentioned in the last chapter. There are also reports on other types of macrocycles, all of them showing biological activity. Most attention has been devoted to the toxic latrunculins A to D (**329–332**) from *Latrunculia magnifica* (*198, 199, 199a*). This sponge is never eaten by fish and when squeezed into an aquarium causes death of the fish within 4–6 minutes. The toxic component caused excitation of the fish in seconds, followed by hemorrhage, loss of balance, and, finally, after a few minutes, death. Extraction of the freeze-dried specimen with petroleum ether furnished the latrunculins (**329–332**) which are members of a new class of 14- and 16-membered macrocycles to which the rare 2-thiazolidinone moiety is attached. SPECTOR *et al.* (*200*) have evaluated the effects of these toxins on cultured mouse neuroblastoma and fibroblast cells. In both types of cells, toxin concentrations as low as 50 ng/mL rapidly induced striking changes in cell morphology that were reversible upon removal of the toxin. Immunofluorescence studies with antibodies specific for cytoskeletal proteins revealed that the toxins caused major alterations in the organization of microfilaments without obvious effects on the organization of the microtubular system.

More recently, the same group (*201*) isolated from *Theonella swinhoei* a novel 22-membered macrolide with antifungal activity, swinholide-A (**333**), whose structure was established after derivatization to the tetraformate on the basis of spectral data using 2D-NMR correlations. Extracts of *Tedania ignis* which showed cytotoxicity and *in vivo* tumor inhibition furnished a potent cytotoxic macrolide designated tedanolide (**334**) (*202*). Because of the extremely low yield of **334** it was suspected that it may be a metabolite of some microorganisms associated with the sponge.

Recently UEMURA *et al.* (*202a*) isolated a new type macrolide with antitumor activity, *nor*halichondrin A (**335**), from *Halichondria okadai*. **335** was the major compound in a series of halichondrins, the most potent antitumor constituent being halichondrin B, the structure of which is not yet published.

(**329**)   (**330**)

References, pp. 320–363

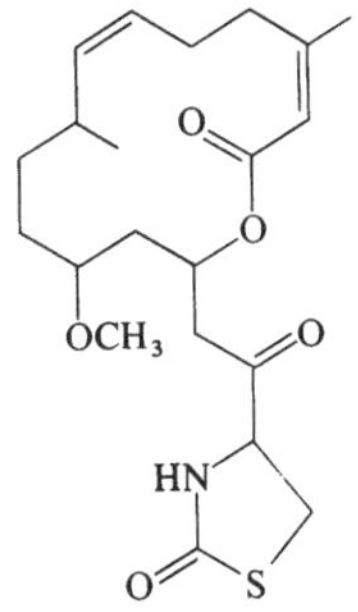

(331)

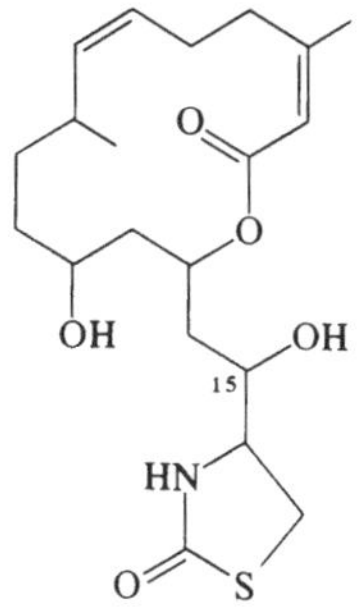

(332)

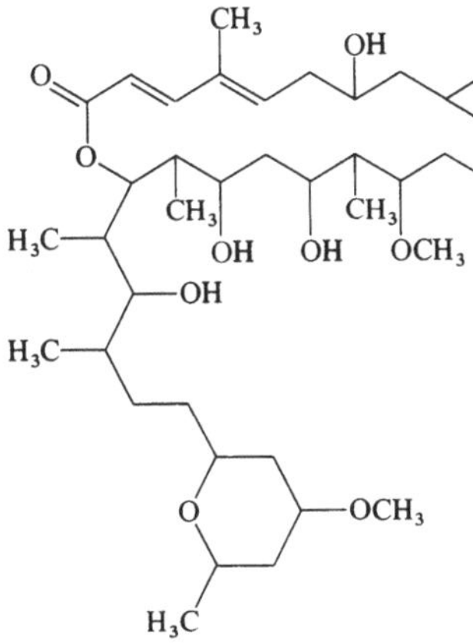

(333)

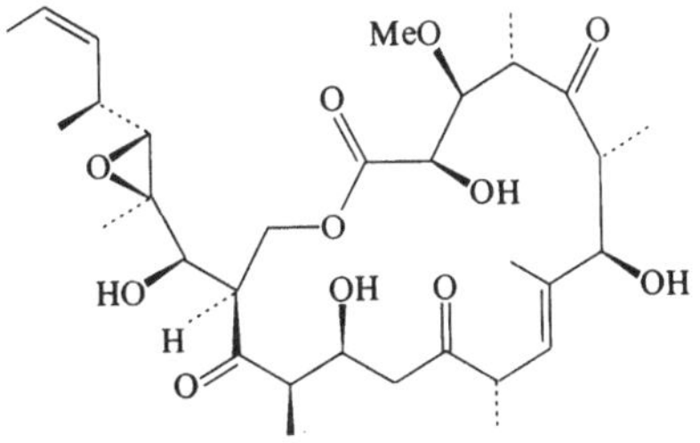

(334)

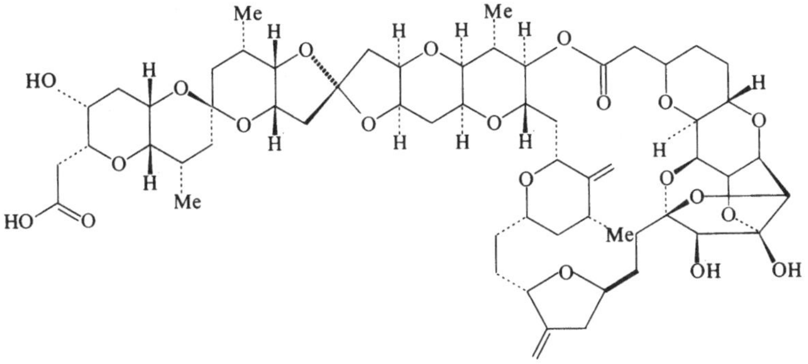

(335)

II.8. Phenols and Aromatic Ethers from Porifera

Three groups have simultaneously investigated halogenated diphenyl ethers from different species of marine sponges (*203–206*), finding eleven compounds (**336–346**) from *Dysidea herbacea*, *D. chlorea*, *Phyllospongia foliascens* and an unidentified species, supposedly a new genus of the family Callyspongiidae (Order Haplosclerida). Although this type of metabolite from Porifera has been known since the 1960's, these compounds were deemed worthy of further investigation because they showed potent antimicrobial activity against gram negative and gram positive bacteria. CARTÉ and FAULKNER (*203*) believe that the isolation of so many different metabolites from *D. herbacea* is in some way due to the presence of symbionts, presumably blue-green algae, in the ectosome of the sponge. NORTON et al. (*204, 205*) described in detail the use of ^{13}C-spin lattice relaxation measurements of quarternary carbons for determination of the structures.

Extracts of *Axinella polycapella* contained 1,2,4-trihydroxybenzene (**347**) and 2,2′,4,4′,5,5′-hexahydroxybiphenyl (**348**) (*207*) as antimicrobial constituents. While **348** could arise through oxidative dimerization of **347**, the authors think that the biphenyl **348** occurs in the living sponge and is not formed *in vitro* after collection.

	R^1	R^2	R^3	R^4	R^5	R^6	R^7	R^8	R^9	isolated from	Lit.
(336)	Br	H	Br	H	Br	Br	Br	H	OH	*D. herbacea*	(203)
(337)	Br	H	Br	H	H	Br	Br	Br	OH	*D. herbacea*	(203)
(338)	Br	H	Br	H	H	Br	H	Br	OH	*D. chlorea*	(203)
(339)	Br	H	Br	H	Br	H	Br	H	OH	*D. herbacea*	(203)
(339)										unidentified	(206)
(340)	OCH$_3$	Br	H	Br	Br	H	Br	H	OCH$_3$	*P. foliascens*	(203)
(340)										*D. herbacea*	(205)
(341)	OH	Br	H	Br	Br	H	Br	Br	OH	*P. foliascens*	(203)
(341)										*D. herbacea*	(204)
(342)	OH	Br	H	Br	Br	Br	Br	Br	OH	*P. foliascens*	(203)
(342)										*D. herbacea*	(204)
(343)	OH	Br	H	Br	Br	H	Br	Br	OCH$_3$	*D. herbacea*	(204)
(344)	OCH$_3$	Br	H	Br	Br	Br	Br	H	OCH$_3$	*D. herbacea*	(204)
(345)	OCH$_3$	Br	H	Br	Br	H	Br	H	OH	*D. herbacea*	(204)
(346)	Br	H	Br	H	Br	Cl	Br	H	OH	unidentified	(206)

(347)   (348)

II.9. Carboxylic Acids from Porifera

Saturated, unsaturated, α-methoxy- and methyl branched fatty acids as well as phospholipid-bound fatty acids from various sponges have been investigated (208–217) and two new brominated acids from phospholipids have been identified (218). The partly unique fatty acids had an unconventional pattern of unsaturation and different substituents superimposed upon a very long hydrocarbon chain. Recently QUINN and TUCKER (219) reported a brominated bisacetylenic acid from *Xestospongia testudinaria*.

Some fatty acids derived metabolites from Porifera have exhibited significant biological activity. Acanthifolicin (**349**) from *Pandaros acanthifolium* was a highly cytotoxic compound which exhibited ED_{50}'s of 2.8×10^{-4}, 2.1×10^{-3}, and 3.9×10^{-3} mcg/mL, respectively, against P 388, KB, and L 1210 cell lines (220). X-ray analysis established that **349** was a novel episulfide-containing member of the polyether antibiotic class of compounds which had been isolated previously only from bacteria. The presence of two fused tetrahydropyran rings was also unique. A similar polyether derivative of a C_{38} fatty acid which did not contain the episulfide group was okadaic acid (**350**) from two sponges, *Halichondria* (syn. *Reniera*) *okadai* and *H. melanodocia* (221). Okadaic acid (**350**) was toxic at doses of less than 0.12 mg/kg (ip. mice) and inhibited growth of KB cells. Furthermore, it exhibited ED_{50} values of 1.7×10^{-3} and 1.7×10^{-2}, respectively, against P 388 and L 1210 cell lines. It was presumed that **350** is a metabolite of an epiphytic microorganism rather than of the sponge. SHIBATA et al. (222) stated that **350** caused a long lasting tonic contraction in human umbilical arteries. They observed that rabbit aortae and guinea pig taenia cecum exhibited a similar contractile response to **350** and reported that the contractile action of **350** was not mediated through a neuronal mechanism.

STIERLE and FAULKNER (223) identified a number of cyclic peroxides and related metabolites (**351–359**) in various samples of *Plakortis hali-*

chondrioides. While the biological activity of these nine compounds was not detailed, a more recent report by PHILLIPSON and RINEHART (*224*) described the isolation of plakinic acid A (**360**) and B (**361**), two new antifungal peroxy acids related to **359**, from a previously undescribed and unnamed genus of the family Plakinidae. *Plakortis zyggompha*, whose extracts were active against *Saccharomyces cerevisiae* and *Penicillium atrovenetum*, gave the active compound, placortic acid (**362**). Plakortin, the methyl ester of **362**, which had been reported earlier as a constituent from the same sponge, was bioinactive.

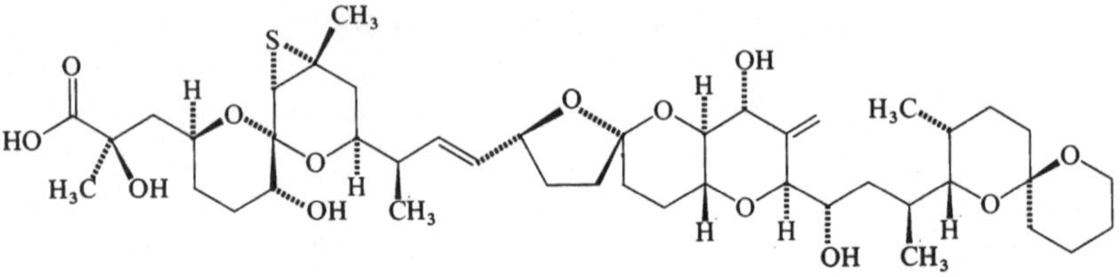

(**349**)

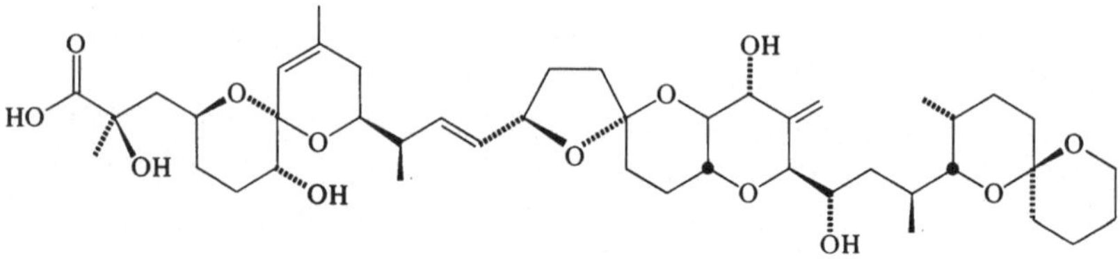

(**350**)

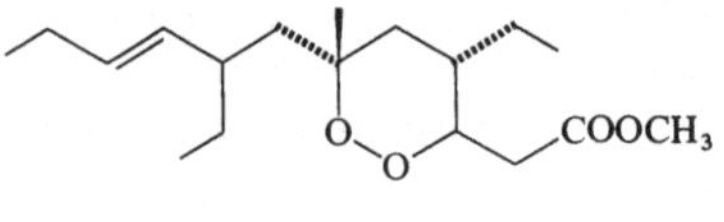

(**351**)

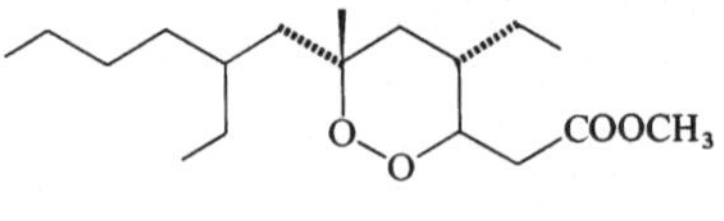

(**352**)

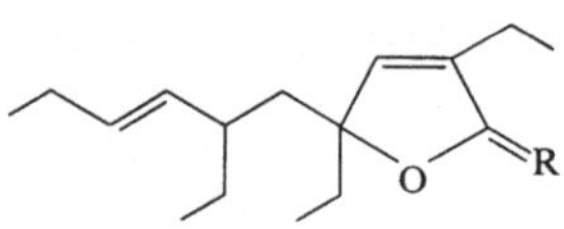

(**353**) R=CH—COOCH$_3$
(**354**) R=O

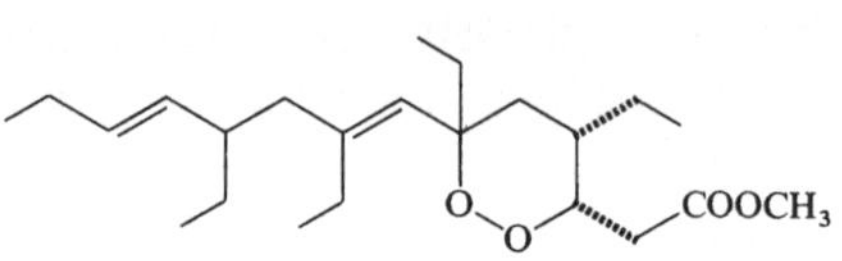

(**355**)

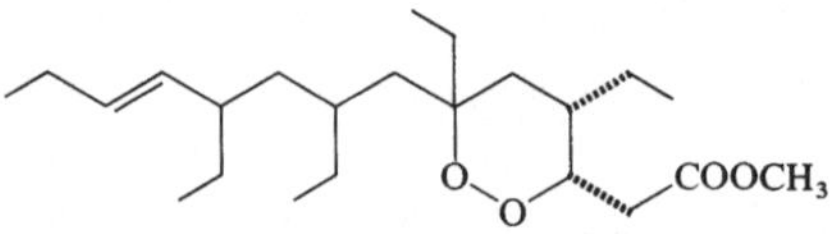

(**356**)

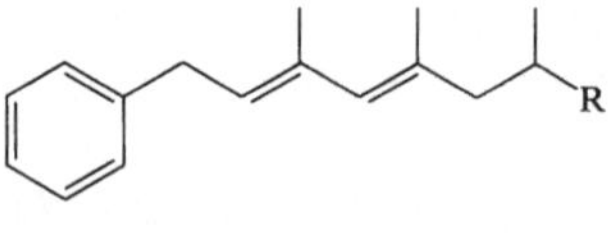

(**357**) R=CH$_3$

References, pp. 320–363

(358) 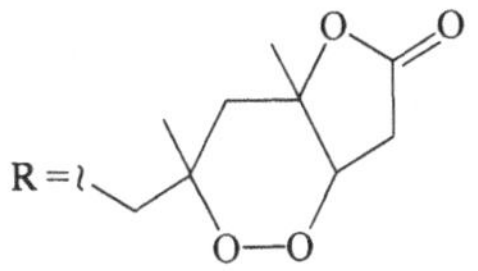 R = ⌊ [structure]

(359) 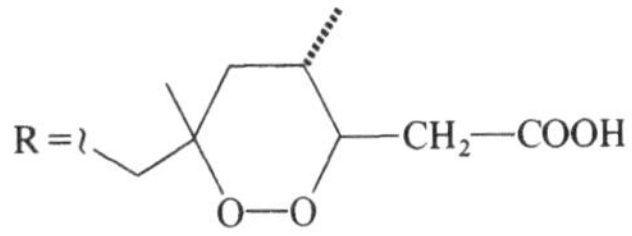 R = ⌊ [structure] —CH$_2$—COOH

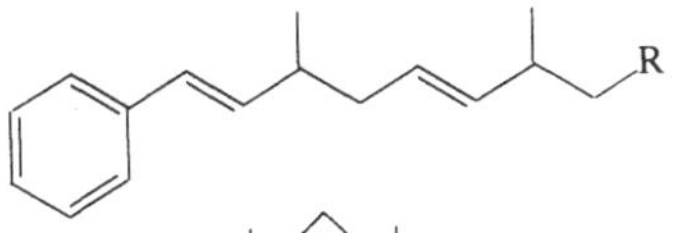

(360) 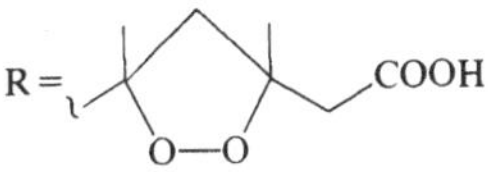 R = ⌊ [structure with COOH]

(361) 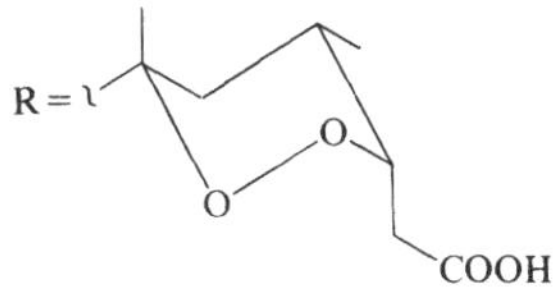 R = ⌊ [structure with COOH]

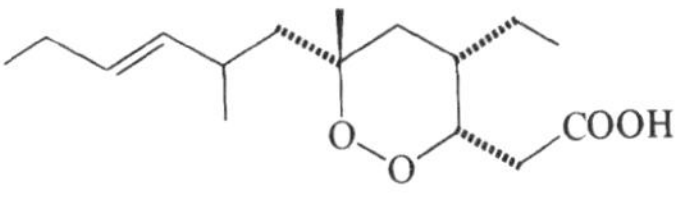

(362) [structure with COOH]

II.10. Miscellaneous Other Compounds from Porifera

Several high molecular weight polyacetylenes have been isolated from sponges. CIMINO et al. (*217*) compared the composition of these metabolites (**363, 364**) from *Petrosia ficiformis* found in dark caves with that from the same species living in its usual habitat. The first mentioned animals lack symbiotic algae and therefore appeared white. Reinvestigation of the more polar metabolites from the etheral extract of *P. ficiformis* furnished **365** and a methyl ester (**366**) (*225*). Experiments to investigate feeding inhibition had not evidenced defensive activity. A new C_{30} linear polyacetylene alcohol (**367**) from a *Tetrosia* species inhibited mitotic cell division in sea urchin eggs and was active against *Penicillium chrysogenum* (*226*). CASTIELLO et al. (*227*) investigated the chemistry of predator-prey relationship and found the same unusual high molecular weight polyacetylenes (**368, 369**) in both invertebrates, the sponge *Petrosia ficiformis* and the nudibranch *Peltodoris atromaculata*.

Glycerol ethers are widely distributed in sponges. 17 Z-Tetracosenyl glycerol-1-yl ether (**370**) has been identified as a major constituent of the dichloromethane-soluble extracts of *Cinachyra alloclada* and *Ulosa ruetzleri* (*228*); it is speculated that the glycerol ethers of sponge tissues may function as growth factors. Examination of an undescribed specimen of the family Plocamiidae had revealed an interesting glyceride, (S)-(+)-1-tridecoxy-2,3-propanediol (**371**), which displayed toxicity to goldfish, leading to death in 70 min at 0.29 mg/mL (*229*). Do and ERICKSON (*230*) isolated two branched chain monoglycerol ethers (**372, 373**) as an inseparable mixture from the alcoholic extracts of a yellow sponge of the genus *Aaptos*. The structure of **372** was confirmed by synthesis.

Halenaquinone (**374**) was a pentacyclic polyketide from *Xestospongia exigua* which possessed *in vitro* antibiotic activity against *Staphylococcus aureus* and *Bacillus subtilis* (*231*). Halenaquinol (**375**) which was readily oxidized either by UV-irradiation or by heating (even at 40° C) in the air to **374** and its sulfate (**376**) could be isolated from the Hawaiian sponge *Xestospongia sapra* (*231a*). CAPON et al. (*232*) obtained two new brominated tetrahydropyrans (**377, 378**) from a *Haliclona* species. Their bioactivity was not described, but the biosynthetic origin has been discussed. HAGADONE et al. (*233*) investigated the origin of the isocyano function in marine sponges of the genus *Hymeniacidon*, using labeled precursors synthesized from 2-isocyanopupukeanane (**379**) which had been described prior to 1980 as a metabolite of the same species. Neither formamide **380** or isothiocyanate **381** was transformed into labeled 2-isocyanopupukeanane (**382**), but, it could be demonstrated unequivocally that the isocyano function is the precursor of the formamide and the isothiocyanate in the sponge. The marine sponge *Plakortis zygompha* which has a sweet odor furnished (Z)-7-methyl-4-octen-3-one (**383**) in addition to several derivatives of 3-hydroxy-4-hydroxymethyl-4-pentenoic acid (**384–387**) and the alcohol **388** (*234*).

$$HC\equiv C-CH(OH)-CH\overset{t}{=}CH-CH_2-R^1-CH_2-CH\overset{t}{=}CH-C\equiv C-CH(OH)-C\equiv C-CH_2-R^2-CH_2-CH\overset{c}{=}CH-C\equiv CH$$

(**363**) $R^1+R^2=C_nH_{2n-4}$, n=26, 29

$$HC\equiv C-CH(OH)-CH\overset{t}{=}CH-CH_2-R^1-CH_2-CH\overset{t}{=}CH-C\equiv C-CH(OH)-C\equiv C-CH_2-R^2-CH_2-C\equiv CH$$

(**364**) $R^1+R^2=C_nH_{2n-4}$, n=28, 31, 34

$$HC\equiv C-CH(OH)-CH=CH-CH(OH)-CH_2-R^1-CH_2-CH=CH-C\equiv C-CH(OH)-$$
$$-C\equiv C-CH_2-R^2-CH_2-CH=CH-CH(OH)-C\equiv CH$$

(**365**) $R^1+R^2=C_{24}H_{44}$

$$HC\equiv C-CH=CH-CH_2-R^1-CH_2-CH=CH-CH_2-CH=CH-CH_2-R^2-CH_2-CH=CH-C\equiv C-COOCH_3$$

(**366**) $R^1+R^2=C_{25}H_{48}$

References, pp. 320–363

HC≡C-CH(OH)-CH=CH-(CH₂)₆-C≡C-CH(OH)-CH=CH-CH(OH)-
                    -C≡C-(CH₂)₆-CH=CH-CH(OH)-C≡CH

(367)

HC≡C-CH(OH)-CH=CH-CH₂-R¹-CH₂-CH=CH-C≡C-CH(OH)-
                    -C≡C-CH₂-R²-CH₂-CH=CH-CH(OH)-C≡CH

(368) $R^1+R^2=C_nH_{2n-6}$, n = 25, 28
(369) $R^1+R^2=C_nH_{2n-4}$, n = 28, 31, 34

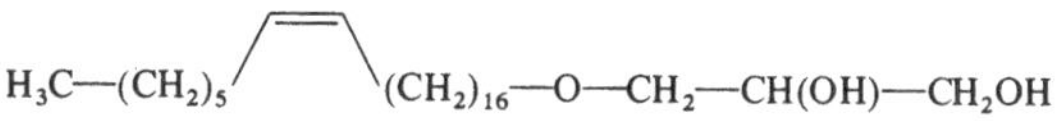

H₃C-(CH₂)₅-CH=CH-(CH₂)₁₆-O-CH₂-CH(OH)-CH₂OH

(370)

H₃C-(CH₂)₁₂-O-CH₂-CH(OH)-CH₂OH

(371)

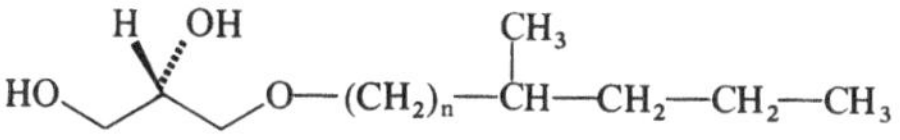

HO-CH₂-C(H)(OH)-CH₂-O-(CH₂)ₙ-CH(CH₃)-CH₂-CH₂-CH₃

(372) n = 12
(373) n = 14

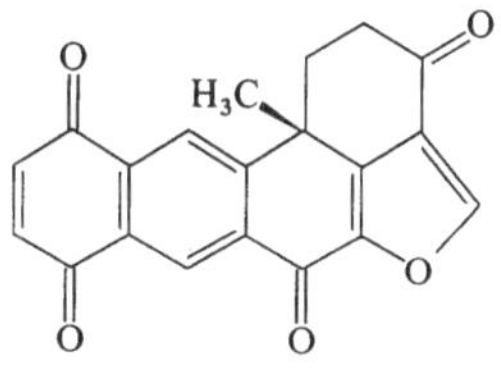

(374)

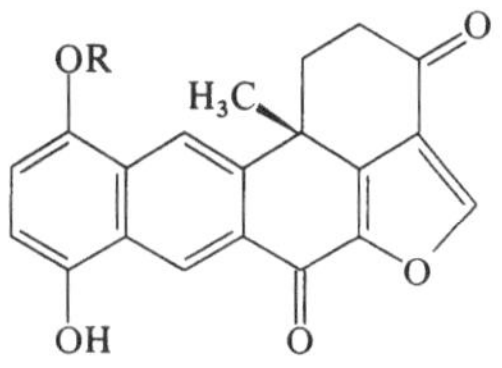

(375) R = H
(376) R = SO₃Na

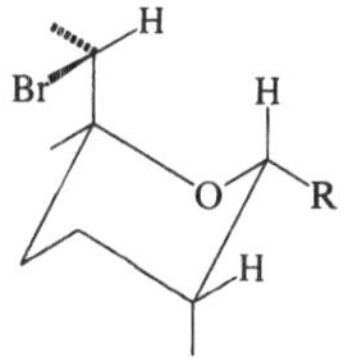

(377) R = CH=CH-CH=CH-CH₃
(378) R = -(CH₂)₃-CH=CH₂

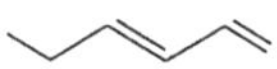

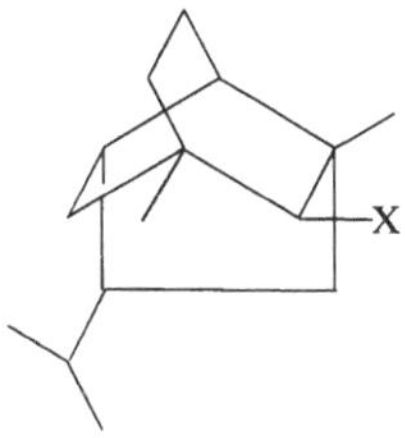

(379) X = NC
(380) X = NH—¹³CHO
(381) X = N=¹³C=S
(382) X = N⁺≡¹³C⁻

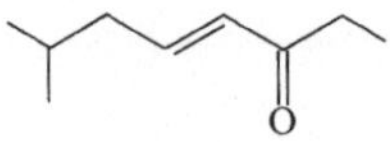

(383)

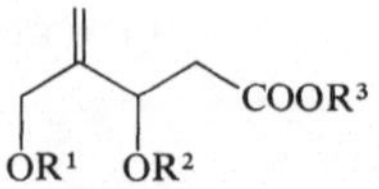

(384) R¹=Et R²=H   R³=Me
(385) R¹=H   R²=H   R³=Et
(386) R¹=Ac R²=Ac R³=Me

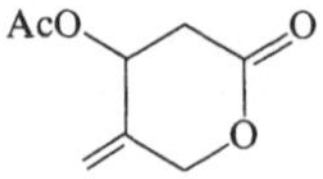

(387)

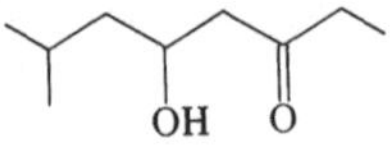

(388)

III. Coelenterata (Cnidaria)

The phylum Coelenterata is subdivided into the classes Hydrozoa, Cubozoa, Scyphozoa, and Anthozoa. The latter class is subdivided into Hexa- and Octocorallia. Many coelenterates produce toxic or other biologically active metabolites (235). For example toxins are used by jellyfish and sea anemones for defence or in order to capture prey.

III.1. Hydrozoa, Cubozoa, and Scyphozoa

This chapter deals mainly with three kinds of jellyfish, the sea nettle (*Chrysaora quinquecirrha*, Scyphozoa), the Portuguese man of war (*Physalia physalis*, Hydrozoa), and the sea wasp (*Chironex fleckeri*, Cubozoa, formerly Scyphozoa). *C. quinquecirrha* is widely distributed in warm and tropic waters. Contact with its tentacles results in painful injuries mediated through the release of a nematocyst venom. The venom of this species is a complex mixture of injurious enzymes and pain producing factors. The nematocyst venom of *P. physalis* also contains a number of toxic proteins and injurious enzymes which exhibit multible actions including mouse lethality, dermonecrosis, neurotoxicity, hemolysis, and cardiotoxicity.

While the chemical structure of none of these high molecular weight substances has been determined, there are some reports dealing with the partial purification and characterization of enzymes from *P. physalis* (236, 237) and *C. quinquecirrha* (238–241) as well as of a hemolytic and toxic protein from the Portuguese man of war (*242*) and toxins from the sea nettle (*243–245*). *C. quinquecirrha* and *P. physalis* both possessed a DNase with approximate molecular weights of 110,000 (*238*) and 75,000 (*236*) respectively, and a collagenase (*239, 237*). The sea nettle nematocyst venom has been shown to have an alkaline (*241*) and an acid (*240*) protease. HARTMAN et al. (*246*) detected a kinin-like factor in sea nettle nematocyst venom using a standard bradykinin radioimmuno assay procedure and compared preparations from the fishing and mesenteric tentacles of *C. quinquecirrha* by means of various toxicological properties (*247*). The same group developed a micro enzyme-linked immunosorbent assay to detect serum antibodies, obtained from guinea-pig, to sea nettle venom (*248*). Monoclonal antibodies against the Portuguese man of war nematocyst lethal factors have also been developed (*249*). The use of these antibodies has allowed partial purification and characterization of lethal factors in the sea nettle fishing and mesenteric tentacles (*245, 250*) as proteins. Recently OLSON et al. (*251*) have studied the cardiotoxicity and the polypeptide content of sea nettle polyps and cysts. The polyp cardiotoxin factor was purified by immunosorbent chromatography.

The nematocyst venom of the sea nettle damages many cellular and subcellular tissue preparations, including nerve and muscle of the rat and frog (*252*), *Aplysia* neurons (*252*), and cultured Chinese hamster ovary K-1 cells (*253, 254*). SHRYOCK and BIANCHI (*255*) studied the cellular mechanism by which this venom injures cardiac and skeletal muscle cells. A lethal toxin from the fishing tentacles of *C. quinquecirrha* affected ion permeability in black lipid membranes by producing monovalent cation channels (*256*). BURNETT et al. (*257*) have undertaken biochemical and pharmacological studies of jellyfish (*C. quinquecirrha* and *P. physalis*) venoms aimed at eventual treatment of envenomations and identification of sensitized patients. FLOWERS and HESSINGER (*258*) described the mast cell histamine release induced by Portuguese man of war venom. The authors concluded that *Physalia* nematocyst venom induces histamine release through a cytolytic mechanism and that this action is antagonized by an intracellular energy requiring process. The role of the jellyfish and its toxicity to humans is reviewed by BEHRENS-BAUMANN (*259*). The local and systemic effect after contact of skin and conjunctiva not only with *P. physalis* and *C. quinquecirrha* but also with *Cyanea capillata* has been reported.

The box jellyfish, *Chironex fleckeri*, has been responsible for numerous fatalities or severe injuries. ENDEAN and RIFKIN (*260*) found that toxic material released from the isolated nematocysts of *C. fleckeri* rapidly produced marked systemic effects in mice when injected intravenously, but no such effects after subcutaneous or intraperitoneal injection. Subsequently, a cardiotoxin from the sea wasp has been purified by immunochromatography on an immobilized mouse monoclonal anti-Portuguese man of war venom antibody column (*261*). The toxin which had a molecular weight of 20,000 caused bradycardia followed by cell lysis.

CIMINO et al. (*262*) found cholest-4-en-4,16β,18,22 R-tetrol-3-one 16,18-diacetate (**389**) in hydroids belonging to the genus *Eudendrium* as well as in its predators, several species of nudibranchs. Another three novel, C-18 oxygenated sterols (**390a–c**) have been isolated in very small amounts from *Eudendrium glomeratum* (*262a,b*). The ether soluble fractions of four species of hydroids, *Aglaophenia pluma*, *Halocordyle disticha*, *Eudendrium glomeratum*, and *Sertularella crassicaulis* contained four acyclic polyhalogenated monoterpenes (**391–394**) (*263*). Compounds **391–393** had been isolated previously from the red alga *Plocamium cartilagineum* and **394** from the digestive gland of the sea hare *Aplysia limacina*. It was presumed that the hydroids utilize these compounds as defensive agents. Garveatin A (**395**) was an antimicrobial 1(4H)-anthracenone derivative from the hydroid *Garveia annulata* (*264*). The structure of **395** was inferred from its spectral data and confirmed by X-ray analysis of an acetylated derivative.

(393)   (394)   (395)

III.2. Hexacorallia: Sea Anemones

The order Actiniaria (sea anemones) of Hexacorallia comprises solitary organisms which lack a skeleton and live in all oceans in the intertidal zone and at depths of more than 10,000 m. Many species live in symbiosis with fishes; thus the clown fish *Amphiprion clarkii* is able to live unharmed amongst the tentacles of the sea anemone *Stichodactyla haddoni* even though the latter has a powerful stinging response (*265*) and is capable of capturing any nonsymbiotic fish that enters its tentacles. *A. clarkii*, however, has managed to produce a thick mucus layer which is apparently not recognized as foreign object by *S. haddoni*. MARETIC and RUSSELL (*266*) reported that stings by sea anemones, particularly *Anemonia sulcata*, have increased manyfold and have become a public health hazard along the coast of Istria because of the recent development of tourism.

Almost all coelenterate toxins are polypeptides or proteins whose amino acid sequences exhibit a high degree of homology. Those of five sea anemone toxins, already known prior to 1980, are ATX I, II, III from *Anemonia sulcata*, APA from *Anthopleura xanthogrammica*, and APC from *A. elegantissima*. A further toxin, ATX IV, has been described as a degradation product of ATX III which lacked the last two amino acids of the latter (*235*). Only two more primary structures have been determined since then, those of toxin V (**396**) from *A. sulcata* (*267*) and cytolysin III (**397**) from *Stoichactis helianthus* (*268*). While toxin V (**396**) with 46 amino acid residues was closely related to ATX I and II, **397** contained 153 amino acid residues. The isolation and primary structure of metridin (**398**), a hemolytic active polypeptide with 36 amino acid residues from *Metridium senile*, has also been reported (*269*). Other toxic polypeptides and proteinase inhibitors have been isolated from *Stoichactis* sp. (*270–273*), *Anthopleura xanthogrammica* (*272–274*), *Actinodendron plumosum* (*272, 273*), *Tealia felina* (*275*), *Condylactis gigantea* (*276*), *Actinia cari* (*277*), *Epiactis prolifera* (*278*), *Actinia equina* (*279*), *Radianthus macrodactylus* (*280*), *R. koseirensis* (*281, 282*), *Gyrostoma helianthus* (*281, 282*), *Rhodactis rhodostoma* (*281, 282*),

Bolocera tuediae (*283*), and *Pseudactinia varia* (*284*). The amino acid composition of some of these toxins have been determined.

BERNHEIMER and AVIGAD (*285*) examined seven kinds of sea anemones for the presence of cytolytic toxins. *Tealia lofotensis* and *Epiactis prolifera* yielded especially powerful preparations, whereas the extracts of the other species showed weaker cytolytic activity. Natural-abundance ^{13}C NMR studies have been performed on toxic polypeptides. These studies of APA and ATX II indicated that their overall conformation was very similar (*286*).

Sea anemone toxins have become very useful tools for studying the voltage-dependent Na^+ channel in nerve, cardiac and muscle cells and the pharmacology as well as the physiological activity of these compounds have been extensively examined (*287–319*). BARHANIN et al. (*320, 321*) established the structure-function relationship of toxin II from *A. sulcata*. SEDMAK et al. (*322*), after observing that cytolysin from the sea anemone *Condylactis aurantiaca* has a significant effect on the survival of mice inoculated with Ehrlich ascites tumor, concluded that coelenterate toxins would be suitable for studies of tumor cell cytolysis *in vitro* and *in vivo*. The isolation of a phosphonomonoesterase which hydrolyzes 4-nitrophenyl(phenyl)phosphonic acid from *Metridium senile* has been described by BURCHAM et al. (*323*).

The sea anemone *Metridium senile* furnished eight known epidioxy sterols (**3–5, 9, 10, 12, 13, 15**) in addition to a new compound (**399**) (*323a*). MILKOVA et al. (*324, 325*) have investigated the sterol composition of Black Sea invertebrates. They examined the sea anemone *Actinia equina* as well as two species of jellyfish.

396 Gly-Val-Pro-Cys-Leu-Cys-Asp-Ser-Asp-Gly-Pro-Ser-Val-Arg-Gly-Asn-Thr-Leu-Ser-Gly-Ile-Leu-Trp-Leu-Ala-Gly-Cys-Pro-Ser-Gly-Trp-His-Asn-Cys-Lys-Lys-His-Lys-$^{Pro}_{Gly}$-Thr-Ile-Gly-Trp-Cys-Cys-Lys

397 Ala-Leu-Ala-Gly-Thr-Ile-Ile-Ala-Gly-Ala-Ser-Leu-Thr-Phe-Gln-Val-Leu-Asp-Lys-Val-Leu-Glu-Glu-Leu-Gly-Lys-Val-Ser-Arg-Ser-Gly-Thr-Thr-Asp-Val-Ile-Leu-Pro-Glu-Phe-Val-Pro-Asn-Thr-Lys-Ala-Leu-Leu-Tyr-Ser-Gly-Arg-Lys-Asp-Thr-Gly-Pro-Val-Ala-Thr-Gly-Ala-Val-Ala-Ala-Phe-Gln-Tyr-Tyr-Met-Ser-Ser-Gly-Asn-Thr-Leu-Gly-Val-Met-Phe-Ser-Val-Pro-Phe-Asp-Tyr-Asn-Trp-Tyr-Ser-Asn-Trp-Trp-Asp-Val-Lys-Ile-Tyr-Ser-Gly-Lys-Arg-Arg-Ala-Asp-Gln-Gly-Met-Tyr-Glu-Asp-Leu-Tyr-Tyr-Gly-Asn-Pro-Tyr-Arg-Gly-Asp-Asn-Gly-His-Trp-Glu-Lys-Asn-Leu-Gly-Tyr-Gly-Leu-Arg-Met-Lys-Gly-Ile-Met-Thr-Ser-Ala-Gly-Glu-Ala-Lys-Met-Gln-Ile-Lys-Ile-Ser-Arg

References, pp. 320–363

398 Asp-Ser-Asp-Cys-Lys-Asp-Lys-Leu-Pro-Ala-Cys-Gly-Glu-Tyr-
Arg-Gly-Ser-Phe-Cys-Lys-Leu-Glu-Lys-Val-Lys-Ser-Asn-Cys-
Glu-Lys-Thr-Cys-Gly-Val-Lys-Cys

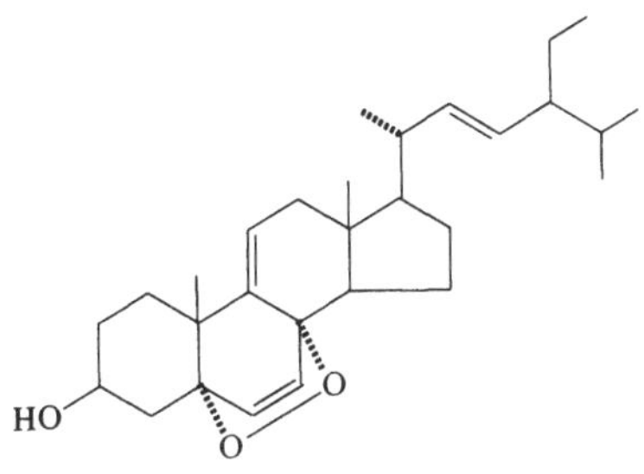

(399)

III.3. Hexacorallia: Other Organisms

Toxins isolated from several *Palythoa* spp. are the most potent known biologically active substances of marine origin. Palytoxin from *P. toxica* and *P. tuberculosa* was isolated number of years ago but its structure was elucidated only recently. The subject has been reviewed in the most recent volume of this series (*326*). The same toxin has since been isolated from a number of other *Palythoa* species, most recently from *P. caribaeorum* by BÉRESS et al. (*326a*) who developed a method for large scale purification. Because of its high molecular weight, low volatility and thermal instability the molecular weight determination of palytoxin required ^{252}Cf plasma desorption mass spectrometry (*327*). Chemical derivatization and degradations, mainly by means of periodate oxidation and ozonolysis followed by identification of the cleavage products, led to the gross structure shown in formula **400** and eventually, with help from synthetic organic chemists to the absolute stereochemistry (*328–338*). The following minor constituents of Okinawan *P. tuberculosa* have been examined recently (*339*), homopalytoxin (**401**), bishomopalytoxin (**402**), neopalytoxin (**403**), and deoxypalytoxin (**404**). Parts of the palytoxin structure have been synthesized (*333–335, 340, 341, 341a*). Palytoxin (**400**) is a potent slow acting cytolysin. Like the polypeptide toxins from sea anemones, it has become an important tool for physiological and pharmacological investigations (*342–353*).

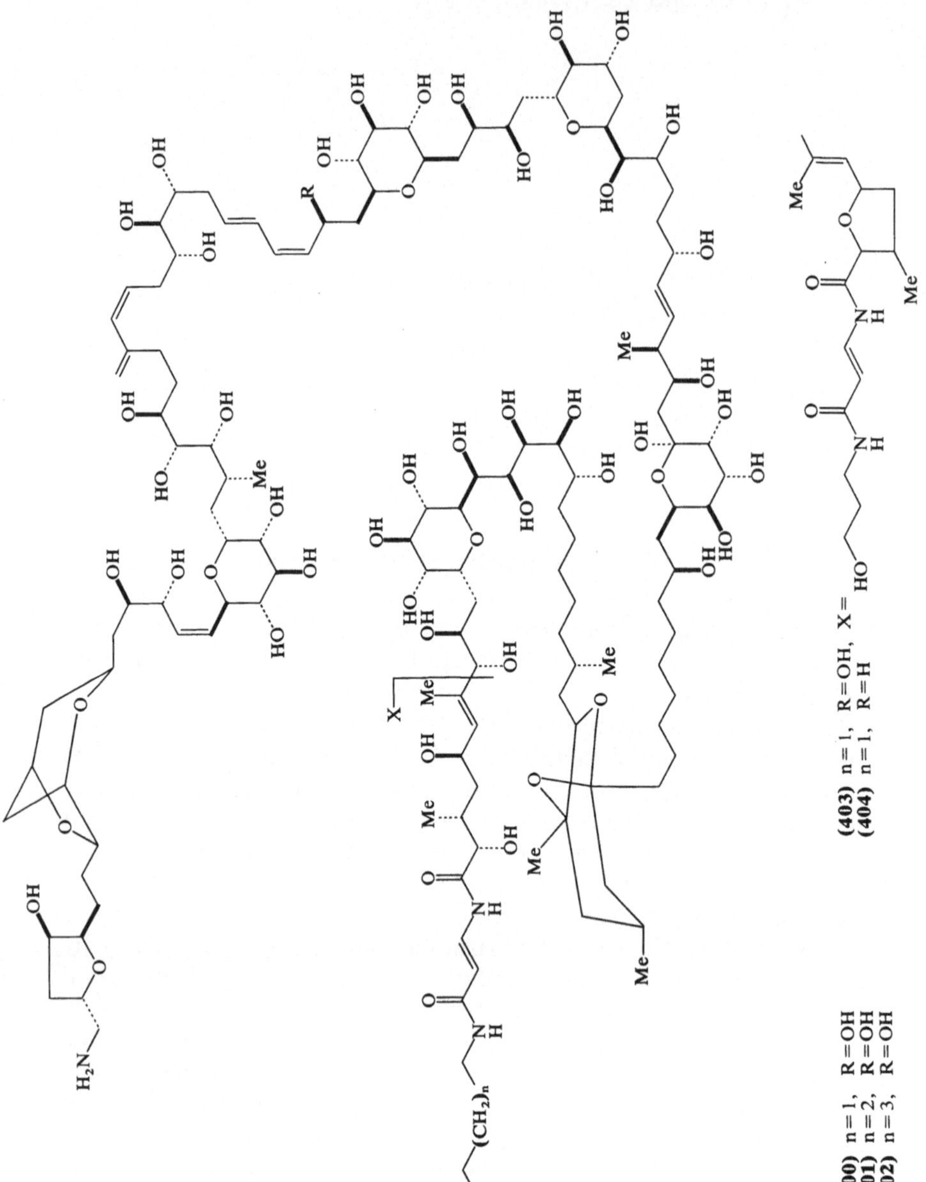

(400) n=1, R=OH
(401) n=2, R=OH
(402) n=3, R=OH

(403) n=1, R=OH, X=
(404) n=1, R=H

Palythoa liscia has furnished several antineoplastic agents. Among these were palystatins 1–3, three proteins with molecular weights ranging from 128,000 to more than 2×10^7 daltons (*354*) which were active in the P388 screen and four relatively low molecular weight (about 3,000 to 5,000) proteins designated palystatins A–D (*355*) which were active in the PS system. Amino acid and carbohydrate analyses of palystatins A–D indicated that A and B were glycopeptides and C and D related peptide conjugates. The ethanol extract of the same species contained two glycerol ethers (**405, 406**) which slightly inhibited Ehrlich ascites tumor (*356*).

Publications on steroids from zoanthids include a comparative study of zoanthid sterols from the genus *Palythoa* (*357*) and an examination of the Brazilian zoanthid *Zoanthus sociatus* (*358*), the latter describing a new marine sterol, 4α-methyl-5α-ergosta-24(28)-en-3β-ol. A further report on the occurrence of large amounts of ecdysterone (**407**) in *Gerardia savaglia* (*359*) represents a new, unexpected marine source of a molting hormone.

KOMODA et al. (*360*) isolated a biologically active and strongly fluorescent marine base, paragracine (**408**), 2-dimethylamino-6-methyl-8-methylamino-1*H*-1,3,7,9-tetrazacyclopent[e]azulene, from an anthozoan, *Parazoanthus gracilis*. Investigations of the pharmacological action of paragracine (**408**) on various autonomic effector organs revealed that **408** has papaverine-like activity and blocks sodium channels of squid axon membranes in a frequency dependent manner.

Colonies of a *Zoanthus* species from the Visakhapatnam coast of India eject jets of water when disturbed. If the spray comes in contact with a victim's eye, it causes tears followed by prolonged redness and pain. The constituents of this species have been investigated by RAO et al. (*361, 361a*); the ether-soluble material from an ethanolic extract of the animals furnished three new alkaloids, zoanthamine (**409**), zoanthenamine (**409a**), and zoanthamide (**409b**).

CH₂—OH
|
CH—OH
|
CH₂—O—CH₂—(CH₂)ₙ—CH₃

(**405**) n = 14
(**406**) n = 16

(**407**)

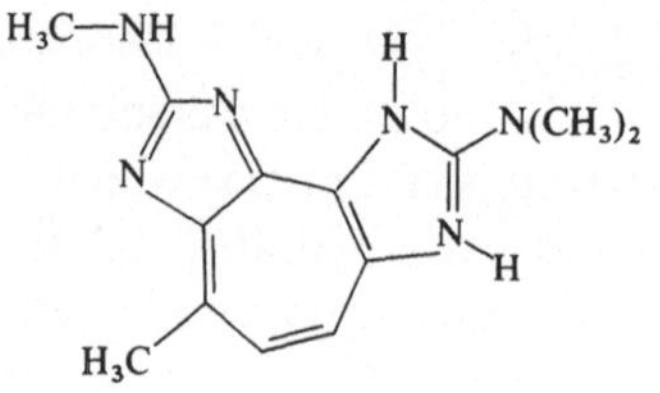

(408)

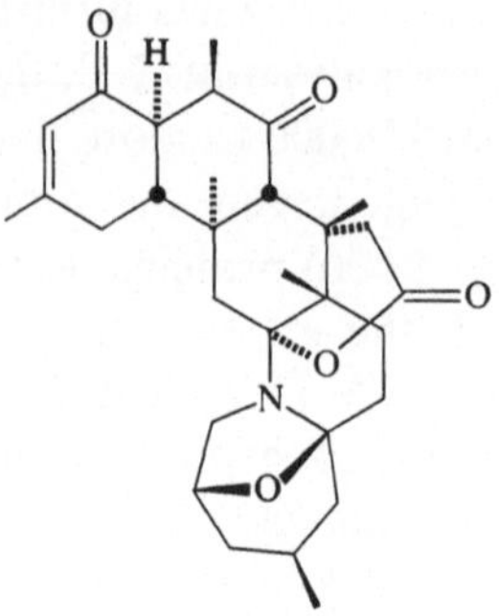

(409)

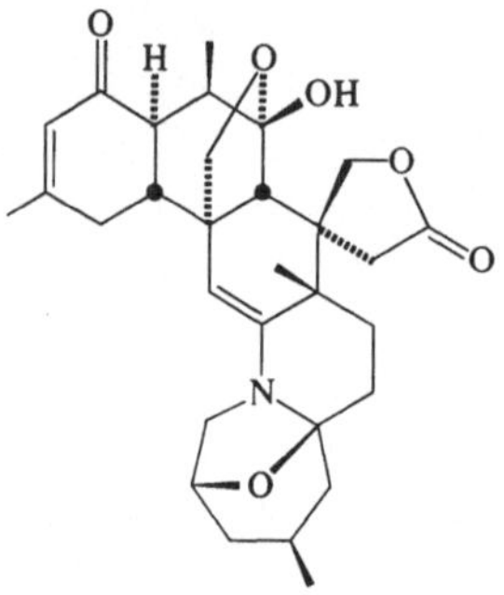

(409 a)

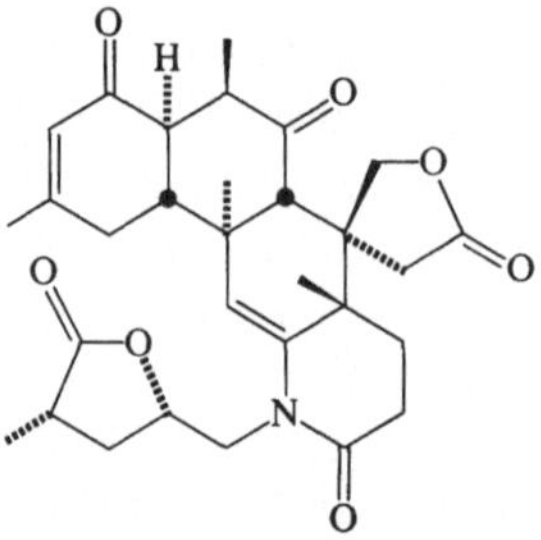

(409 b)

III.4. Octocorallia

III.4.1. Steroids from Octocorallia

The sterol composition of several soft corals and gorgonians as well as the composition of their associated symbiotic dinoflagellates known as zooxanthellae has been investigated (362–369). In addition to well known 3β-hydroxysterols some new 4α-methylated or cyclopropyl-containing 3β-hydroxysterols have been isolated. POPOV et al. (369) reported the first natural occurrence of 19-*nor*cholestenone together with 17 sterols and one other Δ^4-3-ketone in the extracts of the gorgonian *Muricea californica*. KASHMAN and CARMELY (370, 371) isolated a series of sterols with polyoxygenated side chains (**410–414**) from the soft coral *Lobophytum depressum*. Another type of polyhydroxylated sterols (**415–418**) has been obtained from *L. pauciflorum* (372), but while highly oxygenated sterols often exhibit pharmacological activities none have been described for these compounds.

KOBAYASHI et al. (373) isolated a new sterol, 5α-cholestane-1β,3β,5,6β-tetrol (**419**), as well as some known C_{28}-polyhydroxysterols (**420–424**) from *Sarcophyton glaucum*. The configurations at C-24 of **420–424** were shown to be (S) by ^{1}H NMR analysis. Another compound from the same species was the 5α,6β-glycol **425** (374) while *Asterospicularia randalli* yielded 24ξ-methyl-5α-cholestane-3β,5,6β,22R,24-pentol 6-acetate (**426**) (375). *S. glaucum* also furnished a polyoxygenated androstane derivative, 1β,3β,5,6β-tetrahydroxy-5α-androstan-17-one (**427**) (376). The structure of **427** has been confirmed by synthesis. JAGODZINSKA et al. (377) obtained a new group of polyhydroxylated sterols (**428–435**), all of them possessing an 11α-hydroxy substituent, from the soft coral *Sinularia dissecta*. Stoloniferone-A (**436**), -B (**437**), -C (**438**), and -D (**439**) are new cytotoxic steroids from the Okinawan soft coral *Clavularia viridis* (378). Growth inhibition against P-388 leukemia cells was 69% at 1 μg/mL. The Mediterranean gorgonian *Leptogorgia sarmentosa* furnished cholest-4,14-diene-15,20ξ-diol-3,16-dione (**440**), a steroid which easily lost the side chain by a retroaldol reaction (379), in addition to three related polyoxygenated steroids (**441–443**) (380). Steroid **441** is a known plant metabolite from the tree *Commiphora mukul*. The resinous exudate from this tree possesses a variety of pharmacological properties which include antiinflammatory, anti-rheumatic and hypocholesterolemic activity, some of which were due to the presence of steroids.

KINGSTON and FALLIS (381) have reported isolation of two unusual C_{26} steroids, 12β-hydroxy-24-*nor*cholesta-1,4,22-trien-3-one (**444**) and its acetate (**445**), from the sea raspberry, *Gersemia rubiformis*. A novel ergostane (**446**) containing four hydroxy groups in positions 3, 5, 6, and 7 has been isolated in addition to some 3β-hydroxy sterols from the soft coral *Anthelia glauca* (382). Another tetrahydroxysteroid, whose oxygenation pattern differed from that of **446** was identified as gorgost-5-ene-3β,7α,11α,12β-tetraol 12-monoacetate (**447**) (383). The same species also furnished six steroids (**448–453**) containing a spiroketal function (384, 385). Two of these, hippuristanol (**450**) and 2α-hydroxyhippuristanol (**453**), were stated to have potent anticancer activity showing 50% inhibition (*in vitro*) of DBA/MC fibrosarcoma at 0.8 and 0.1 μg/mL, respectively, while **451** and **452** seemed to be artifacts of the isolation procedure.

Several 5α,8α-epidioxy sterols (**3–12, 14–16, 454**) have been isolated from the common pillar coral, *Dendrogyra cylindrus* (67) and (**3–5, 6** and/or **7, 8–10**) from the sea pen, *Virgularia* sp. (386). The probable biological significance of these epidioxy sterols has been discussed by GUNATILAKA et al. (67) with special reference to sterol biosynthesis. A soft coral of the genus *Sinularia* contained a series of sterols with the unique 9,11-*seco*cholestane system (**455–460**) (*387, 388*). It was suggested that the animal probably cleaves ring C of exogenously provided sterols but neither the mechanism of this cleavage nor the biological function of the *seco* sterols has so far been established. Five new pregnene-type steroidal glycosides, named pregnedioside-A (**461**), -B (**462**), and their monoacetates **463–465** have been isolated from an Okinawan soft coral of *Alcyonium* sp. (*389*). BANDURRAGA and FENICAL (*390*) investigated the closely-related Pacific gorgonians *Muricea californica* and *M. fruticosa* and have shown that the much less fouling *M. fruticosa* contained four new esterified aminogalactose saponins, muricins-1 through -4 (**466–469**). Structures of **466–469** have been determined spectroscopically and by chemical degradation to the aglycone pregna-5,20-dien-3β-ol. The muricins, at 100 ppm concentrations, were found to effectively inhibit growth (70% inhibition) of the pennate diatom *Phaeodactylum tricornutum*. This activity may play a significant role in reducing fouling on *M. fruticosa*.

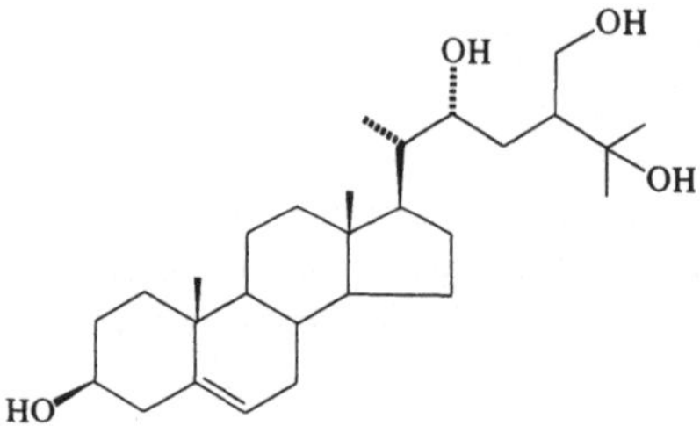

(**410**)

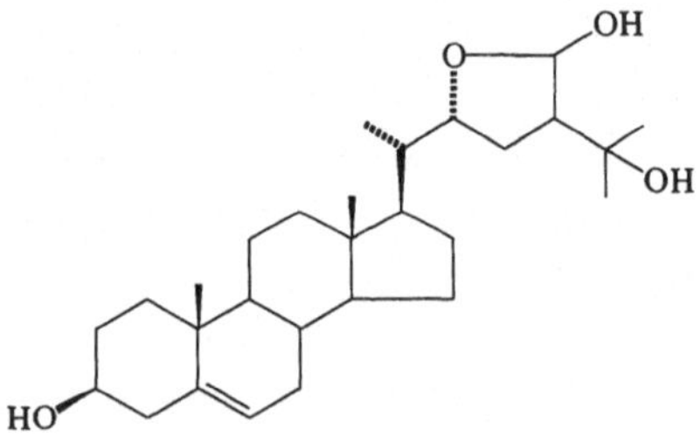

(**411**)

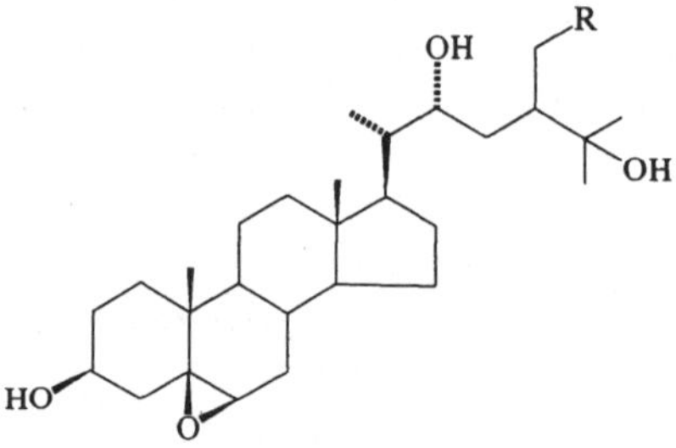

(**412**) R = H
(**413**) R = OH

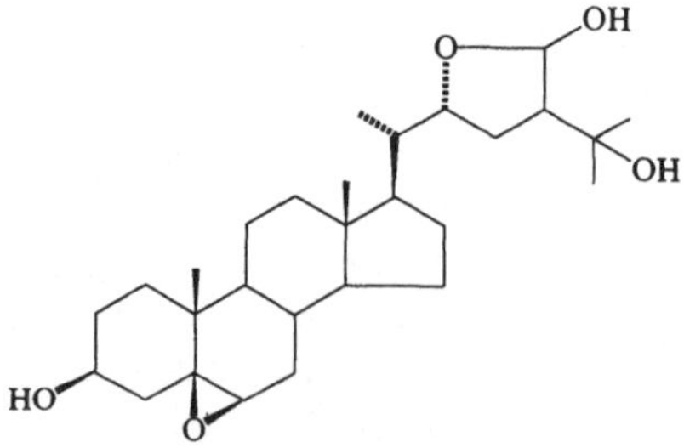

(**414**)

References, pp. 320–363

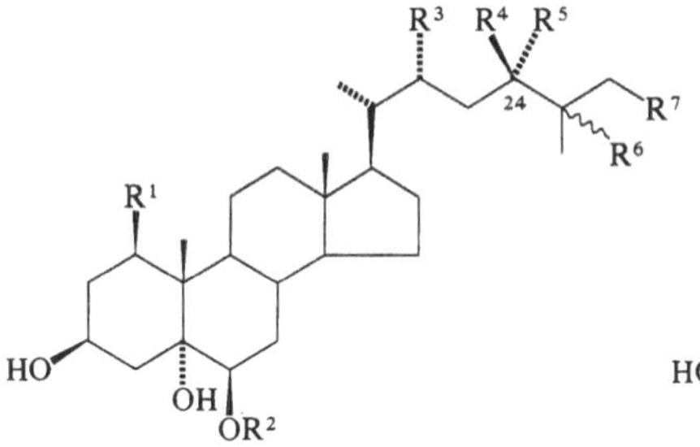

	R¹	R²	R³	R⁴	R⁵	R⁶	R⁷
(415)	H	H	H	H/CH₃*		OAc	H
(416)	H	H	H	H/CH₃*		OH	H
(417)	OH	H	H	H/CH₃*		H	H
(418)	OH	H	H	CH₂		H	H
(419)	OH	H	H	H	H	H	H
(420)	OH	H	H	H	CH₃	H	H
(422)	H	H	H	H	CH₃	OAc	H
(423)	H	H	H	H	CH₃	OH	H
(424)	OH	H	H	H	CH₃	OAc	H
(425)	H	H	H	H	CH₃	OH	OH
(426)	H	Ac	OH	CH₃/OH*		H	H

* Stereochemistry not assigned.

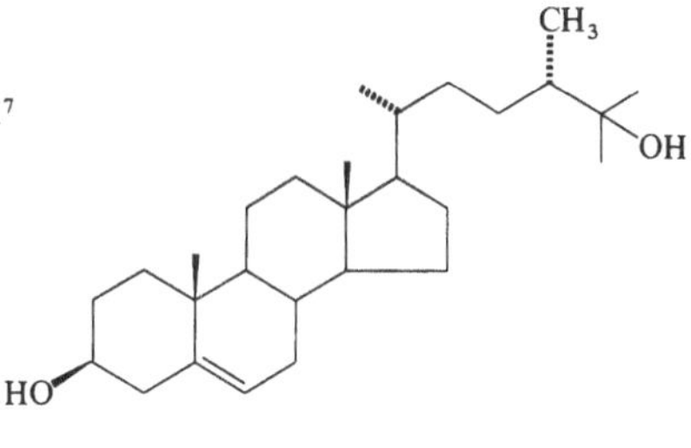

(415)–(426), (421), 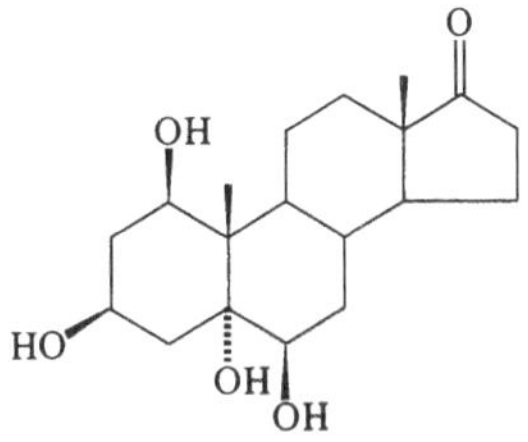 (427)

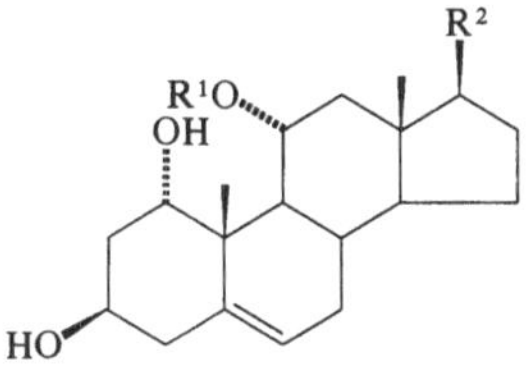

(428) R¹=H R² = 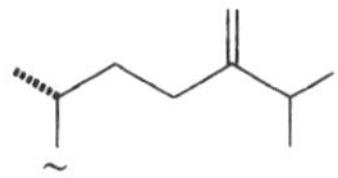
(429) R¹=H R² = 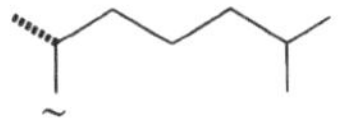
(430) R¹=H R² = 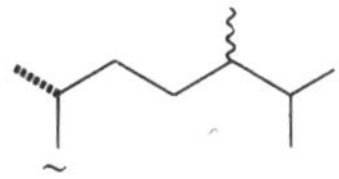
(431) R¹=H R² = 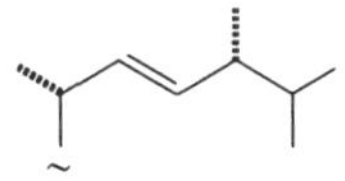
(432) R¹=H R² = 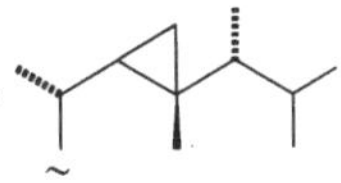
(433) R¹=H R² = 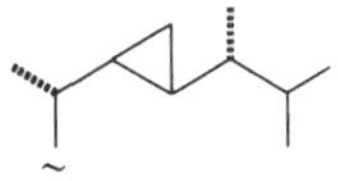
(434) R¹=Ac R² =

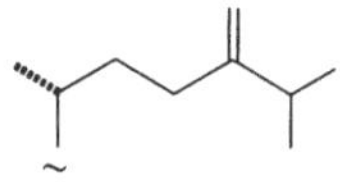

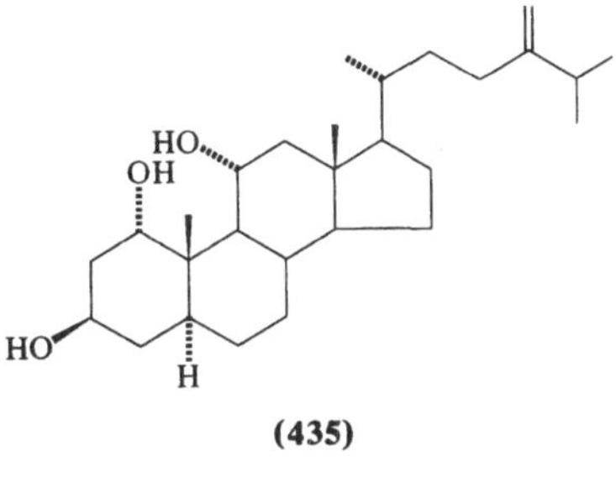

(435)

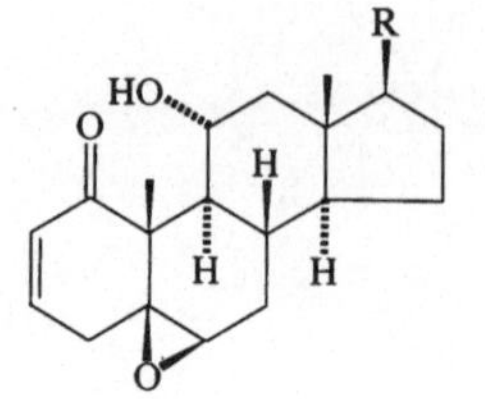

(436) R = [side chain with exocyclic methylene]

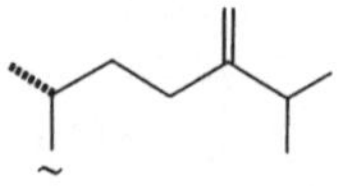

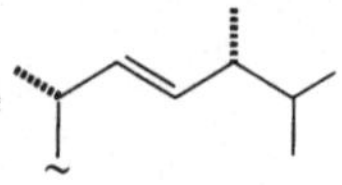

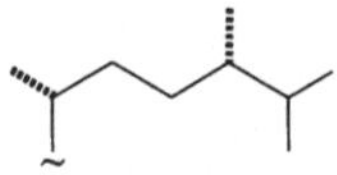

(437) R = [side chain with trans double bond]

(438) R = [saturated side chain]

(439) R = [cyclopropane-containing side chain]

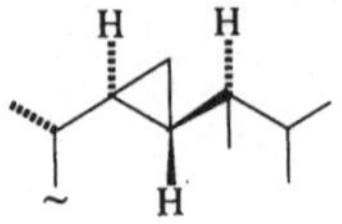

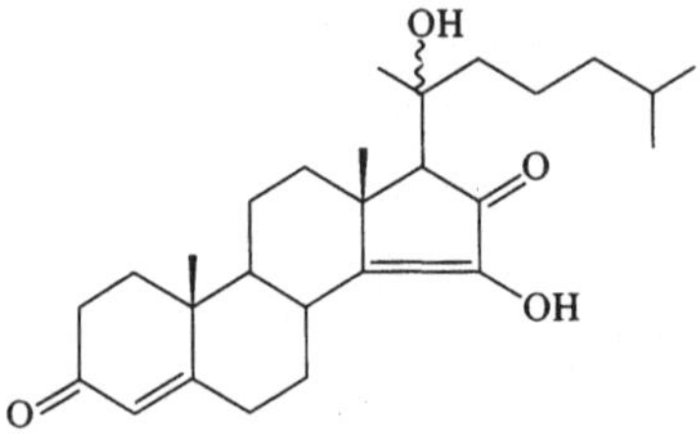

(440)

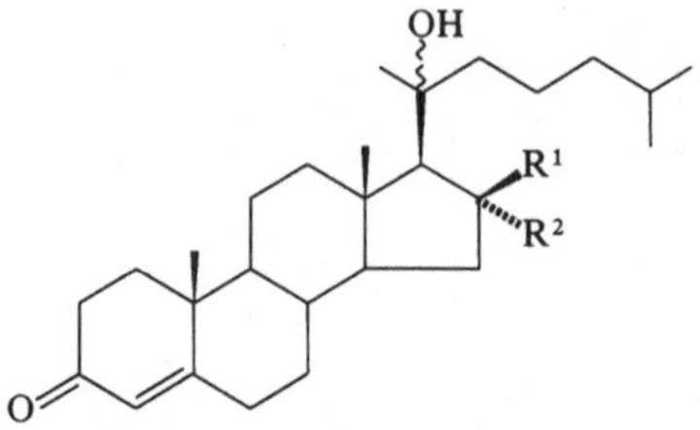

(441) R¹ = OH R² = H
(442) R¹/R² = O
(443) R¹ = H R² = OH

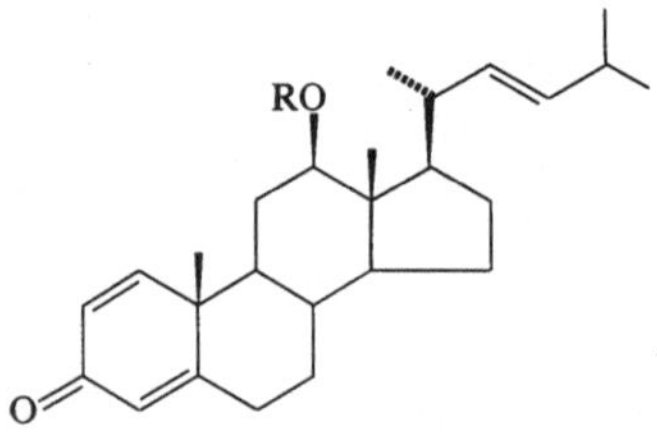

(444) R = H
(445) R = Ac

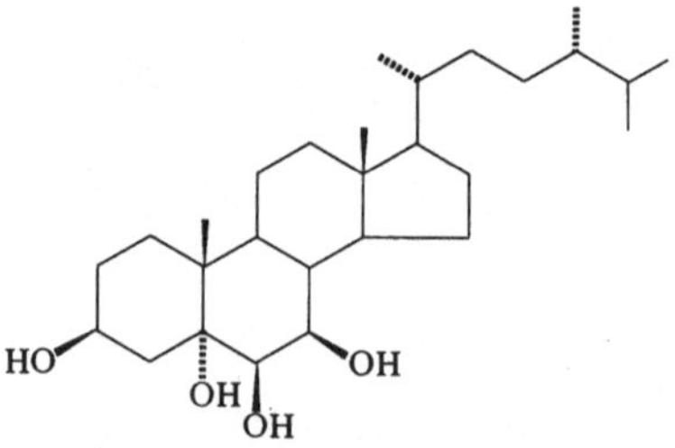

(446)

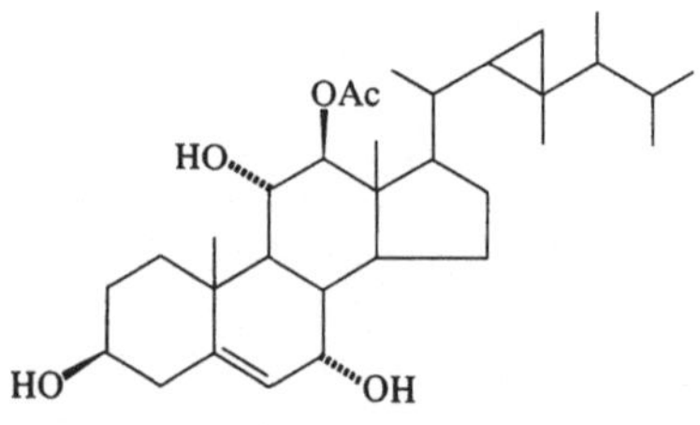

(447)

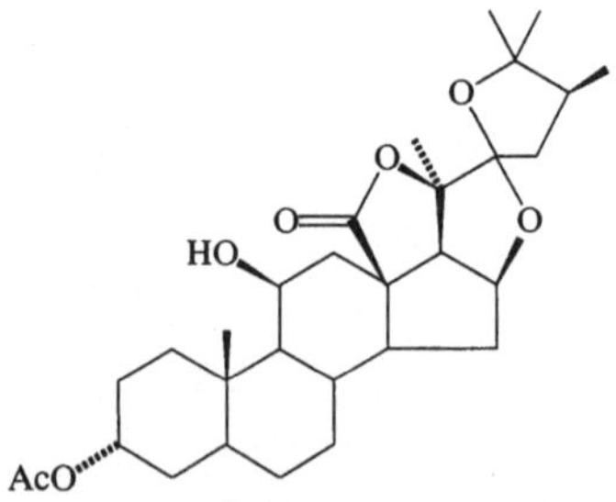

(448)

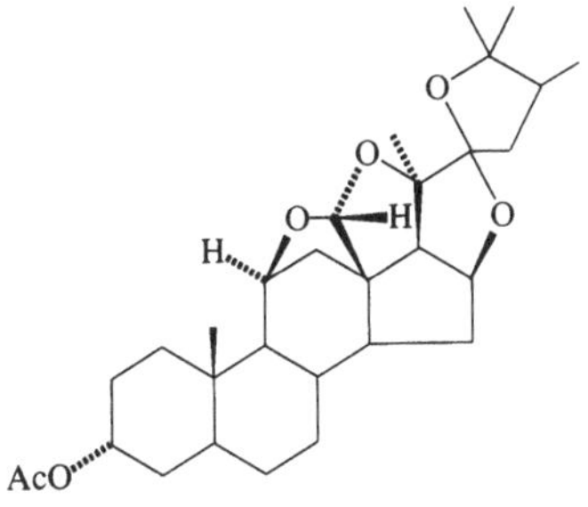

(449)

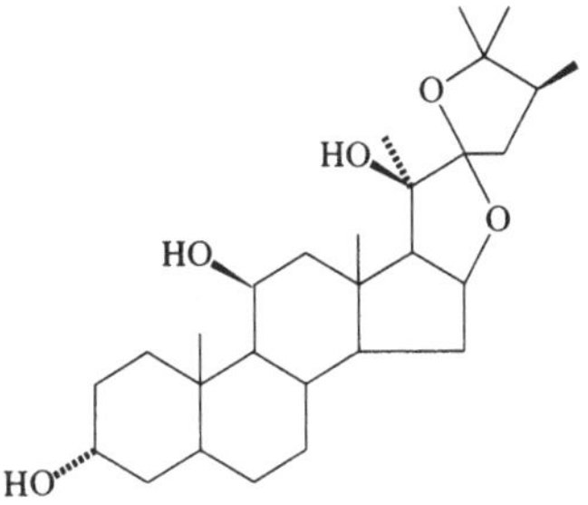

(450) 22 R
(451) 22 S

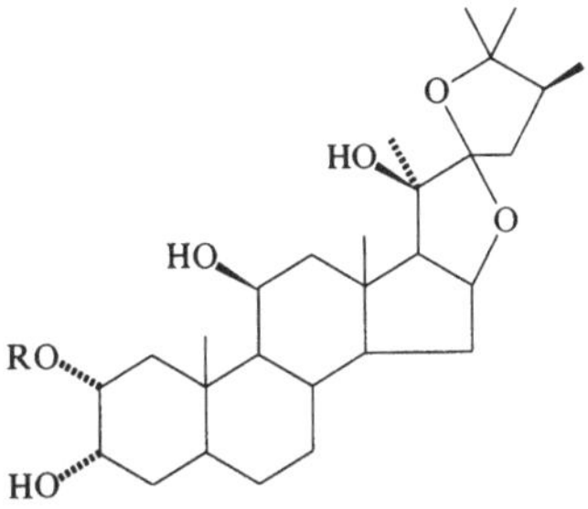

(452) R=Ac 22 S
(453) R=H   22 R

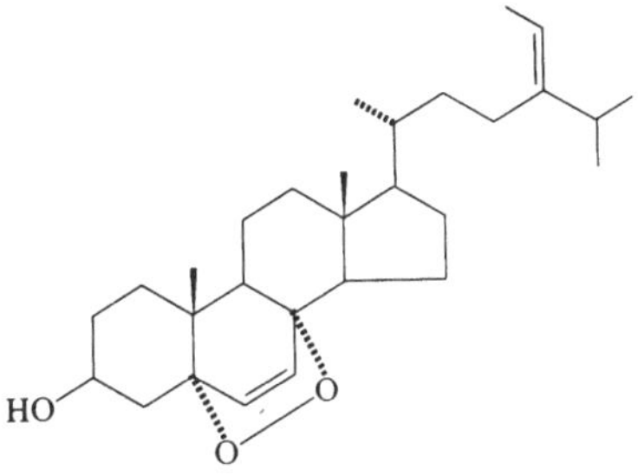

(454)

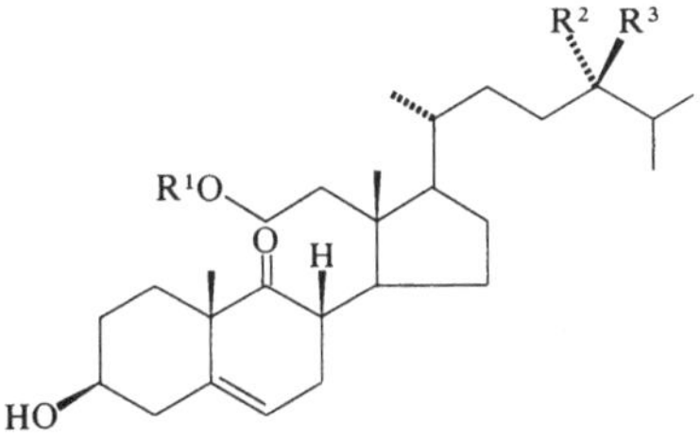

(455) R¹=H   R²/R³=CH₂
(456) R¹=H   R²=CH₃ R³=H
(457) R¹=R²=R³=H
(459) R¹=Ac R²/R³=CH₂
(460) R¹=Ac R²=CH₃ R³=H

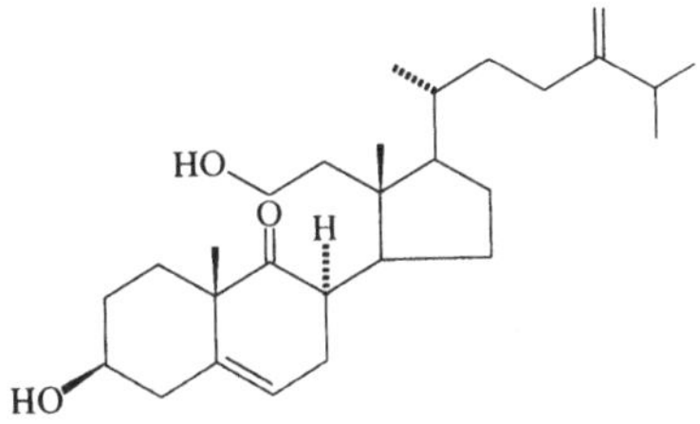

(458)

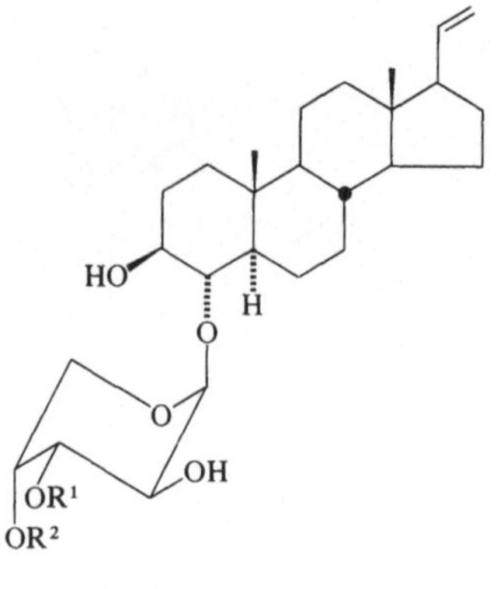

(461) R¹=R²=H
(463) R¹=H R²=Ac
(464) R¹=Ac R²=H

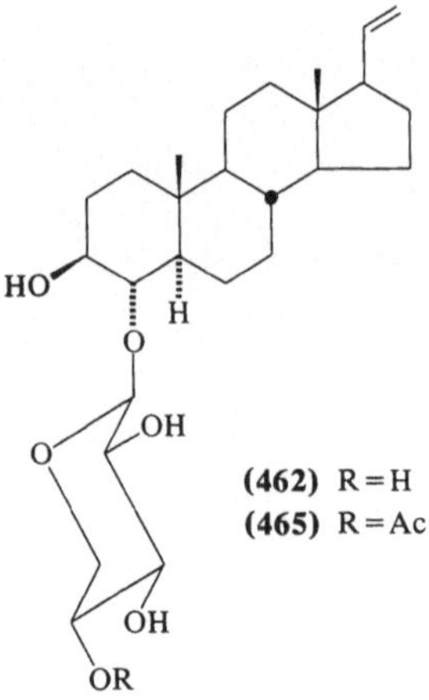

(462) R=H
(465) R=Ac

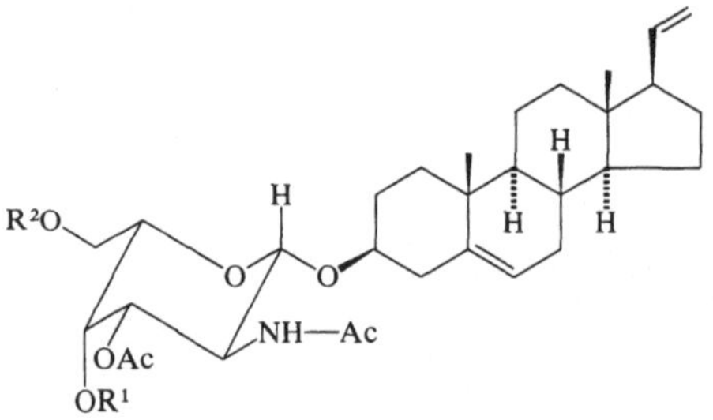

(466) R¹=R²=Ac
(467) R¹=Ac   R²=—CO—CH₂—CH₂—CH₃
(468) R¹=—CO—CH₂—CH₂—CH₃   R²=Ac
(469) R¹=R²=—CO—CH₂—CH₂—CH₃

III.4.2. Terpenes from Octocorallia

Alcyonarian soft corals have produced a vast range of sesquiterpenoid and diterpenoid metabolites several of which have been shown to be toxic. COLL and SAMMARCO (*391*) examined 67 soft corals for ichthyotoxicity and found that levels of toxicity varied greatly. COLL et al. (*392*) in studying the sesquiterpenoids from a *Nephthea* species and discussing the terpenoid chemistry of alcyonacean soft corals in general found that mevalonolactone was incorporated specifically into the terpene portion of a sesquiterpene hydroquinone and proposed that toxic terpenes in soft coral tissue ingested by specialist predators serve as stored defensive substances. Certain soft coral toxins can kill the scleractinian corals (*391*) and show significant effects on photosynthesis and respiration of these species (*393*).

References, pp. 320–363

Tris*nor*sesquiterpenes have been isolated from a *Cespitularia* and two *Clavularia* species. 1,4-Dimethyl-2,3,3a,4,5,6-hexahydroazulene (**470**) was a metabolite of the former (*394*). The soft coral *Clavularia koellikeri* furnished clavukerin A (**471**) (*395*), B (**472**), and C (**473**) (*396*). The structure reported for clavukerin B (**472**) was identical with that of inflatene from *Clavularia inflata* (*397*). Inflatene (**472**) has shown ichthyotoxicity toward the Pacific damselfish *Pomacentrus coeruleus* at 10 µg/mL. $\Delta^{9(15)}$-Africanene (**474**), a tricyclic sesquiterpene, has been isolated from *Sinularia erecta* (*398*) and *S. polydactyla* (*399*).

Extraction of the soft coral *Nephthea chabrolii* afforded a new guaiane-based alcohol (**475**) as well as two known terpenoids (*400*). The ether soluble material from a methanol extract of the gorgonian *Euplexaura erecta* showed mild activity against *Pseudomonas aeruginosa*, with guaiazulene (**476**) being responsible for the activity and apparently also for the blue color of the gorgonian (*401*). Guaiazulene (**476**) is frequently found in essential oils of terrestrial plants and has also been found in a marine red alga. Three halogenated blue pigments (**477–479**), derivatives of **476**, have been isolated from a deep sea gorgonian (*402*). The same species also furnished guaiazulene (**476**) itself and N,N-dimethylamino-3-guaiazulenylmethane (**480**) (*403*). The structure of **480** has been confirmed by synthesis. The azulenofuran, linderazulene (**481**), first obtained by zinc dust distillation of linderene from the roots of *Lindera strychnifolia*, is a pigment of the gorgonian *Paramuricea chamaeleon* (*404*), whereas an olive-colored gorgonian, *Placogorgia* sp., furnished **481** together with a yellow pigment (**482**) (*405*). Linderazulene (**481**) could be oxidized to a ketolactone identical with the natural product **482** in all respects and ALPERTUNGA *et al.* (*406*) have shown that **482** could also be obtained by exposure of an ethanolic solution of linderazulene (**481**) to direct sunlight.

GOPICHAND *et al.* (*407*) isolated two new acyclic sesquiterpene hydrocarbons (**483**) and (**484**) from the gorgonian *Plexaurella grisea* as well as (+)-α-santalene (**485**), which is well-known from terrestrial source. BOWDEN *et al.* (*408*) described new sesquiterpene furans from *Sinularia capillosa* and *S. firma*, all of which were structurally related. The furanoquinol (**486**) was the predominant isolate and was accompanied by the related furanoquinone (**487**) which may be an artifact. In addition they found the methylfurans (**488–492**), the furan methyl esters (**493–496**), and the furan carboxylic acids (**497–500**). A related methylfuran (**501**) has been obtained from the Mediterranean alcyonacean, *Alcyonum palmatum* (*409*).

Extraction of several soft corals of the genera *Cespitularia* and *Efflatounaria* afforded in addition to two previously reported epimeric diterpene alcohols varying amounts of three sesquiterpene furans, atractylone (**502**), isofuranodiene (**503**), and the new but related furan (**504**) (*410*). Examinations of the extracts of 8 taxonomically diverse marine soft corals of the orders Gorgonaceae, Alcyonaceae and Stolonifera, resulted in the isolation of four previously known furanogermacradienes (**503, 505–507**) and four new metabolites (**508–511**) (*411*). Furanodiene (**503**) from the Pacific gorgonians *Pacifigorgia pulchra exilis* and *P. media*, as well as from the stoloniferan coral *Tubipora musica* is a compound originally isolated from the terrestrial plant *Curcuma zedoaria*. A compound similar to **509**, germacrene-C (**512**), has been isolated from the soft coral *Nephthea chabrolii* (*400*) and may be derived from **509** by elimination of water. *Parerythropodium fulvum* furnished two oxigenated bicyclogermacrenes (**513, 514**) in addition to the previously known lemnacarnol (**515**) and its 2-keto derivative (**516**) (*412*).

Three new sesquiterpenes named valerenenol (**517**), isovalerenenol (**518**), and anhydrovalerenenol (**519**) from an unidentified Okinawan soft coral are based on the bicarbocyclic carbon skeleton of *ent*-valerenane (*413*) and pacifigorgiol (**520**) is a related sesquiterpene from *Pacifigorgia adamsii* (*414*). The structure of **520** which showed modest activity as a fish toxin was established by a combination of spectral and X-ray crystallographic techniques and was confirmed by an eleven step synthesis (*415*).

The isolation and interconversion of two new sesquiterpenes containing the aristolane skeleton, $(4R,5S,6R,7S)$-aristol-9-en-3-one (**521**) and $(3S,4R,5S,6R,7S)$-aristol-9-en-3-ol (**522**), from the soft coral *Lemnalia humesi* has been reported (*416*). Another aristolane sesquiterpenoid, β-(−)-1,10-epoxyaristolane (**523**), has been reported as a metabolite of the sea pen *Scytalium splendens* (*417*). Two new sesquiterpenes (**524, 525**) and 2-deoxylemnacarnol (**526**) have been isolated from *Paralemnalia thyrsoides*; the structures, deduced from spectroscopic data, were confirmed by conversion into known sesquiterpenes (*418*). The closely related soft coral *Lemnalia africana* furnished several novel nardosinane-based sesquiterpenes and a *nor*-sesquiterpene (*419*). The sesquiterpenes were identified as $(2R,7S,11R)$-7-acetoxy-2-hydroxynardosin-1(10)-en-12-al (**527**) and $(2R,11S,12R)$-lemnal-1(10)-ene-2,12-diol (**528**) while the *nor*-sesquiterpene provisionally assigned the formula of $(2R,7S)$-7-formyloxy-2-hydroxy-12-*nor*nardosin-1(10)-en-11-one (**529**), although the substance could not be converted into a known nardosinane or lemnalane derivative. Two further *nor*-sesquiterpenoids (**530, 531**) of apparent nardosinane origin were reported as constituents of the Pacific soft coral *Paralemnalia thyrsoides* (*420*). The diketone **530** could be converted to **531** in two steps. IZAC et al. (*421*) investigated *Lemnalia africana* and found the diol **532** as well as two new sesquiterpenoid ketones **533** and **534** which based on a new irregular terpenoid carbon skeleton for which the semisystematic name "neolemnane" was suggested. The structures were confirmed by X-ray crystallography.

(+)-Cubetol (**535**) was a tricyclic monohydric alcohol from a *Cespitularia* species (*394*). A sample of *Clavularia inflata* yielded two sesquiterpenes closely related to sinularene, 12-acetoxycyclosinularane (**536**) and 12-acetoxysinularene (**537**) (*422*). Compound **537** could be derived from **536** by opening of the cyclopropane ring. Another species of the order Stolonifera, *C. koellikeri* yielded the methyl ester (**538**) (*422*).

The structure and absolute stereochemistry of lemnalol (**539**), a new ylangene-type sesquiterpenoid with cytotoxic activity from the soft coral *Lemnalia tenuis*, has been established by spectroscopic, chemical, and X-ray crystallographic analysis (*423*). Ylangene-type sesquiterpenoids are commonly produced by plants and **539** was the first oxygenated compound of this type from any source. Lemnalol (**539**) exhibited strong antitumor activity (*424*).

A study of *Capnella imbricata*, a widely distributed member of the Alcyonacea, by KAISIN *et al.* (*425*) afforded an array of eight new nonisoprenoid sesquiterpenes (**540–547**) based on the new tricyclo-[6.3.0.0^{2,6}]undecane or triquinane skeleton named capnellane. By extracting fresh colonies of *C. imbricata* with acetone immediately after collection the authors showed that the previously reported capnellene polyols, isolated by methylene chloride or hexane extraction of sun-dried colonies (*425a*), were artifacts produced by a substrate-specific hydrolase present in the soft coral. The hydrolase containing fraction was also able to catalyze the transfer of acetyl groups from several alkyl acetates to the capnellene polyols. The capnellane sesquiterpenes had feeding deterrent activities against *Lebistes reticulatus*. Recently GROWEISS *et al.* (*426*) reported isolation of subergorgic acid (**548**), a cardiotoxic sesquiterpene with the tricyclo[6.3.0.0^{1,5}]undecane skeleton of silphiperfdene from the Pacific gorgonian coral *Subergorgia suberosa*. In the isolated guinea pig heart essay, acid **548** inhibited neuromuscular transmission at threshold levels as low as 0.16 μg/mL.

(470)   (471)   (472)

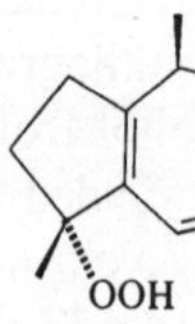

(473) (474) (475)

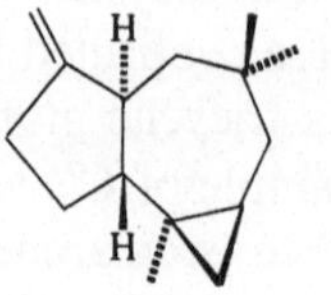

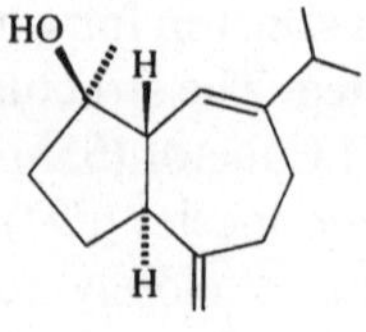

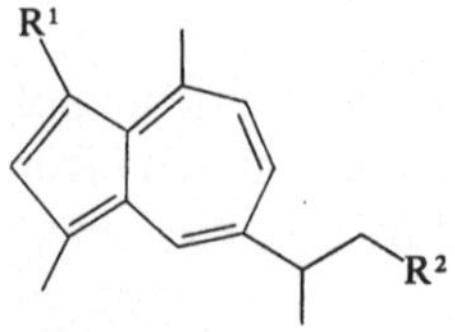

(476) R¹=R²=H
(477) R¹=Cl   R²=H
(478) R¹=Br   R²=H
(479) R¹=H   R²=Br
(480) R¹=CH₂NMe₂ R²=H

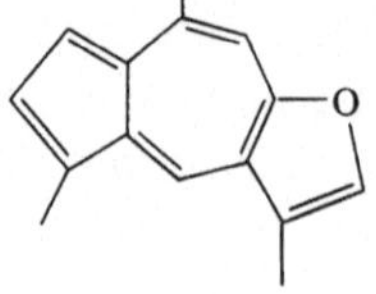

(481)

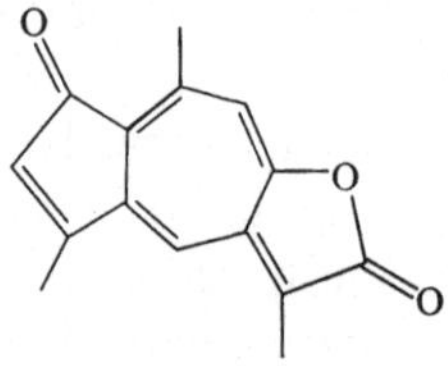

(482) (483)

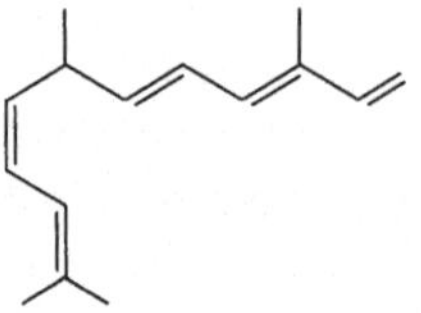

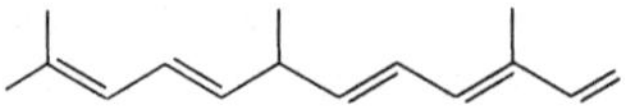

(484) (485)

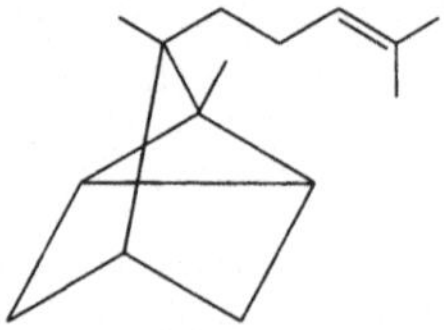

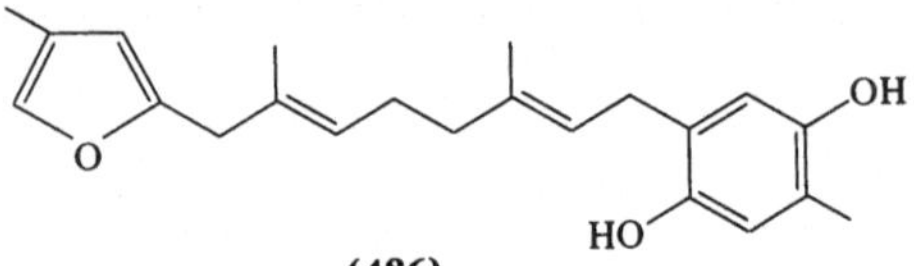

(486)

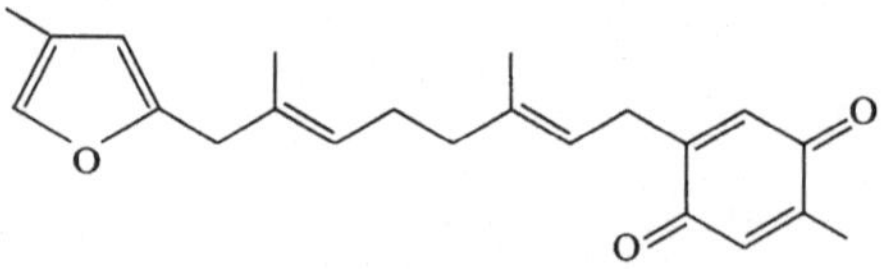

(487)

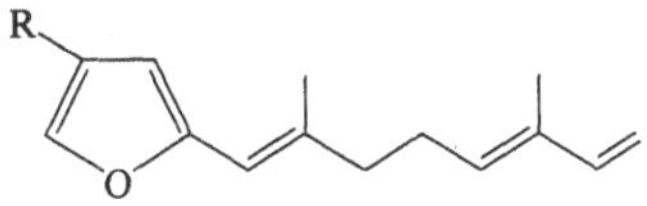

(488) R=CH₃
(493) R=COOMe
(498) R=COOH

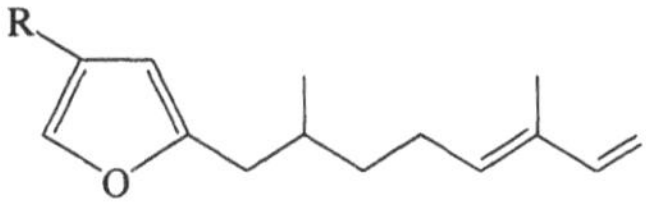

(489) R=CH₃
(495) R=COOMe
(499) R=COOH

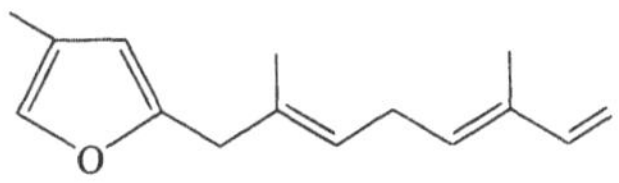

(490)

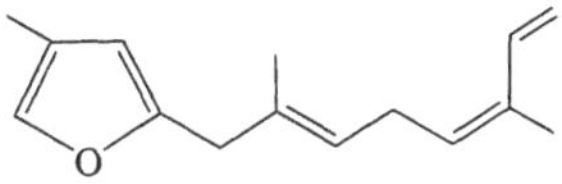

(491)

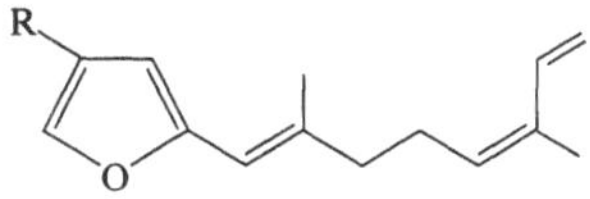

(492) R=CH₃
(494) R=COOMe
(497) R=COOH

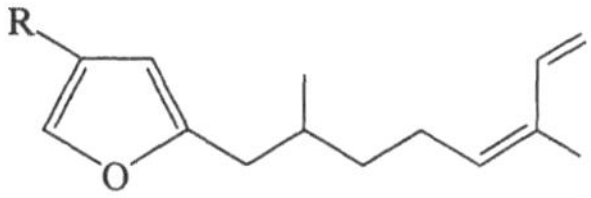

(496) R=COOMe
(500) R=COOH

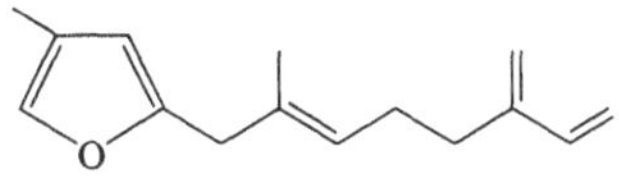

(501)

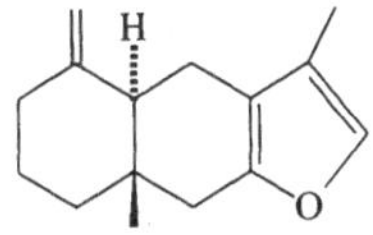

(502)

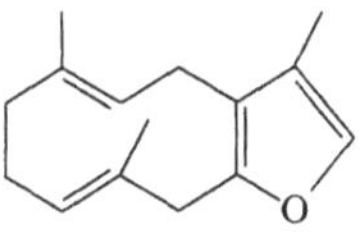

(503)

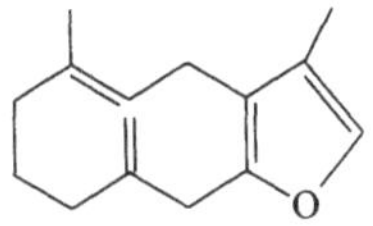

(504)

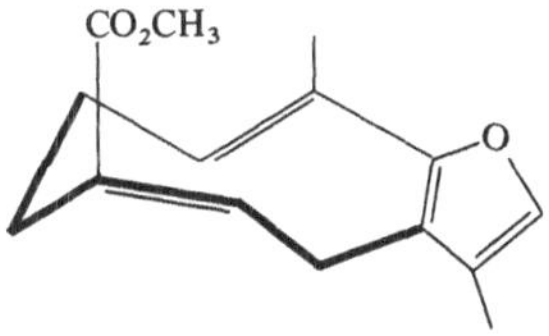

(505)

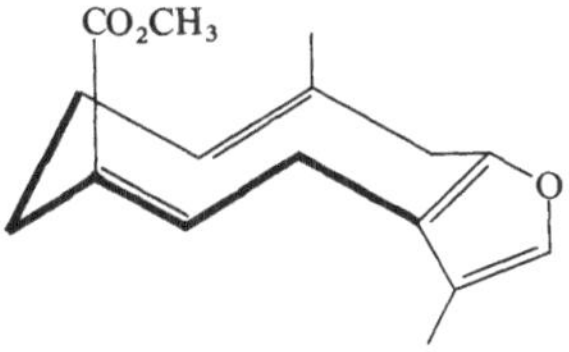

(506)

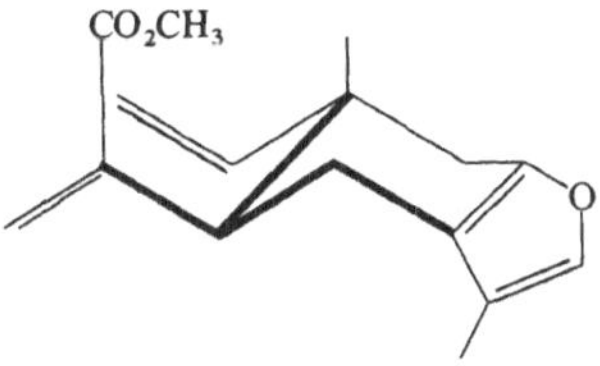

(507)

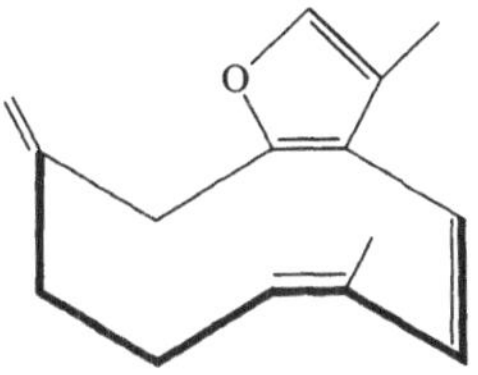

(508)

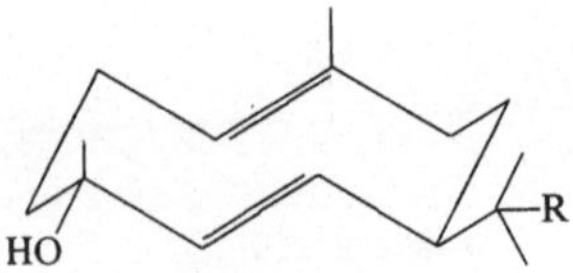

(509) R=H
(510) R=OH

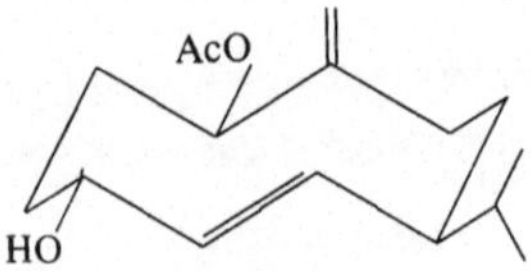

(511)

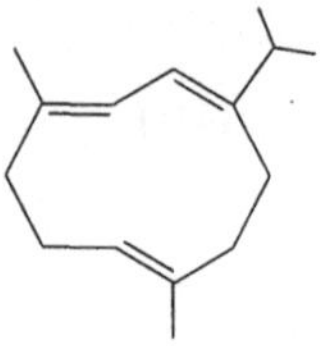

(512)

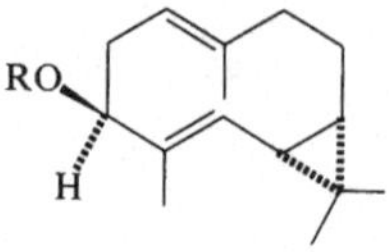

(513) R=Ac
(514) R=H

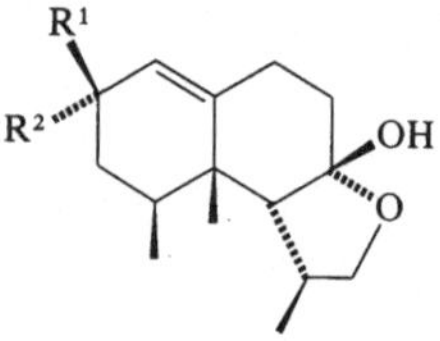

(515) R¹=H R²=OH
(516) R¹/R²=O

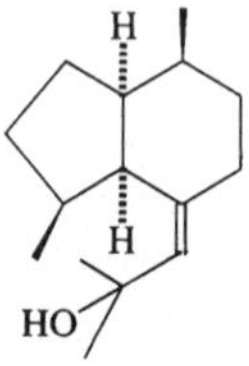

(517)

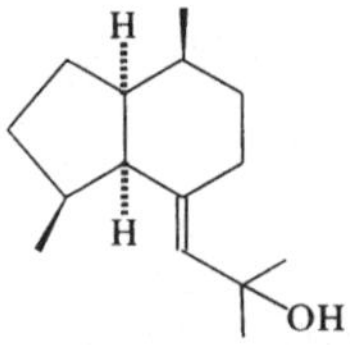

(518)

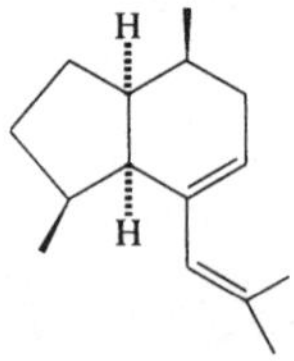

(519)

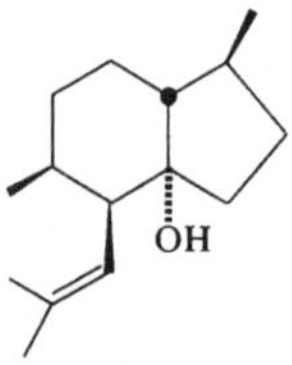

(520)

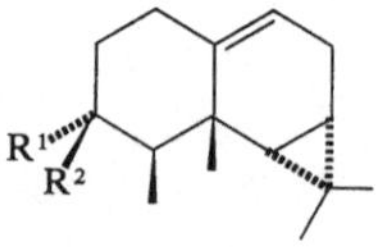

(521) R¹/R²=O
(522) R¹=H R²=OH

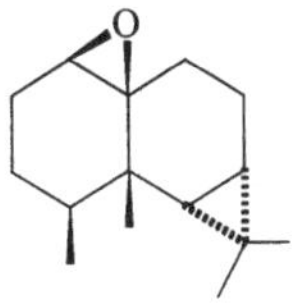

(523)

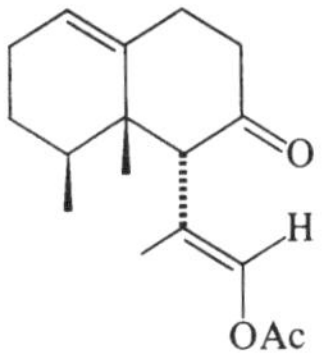

(524)

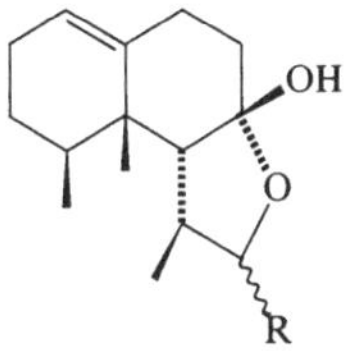

(525) R=OH
(526) R=H

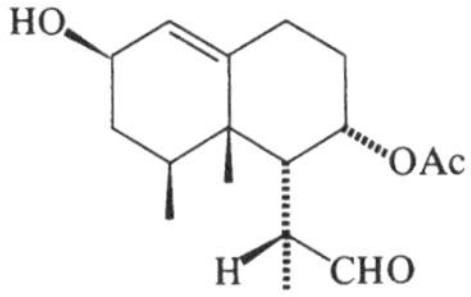

(527)

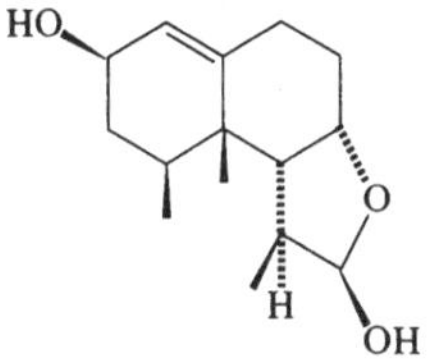

(528)

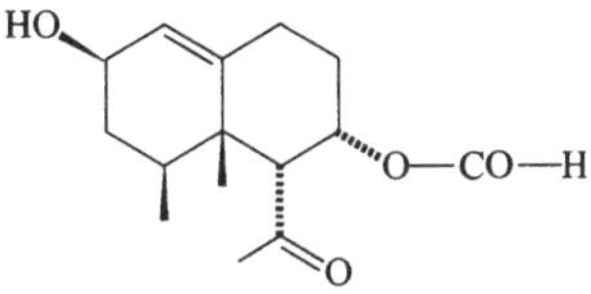

(529)

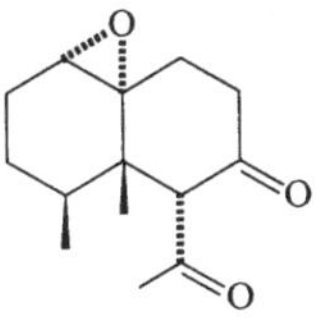

(530)

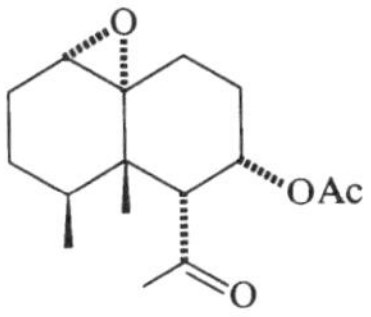

(531)

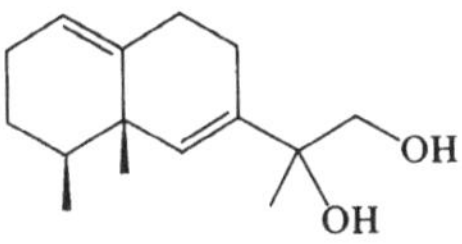

(532)

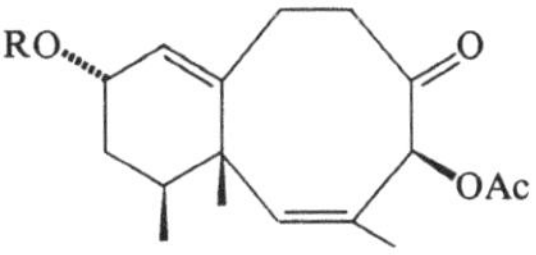

(533) R=H
(534) R=Ac

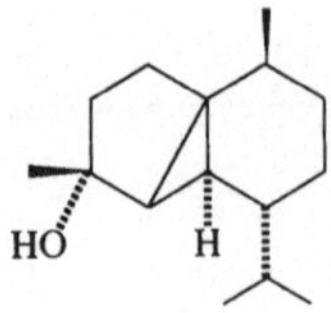

(535)

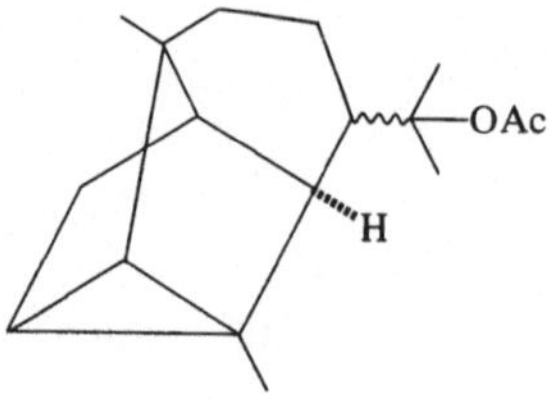

(536)

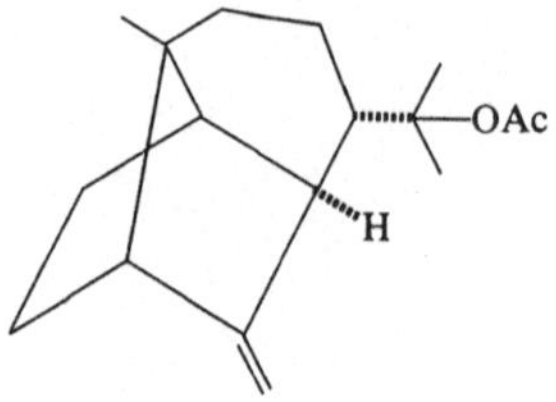

(537)

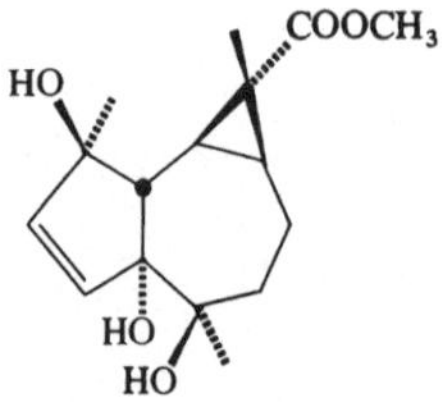

(538)

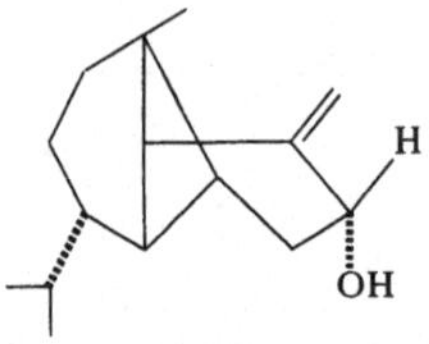

(539)

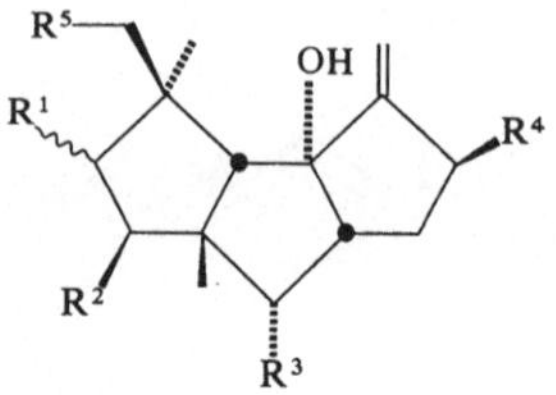

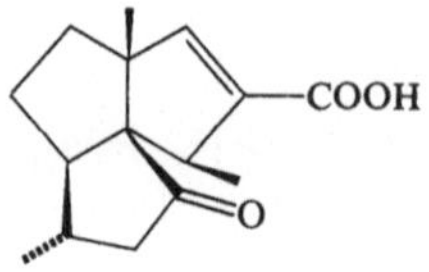

(548)

	R¹	R²	R³	R⁴	R⁵
(540)	H	H	H	OAc	H
(541)	H	OAc	H	OH	H
(542)	H	OAc	H	OAc	H
(543)	H	H	OAc	OH	H
(544)	H	H	OAc	OAc	H
(545)	H	OAc	H	OH	OAc
(546)	β-OAc	H	OAc	OH	H
(547)	β-OAc	H	OAc	OAc	H

References, pp. 320–363

Among the diterpenes isolated from gorgonian and soft corals, the cembranes so far are the largest and most abundant group of compounds. Cembrane diterpenes were originally obtained from terrestrial plants, but the occurrence of these metabolites in octocorals was first noted in the 50's. Some of these were reported to have anticancer or other biological activities and several are toxic to fish and are believed to play a role in the protection of the soft corals from predators. Although many new cembranoide terpenes have been described in the last five years biological activities have been reported for only a few. Lophotoxin (**549**) was a neuromuscular toxin isolated from several Pacific gorgonians of the genus *Lophogorgia* (*427, 428*) together with the previously known pukalide (**550**) (*427*). Both metabolites were correlated chemically, with **549** being lethal to mice on subcutaneous injection with an LD_{50} of 8.0 µg/g. The prominent signs displayed by lophotoxin-treated mice were indicative of muscle paralysis, an effect which has been studied further by CULVER and JACOBS in isolated skeletal muscle preparations (*429*).

KAZLAUSKAS et al. (*430*) found that the crude extract of a *Sarcophyton* species caused a transient convulsant response in mice when administered intraperitoneally and succeeded in isolating two new cembranes (**551, 552**), of which **552** was the compound responsible for the effect on the central nervous system demonstrated by the crude extract. The cembranoid also potentiated the sedative effects of barbiturates. Another new cembranoid (**553**) from a soft coral of the genus *Sarcophyton* which possessed Ca-antagonistic action on the isolated rabbit aorta (*431*) was a stereoisomer of the previously known sarcophytoxide (**554**) (*432*). Lobohedleolide (**555**) and (7Z)-lobohedleolide (**556**) were two new cembranolides from *Lobophytum hedleyi*. Lobohedleolide (**555**) produced *in vitro* growth inhibition of Hella cells at 5 µg/mL (*433*). ^{13}C spin-lattice relaxation measurements were used to establish the structure of flexibilide (**557**), an antiinflammatory agent from *Sinularia flexibilis* (*434*). The same soft coral gave a cytotoxic compound (**558**) together with two other cembranolides (**559, 560**) (*435*). Recently UCHIO et al. (*435a*) isolated denticulatolide (**561**), an ichthyotoxic cembranoid diterpene bearing a cyclic peroxide function, from *Lobophytum denticulatum*.

All other cembranoids of marine origin described in the past five years are listed in Table 1 below. Of special interest is the observed variation in cembrane content for a given soft coral.

Table 1. *Cembrane-type Diterpenes Isolated in the Last Five Years*

Compound	Source	Ref.	Source	Ref.
562	*Nephthea chabrolii*	*(400)*	*Alcyonium flaccidum*	*(436)*
	Nephthea sp.	*(437)*	*Nephthea brassica*	*(438)*
	Alcyonium utinomii	*(439)*	*Lobophytum* sp.	*(440)*
563	*Sarcophyton glaucum*	*(441)*	*Sarcophyton* sp.	*(438)*
564	*Sarcophyton glaucum*	*(441)*	*Sarcophyton* sp.	*(438)*
565	*Sarcophyton glaucum*	*(441)*	*Alcyonium flaccidum*	*(436)*
566	*Nephthea* sp.	*(437, 438)*		
567	*Nephthea* sp.	*(437, 438)*		
568	*Sinularia mayi*	*(442)*		
569	*Nephthea brassica*	*(438)*		
570	*Nephthea brassica*	*(438)*		
571	*Nephthea brassica*	*(438)*		
572	*Sinularia facile*	*(443)*		
573	*Sarcophyton decaryi*	*(444)*		
574	*Lobophytum pauciflorum*	*(439)*		
575	*Lobophytum* sp.	*(440)*		
576	*Sarcophyton glaucum*	*(441)*		
577	*Alcyonium utinomii*	*(439)*		
578	*Sarcophyton glaucum*	*(441)*		
579	*Alcyonium utinomii*	*(439)*		
580	*Sinularia mayi*	*(442)*		
581	*Sinularia mayi*	*(442)*		
582	*Sarcophyta elegans*	*(445)*		
583	*Sarcophyta elegans*	*(445)*		
584	*Sarcophyta elegans*	*(445)*		
585	*Lobophytum pauciflorum*	*(439)*		
586	*Lobophytum* sp.	*(440)*		
587	*Sarcophyton crassocaule*	*(432)*		
588	*Sarcophyton crassocaule*	*(432)*		
589	*Sarcophyton crassocaule*	*(432)*		
590	*Nephthea brassica*	*(438)*		
591	*Nephthea brassica*	*(438)*		
592	*Nephthea brassica*	*(438)*		
593	*Sinularia facile*	*(443)*		
594	*Sarcophyton decaryi*	*(444)*		
595	*Nephthea brassica*	*(438)*		
596	*Sarcophyton glaucum*	*(441)*		
597	*Alcyonium flaccidum*	*(436)*		
598	*Nephthea* sp.	*(437)*		
599	*Lobophytum* sp.	*(440)*		
600	*Lobophytum* sp.	*(440)*		
601	*Lobophytum* sp.	*(440)*		
602	*Alcyonium utinomii*	*(439)*		
603	*Sinularia mayi*	*(446)*		
604	*Sinularia mayi*	*(446)*		
605	*Lobophytum pauciflorum*	*(447)*		
606	*Lobophytum pauciflorum*	*(447)*		
607	*Sarcophyton crassocaule*	*(432)*		

References, pp. 320–363

Table 1 (*continued*)

Compound	Source	Ref.	Source	Ref.
608	*Sarcophyton* sp.	*(448)*		
609	*Lobophytum pauciflorum*	*(449)*		
610	*Lobophytum pauciflorum*	*(449)*		
611	*Lobophytum crassum*	*(436)*		
612	*Lobophytum crassum*	*(436)*		
613	*Lobophytum crassum*	*(436)*		
614	*Lobophytum crassum*	*(439)*		
615	*Cespitularia* sp.	*(450)*		
616	*Cespitularia* sp.	*(450)*		
617	*Lobophytum crassum*	*(439)*		
618	*Lobophytum crassum*	*(439)*		
619	*Eunicea succinea*	*(451)*		
620	*Eunicea succinea*	*(451)*		
621	*Eunicea succinea*	*(451)*		
622	*Eunicea succinea*	*(451)*		
623	*Lophogorgia alba*	*(428)*		
624	*Lophogorgia alba*	*(428)*		
625	*Lophogorgia alba*	*(428)*		
626	*Sarcophyton glaucum*	*(452)*		
627	*Sarcophyton glaucum*	*(453)*	*Sarcophyta elegans*	*(454)*
628	*Sarcophyton glaucum*	*(453)*		
629	*Sarcophyta elegans*	*(454)*		
630	*Sarcophyton decaryi*	*(444)*		

Recently deoxosarcophine (**608**) has been decribed as a potentiator of directly stimulated contraction of skeletal muscle in the isolated rat hemidiaphragm *(455)*.

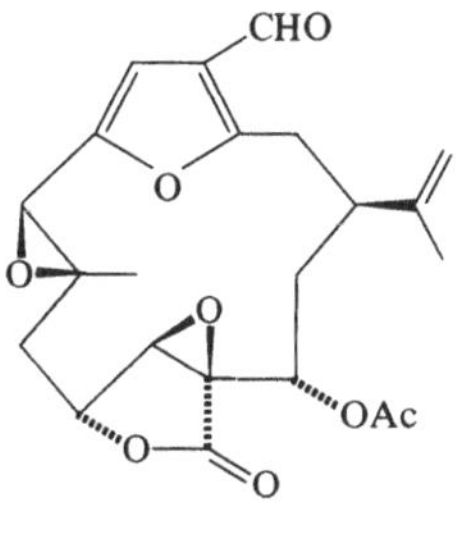

(**549**)

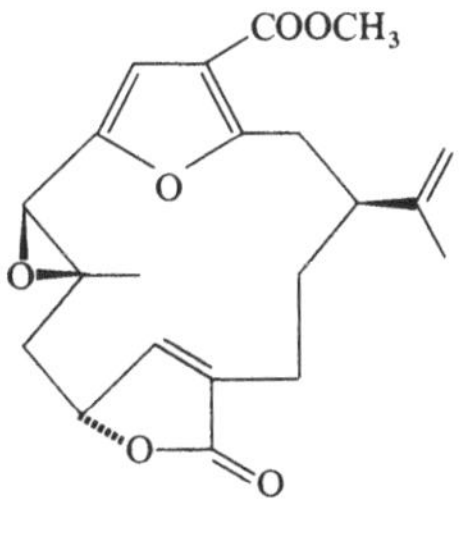

(**550**)

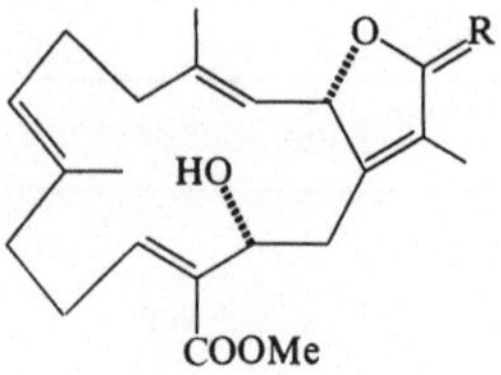

(551) R=H₂
(552) R=O

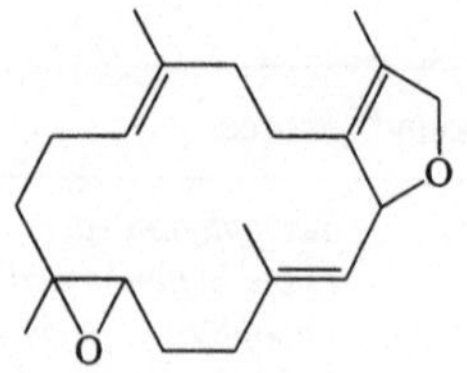

(553)

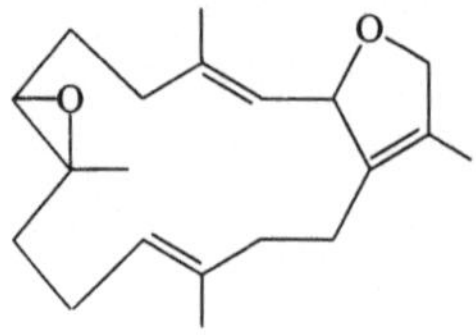

(554)

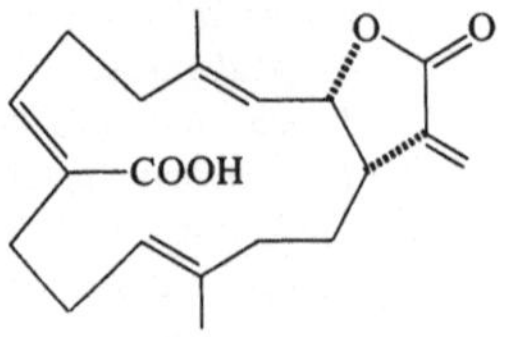

(555)

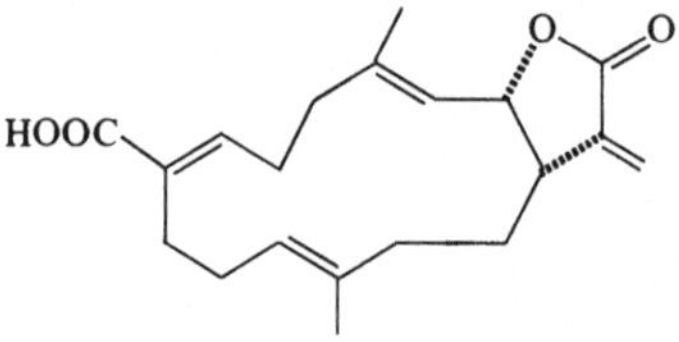

(556)

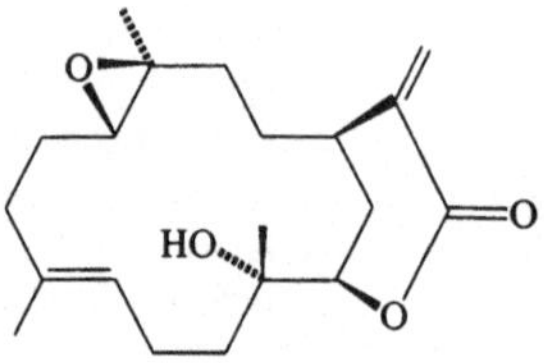

(557)

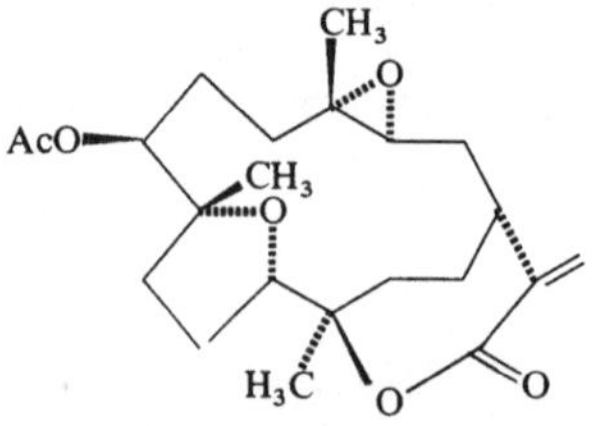

(558)

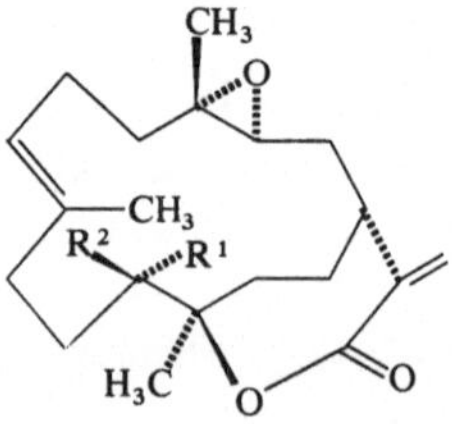

(559) R¹=H   R²=OH
(560) R¹=OH R²=H

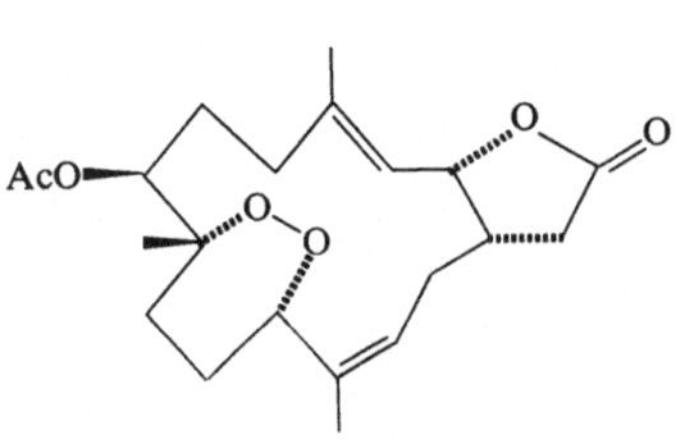

(561)

References, pp. 320–363

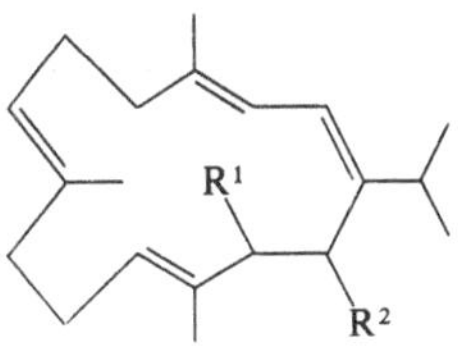

(562) R¹=R²=H
(563) R¹=H   R²=OH
(564) R¹=H   R²=OAc
(565) R¹=R²=OH
(566) R¹=(S)—OH R²=H
(567) R¹=(S)—OAc R²=H

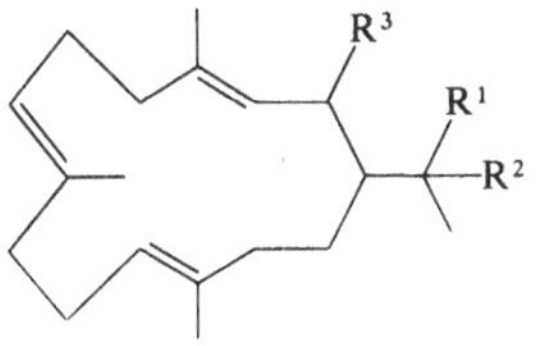

(568) R¹/R²=CH₂   R³=OH
(569) R¹/R²=CH₂   R³=H
(570) R¹=CH₃ R²=OH R³=H
(571) R¹=CH₃ R²=OAc R³=H

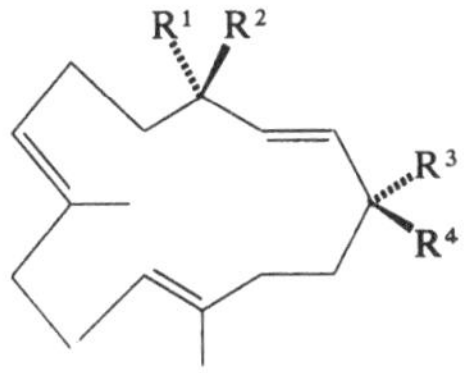

(572) R¹=CH₃ R²=OH R³=H   R⁴=CH(CH₃)₂
(573) R¹=CH₃ R²=OH R³=CH(CH₃)₂ R⁴=H
(574) R¹/R²=CH₃/OH R³/R⁴=H/C(OH)(CH₃)₂
(575) R¹=OH R²=CH₃ R³=CH(CH₃)₂ R⁴=H

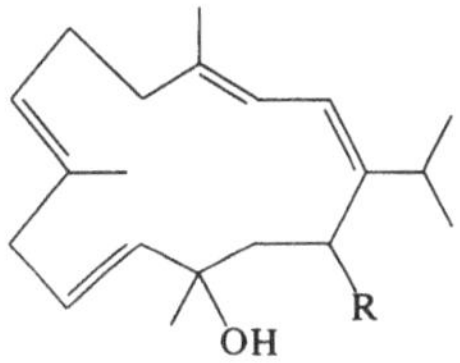

(576) R=OH
(577) R=H

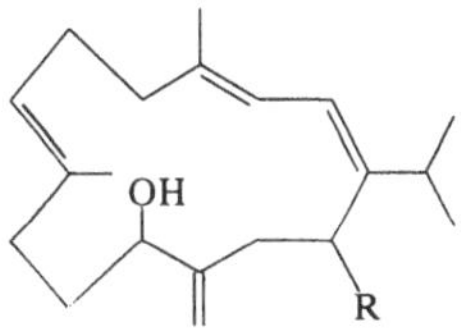

(578) R=OH
(579) R=H

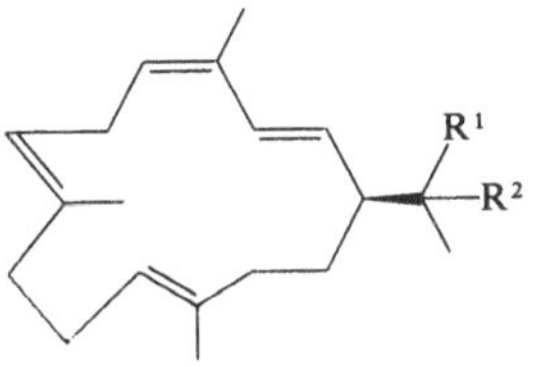

(580) R¹/R²=CH₂
(581) R¹=CH₃ R²=H

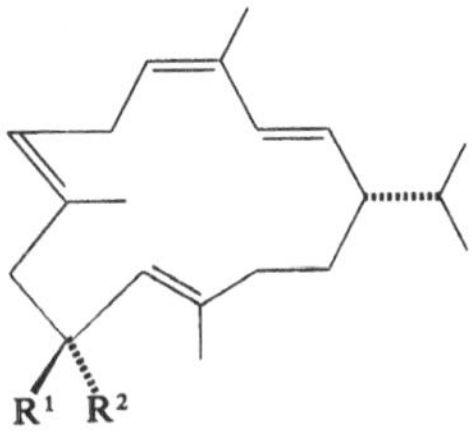

(582) R¹/R²=O
(583) R¹=H R²=OH
(584) R¹=H R²=OCH₃

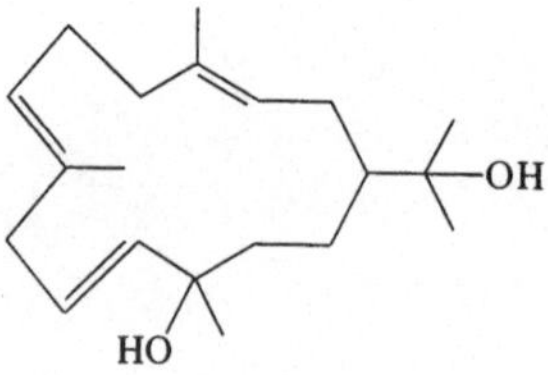

(585)

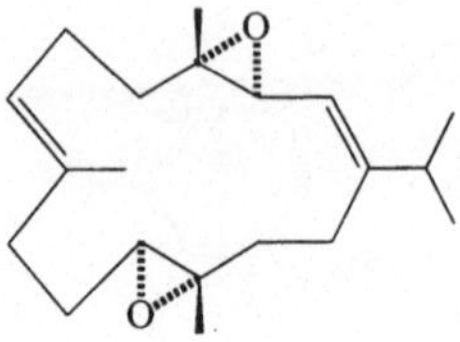

(586)

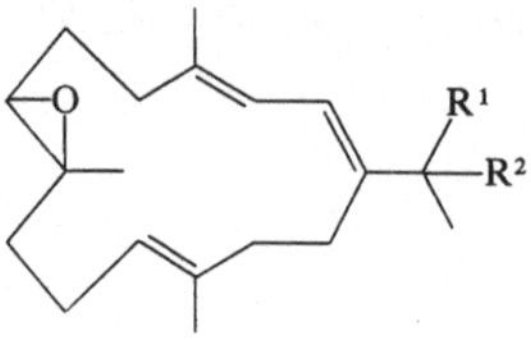

(587) R¹=H   R²=CH₃
(588) R¹=OH   R²=CH₃
(589) R¹/R²=CH₂

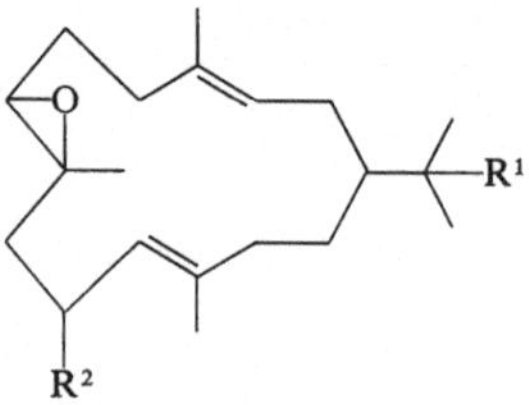

(590) R¹=OAc R²=H
(591) R¹=R²=OAc
(592) R¹=OH   R²=OAc

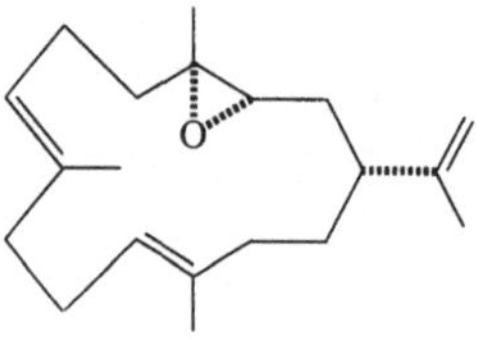

(593)

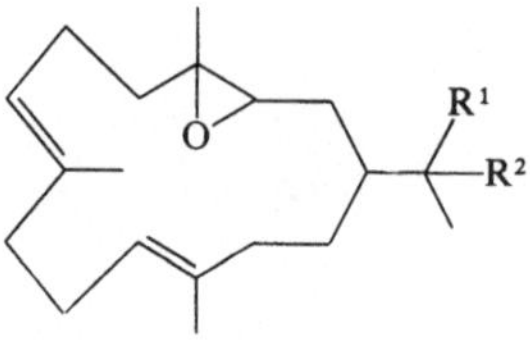

(594) R¹=CH₃ R²=OH
(595) R¹/R²=CH₂

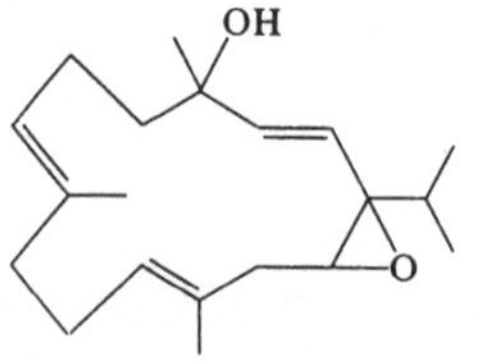

(596)

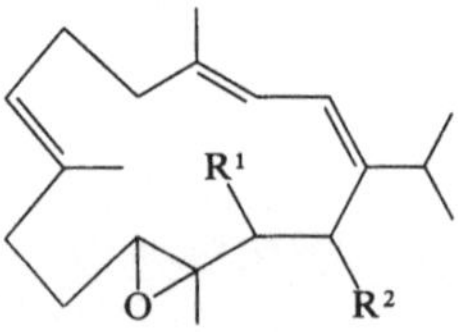

(597) R¹=OH   R²=OAc
(598) R¹=R²=H

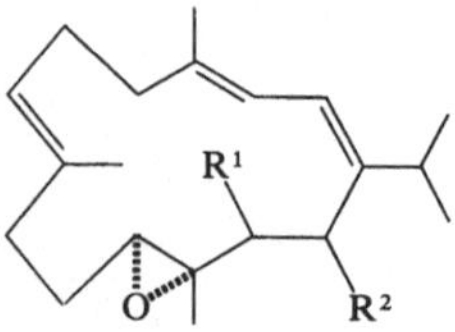

(599) R¹=R²=H
(600) R¹=H   R²=(S)—OAc
(601) R¹=H   R²=(S)—OH

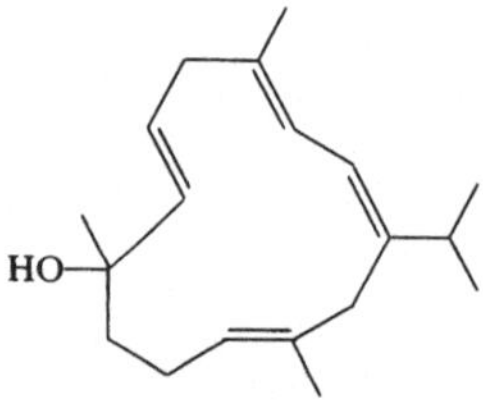

(602)

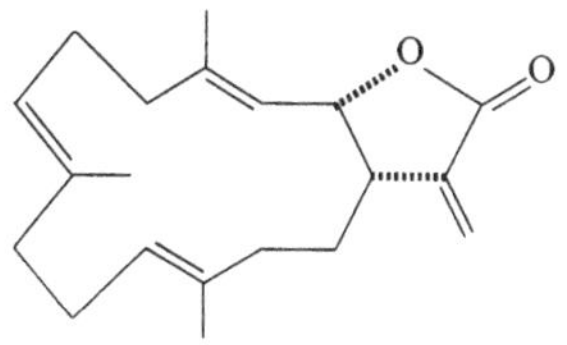

(603)

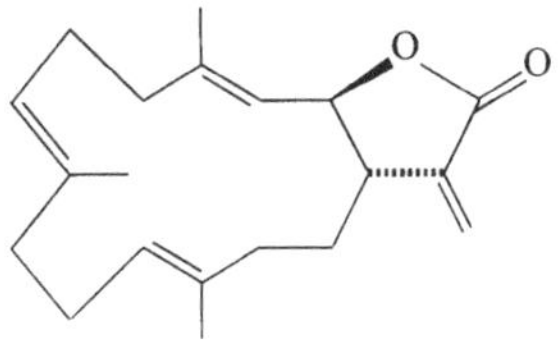

(604)

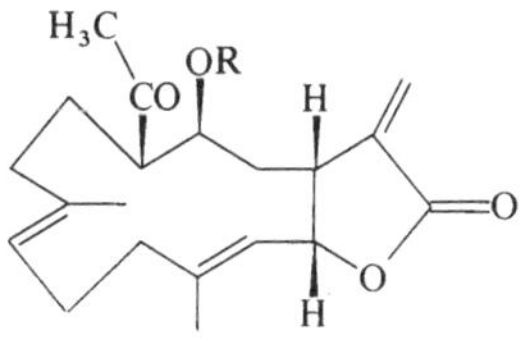

(605) R = Ac
(606) R = H

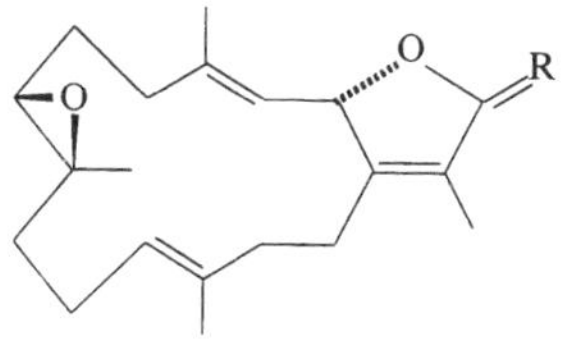

(607) R = O
(608) R = H$_2$

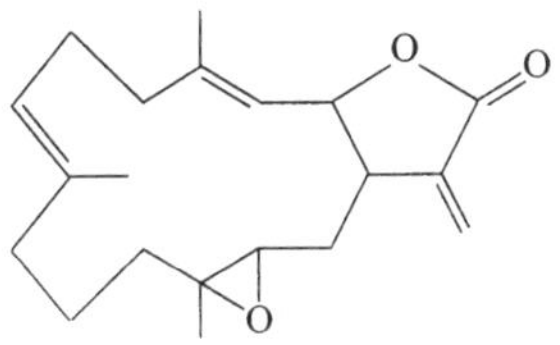

(609)

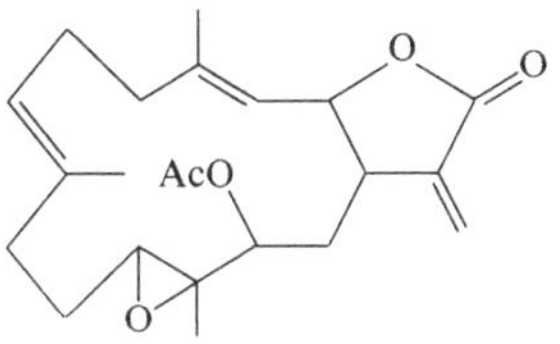

(610)

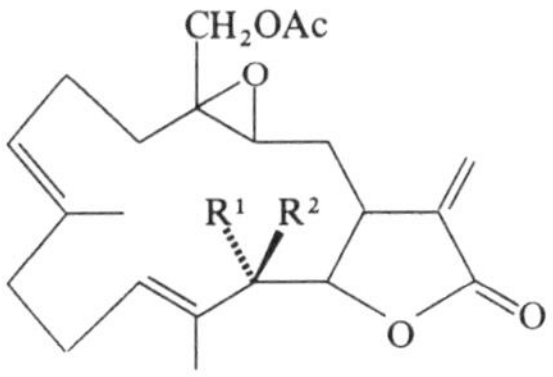

(611) R^1 = R^2 = H
(612) R^1 = H   R^2 = OH
(613) R^1 = OH   R^2 = H

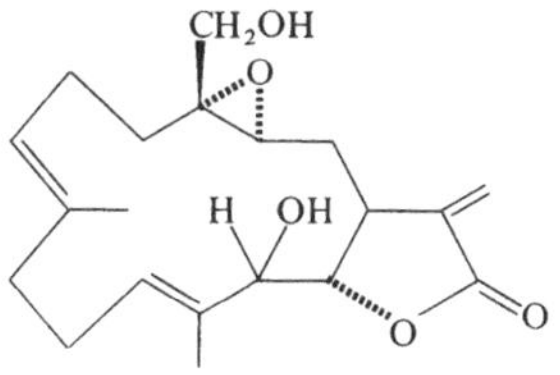

(614)

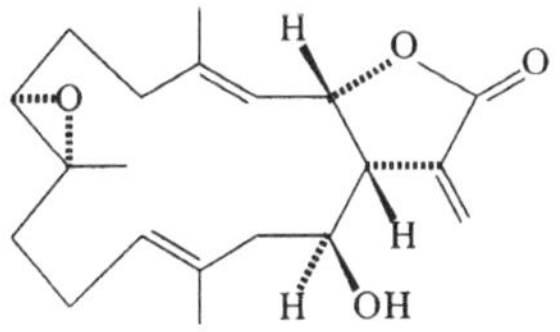

(615)

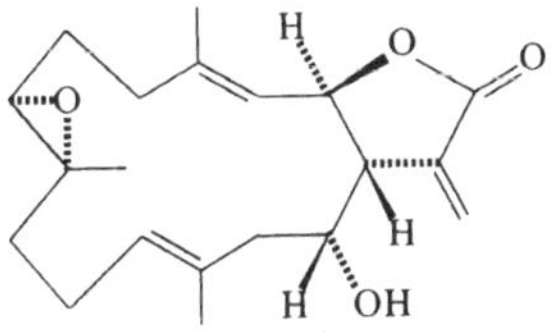

(616)

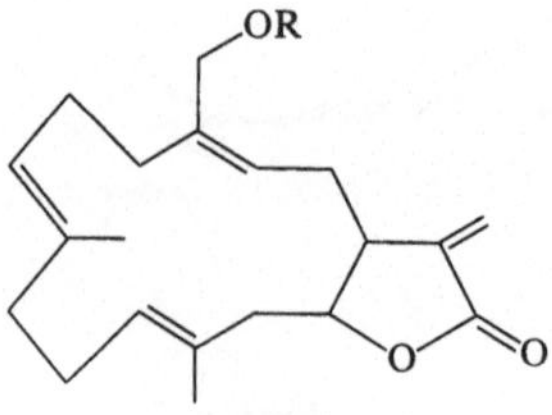

(617) R=H
(618) R=Ac

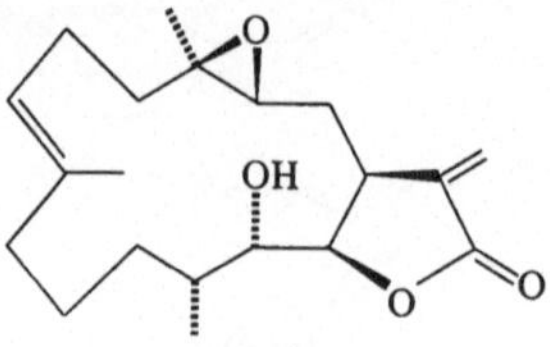

(619)

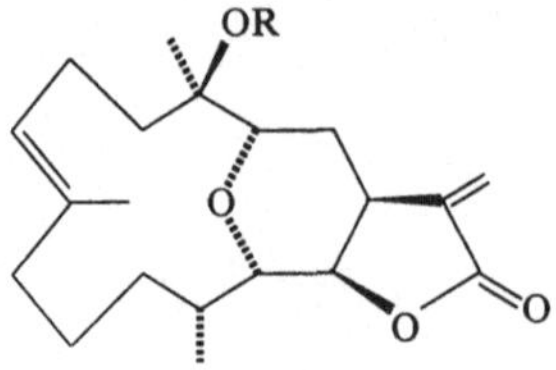

(620) R=H
(621) R=Ac

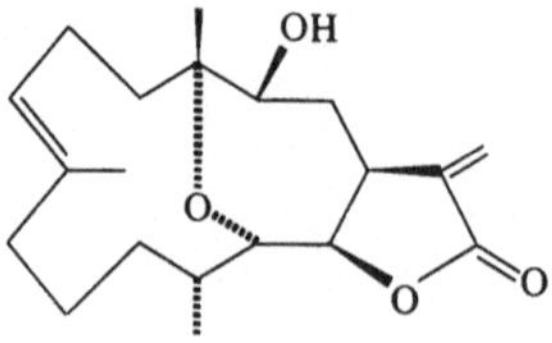

(622)

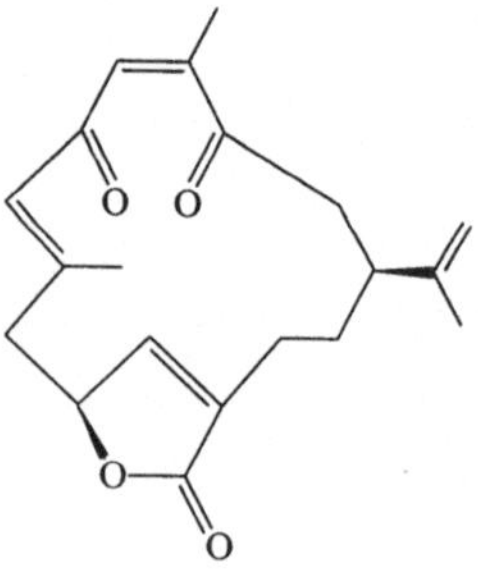

(623)

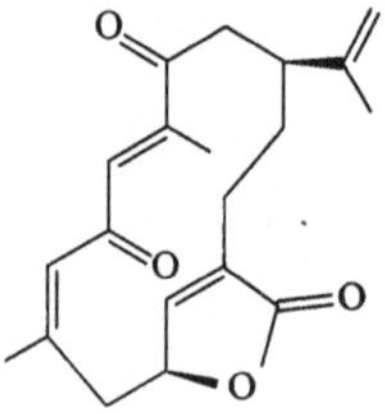

(624)

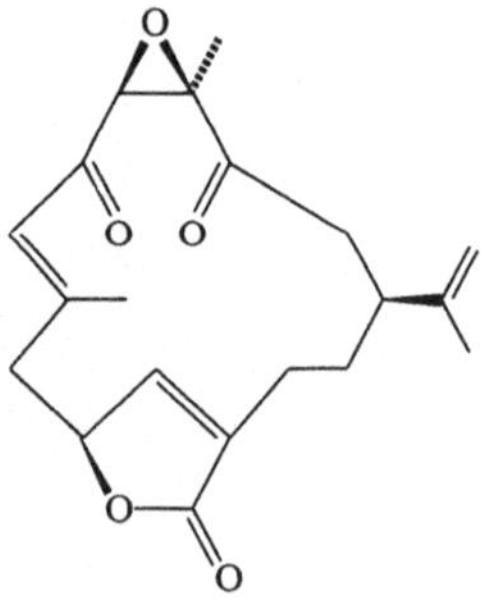

(625)

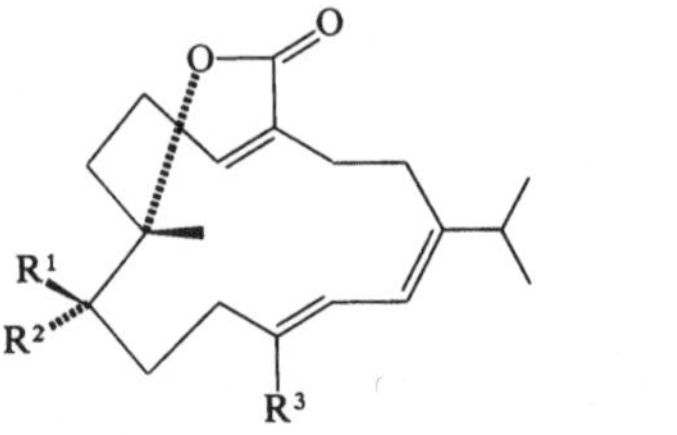

(626) R¹=OAc R²=H R³=COOCH₃
(627) R¹/R²=O   R³=CH₃
(628) R¹=OH   R²=H R³=CH₃
(629) R¹/R²=O   R³=COOCH₃

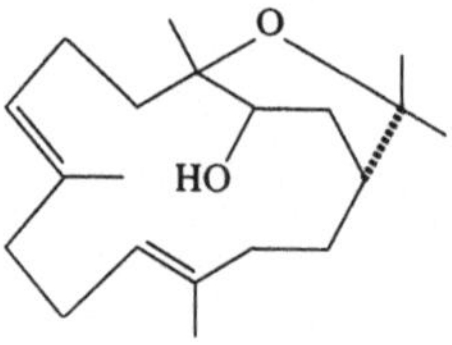

(630)

Pseudopterolide (**631**), an irregular constituted diterpenoid from the Caribbean gorgonian *Pseudopterogorgia acerosa*, with a 12-membered ring system bearing two isopropenyl groups (*456*) showed unusual cytotoxic properties. It inhibited overall cell cleavage but did not inhibit nuclear division in the fertilized sea urchin egg essay. Look et al. (*457*) reported three new such compounds, the calyculones A–C (**632–634**), as minor constituents of *Eunicea calyculata* which were new examples of the cubitane class of rearranged diterpenoids. A compound with the cubitane skeleton was first isolated from the defensive secretion of a termite, *Cubitermes umbratus*. The same group isolated two new dolabellanes **635** and **636** from *E. calyculata* (*458*). Dolabellanes were reported earlier as metabolites of soft corals, e.g. the epimeric diterpene alcohols **637** and **638** from the genera *Cespitularia*, *Efflatounaria*, and *Clavularia* (*410*), as well as from molluscs and sea weeds. Neodolabelline (**639**), which possesses a methyl migrated dolabellane-type skeleton, in which a methyl group has undergone migration, has been isolated from *Clavularia koellikeri* (*459*).

Two new chlorinated diterpenoids **640** and **641** from *Erythropodium caribaeorum* have been investigated by Look et al. (*460*). The structure of erythrolide A (**640**) was deduced by spectroscopic methods and X-ray crystallography and is related formally to the second compound, erythrolide B (**641**), by a di-π-methane rearrangement, thus leading to the claim that this is the first observation of this photochemical rearrangement in the production of natural products. Indeed, irradiation of **641** under a variety of conditions yielded **640** as the sole product. Brianthein X (**642**), Y (**643**) (*461*) and Z (**644**) (*462*) from *Briareum polyanthes* are further examples of chlorinated diterpenoids which contain a ten membered ring-system. Brianthein Y (**643**) is toxic to grasshoppers and exhibited no mutagenicity in *Salmonella* strains TA 98 and TA 100. Concentrations approaching 7 µg/mL initiated toxicity to the bacteria but still showed no sign of mutagenicity (*461*). Brianthein W (**645**), another minor metabolite of *B. polyanthes* had the same carbon skeleton as **642–644** but contains no chlorine (*463*). Compounds related to brianthein W (**645**) were isolated by Clastres et al. from two species of sea pens, *Pteroides laboutei* (*464*) and *Cavernulina grandiflora* (*465*). The former gave two ichthyotoxic compounds, pteroidine (**646**) and O-deacetyl-12-O-benzoyl-12-pteroidine (**647**) in addition to labouteine (**648**), and the latter cavernuline (**649**), O-deacetyl-propionyl cavernuline (**650**), and cavernulinine (**651**). Three related diterpenoids (**652–654**) were isolated by Ravi et al. (*466*) from the crude extract of the sea pen *Scytalium tentaculatum* which showed significant *in vitro* activity in the isolated guinea pig atrium.

Repeated silica gel column chromatography of the petroleum ether extract of a freeze dried *Muricella* species afforded ophirin (**655**) (*467*) which has the carbon skeleton of eunicellin and a diterpenoid alcohol (**656**) of the same type has been isolated from an unidentified soft coral collected at the Majuro Atoll (*468*). The gorgonian *Briareum asbestinum* furnished a series of compounds of a slightly different type, asbestinin-1 through -5 (**657–661**), asbestinin epoxide (**662**), and asbestinin-5 acetate (**663**) (*469, 470*). Investigation of this species was stimulated by the finding that the crude extract was highly toxic to goldfish. Compounds **657** and **661** showed the ability to antagonize the effects of acetylcholine on guinea pig ileum preparations at levels of 13% and 38%, respectively, at a concentration of 16 µg/mL. At the same concentration, asbestinin-5 (**661**) also exhibited histamine antagonism in the same assay at a level of 40%.

Obscuronatin (**664**) is a diterpenoid isoprenologue of a germacradiene which has been isolated from the soft coral *Xenia obscuronata* (*471*). This is a member of a genus which frequently affords diterpenes containing a cyclononane ring whose biogenesis derives from a relatively unusual cyclization of geranylgeraniol and is similar to the biogenesis of the lower isoprenologue caryophyllene. The first representative of this type was xenicin (**665**) (*471a*) and since then many other xenicins containing the 2-oxabicyclo[7.4.0]tridecane ring system of **665**, xeniolides, the corresponding lactones, and xeniaphyllanes, which have a bicyclo[7.2.0]undecane skeleton, have been isolated not only from *Xenia* species (*471, 472, 473*), but also from other soft corals of the genera *Nephthea* (*474*) and *Anthelia* (*475*), as well as from gorgonians (*476*). No biological activity has been described for any of these metabolites which are listed in Table 2.

From an Okinawan soft coral of the genus *Alcyonium*, KOBAYASHI *et al.* (*478*) have isolated a new diterpenoid acetate, alcyonolide (**701**), whose carbon skeleton is that of *seco*-xenicin. Three compounds (**702–704**) of this type have been isolated from several *Efflatounaria* species (*479*).

The diterpene **705** from a *Nephthea* sp. has been shown to be a eudesmane isoprenologue (*437*). Dichloromethane extraction of a soft coral of the genus *Lobophytum* yielded the elemadiene isoprenologue (**706**) with anticonvulsant activity in addition to four analogous but inactive novel diterpenes (**707–710**) (*480*). The anticonvulsant activity of some synthetic derivatives was also investigated, but only a monoacetate of **706** showed activity against pentetrazole-induced convulsion, and protected, like **706**, against the tonic-extensor phase of a convulsion produced by maximal electroshock.

RAVI and WELLS (*481*) isolated the quinone **711** and the chromene **712** in addition to several new lipids from the crude dichloromethane

Table 2. *Isolated Xenicins, Xeniolides and Xeniaphyllanes with Sources*

Compound	Source	Ref.
666	*Xenia macrospiculata*	*(471)*
667	*Xenia obscuronata*	*(471)*
668	*Xenia crassa*	*(472)*
669	*Anthelia edmondsoni*	*(475)*
670	*Anthelia edmondsoni*	*(475)*
671	*Xenia obscuronata*	*(471)*
672	*Xenia macrospiculata, X. obscuronata*	*(471)*
673	*Xenia macrospiculata, X. obscuronata*	*(471)*
674	*Xenia macrospiculata*	*(471)*
675	*Xenia macrospiculata, X. obscuronata*	*(471)*
676	*Corallium* sp.	*(476)*
677	*Corallium* sp.	*(476)*
678	*Corallium* sp.	*(476)*
679	*Xenia obscuronata, X. lilielae*	*(473)*
680	*Xenia obscuronata*	*(473)*
681	*Xenia macrospiculata*	*(471)*
682	*Corallium* sp.	*(476)*
683	*Corallium* sp.	*(476)*
684	*Xenia viridis*	*(472)*
685	*Efflatounaria* sp.	*(477)*
686	*Xenia macrospiculata, X. obscuronata*	*(471)*
687	*Xenia macrospiculata*	*(471)*
688	*Xenia macrospiculata, X. obscuronata*	*(471)*
689	*Xenia macrospiculata*	*(471)*
690	*Nephthea chabrolii*	*(474)*
691	*Nephthea chabrolii*	*(474)*
692	*Xenia macrospiculata*	*(471)*
693	*Xenia macrospiculata, X. obscuronata*	*(471)*
694	*Xenia macrospiculata*	*(471)*
695	*Xenia lilielae*	*(473)*
696	*Xenia macrospiculata, X. obscuronata, X. lilielae*	*(473)*
697	*Xenia lilielae*	*(473)*
698	*Xenia macrospiculata*	*(473)*
699	*Xenia macrospiculata*	*(473)*
700	*Xenia obscuronata*	*(473)*

extract of freeze-dried specimens of *Plexaura flava*. Both **711** and **712** were also metabolites of a *Nephthea* species, together with **713** which has been shown to be a dihydro derivative of **712** (*482*).

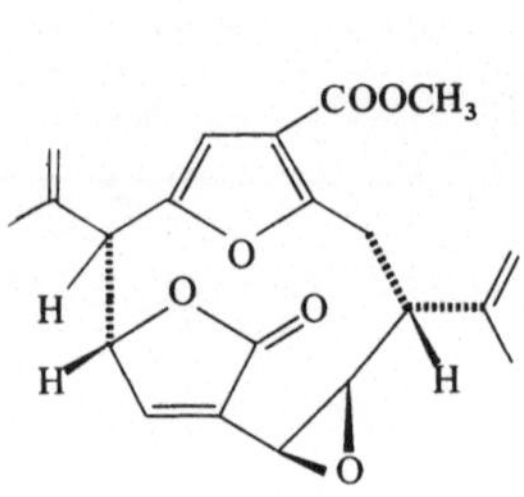

(631)

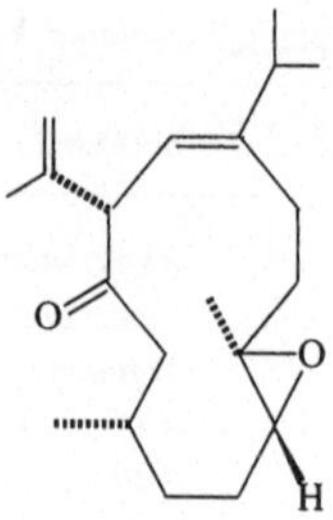

(632)

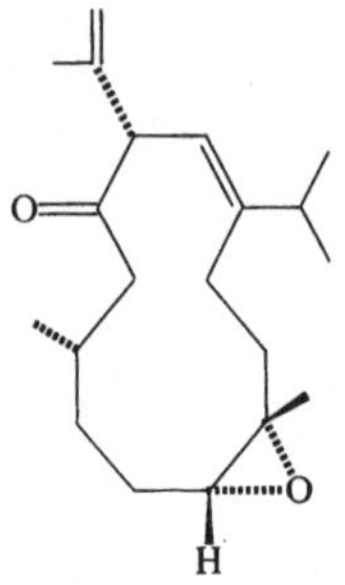

(633)

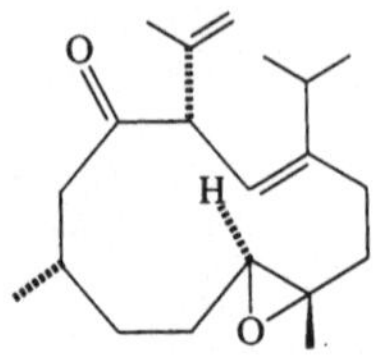

(634)

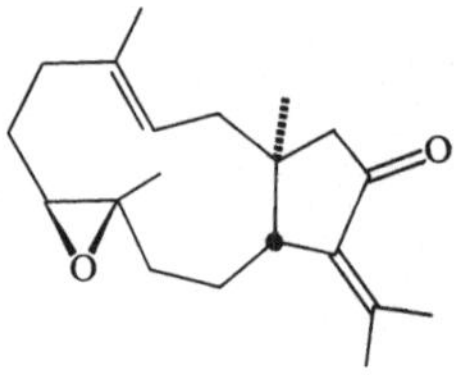

(635)

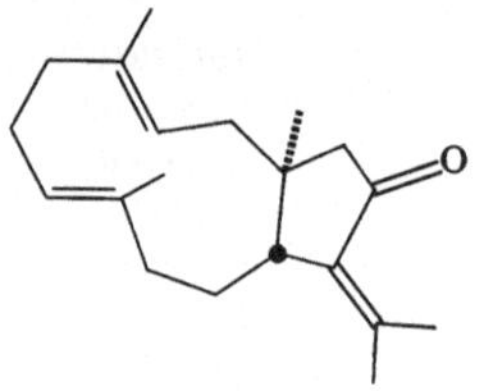

(636)

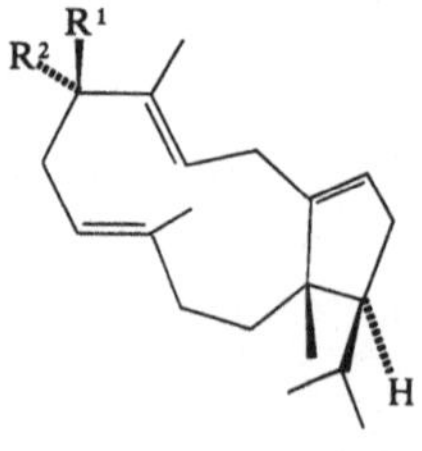

(637) R¹=H
R²=OH
(638) R¹=OH
R²=H

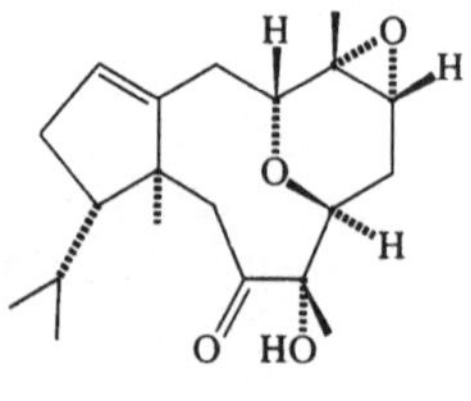

(639)

References, pp. 320–363

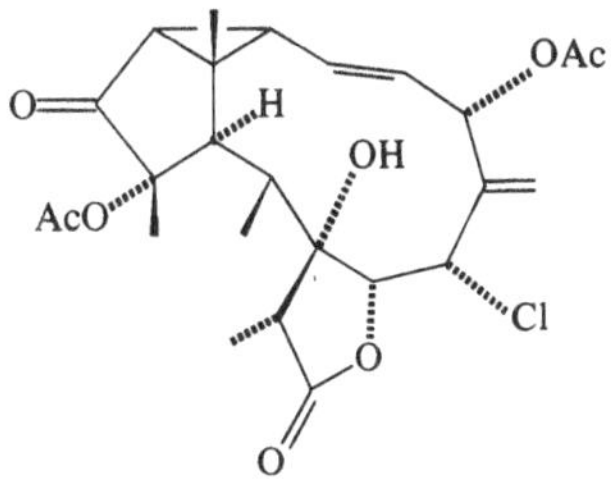

(640)

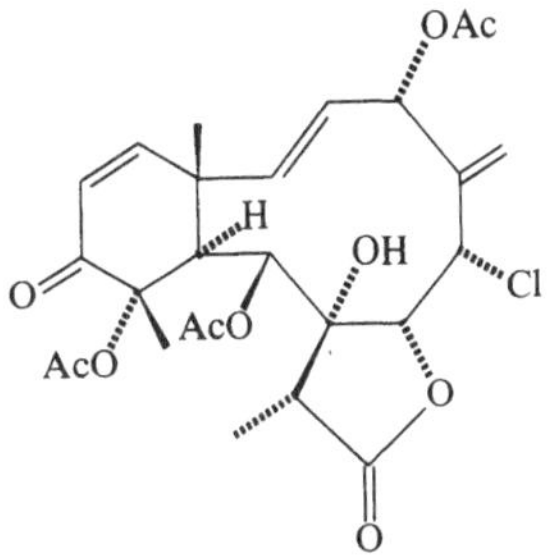

(641)

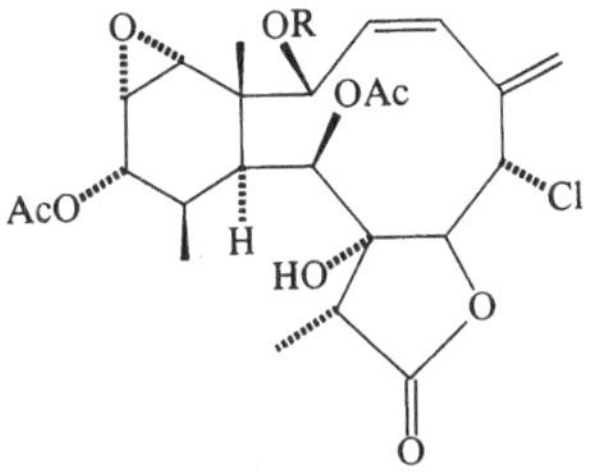

(642) R = H
(643) R = CO—CH$_2$—CH$_2$—CH$_3$
(644) R = Ac

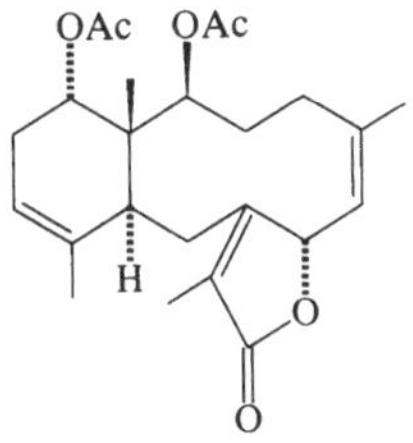

(645)

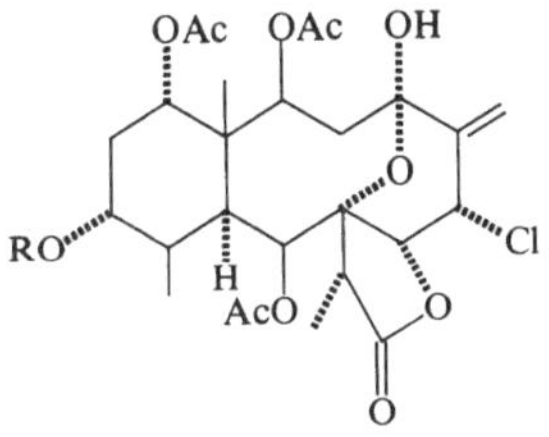

(646) R = Ac
(647) R = Bz

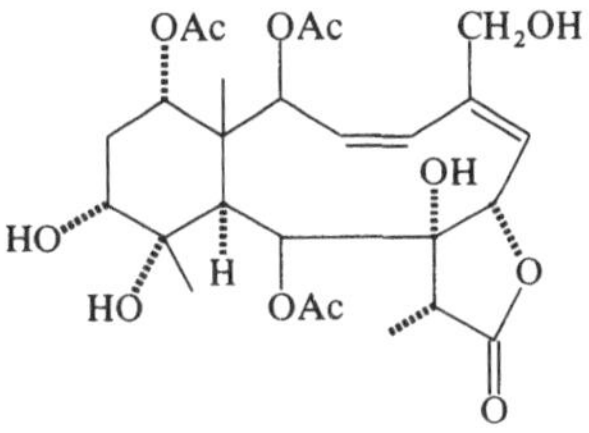

(648)

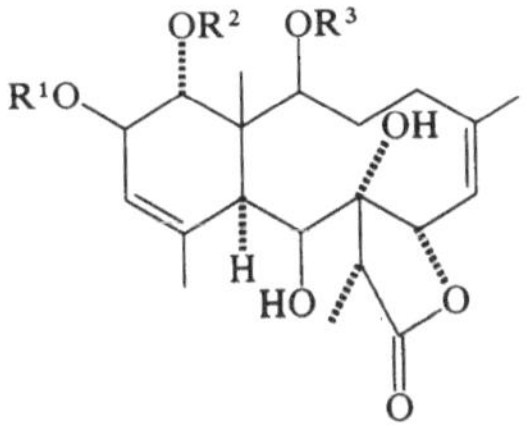

(649) R^1, R^2, R^3 = Ac, Ac, n-C$_5$H$_{11}$CO
(650) R^1, R^2, R^3 = Ac, C$_2$H$_5$CO, n-C$_5$H$_{11}$CO
(651) R^1, R^2, R^3 = C$_2$H$_5$CO, n-C$_3$H$_7$CO, n-C$_3$H$_7$CO

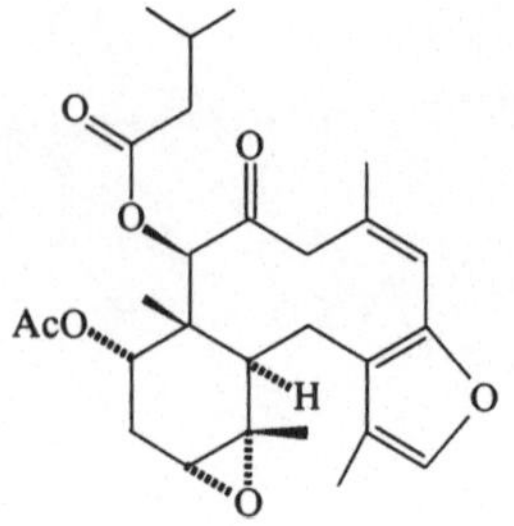

(652)

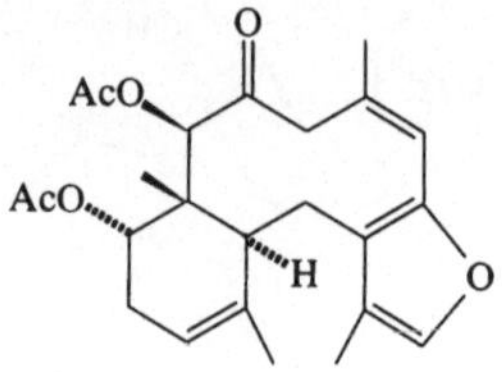

(653)

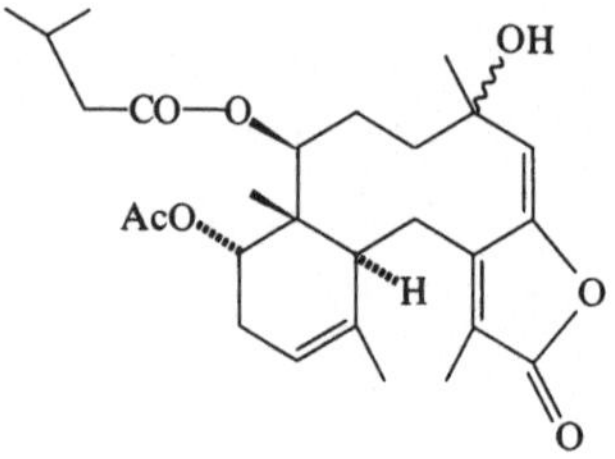

(654)

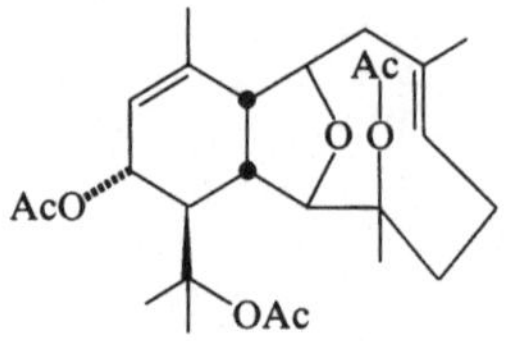

(655)

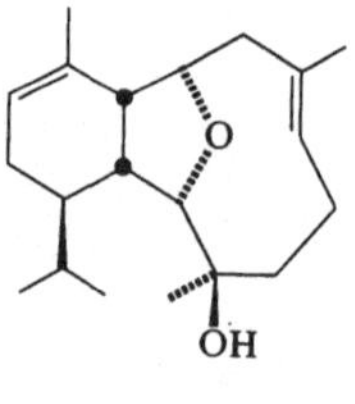

(656)

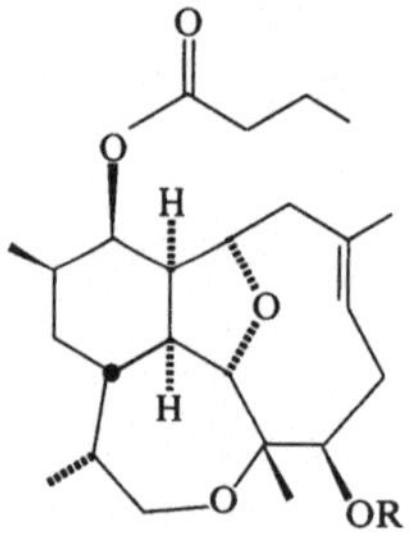

(657) R = Ac
(659) R = H

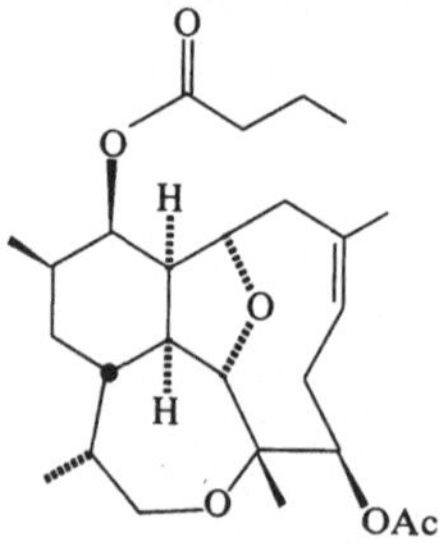

(658)

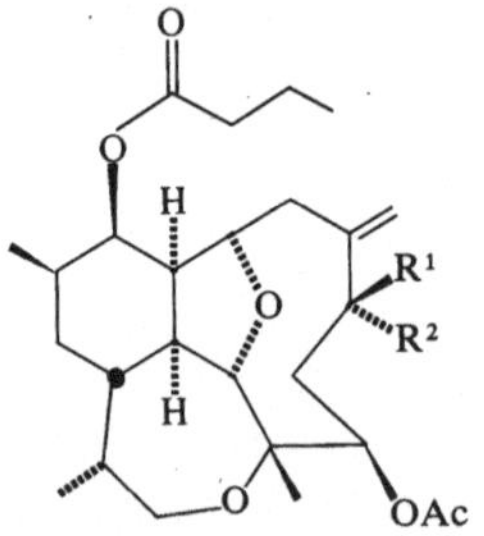

(660) $R^1/R^2 = O$
(661) $R^1 = H$ $R^2 = OH$
(663) $R^1 = H$ $R^2 = OAc$

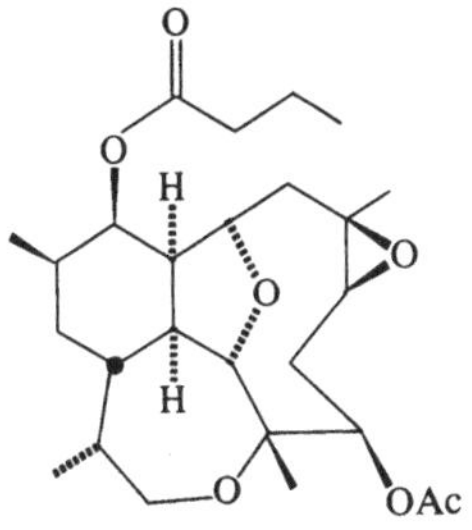

(662)

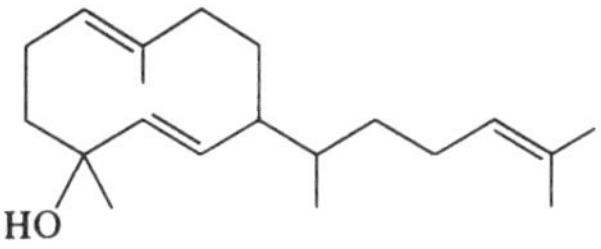

(664)

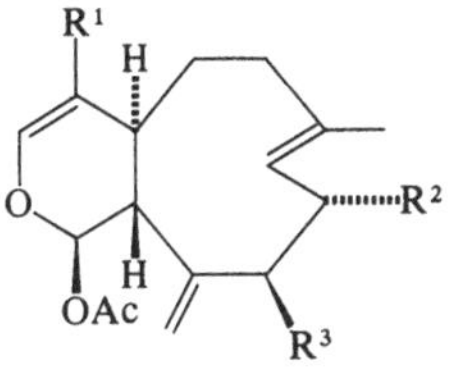

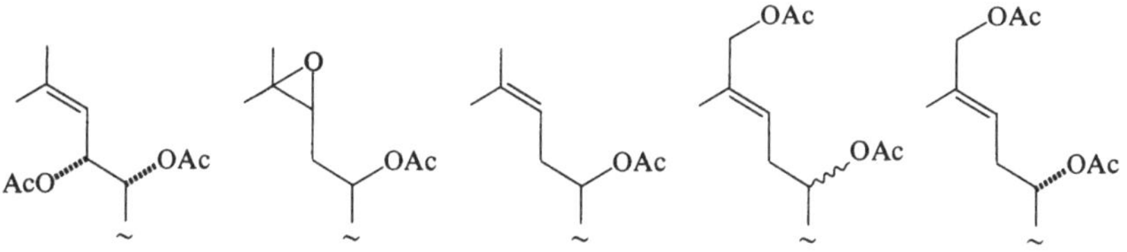

A B C D E

(665)	R^1=A	R^2=OAc	R^3=H
(666)	R^1=B	R^2=OAc	R^3=H
(667)	R^1=C	R^2=H	R^3=H
(668)	R^1=A	R^2=H	R^3=H
(669)	R^1=D	R^2=H	R^3=H
(670)	R^1=E	R^2=H	R^3=OH

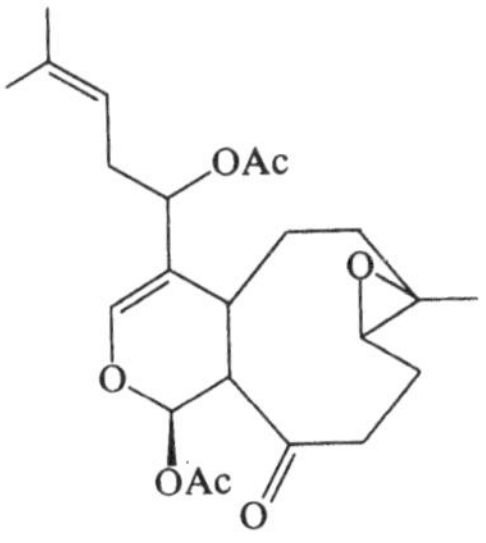

(671)

	R¹	R²	R³	R⁴	R⁵
(672)	H	A	O	H₂	OH
(673)	H	A	H₂	O	OH
(674)	H	A	H₂	O	OAc
(675)	H	A	H₂	H/OH	OH
(676)	H	B	O	H₂	H
(677)	C	H	O	H₂	H
(678)	H	C	O	H₂	H
(679)	H	D	H₂	H, OH	⋯OH
(680)	H	D	H₂	O	⋯OH

References, pp. 320–363

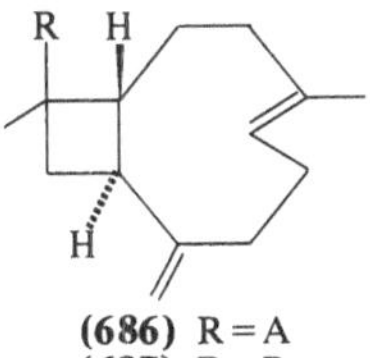

A=—CH₂—CH(OH)—CH=C(CH₃)₂
B=—CH₂—CH=ᵗCH—C(OH)(CH₃)₂
C=—CH₂—CH₂—CH(OH)—C(OH)(CH₃)₂
D=—CH₂—CH₂—CH(OAc)—C(OH)(CH₃)₂
E=—CH₂—CH₂—CH—C(CH₃)₂
                    \O/
F=—CH₂—CH₂—CH(OH)—CCl(CH₃)₂
G=—CH₂—CH(OH)—CH—C(CH₃)₂
                    \O/

(686) R=A
(687) R=B
(688) R=C
(689) R=D
(690) R=E
(691) R=F

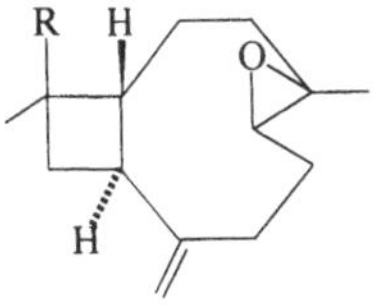

(692) R=A
(693) R=B
(694) R=D
(695) R=—E, *
(696) R=—C, *
(697) R=—G, *

*=β-epoxide

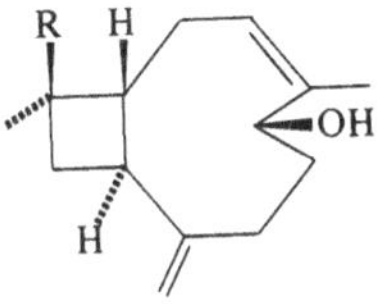

(698) R=—A

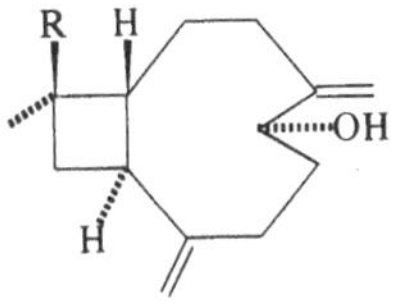

(699) R=—A
(700) R=—C

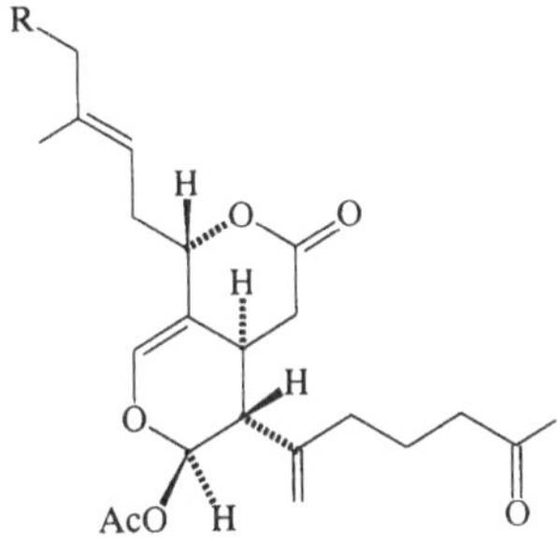

(701) R=H
(702) R=OAc

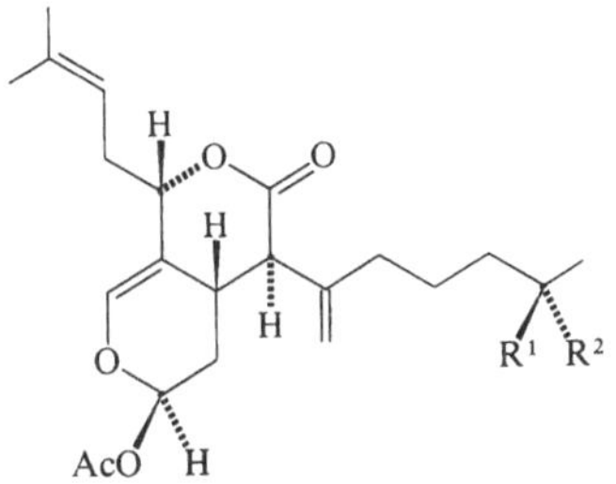

(703) R¹=OAc R²=H
(704) R¹/R²=O

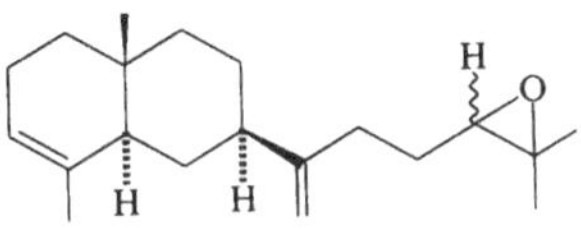

(705)

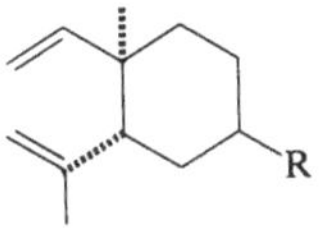

(706) R=

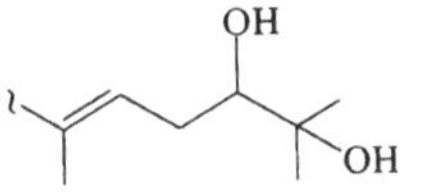

(707) R=

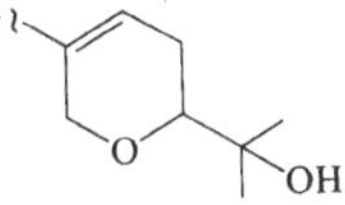

(708) R= [structure with epoxide]

(709) R= [structure with OAc groups and OH]

(710) R= [structure with OH, OH, OAc]

(711) [quinone with polyprenyl chain]

(712) [chromene with polyprenyl chain]

(713) [chroman with polyprenyl chain]

III.4.3. Miscellaneous Compounds from Octocorallia

Prostaglandins were known to occur in the gorgonians *Plexaura homomalla* and *Euplexaura erecta* already prior to 1980. Since then, the Red Sea soft coral *Lobophyton depressum* has been shown to contain four PGF derivatives, (15 S)-PGF$_{2\alpha}$-11-acetate methyl ester (**714**), the 18-acetoxy (**715**) derivative of **714** as well as the two corresponding free carboxylic acids (**716**) and (**717**) (*483*). Recently COREY *et al.* (*483a*) reported on the enzymatic transformation of arachidonic acid by a cell-free extract from the stolonifer *Clavularia viridis* to a new metabolite (**718**). Two Japanese groups independently investigated the prostanoids from *C. viridis*. KITAGAWA *et al.* isolated claviridenone-a (**719**), -b (**720**), -c (**721**), -d (**722**) (*484, 485*), 20-acetoxy-claviridenone-b (**723**), and 20-acetoxy-claviridenone-c (**724**) (*486*) from the acetone extract of this species. Compounds **719–722** were described as anti-tumor active and the claviridenones were also active against L 1210 leukemia cells. KIKUCHI *et al.* obtained what they called clavulones I to III (*487, 488*) and three 20-acetoxyclavulones (*489*) from *C. viridis*. Clavulones I–III which exhibited significant antiinflammatory activity were identical with **722, 721,** and **720,** respectively. Two of the 20-acetoxyclavulones were identical with **723** and **724** and the third was isomer **725**.

Recently BAKER *et al.* (*490*) described four punaglandins (**726–729**) from *Telesto riisei* which were characterized by unprecedented C-10 chlorine functions. Punaglandin 3 (**728**) inhibited L 1210 leukemia cell proliferation with an IC_{50} value of 0.02 µg/mL. Another four new halogenated marine prostanoids (**729a–d**), **729d** as the corresponding acetylated product (**729e**), with antitumor activity have been isolated from *Clavularia viridis* (*490a*). Chlorovulone I (**729a**) showed strong antiproliferative and cytotoxic activity in human promyelocytic leukemia (HL-60) cells *in vitro* with an IC_{50} value of 0.01 µg/mL. The biological properties of the clavulones/claviridenones have prompted work on their total synthesis (*491–493*).

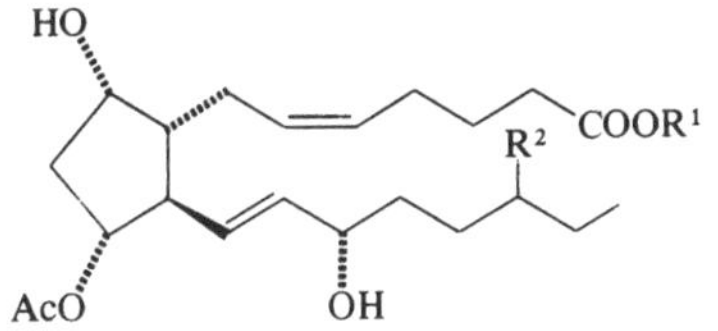

(**714**) $R^1 = CH_3$ $R^2 = H$
(**715**) $R^1 = CH_3$ $R^2 = OAc$
(**716**) $R^1 = H$ $R^2 = H$
(**717**) $R^1 = H$ $R^2 = OAc$

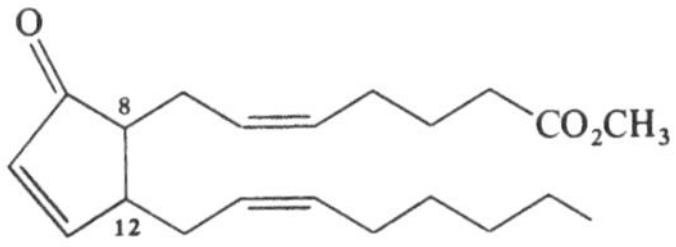

(**718**) H-8/H-12 cis

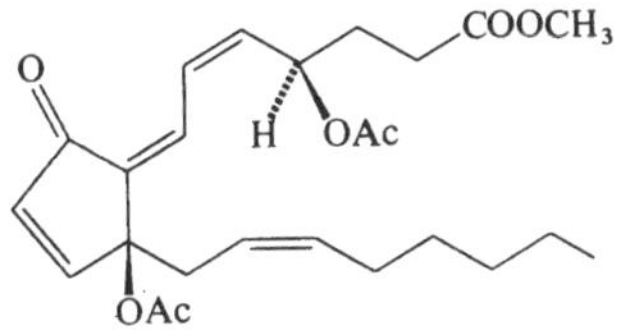

(**719**)

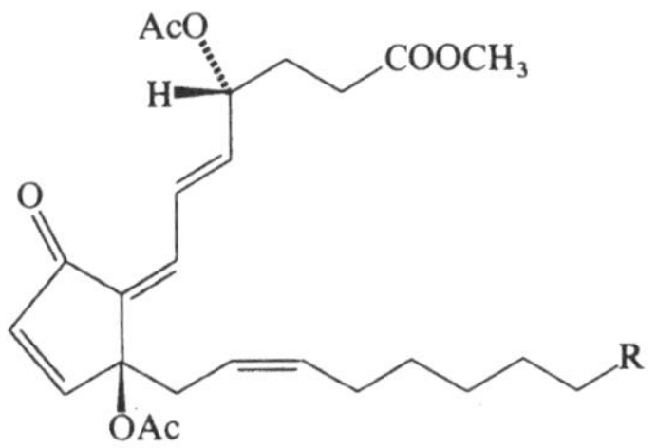

(**720**) R = H
(**723**) R = OAc

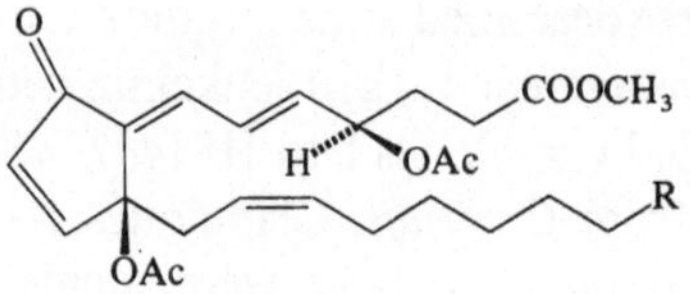

(721) R=H
(724) R=OAc

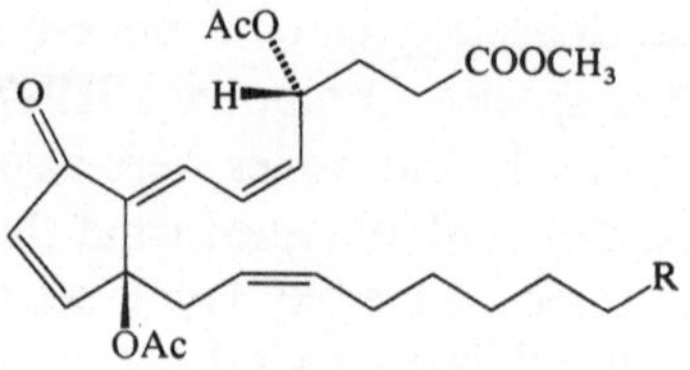

(722) R=H
(725) R=OAc

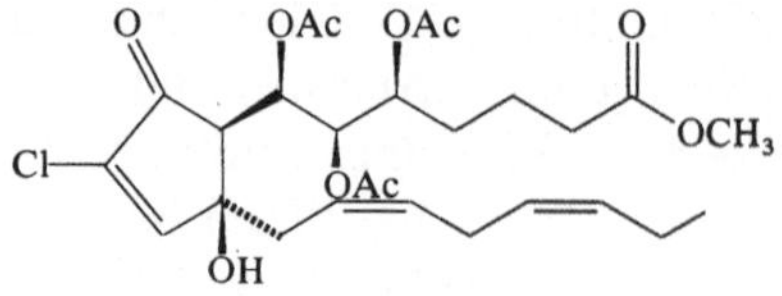

(726)

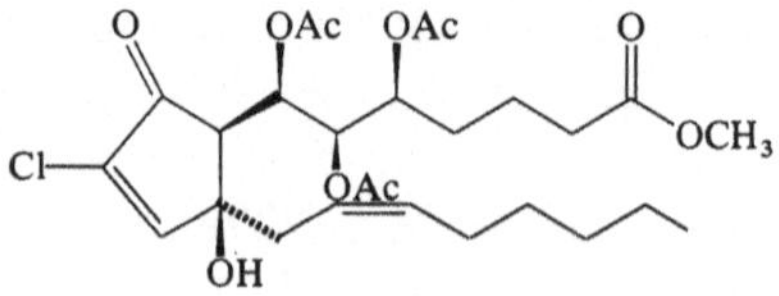

(727)

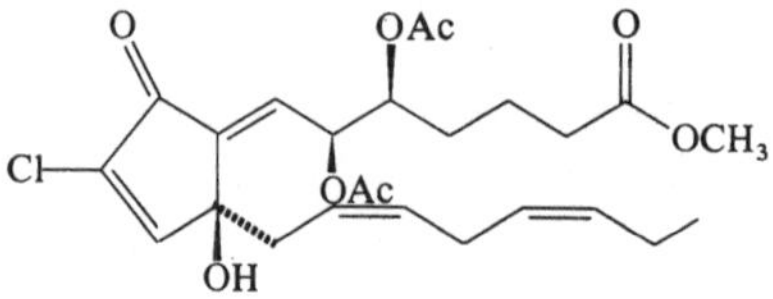

(728)

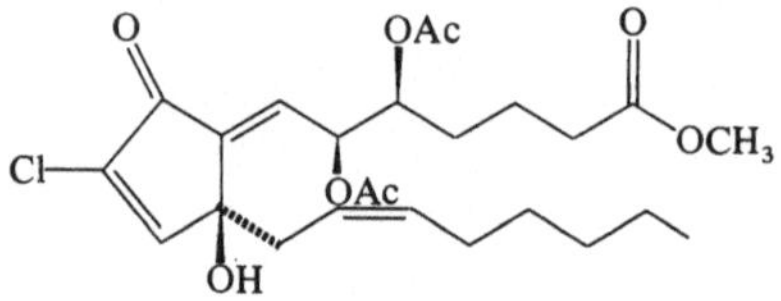

(729)

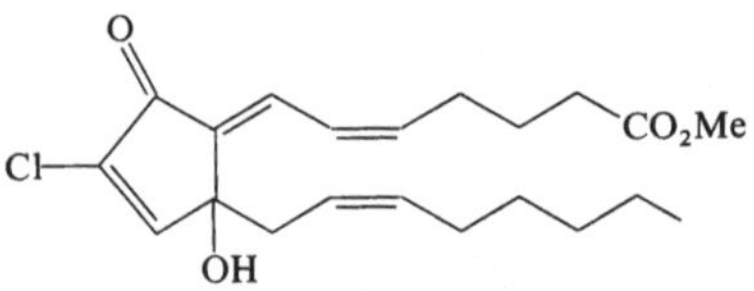

(729 a)

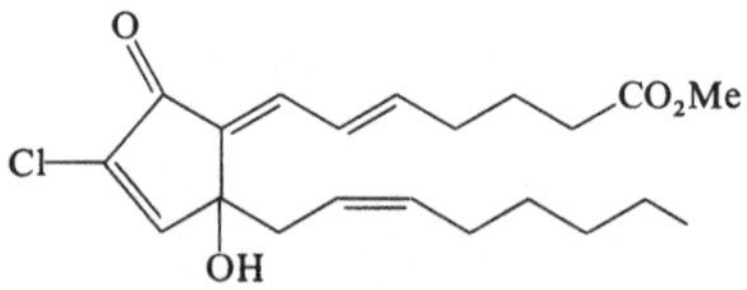

(729 b)

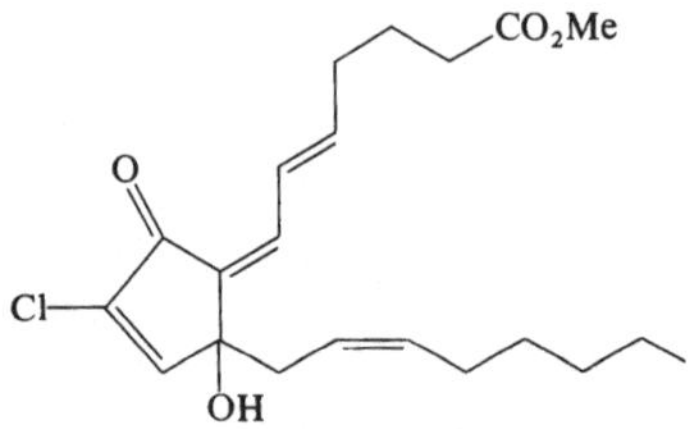

(729 c)

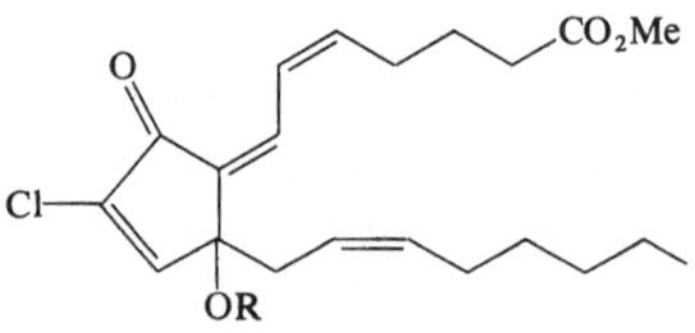

(729 d) R=H
(729 e) R=Ac

RAVI and WELLS (*481*) isolated six novel branched chain lipids (**730–735**) containing a γ-lactone ring from the gorgonian *Plexaura flava*. Similar compounds (**736–739**) have been found in the methanol-acetone extract of *Euplexaura flava* (*494*). The butenolides **736–739** showed a significant antiinflammatory effect at 100 μg/mL in the fertile egg test (*495*). The soft coral *Clavularia koellikeri* furnished clavularin A (**740**) and B (**741**), two metabolites belonging to a new class of cytotoxic compounds (*496*). It may be assumed that **740** either is formed from clavukerin C (**473**) (*396*) biogenetically or chemically during the isolation procedure.

Spermidine derivatives are known to have cytotoxic properties. KAZLAUSKAS *et al.* (*387*) isolated two new N-methylated amides (**742**) and (**743**) from a species of *Sinularia* after the methanol extract had shown potent *in vitro* and *in vivo* activity against a human pathogenic bacterium *Pseudomonas aeruginosa*. The amides **742** and **743** which accounted for the antimicrobial activity were toxic and therefore unlikely to be of value as antibiotics. Some efforts have been made toward the synthesis of such spermidine derivatives derived from corals (*387, 497*). Three amides (**744–746**) different from those mentioned earlier and derived from tyramine, have been reported as metabolites from *Sinularia flexibilis* (*498*). Structures of **744, 745**, and **746** were deduced spectroscopically and that of **744** has been confirmed by synthesis. The amides (**744–746**) accounted for the cardiotonic activity which has been observed for the crude dichloromethane extract of *S. flexibilis*.

BOWDEN *et al.* (*499*) in investigating a coral of the genus *Telesto* found that the coral was encrusted with epibiotic sponges and that it was almost impossible to separate the two organisms. Extraction of the freeze-dried material furnished 5-nonylpyrrole-2-carbaldehyde (**747**). Similar metabolites have been reported from sponges previously.

$CH_3-(CH_2)_n$

(**730**) n=15	R^1=OH	R^2=H	R^3/R^4=CH$_2$
(**732**) n=15	R^1=H	R^2=OAc	R^3=H R^4=CH$_3$
(**733**) n=13	R^1=H	R^2=OAc	R^3=H R^4=CH$_3$
(**734**) n=15	R^1=OAc	R^2=H	R^3=H R^4=CH$_3$
(**735**) n=13	R^1=OAc	R^2=H	R^3=H R^4=CH$_3$

(**731**)

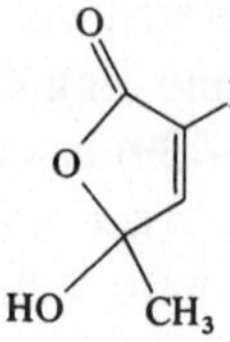

(736) R=—(CH$_2$)$_{15}$—CH$_3$
(737) R=—(CH$_2$)$_4$—CH=CH—CH=CH—CH=CH—(CH$_2$)$_8$—CH$_3$
(738) R=—(CH$_2$)$_4$—CH=CH—CH=CH—CH=CH—CH=CH—(CH$_2$)$_7$—CH$_3$
(739) R=—CH$_2$—CH=CH—CH=CH—CH=CH—CH=CH—CH=CH—(CH$_2$)$_7$—CH$_3$

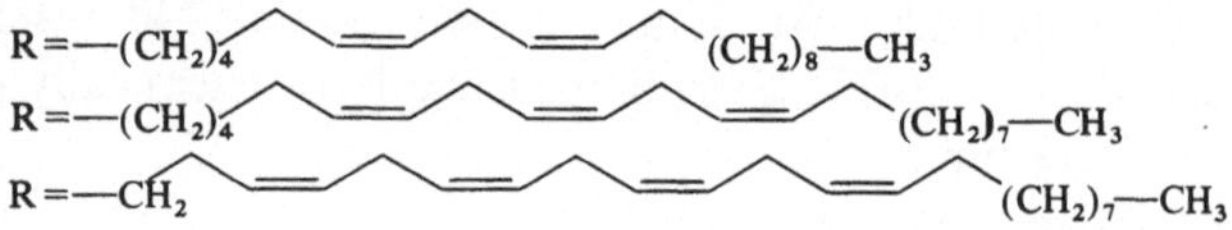

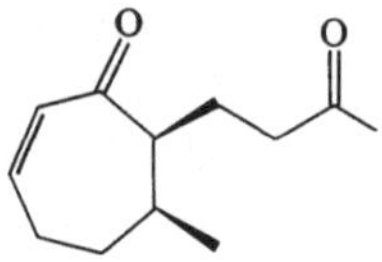

(740)

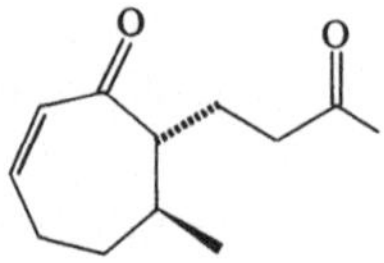

(741)

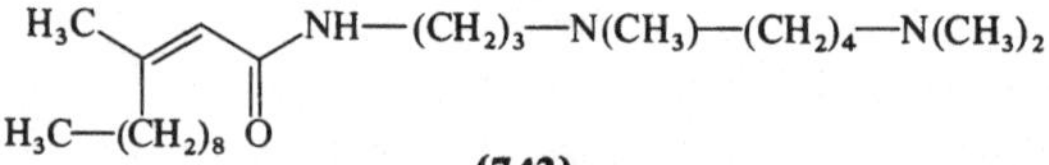

(742)

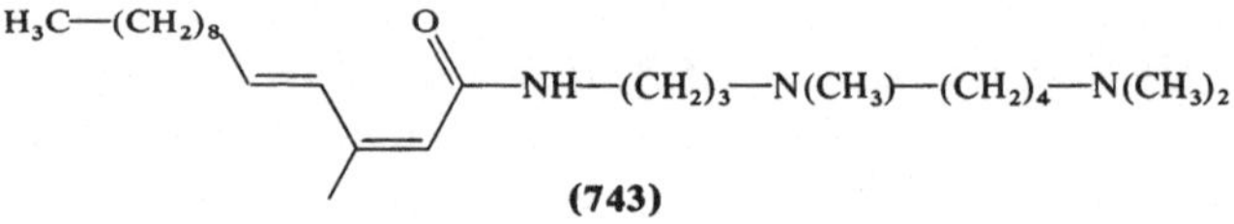

(743)

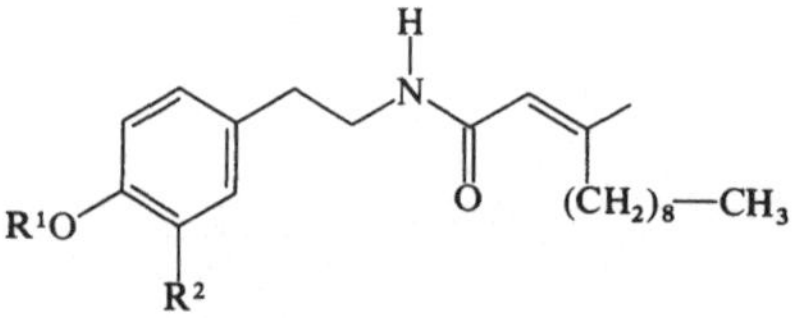

(744) R^1=H   R^2=H
(745) R^1=H   R^2=OCH$_3$
(746) R^1=CH$_3$ R^2=OH

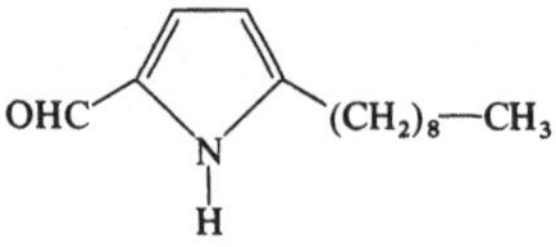

(747)

Goniopora toxin (GPT), a marine polypeptide composed of 105 amino acids which has been isolated from a coral belonging to the genus *Goniopora*, has been known for more than ten years. Recently a Japanese group examined its effects on guinea pig blood vessels (*500*), on non-adrenergic, non-cholinergic response and purine nucleotide release in guinea pig taenia coli (*501*), and also investigated the nicotine-induced response in guinea pig aorta enhanced by this toxin (*502*).

LESNIAK and LIU (*503*) have reported that extracts of the marine coral *Leptogorgia virgulata* possessed potent hemagglutination activity. This activity was apparently due to the presence of a lectin.

IV. Bryozoa

Marine animals of the phylum Bryozoa are colonial filter-feeders which contain biological active alkaloids and macrolides. In addition to these two types of metabolites, an agent that caused an eczematous allergic contact dermatitis known as the "Dogger Bank Itch" has been decribed (*504–506*). The dermatitis, which is a severe occupational disease is especially widely distributed among trawlermen working in the Dogger Bank area in the North Sea. It has been shown that the allergy is due to repeated exposure to the marine bryozoan *Alcyonidium gelatinosum*. Alcohol extraction followed by gel permeation and ion exchange chromatography furnished the allergen which was shown spectroscopically to be (2-hydroxyethyl)dimethylsulfoxonium ion (**748**). The structure has been confirmed by synthesis.

The marine bryozoan *Flustra foliacea* furnished a series of brominated alkaloids called flustramines A (**749**), B (**750**) (*507*), C (**751**), flustraminol A (**752**), B (**753**) (*508*), dihydroflustramine C (**754**) (*509*), flustrabromine (**755**) (*510*), and 7-bromo-4-(2-ethoxyethyl)quinoline (**756**) (*511*). Structures **749–754** are all 6-bromo-substituted prenylated analogs of the physostigmines, a group of alkaloids from the Calabar bean. While no biological activity has been reported for **749–753**, dihydroflustramine C (**754**) show strong activity against *Bacillus subtilis* (*509*). Flustrabromine (**755**) was a 6-bromotryptamine derivative and **756** has been described as the first naturally occurring bromoquinoline alkaloid. Two further new brominated alkaloids, 2,5,6-tribromo-N-methylgramine (**757**) and its derived side chain N-oxide (**758**), have been isolated from *Zoobotryon verticillatum* (*512*). For confirmation of structures **757** and **758**, both compounds have been synthesized. Initial tests indicated that **757** inhibits cell division of the fertilized sea urchin egg. AYER *et al.* (*513*) isolated phidolopin (**759**), a purine derivative from the bryozoan *Phidolopora pacifica*. The substance, which has been synthesized by another group (*514*), was largely responsible for the strong *in vitro* antifungal and antialgal activity of the extract from this species.

In 1982 PETTIT et al. *(515)* reported the first structure of a series of macrolides derived from bryozoans. Bryostatin 1 **(760)** from *Bugula neritina* showed antineoplastic activity at an extremely low dose level. The complete structure has been deduced by crystallographic and spectroscopic techniques. The second member of this group, bryostatin 2 **(761)**, has been shown to be a 20-membered ring lactone, too *(516)*. This very potent anticancer constituent has been isolated together with microgram quantities of 15 more highly active bryostatins from *B. neritina*. Bryostatins 3 **(762)** *(517)* and 4 **(763)** *(518)* were two other active macrolides from the same species. The discovery of **763** in geographically distant collections of this bryozoan suggested that these compounds might be true biosynthetic products of the animal. Another marine bryozoan species, *Amathia convoluta*, furnished bryostatins 4 **(763)**, 5 **(764)**, 6 **(765)**, and 8 **(766)** *(519)*. The authors presumed that **766** was a genuine constituent of *A. convoluta*, while **763–765** appeared to be derived from the closely related *B. neritina*.

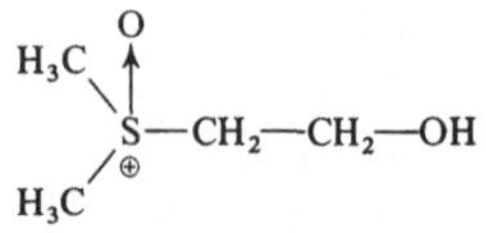

(748)

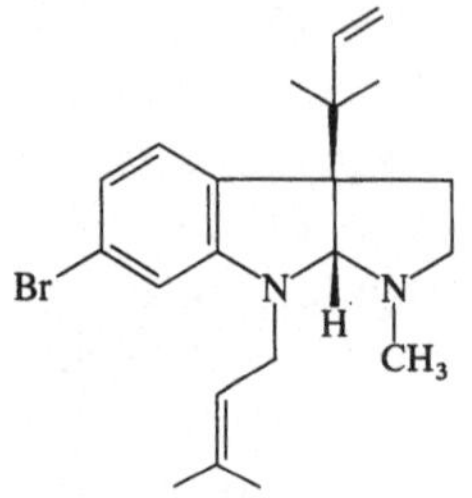

(749)

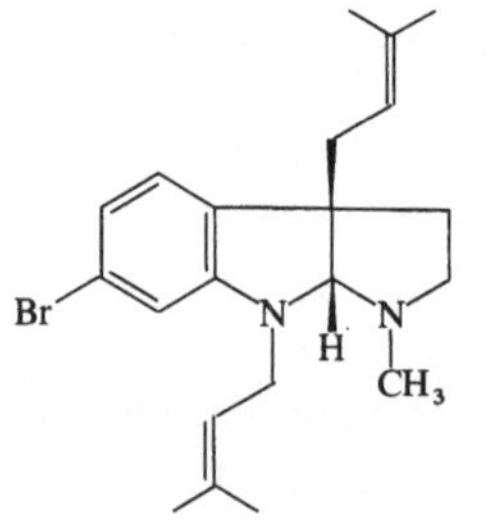

(750)

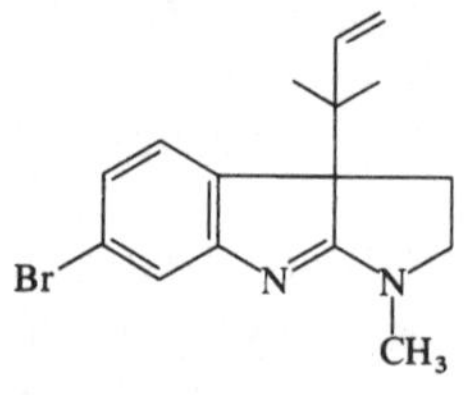

(751)

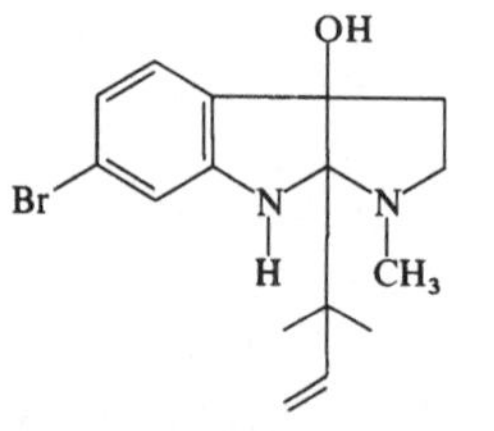

(752)

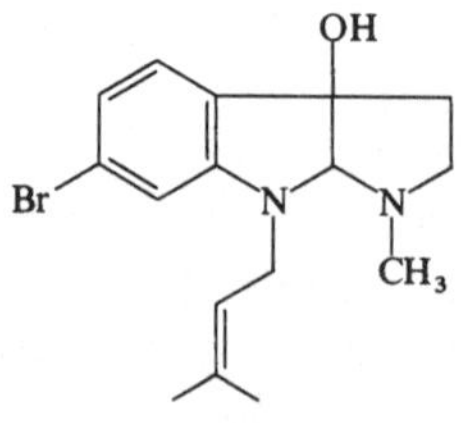

(753)

References, pp. 320–363

(760) R=CO—CH=CH—CH=CH—CH₂—CH₂—CH₃   R₁=CO—CH₃
(761) R=CO—CH=CH—CH=CH—CH₂—CH₂—CH₃   R₁=H
(763) R=CO—CH₂—CH₂—CH₃   R₁=CO—CH₂—CH(CH₃)₂
(764) R=CO—CH₃   R₁=CO—CH₂—CH(CH₃)₂
(765) R=CO—CH₃   R₁=CO—CH₂—CH₂—CH₃

(762)

(766)

V. Mollusca

V.1. Gastropoda

Marine snails contain many different chemical constituents with various biological activities. There are more than 3,000 species of opisthobranchs, a subclass of the gastropod molluscs. Most opisthobranchs lack a protective shell, but are protected by chemical substances with toxic or antifeedant activities. The nudibranchs which lack the shell form the largest order in the subclass Opisthobranchia. They are often beautifully colored and it has been pointed out that this fact may provide camouflage against the background of a similarly colored food source. Nudibranchs feed on sponges and most of the compounds isolated from them are derived from dietary sources.

Many unusual and potentially noxious compounds have been isolated from another kind of opisthobranch molluscs, the sea hares or Aplysiidae. These animals can be easily collected in the intertidal zone and contain large quantities of metabolites which are identical with or closely related to metabolites of algae.

Marine snails belonging to the genus *Conus* are venomous predators that immobilize their prey by a highly specialized venom apparatus. The venom which is of proteinaceous or peptide nature is injected into the prey by means of a disposable hollow tooth which serves both as harpoon and hypodermic needle. A number of human fatalities have resulted from the sting of *Conus geographus*.

As the nudibranchs and the sea hares derive most of the substances which have been isolated from them from their food, the chemistry of these molluscs will be reviewed in two separate chapters. A third chapter deals with the constituents of Conidae because of their specialized venom apparatus which contains proteins and peptides.

V.1.1. Nudibranchia

The brightly colored and delicately shaped nudibranchs form the largest order in the subclass Opisthobranchia. As they lack an external shell these animals are not protected physically from their predators, but nevertheless the nudibranchs are rarely eaten by fish. There have been many studies of their defence mechanism. Some of the animals exude a strongly acidic skin secretion, other species are reported to secrete toxic or noxious substances. Nudibranchs feed on sponges as well as on coelenterates and bryozoans, and it has been demonstrated that they selectively accumulate chemical defensive agents from their food. Indeed, many metabolites have been isolated from nudibranches and from sponges. CIMINO et al. (520) investigated novel metabolites from some predator-prey pairs and found terpenes and acetylenic compounds. Although most of the substances isolated from both kinds of animals are terpenes or steroids, mention should be made of doridosine which is identical with 1-methylisoguanosine (**290**) and has been found in the sponge *Tedania digitata* (*176, 177*) as well as in the nudibranch *Anisodoris nobilis* (*521, 522*). The pharmacological properties of **290** are described in chapter II.5. Isoguanosine has also been isolated from a marine nudibranch mollusk, *Diaulula sandiegensis* (*523*). It produces hypotension in mammals, bradycardia, and relaxation of smooth muscle. Compounds which have been found in Porifera as well as in Nudibranchia are summarized in Table 3.

Table 3. *Compounds Isolated from Porifera and Nudibranchia*

Compound	Porifera	Ref.	Mollusca	Ref.
3	*Hyrtios* sp.	(*62*)	*Adalaria* sp.	(*386*)
	Thalysia junipertina	(*67*)		
4	*Hyrtios* sp.	(*62*)	*Adalaria* sp.	(*386*)
	Thalysia junipertina	(*67*)		
	Axinella cannabina	(*386*)*		
5	*Hyrtios* sp.	(*62*)	*Adalaria* sp.	(*386*)
	Thalysia junipertina	(*67*)		
	Axinella cannabina	(*386*)*		
	Tethya aurantia	(*386*)*		
6/7	*Hyrtios* sp.	(*62*)	*Adalaria* sp.	(*386*)
	Thalysia junipertina	(*67*)		
	Axinella cannabina	(*386*)*		
	Tethya aurantia	(*386*)*		
9	*Hyrtios* sp.	(*62*)	*Adalaria* sp.	(*386*)
	Thalysia junipertina	(*67*)		
	Axinella cannabina	(*386*)*		
10	*Hyrtios* sp.	(*62*)	*Adalaria* sp.	(*386*)
	Thalysia junipertina	(*67*)		
	Axinella cannabina	(*386*)*		
	Tethya aurantia	(*386*)*		
13	*Thalysia junipertina*	(*67*)	*Adalaria* sp.	(*386*)
	Tethya aurantia	(*386*)*		
35	*Dysidea fragilis*	(*74*)	*Hypselodoris godeffroyana*	(*74*)
			Chromodoris maridadilus	(*74*)
			Hypselodoris californiensis	(*78*)

References, pp. 320–363

Table 3 (*continued*)

Compounds	Porifera	Ref.	Mollusca	Ref.
36	*Dysidea fragilis*	*(74)*	*Hypselodoris godeffroyana*	*(74)*
			Chromodoris maridadilus	*(74)*
			Hypselodoris ghiselini	*(78)*
44	*Dysidea fragilis*	*(76)*	*Cadlina luteomarginata*	*(524)*
50	*Dysidea amblia*	*(77)*	*Cadlina luteomarginata*	*(524)*
			Hypselodoris californiensis	*(78)*
52	*Euryspongia* sp.	*(78)*	*Hypselodoris californiensis*	*(78)*
	Dysidea herbacea	*(79)*	*Hypselodoris porterae*	*(78)*
			Hypselodoris zebra	*(80)*
60	*Dysidea amblia*	*(524)*	*Cadlina luteomarginata*	*(524)*
	Dysidea etheria	*(80)*	*Cadlina luteomarginata*	*(525)*
			Hypselodoris californiensis	*(78)*
			Hypselodoris porterae	*(78)*
			Hypselodoris zebra	*(80)*
119	*Dendrilla* sp.	*(94)*	*Chromodoris norrisi*	*(102)*
156	*Spongia idia*	*(115)*	*Cadlina luteomarginata*	*(524)*
290	*Tedania digitata*	*(176)*	*Anisodoris nobilis*	*(521)*
	Tedania digitata	*(177)*	*Anisodoris nobilis*	*(522)*
368	*Petrosia ficiformis*	*(227)*	*Peltodoris atromaculata*	*(227)*
369	*Petrosia ficiformis*	*(227)*	*Peltodoris atromaculata*	*(227)*
767	*Ircinia fasciculata*	*(520)**	*Dendrodoris grandiflora*	*(520)*
768	*Pleraplysilla spinifera*	*(520)**	*Glossodoris gracilis*	*(520)*
	Pleraplysilla spinifera	*(526)**	*Glossodoris valenciennesi*	*(526)*
769	*Axinella* sp.	*(524)*	*Cadlina luteomarginata*	*(524)*
770	*Axinella* sp.	*(524)*	*Cadlina luteomarginata*	*(524)*
771	*Axinella cannabina*	*(526)**	*Phyllidia pulitzeri*	*(526)*
772	*Cacospongia mollior*	*(526)**	*Glossodoris tricolor*	*(526)*
773	*Spongia officinalis*	*(526)**	*Glossodoris tricolor*	*(526)*
774	*Spongia* sp.	*(527)**	*Casella atromarginata*	*(527)*
775	*Spongia* sp.	*(527)**	*Casella atromarginata*	*(527)*
776	*Dysidea* sp.	*(525)**	*Cadlina luteomarginata*	*(525)*
777	*Microciona toxystila*	*(525)**	*Cadlina luteomarginata*	*(525)*
			Dendrodoris grandiflora	*(528)*
778	*Pleraplysilla spinifera*	*(529)**	*Hypselodoris daniellae*	*(529)*

(*XXX*)* = see references in (*XXX*)

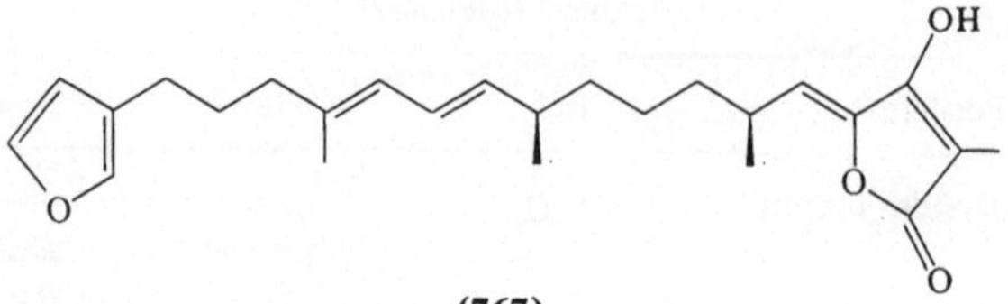

(767)

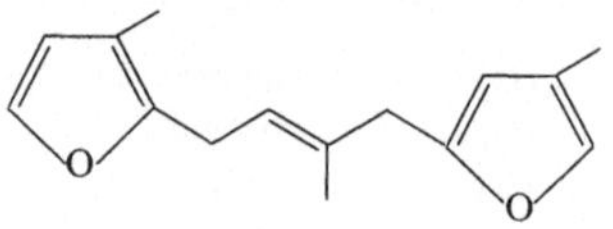

(768)

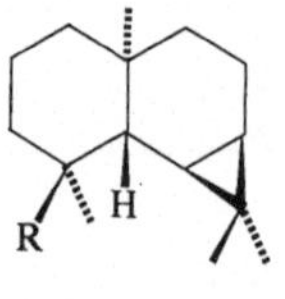

(769) R=—NC
(770) R=—NCS

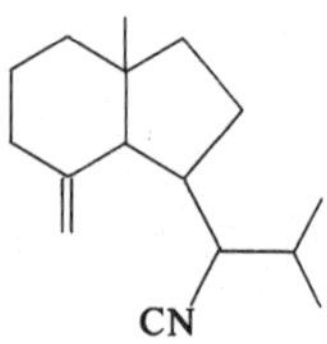

(771)

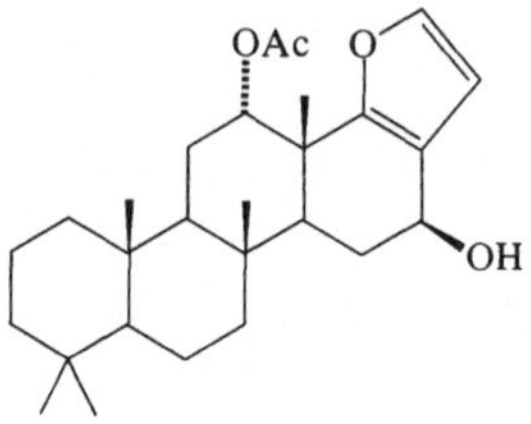

(772)

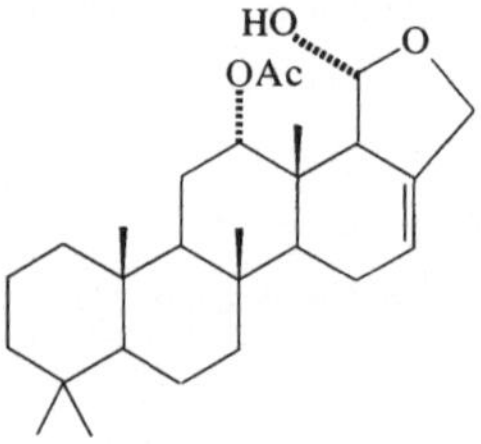

(773)

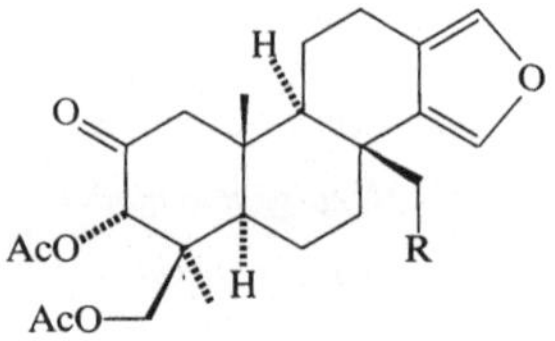

(774) R=OAc
(775) R=H

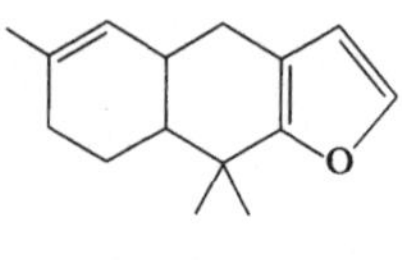

(776)

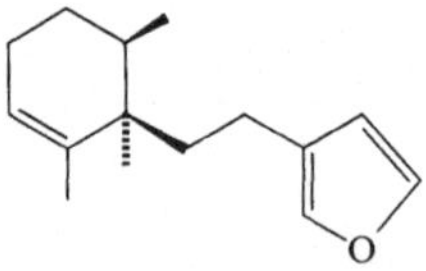

(777)

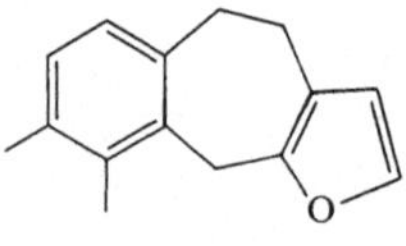

(778)

Furodysinine (**60**) and euryfuran (**52**) have been identified as feeding deterrents in nudibranchs (*78, 80, 524, 525*). Deterrent activity has also been reported for pallescensin (**50**), idiadione (**156**), the isonitrile **769**, and the isothiocyanate **770** (*524*). These compounds were also toxic to goldfish at 0.1 mg/mL. Axisonitrile (**771**) was ineffective as antifeedant, but showed strong ichthyotoxicity, the minimum active concentration being at 8 ppm. Furoscalarol (**772**) and deoxoscalarin (**773**) were active in the feeding inhibition bioassay (*526*).

V.1.1.1. Steroids from Nudibranchia

In addition to the epidioxysteroids listed in Table 3 several other steroids have been isolated from nudibranchs. CIMINO *et al.* (*262, 520*) investigated another predator-prey pair, the hydroid *Eudendrium* spp. which are the prey of several nudibranchs (*Hervia peregrina, Flabellina affinis, Coryphella lineata*) and on which the same nudibranchs deposit their eggs. The same group of polyhydroxylated steroids was found in the hydroid and the nudibranchs, the major compound being **389**. Two steroidal acids, 3-oxo-chol-4-ene-24-oic acid (**779**) and its unsaturated analog **780** were isolated from the dorid nudibranch *Aldisa sanguinea cooperi* (*530*). The major compound **779** displayed significant antifeedant activity. *A. sanguinea* were usually found deeply embedded in the sponge *Anthoarcuata graceae*. The sponge, however, contained neither **779** nor **780**, but only cholestenone, which had no antifeedant properties in this animal as well as in the nudibranch. Apparently the nudibranch obtains an inactive metabolite from its diet and modifies it chemically to produce an active antifeedant.

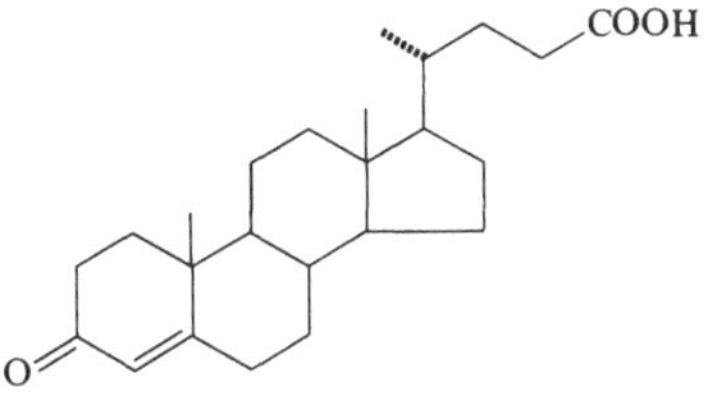

(**779**)

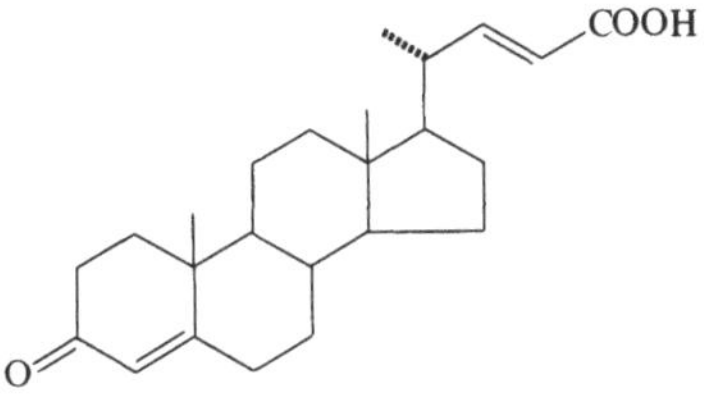

(**780**)

V.1.1.2. Terpenes from Nudibranchia

Although many terpenes isolated from nudibranchs were demonstrably sponge metabolites, other compounds not present in sponges can also be isolated from this kind of molluscs. Thus, 2,6-dimethyl-5-heptenal and 2,6-dimethyl-5-heptenoic acid were found in the skin extracts of *Melibe leonina* (*531*). The aldehyde was responsible for the pleasant odour of the animals.

Most terpenes from nudibranchs belong to the group of sesquiterpenes. (All-*E*)-2,3-dihydroxypropylfarnesate (**781**) and its two monoacetoxy derivatives **782** and **783** were metabolites of *Archidoris odhneri* (*532*). While **781** showed moderate *in vitro* antibiotic activity against *Staphylococcus aureus*, the monoacetates **782** and **783** were totally inactive. A monocyclofarnesic acid glyceride (**784**) has been obtained from the extracts of *Archidoris montereyensis* (*533*). *Chromodoris marislae* furnished marislin (**785**) and four minor metabolites **786–789** (*534*). Marislin (**785**) is closely related structurally to pleraplysillin-2 (**790**), a metabolite of the Mediterranean sponge *Pleraplysilla spinifera*.

Several terpenoid furans have been isolated from *Cadlina luteomarginata*, most of them cited previously as metabolites from Porifera. Two more compounds from this species were dendrolasin (**791**) and pleraplysillin-1 (**792**) (*524*). Dendrolasin (**791**) was also a constituent of *Hypselodoris ghiselini* and *H. californiensis* (*78*). The digestive glands of *Dendrodoris grandiflora* contained the previously known sesquiterpenes microcionin-1 (**793**), -2 (**777**), -3 (**794**), and -4 (**795**) (*528*).

CIMINO *et al.* isolated a mixture of previously unknown sesquiterpenoid esters (**796**) from the digestive glands of *Dendrodoris limbata* (*520, 526, 535*) and *D. grandiflora* (*528*). The components of this mixture differed from each other in the acyl components which were fatty acids containing different degrees of unsaturation. Heating of an hexane solution of **796** for a few minutes resulted in elimination of the acyl groups to give euryfuran (**52**). While mixture **796** was inactive as an antifeedant, the mantles of both species yielded the dialdehyde polygodial (**797**) (*526, 528*), a worm antifeedant from East African *Warburgia* plants. When polygodial (**797**) was tested as feeding inhibitor against fish, it was observed that treated food particles were mouthed by the fish but immediately rejected, the minimum active concentration being 30 µg/mL. Neither **796** nor **797** could be found in the extracts of several sponges (*526*). *In vitro* experiments suggested that the biological activity of **797** was primarily related to its ability to form adducts with -NH$_2$ groups (*536*). It has been shown by biosynthetic experiments with mevalonic acid labeled with carbon-14 that the nudibranch biosynthesizes **797** as well as the mixture **796** (*537*) and it was hypothesized that esters **796** might be detoxication products of polygodial (**797**). The enol acetate of **797**, named pu'ulenal (**798**), could be obtained from *Chromodoris albonotato* (*529*).

OKUDA *et al.* (*538*) investigated eight species of *Dendrodoris* and

Doriopsilla, and found **796** and **797** as well as two related compounds, olepupuane (**799**) and the methoxy acetal **800**. Since the nudibranch was extracted with cold methanol, **800** is probably an artifact. Olepupuane (**799**) has been shown to inhibit feeding of the Pacific damsel fish (*Dascyllus aruanus*) on food pellets that were impregnated with various concentrations (5–50 µg/mg) of the terpenoid. 6β-Acethoxyolepupuane (**801**) from *Dendrodoris grandiflora* also showed antifeedant properties (*528*).

Extracts of the dorid nudibranch *Acanthodoris nanaimoensis* contained three unusual sesquiterpenoids, nanaimoal (**802**), the major product (*539*), as well as acanthodoral (**803**) and isoacanthadoral (**804**) (*540*). Nanaimoal (**802**) is a fragrant sesquiterpenoid aldehyde whose structure has been confirmed by an unambiguous synthesis of its *p*-bromophenyluretane derivative. *Archidoris montereyensis* furnished a sesquiterpene acid glyceride (**805**) as well as its monoacetate (**806**), the structures of which have been determined by chemical correlations (*533, 541*). Albicanyl acetate (**807**) and albicanol (**808**) were reported as constituents of *Cadlina luteomarginata* (*525*). The latter was first isolated from the liverwort *Diplophyllum albicans* (*541a*); the former showed strong antifeedant activity at 5 µg/mg. Agassizin (**809**) was a new sesquiterpene furan from *Hypselodoris agassizi* (*78*). The same authors isolated a methoxy butenolide (**810**) related to nakafuran 9 (**36**) from *Hypselodoris ghiselini* (*78*) and suggested that **809** and **810** might function as predators repellents. Gel permeation chromatography of an extract of *Hypselodoris zebra* furnished 5-acetoxy-nakafuran-8 (**811**) and 5-hydroxy-nakafuran-8 (**812**) in addition to the sponge metabolites furodysinin (**60**) and euryfuran (**52**) (*80*).

(**781**) $R^1 = R^2 = H$
(**782**) $R^1 = H$ $R^2 = Ac$
(**783**) $R^1 = Ac$ $R^2 = H$

(**784**)

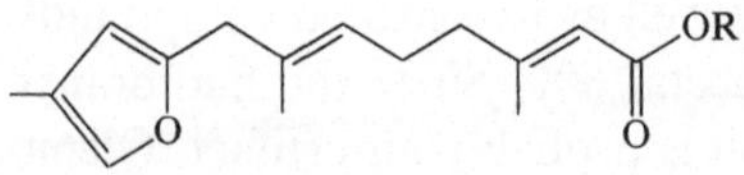

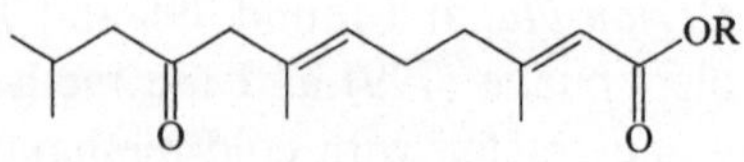

(785) R = [2-methylene-2,5-dihydrofuran-5-yl]

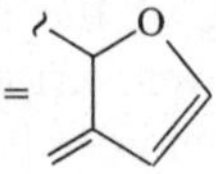

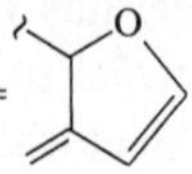

(790) R = [furan-3-ylmethyl]

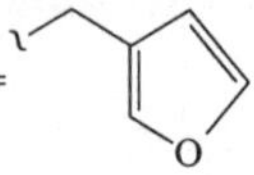

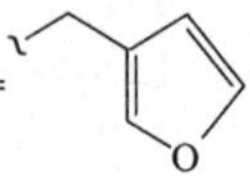

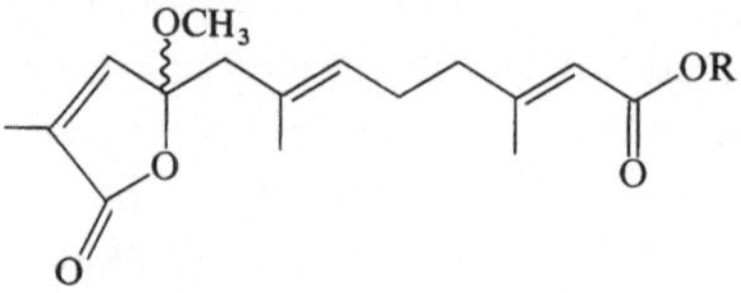

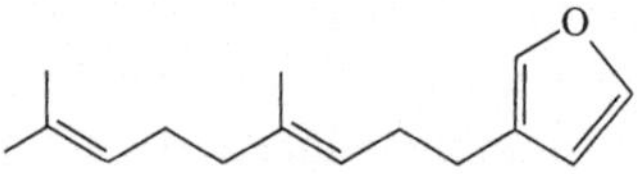

(786) R = [2-methylene-2,5-dihydrofuran-5-yl]

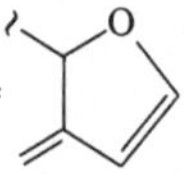

(787) R = [furan-3-ylmethyl]

(788) R = [2-methylene-2,5-dihydrofuran-5-yl]   (789) R = [furan-3-ylmethyl]

(791)

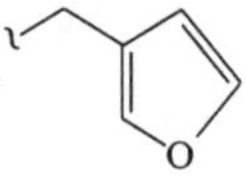

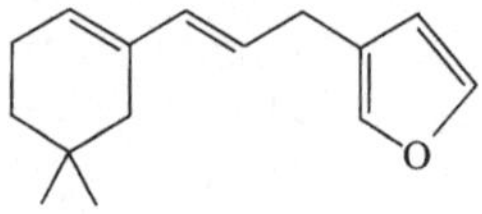

(792)

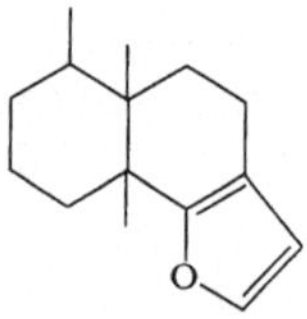

(793)

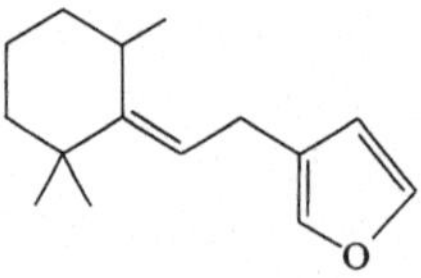

(794)

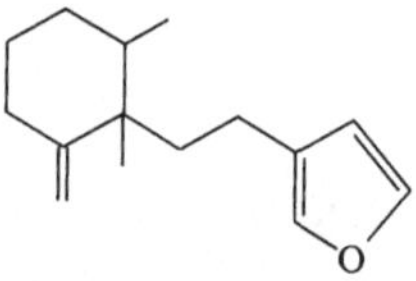

(795)

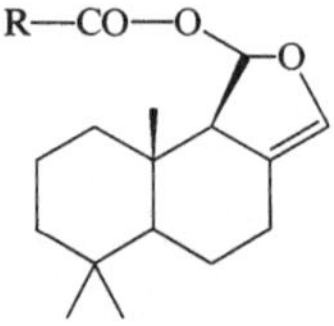

(796)

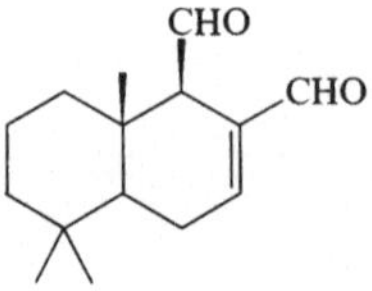

(797)

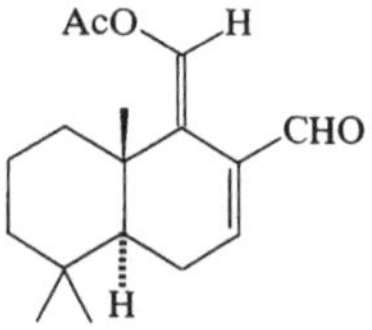

(798)

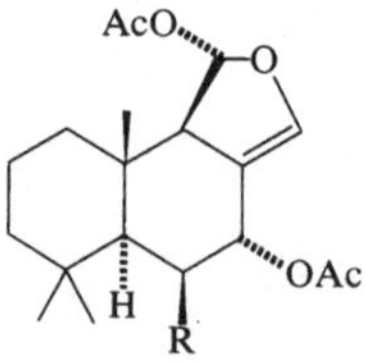

(799) R = H
(801) R = OAc

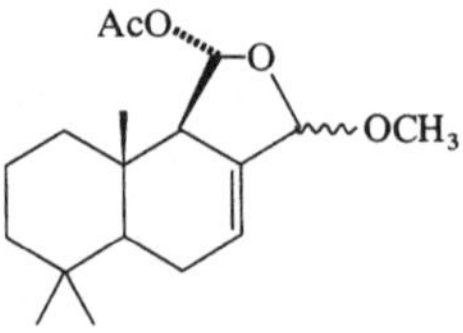

(800)

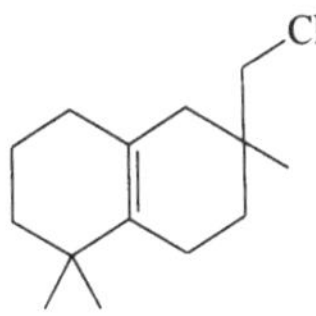

(802) 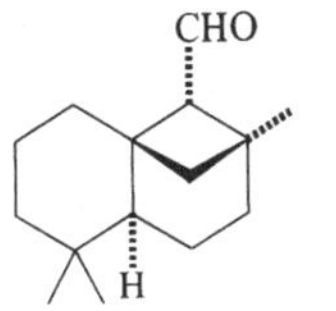 (803) 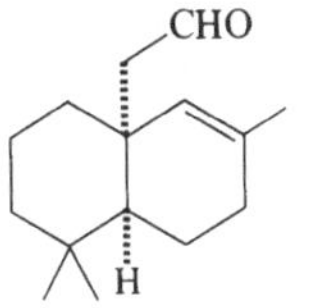 (804)

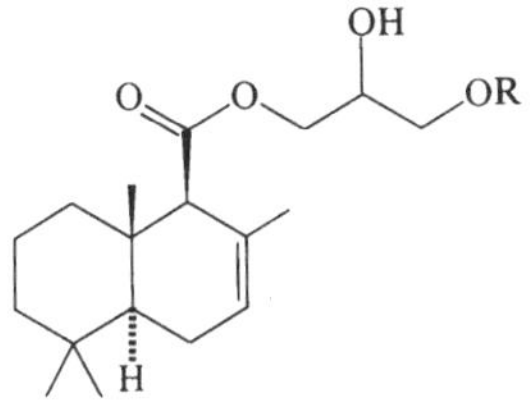

(805) R=H
(806) R=Ac

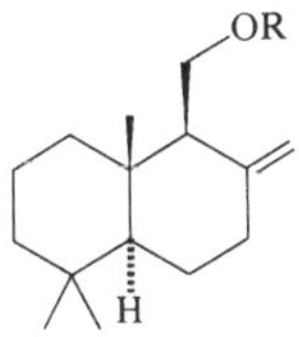

(807) R=Ac
(808) R=H

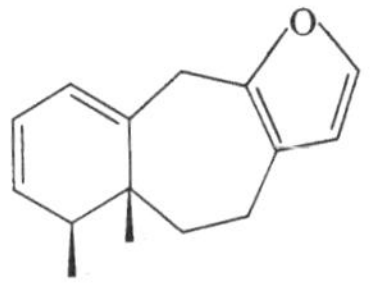

(809)

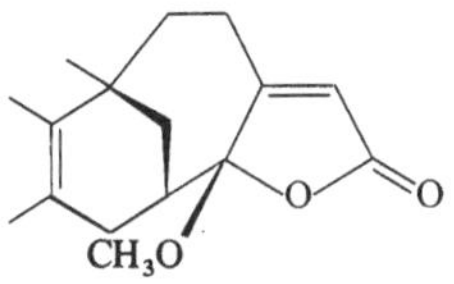

(810)

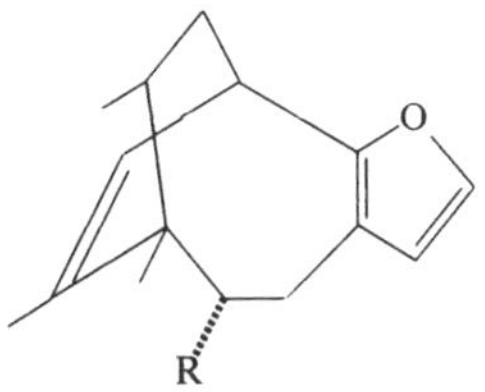

(811) R=OAc
(812) R=OH

In addition to the sesquiterpenes mentioned in the previous paragraph, *Hypselodoris ghiselini* also contained a diterpene epoxide, the labdane ghiselinin (813) (78). An isoagathane diterpene (814), together with its two monoacetates (815) and (816), has been reported as a constituent of *Archidoris montereyensis* (533, 541). The glyceride 814 has also been isolated from extracts of *Archidoris odhneri* (533). The structure of 814 was established by X-ray analysis. The structures of six related furanoditerpenoids (774, 775, 817–820) from *Casella atromarginata* have been determined by DE SILVA and SCHEUER (527). Marginatafuran (821) from *Cadlina luteomarginata* represents a new mode of cyclization of a labdane precursor; its structure was solved by X-ray diffraction (542).

Luteone (822) was a C_{23}-terpenoid from the dorid nudibranch *Cadlina luteomarginata* (525, 543) whose purification was achieved through a crystalline 2,4-dinitrophenylhydrazone. It appears to be a degraded sesterterpene of the cheilanthane class (110c). As mentioned earlier other degraded sesterterpenoids with twentyone carbons have been reported from marine sponges, and furospongin-1 acetate (**823**) as well as the furanoterpene **824** have been obtained from the digestive glands of *Dendrodoris grandiflora*, together with the scalarane fasciculatin (**825**) (*528*). Five C_{26} scalaranes have been isolated from the dorid nudibranch *Chromodoris sedna* (*544*). 12-Deacetyl-20-methyldeoxoscalarin (**826**), 23-hydroxy-20-methyldeoxoscalarin (**827**), 23-hydroxy-20-methylscalarolide (**828**), sednolide (**829**), and sednolide 23-acetate (**830**) were all 24-methylscalarins. This type of compound has been found in sponges of the genus *Phyllospongia*, but the authors could not find a sponge source for the metabolites **826** to **830**. **826** and **829** inhibited growth of the marine bacterium *Vibrio angiullarum* at 0.1 mg/disk. *Dendrodoris grandiflora* contained a mixture of prenylated chromanols (**831**) (*528*).

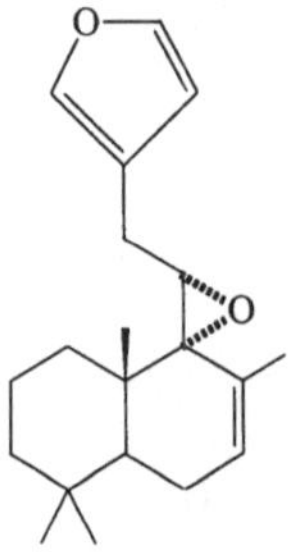

(**813**)

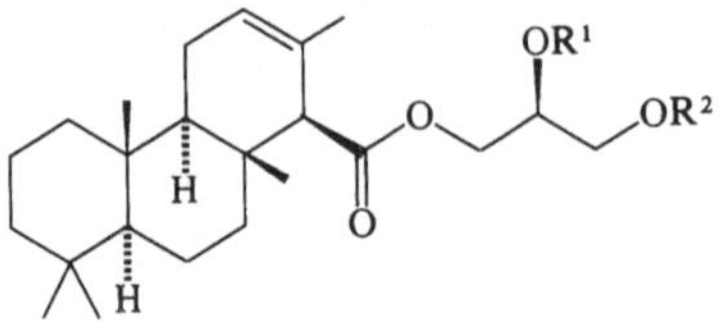

(**814**) $R^1=R^2=H$
(**815**) $R^1=H$ $R^2=Ac$
(**816**) $R^1=Ac$ $R^2=H$

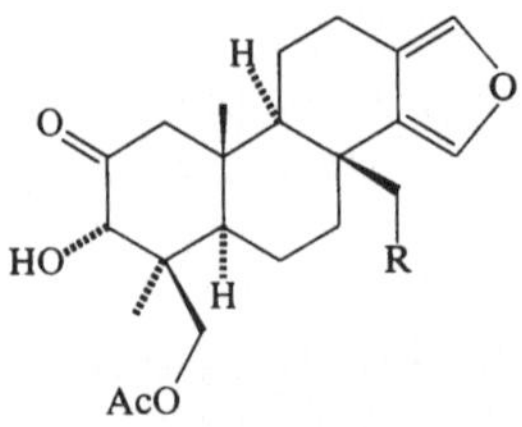

(**817**) R=OAc
(**818**) R=H

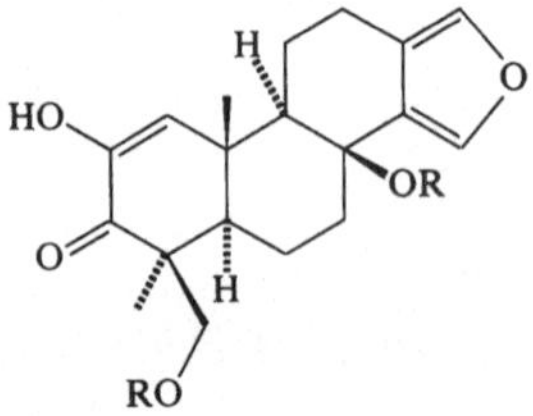

(**819**) R=Ac
(**820**) R=H

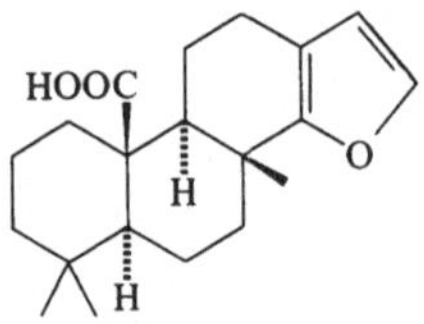

(**821**)

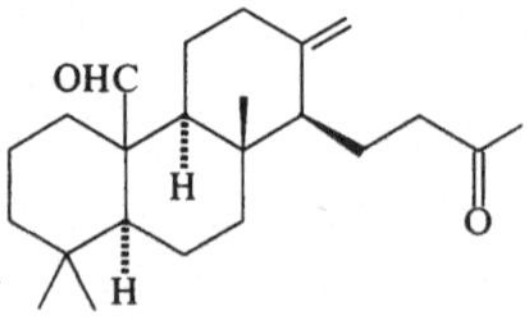

(**822**)

References, pp. 320–363

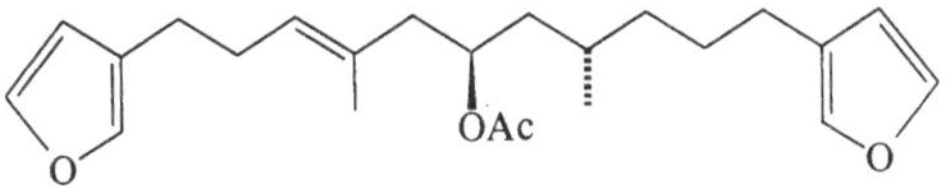

(823)

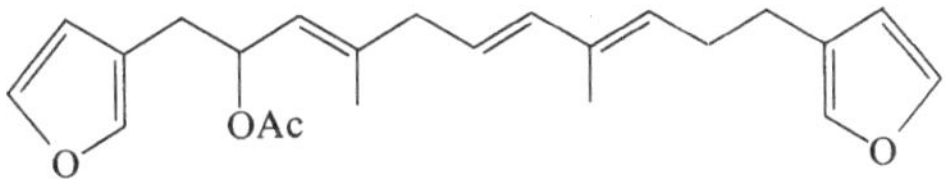

(824)

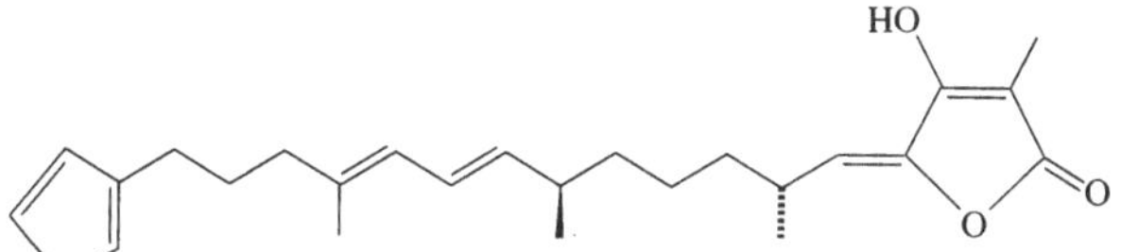

(825)

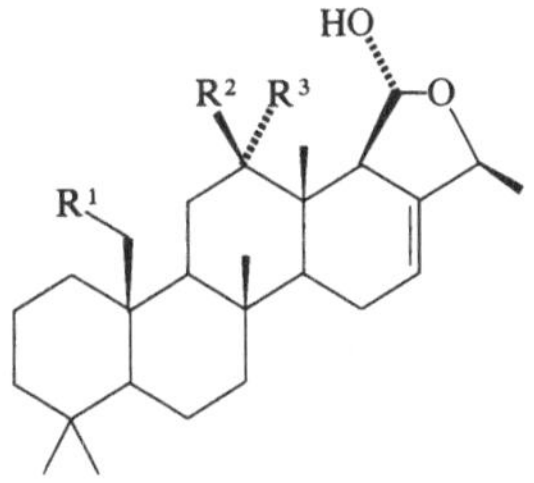

(826) R¹=H   R²=OH   R³=H
(827) R¹=OH R²=H   R³=OAc

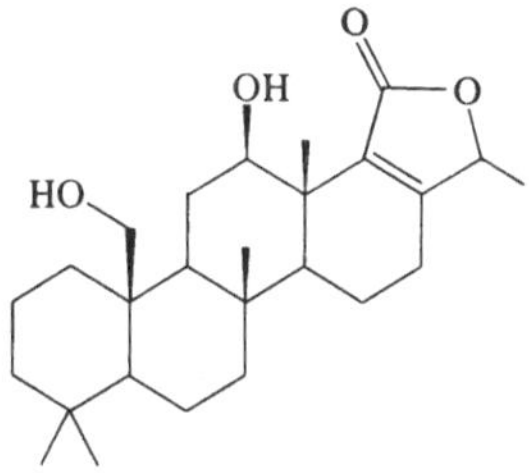

(828)

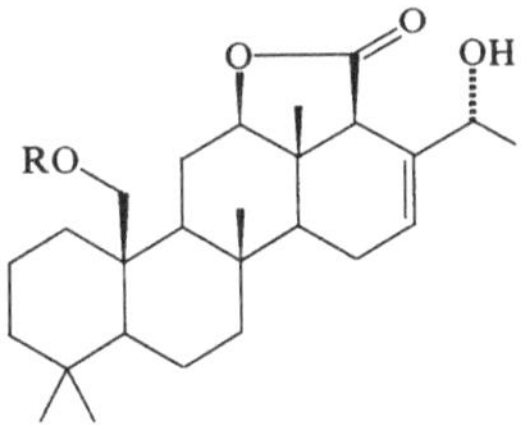

(829) R=H
(830) R=Ac

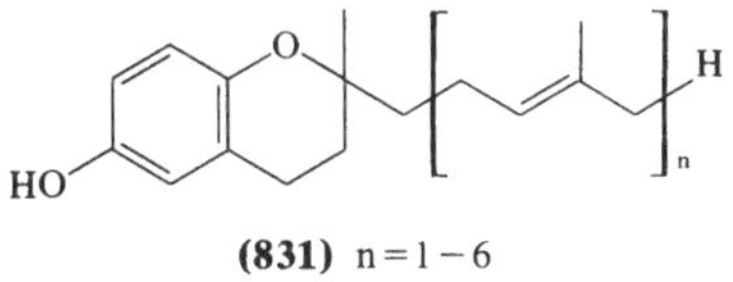

(831) n=1–6

V.1.1.3. Miscellaneous Compounds from Nudibranchia

WALKER and FAULKNER (545) investigated the nudibranch *Diaulula sandiegensis* and isolated nine chlorinated acetylenes (**832–840**) which were believed to be involved in the chemical defence mechanism of the animal. *D. sandiegensis* has been observed to feed on a variety of sponges, but none of these species contained the chlorinated acetylenes **832–840** or an obvious precursor. **832** exhibited mild antimicrobial activity against several bacteria and a yeast. Although the authors could not perform toxicity assays on these compounds due to their instability, closely related acetylenes are often highly toxic to fish and invertebrates. An antibiotic compound, 1-O-hexadecylglycerol (**405**), which had previously been isolated from *Palythoa liscia* (*356*), has also been obtained from *Archidoris montereyensis* and *Aldisia sanguinea cooperi* (*533*). **405** showed potent *in vitro* activity against *Staphylococcus aureus* and *Bacillus subtilis*.

The nembrothid nudibranchs *Tambje abdere*, *T. eliora*, and *Roboastra tigris* all contained the tambjamines A to D (**841–844**) (*546*) which were traced to a dietary source, the bryozoan *Sessibugula translucens*, and which were implicated in the chemical defence mechanism of the *Tambje* species. All four compounds were enamines easily hydrolysed to the corresponding aldehydes and showed antimicrobial activity. In addition they inhibited cell division of fertilized sea urchin eggs as well as the artifical aldehydes. GUSTAFSON and ANDERSEN (*547*) reported the structure of triophamine (**845**), a novel diacylguanidine which was the major component of *Triopha catalinae* and has also been obtained from *Polycera tricolor* which feeds like *T. catalinae* on bryozoans (*533*). A total synthesis of **845** established the alkene geometry and verified the overall structure (*548*).

$$Cl-CH=CH-CH=CH-C\equiv C-C\equiv C-CH=CH-(CH_2)_4-CO-CH_3$$

(**832**) (1 Z, 3 E, 9 Z)
(**833**) (1 Z, 3 Z, 9 Z)

$$Cl-CH=CH-CH=CH-C\equiv C-C\equiv C-CH=CH-(CH_2)_4-CH(OH)-CH_3$$

(**834**) (1 Z, 3 E, 9 Z)
(**835**) (1 Z, 3 Z, 9 Z)
(**836**) (1 E, 3 E, 9 Z)

$$Cl-CH=CH-CH=CH-C\equiv C-C\equiv C-CH=CH-(CH_2)_3-CH(OH)-CH_2-CH_3$$

(**837**) (1 Z, 3 E, 9 Z)
(**838**) (1 Z, 3 Z, 9 Z)
(**839**) (1 E, 3 E, 9 Z)

$$Cl-CH=CH-CH=CH-C\equiv C-C\equiv C-(CH_2)_5-CH(OH)-CH_2-CH_3$$

(**840**) (1 Z, 3 E)

References, pp. 320–363

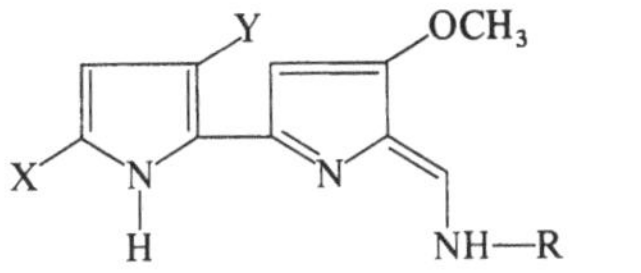

(841) X=H   Y=H   R=H
(842) X=Br Y=H   R=H
(843) X=H   Y=H   R=iso-Butyl
(844) X=H   Y=Br R=iso-Butyl

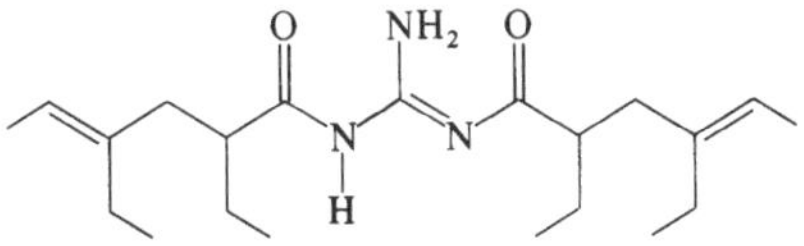

(845)

V.1.2. Aplysiidae

The first natural product research on opisthobranchs dates from the early 1960's and resulted in the isolation of brominated terpenes from a sea hare (*548a*). These animals, which can be collected easily, contain large quantities of metabolites which are found predominantly in the digestive gland and the skin and often contain halogen. The sea hares obtain their metabolites from their food, the algae. Aplysiatoxin (**846**) is the most prominent toxic compound isolated from the Hawaiian sea hare *Stylocheilus longicauda* and from the blue-green algae on which it feeds. Aplysistatin (**847**) is another well known antileucemic metabolite from a sea hare, the South Pacific Ocean *Aplysia angasi*. The structure elucidation of this compound involved some synthetic work (*549*). Extracts of *Aplysia dactylomela* are reported to show cytotoxic and *in vivo* antitumor activity. The constituents of this species have been reviewed by SCHMITZ et al. (*550*).

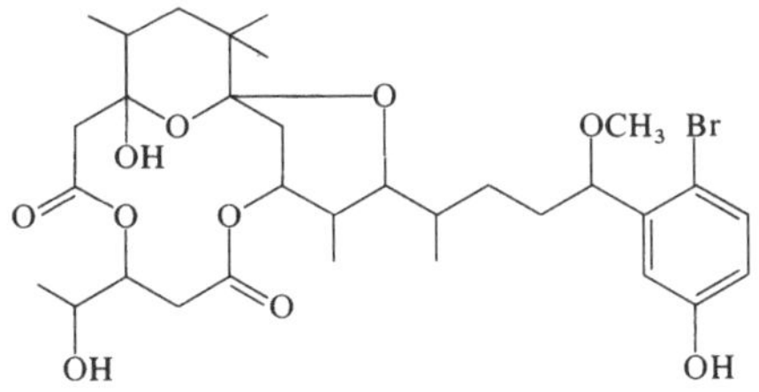

(846)

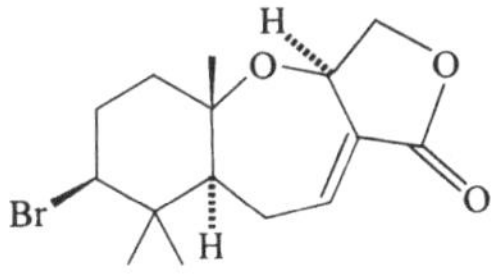

(847)

V.1.2.1. Terpenes from Aplysiidae

Two new sesquiterpenoids, dihydroxydeodactol monoacetate (**848**) (*551*) and isodeodactol (**849**) (*552*), have been isolated from alcohol extracts of the digestive glands of the sea hare *Aplysia dactylomela*. Similar compounds were shown earlier to have mildly cytotoxic activity. GONZÁLEZ et al. (553) isolated two more new brominated labile sesquiterpenes (**850, 851**) from the same species.

A. dactylomela also furnished a series of diterpenes. Five new brominated compounds with modified pimarane skeletons were isolated by SCHMITZ et al. (554). Parguerol (**852**), its 16-acetate (**853**), and deoxyparguerol (**854**) contained a cyclopropane ring, while isoparguerol (**855**) and its 16-acetate (**856**) had a cyclobutane ring. All five compounds were said to be cytotoxic. Spectroscopic data and several chemical transformations led to the structure of the main compound **852**. In addition, the authors found several known algal metabolites in the extracts of the digestive glands and they assumed that **852–856** were also of algal origin. Another new compound from the digestive gland of the same species was a trioxygenated diterpene (**857**) possessing the dolabellane skeleton (555). The structure has been solved spectroscopically and by means of an X-ray analysis, but the biological activity was not mentioned. PETTIT et al. (556) isolated from *Dolabella ecaudata* dolatriol (**858**), which they had previously obtained from *D. auricularia*. In addition they found a well known plant metabolite, (−)-loliolide (**859**), which is a degraded carotenoid and showed cytotoxic activity. MIDLAND et al. (557) described the isolation of three new crenulides (**860–862**) and an acetoxycrenulide (**863**) from the digestive glands of *Aplysia vaccaria*. The crenulides are brown algal metabolites and **863** has been reported to be severely debilitating and ultimately toxic at levels of 10 µg/mL toward the herbivorous reef-dwelling fish *Eupomacentrus leucostictus* (558).

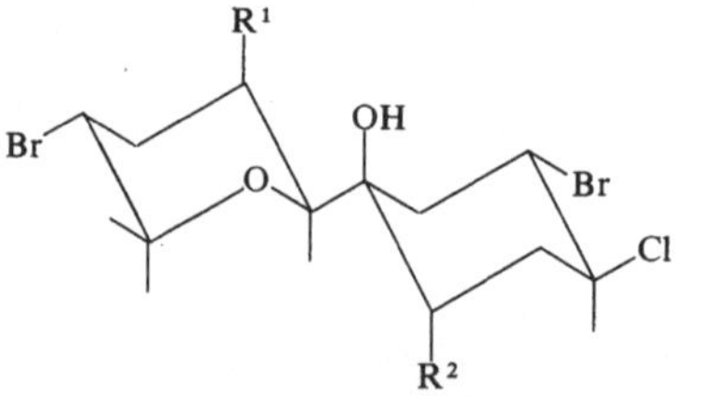

(**848**) R^1=OAc R^2=OH
(**849**) R^1=R^2=H

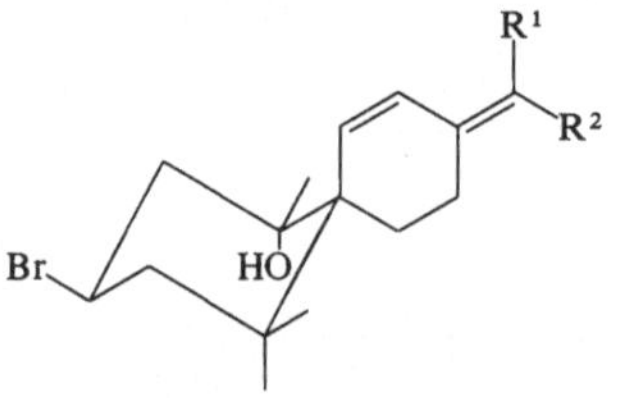

(**850**) R^1=H R^2=Br
(**851**) R^1=Br R^2=H

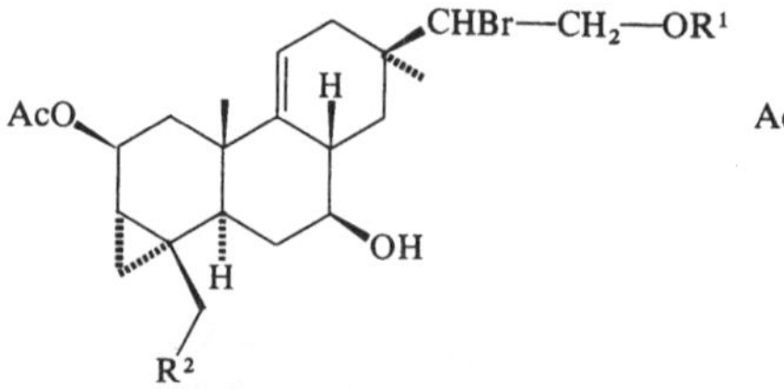

(**852**) R^1=H R^2=OH
(**853**) R^1=Ac R^2=OH
(**854**) R^1=H R^2=H

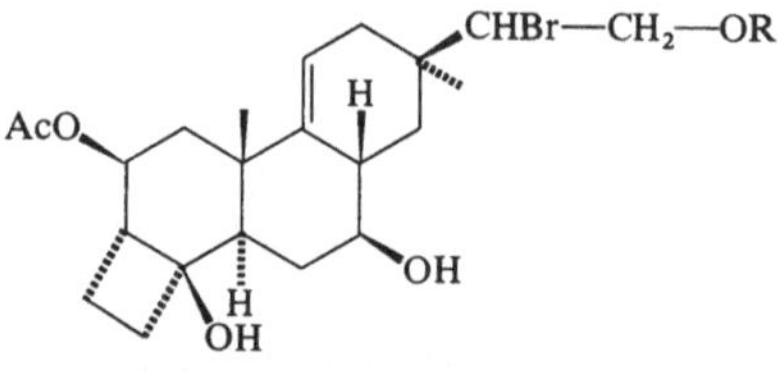

(**855**) R=H
(**856**) R=Ac

References, pp. 320–363

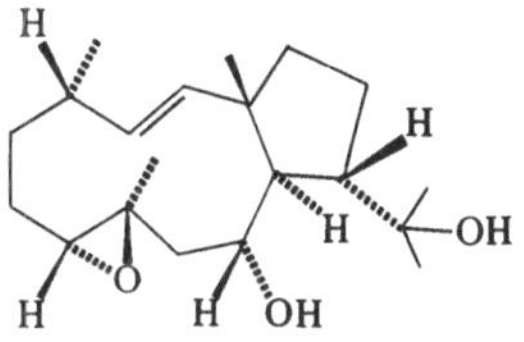

(857)

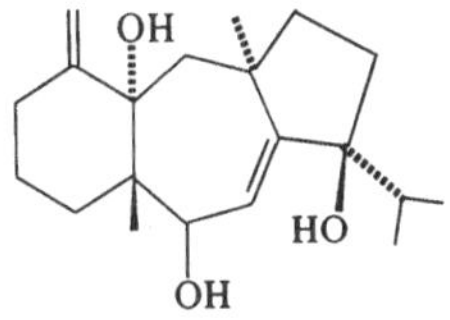

(858)

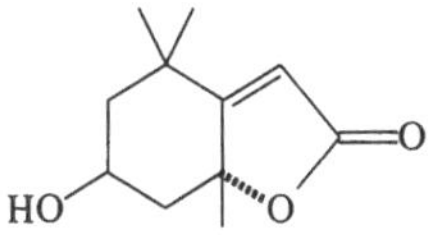

(859)

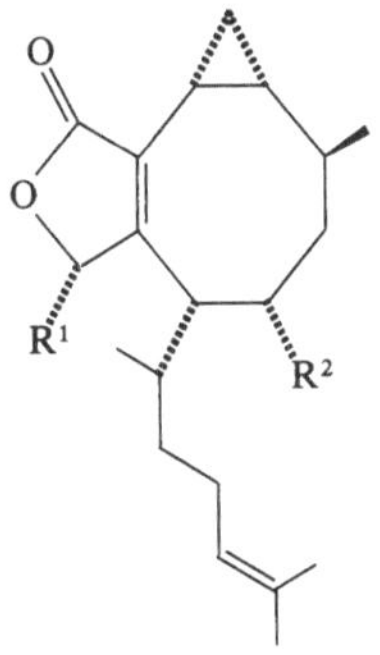

(860) R¹=OH R²=OH
(861) R¹=OH R²=OAc
(862) R¹=H   R²=OH
(863) R¹=H   R²=OAc

V.1.2.2. Miscellaneous Compounds from Aplysiidae

In addition to the compounds mentioned in the previous section, *Aplysia dactylomela* also contains four epidioxysterols (**5, 8, 10, 12**) (*67*), compounds which have been found in tunicates, sponges, and corals as well. But while the latter animals contained a variety of these compounds, only these four have been isolated from the sea hare. This may be attributed to the food habits of these animals.

Extracts of *A. dactylomela* are rich in other organic metabolites as has been demonstrated earlier. GOPICHAND et al. (*552*) isolated four new halogenated acetylenic ethers (**864–867**) from the digestive glands. These ethers which are fatty acid derived were two isomeric pairs named (3 *E*)-12-epiobtusenyne (**864**), 12-*epi*-obtusenyne (**865**), (3 *E*)- and (3 *Z*)-dactomelyne (**866** and **867**), respectively. Another isomeric pair, (*E*)- and (*Z*)-ocellenyne (**868** and **869**), has been obtained from *Aplysia oculifera* (*559*). No biological activity has been described for any of these compounds.

The extracts of *Bursatella leachii pleii* consisted mainly of aliphatic lipids, but among the minor components GOPICHAND and SCHMITZ (*560*) had isolated a novel dinitrile, bursatellin (**870**).

The Indian ocean sea hare *Dolabella auricularia* has been found to be an exceptionally productive source of new bioactive products. PETTIT *et al.* (*561, 562*) reported discovery of nine antineoplastic and/or cytotoxic substances in *D. auricularia*. Some 100 kg of the wet sea hare afforded about 1 mg each of these substances designated dolastatins 1–9. The primary structure of Dolastatin 3 was proposed as *cyclo* [Pro-Leu-Val-(gln)Thz-(gly)Thz] (**871**) (*562*) containing two unusual thiazole amino acids and the absolute configurations of each amino acid unit were speculated to bear the usual L-configuration on biosynthetic grounds. However, synthetic studies showed that structure **871** requires revision (*562a, 562b*).

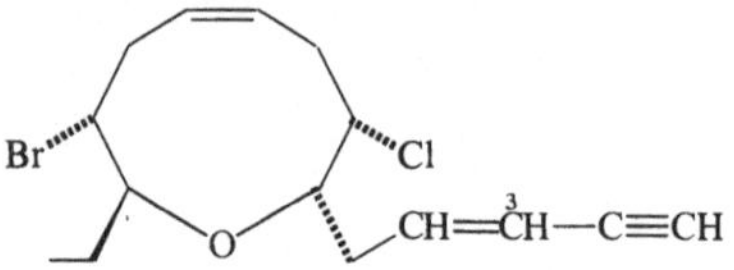

(**864**) (3 E)
(**865**) (3 Z)

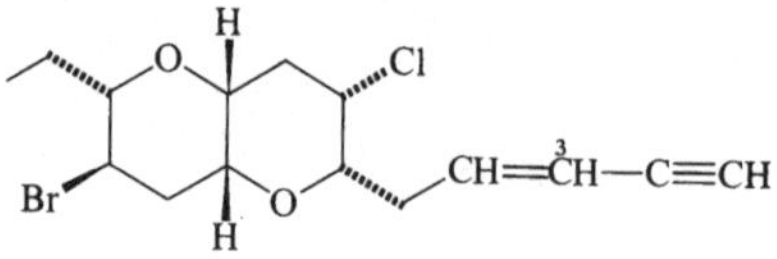

(**866**) (3 E)
(**867**) (3 Z)

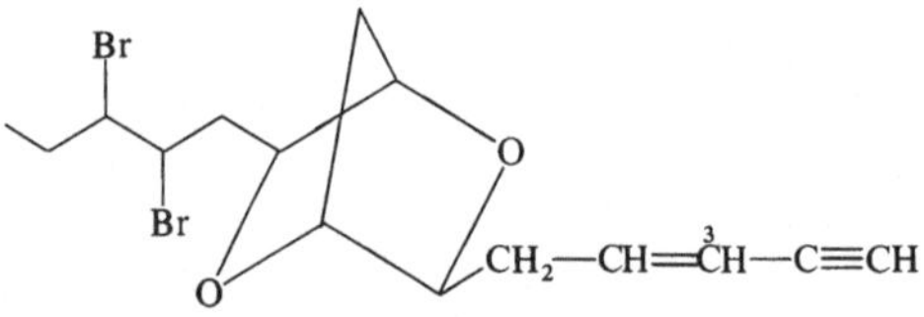

(**868**) (3 E)
(**869**) (3 Z)

(**870**)

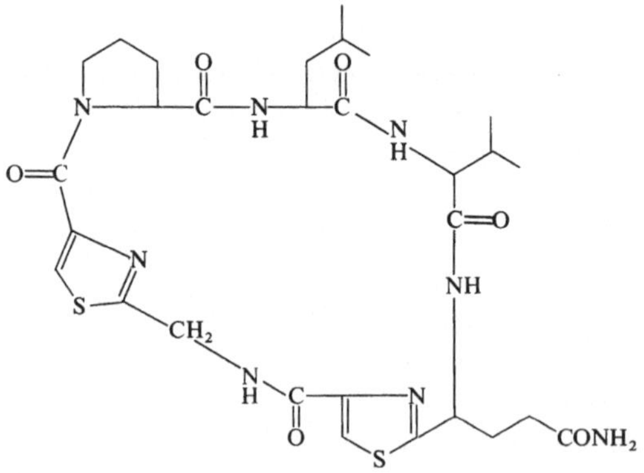

(**871**)

V.1.3. Conidae

Marine snails of the family Conidae possess potent venoms which are used in immobilizing prey. With their highly specialized venom apparatus, the animals inject their toxin into the prey by means of a disposable hollow tooth. There are three major feeding types in the genus, worm eaters (vermivorous), mollusc eaters (molluscivorous) and fish eaters (piscivorous). The venoms of the piscivorous Conidae are active against vertebrates and the venoms of several of the species have been responsible for human injuries and fatalities. YOSHIBA and SAKURAI (*563*) compared the toxicities of the venoms of five species. All venoms act as neurotoxins with the venom of *Conus geographus* being regarded as the most dangerous to man. KOBAYASHI *et al.* (*564*) investigated the effects of the venoms of 29 species on the mouse isolated diaphragm, the guinea pig isolated left atria and ileum and the rabbit isolated aorta. While the active principles in the venoms of *C. geographus*, *C. textile*, and *C. imperialis* were stable at 100° C for 15 min, an observation which suggested molecular weights lower than 10,000, those of *C. magus*, *C. striatus*, *C. tessulatus*, and *C. eburneus* were partially destroyed by heating. In the atria, the venoms of *C. magus* and *C. striatus* caused powerful, positive inotropic actions. The venoms of *C. eburneus* and *C. tessulatus* caused marked contractions in the rabbit aorta and the venoms from piscivorous Conidae, *C. geographus*, *C. striatus*, *C. magus*, and *C. fulmen* markedly inhibited the contractile response of mouse skeletal musculature to electrical stimulation.

The conotoxins GI (**872**), GIA (**873**), and GII (**874**) are the first toxins from the more than 300 species of venomous Conidae whose structures have been elucidated (*565*). These three homologous toxic peptides from *C. geographus* cause postsynaptic inhibition at the vertebrate neuromuscular junction and a number of human fatalities have resulted from the sting of this species. The biological active peptides were monomeric, with internal disulfide bridges which were not assigned. The toxins may be breakdown products from larger precursors and are reported to be equally sensitive to reagents which cleave disulfide bonds. As four cysteine units and two disulfide bonds are present in each molecule, three different modes of disulfide bond formation could be considered for each peptide. NISHIUCHI and SAKAKIBARA (*566*) synthesized all peptides having the sequence proposed for conotoxin GI (**872**) and showed that the toxicity of the peptide with bridges between positions 2–7 and 3–13, respectively, was highest and compared with the value reported for native **872**. A 22-residue peptide toxin from the same species, named conotoxin GIII, was extremely basic and contained three disulfide bridges as well as three residues of hydroxyproline *(567)*. NAKAMURA et al. *(568)* isolated two peptides, geographutoxin I and II, from *C. geographus* and determined their amino acid compositions. Conotoxin GIII seemed to be identical with geographutoxin I. The amino acid sequences of geographutoxin I (**875**) and II (**876**) were subsequently determined by the same group *(569)*. Both peptides inhibited the contractile response of directly stimulated mouse diaphragm. CLARK et al. *(570)* described the purification to homogeneity of a central nervous system toxin, conotoxin GIV, from *C. geographus* venom. This toxin which was a protein with an apparent monomer molecular weight of 13,000 had no activity when injected intraperitoneally into mice. Conotoxin GIV was relatively acidic but its amino acid composition has not been reported.

A 14-residue peptide toxin homologous to conotoxins GI, GIA, and GII has been isolated from the venom of *C. magus (571)*. This peptide, conotoxin MI (**877**), was two to three times more active than GI, GIA, and GII, and was presumed to act at the acetylcholine receptor of vertebrate neuromuscular junctions. A successful chemical synthesis of **877** has been achieved *(571)*. A myotoxin from *C. magus* venom and apparently a peptide of molecular weight 1,500 elicited complete loss of contractile response of mouse diaphragm to electrical stimulation followed by a gradual rise in the base line, an increase in the contractile force of guinea pig left atria, a tonic contraction of guinea pig taenia caeci and powerful rhythmic contractions of guinea pig ileum and vas deferens *(572)*. The myotoxin-induced contraction of the diaphragm was abolished by a specific Na^+ channel blocker (tetrodotoxin), but not affected by a nicotinic blocker (d-tubocurarine).

An aqueous extract of the venom duct of *C. striatus* produced repetitive impulse firing in response to a single stimulus applied to a frog sciatic nerve *(573)*. KOBAYASHI et al. *(574)* have shown that venom extract of *C. striatus* elicited rhythmic, transient contraction of guinea pig ileum followed by relaxations. The chemical properties of the active principle in the venom suggested a protein with a molecular weight of about 10,000 or more. Later, the same group isolated a cardiotonic glycoprotein, striatoxin, from the venom of the same species *(575)*. The molecular weight has been estimated to be 25,000 by gel filtration. The toxin was found to have long-lasting inotropic activity on guinea pig left atria. Striatoxin may be one of the major principles in the venom toxic to fish. The minimum lethal dose in the fish *Rhodeus ocellatus smithi* was 1 µg/g body weight. JIMENEZ et al. *(576)* investigated the localization of enzymes and possible toxin precursors in granules from *C. striatus* venom and argued that the granules could be considered as storage packets not only for hydrolytic or digestive en-

References, pp. 320–363

zymes but also for the precursors of the conotoxins. The mode of action of the crude venom of *C. textile* on mammalian skeletal, smooth and cardiac musculature has been elucidated by KOBAYASHI *et al.* (*577*). The active principle(s) was (were) stable to heating, to acid and base, and apparently substance(s) of relatively low molecular weight. The venom caused a marked, transient contraction in guinea pig isolated ileum which was abolished by atropine and tetrodotoxin. The venom exhibited a positive inotropic effect on guinea pig left atria which was almost completely inhibited by verapamil. The extract of the venom has been fractionated and arachidonic acid has been shown to be present as an active substance (*578*), having a contractile effect on the ileum.

Tessulatoxin, a powerful vasoactive protein from *C. tessulatus*, has been purified by affinity and electrofocusing chromatography (*579*). This protein with a molecular weight of 26,000 was lethal to the fish *Rhodeus ocellatus smithi* at a dose of 1 µg/g and caused a marked contraction of rabbit isolated aorta which was inhibited by verapamil. The same group isolated another vasoactive substance, eburnetoxin from *C. eburneus* (*580*). The molecular weight of this peptide has been estimated to be 28,000 by gel permeation chromatography and slab gel electrophoresis. Eburnetoxin had the same biological activities as tessulatoxin. The venom of *Chelyconus fulmen* which causes promptly convulsions and paralysis accompanied by death in various species of vertebrates is probably a protein (*581*). OLIVERA *et al.* (*582*) described a novel peptide from *Conus geographus* which induced a comatose state in mice upon intracerebral injection. This peptide containing 17 amino acid residues elicited no obvious effects when injected intraperitoneally into either mice or fish. In contrast to the conotoxins the purified compound was highly acidic and did not contain a cystine residue.

872 Glu-Cys-Cys-Asn-Pro-Ala-Cys-Gly-Arg-His-Tyr-Ser-Cys-NH_2
873 Glu-Cys-Cys-Asn-Pro-Ala-Cys-Gly-Arg-His-Tyr-Ser-Cys-Gly-Lys-NH_2
874 Glu-Cys-Cys-His-Pro-Ala-Cys-Gly-Lys-His-Phe-Ser-Cys-NH_2
875 Arg-Asp-Cys-Cys-Thr-Hyp-Hyp-Lys-Lys-Cys-Lys-Asp-Arg-Gln-Cys-Lys-Hyp-Gln-Arg-Cys-Cys-Ala-NH_2
876 Arg-Asp-Cys-Cys-Thr-Hyp-Hyp-Arg-Lys-Cys-Lys-Asp-Arg-Arg-Cys-Lys-Hyp-Met-Lys-Cys-Cys-Ala-NH_2
877 Gly-Arg-Cys-Cys-His-Pro-Ala-Cys-Gly-Lys-Asn-Tyr-Ser-Cys-NH_2

V.1.4. Miscellaneous Compounds from Other Marine Snails

Two novel tetraesters, kelletinin I (**878**) and II (**879**), which inhibited the growth of *Bacillus subtilis* and L1210 leukemia cells have been isolated from the marine mollusc *Kelletia kelletii* (*583*). They are said to be rare examples of bioactive compounds from hard-shelled molluscs and to represent a new class of antibacterial marine natural products. Isolation of **878** and **879** from the kidney and the hypobranchial gland and detection in *Kelletia* specimens over a long time span supported the hypothesis that they are intrinsic metabolites.

It has been found that non-poisonous animals of the marine snail *Babylonia japonica* (Japanese ivory shell), which is used as food in Japan, became inedible after living for a short time in Suruga Bay. On the other hand poisonous animals from Suruga Bay lost their toxicity after living in another area for a few months. Following up these observations KOSUGE et al. (*584–586*) isolated in addition to the known surugatoxin (**880**), a low molecular weight compound of unique structure, a new marine toxin, neosurugatoxin (**881**), which had about hundred times greater antinicotinic activity than **880**. The structure of this glycoside was determined by X-ray crystallography. At a concentration of 1×10^{-9} g/mL **881** inhibited the contractile response of isolated guinea pig ileum to 3×10^{-5} g/mL of nicotine and evoked mydriasis in mice at a minimum dose of 3 ng/g. NOGUCHI et al. (*587*) as well as YASUMOTO et al. (*588*) also detected tetrodotoxin in the Japanese ivory shell specimens collected in Sakajiri Bay, Fukui Pref., Japan at the same time in 1980. Tetrodotoxin has also been found in a trumpet shell *Charonia sauliae* (*589*) and in the frog shell *Tutufa lissostoma* (*590*).

Marine pulmonates of the genus *Siphonaria* are air breathing molluscs that resemble limpets with which they co-occur in the intertidal zone. Nearly all specimens examined contained "polypropionate" metabolites that are believed to be employed in chemical defense against predators. *Siphonaria diemenensis* collected intertidally in southeastern Australia contained two antibiotics, diemenensin-A (**882**) and -B (**883**) (*591*). Diemenensin-A (**882**) inhibited *Staphylococcus aureus* and *Bacillus subtilis* at 1 µg/disc and 5 µg/disc, respectively, and inhibited cell division in the fertilized sea urchin egg assay at 1 µg/mL. Another new "polypropionate" antibiotic with a γ-hydroxy-α-pyrone ring, pectinatone (**884**) was isolated from *Siphonaria pectinata* (*592*). *S. denticulata* furnished two isomeric "polypropionate" metabolites, denticulatin A (**885**) and B (**886**) (*593*). Denticulatin A (**885**) was ichthyotoxic at 10 µg/mL while **886** was reported to be toxic at 30 µg/mL. Norpectinatone (**887**) and the *E*- and *Z*-isomers **888** and **889** of a related furanone were constituents of the Chilean pulmonate *S. lessoni* (*594*), and a cyclic hemiacetal (**890**) as well as an associated ketol ester (**891**) were reported as metabolites of *S. australis* (*595*). HOCHLOWSKI et al. (*596*) investigated four other *Siphonaria* species and isolated siphonarin-A (**892**) and -B (**893**) from *S. zelandica* and *S. atra*. The corresponding dihydrosiphonarins **894** and **895** have been obtained from *S. normalis* and *S. laciniosa*.

"Polypropionate" metabolites have also been isolated from the sacoglossans *Tridachiella diomedea* and *Tridachia crispata* (*597*) as well as from *Aglaja depicta* (*597a*). *T. diomedea* from the Gulf of California furnished tridachione (**896**) and 9,10-deoxytridachione (**897**), while the corresponding mollusc from the Caribbean, *T. crispata* contained crispatone (**898**) and crispatene (**899**). The structures of these four metabolites were elucidated by X-ray crystallography, by spectral data and by chemical interconversion. Aglajne-1 (**900**) from *A. depicta* is a representative of acyclic polypropionates. IRELAND et al. (*598*) reported the isolation of a mixture of esters, all based on the bispyrone alcohol ilikonapyrone (**901**), as probable defense allomones of the Hawaiian onchid *Onchidium verruculatum*. The mixture could only be partially resolved chromatographically to give small amounts of pure **902**. Subsequently, the same group (*598a*) obtained peroniatriols I (**903**) and II (**904**), two cytotoxic metabolites from saponified extracts of *Peronia peronii*.

SANDUJA et al. (*599*) described the occurrence of fulvoplumierin (**905**), an antibacterial pigment originally isolated from terrestrial *Plumeria* species (Apocynaceae), in the marine mollusc *Nerita albicilla* which was also found to contain an unusually high concentration of benzoic acid. The opisthobranch *Navanax inermis* produces alarm pheromons which have been investigated by SLEEPER et al. (*600*). Three major compounds were found, navenone A (**906**), B (**907**), and C (**908**), as well as four minor constituents (**909–912**). A related caphalaspidean mollusc, *Philinopsis speciosa*, furnished another pyridine derivative, pulo'upone (**913**), substituted at C-2 by a bicyclic C_{16}-polyketide (*600a*).

The intertidal limpet *Collisella limatula* contains limatulone (**914**) which inhibits fish and crab predation (*600b*). Food pellets containing **914** at levels of 0.05% dry weight or more induced regurgitation in the intertidal fish *Gibbonsia elegans*, a known limped predator. The marine prosobranch mollusc, *Lamellaria* sp. furnished four aromatic metabolites, lamellarins A–D (**915–918**) (*600c*). At concentrations of 19 μg/mL, **918** caused a 78% inhibition of cell division in the fertilized sea urchin egg assay while **917** caused 15% inhibition and **915** as well as **916** were inactive.

MILKOVA et al. (*325*) investigated the sterol composition of Black sea organisms belonging to the phyla Coelenterata and Mollusca, in order to study food chains in this ocean and to shed light on the origin of sterols in marine organisms.

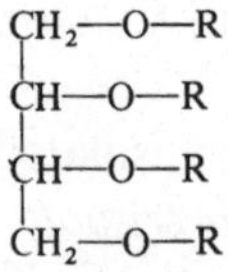

(878)

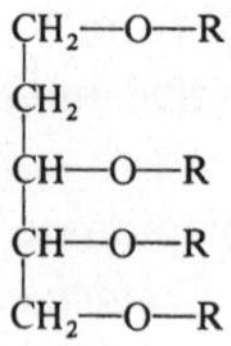

(879)

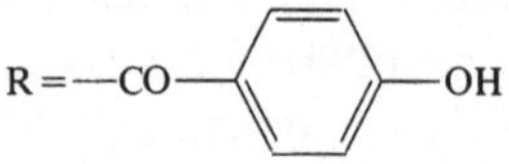

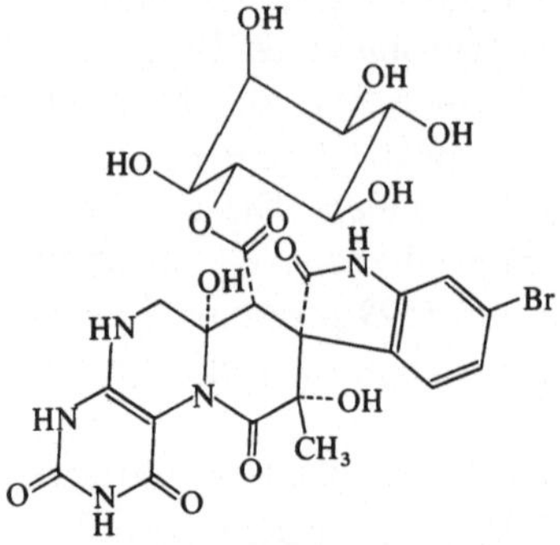

R = —CO—⟨C₆H₄⟩—OH

(880)

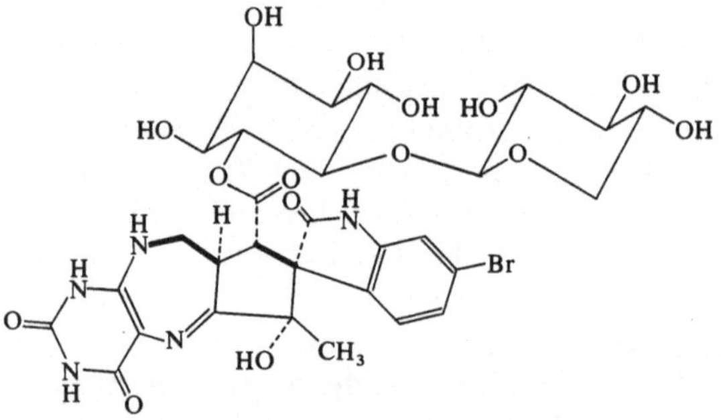

(881)

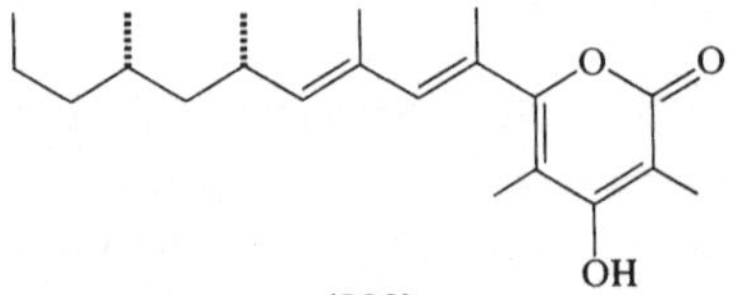

(882)

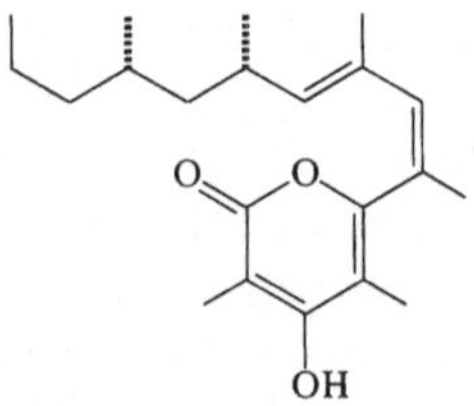

(883)

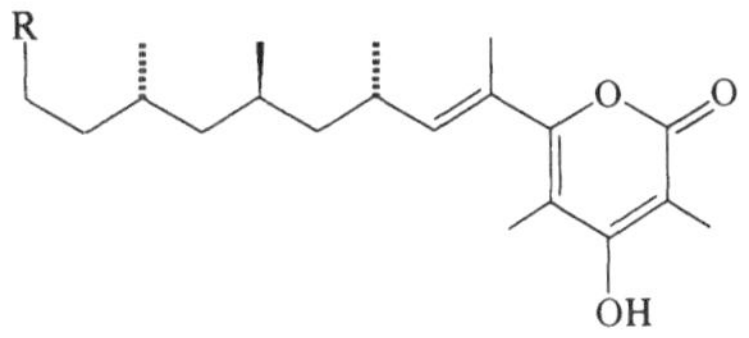

(884) R=CH$_3$
(887) R=H

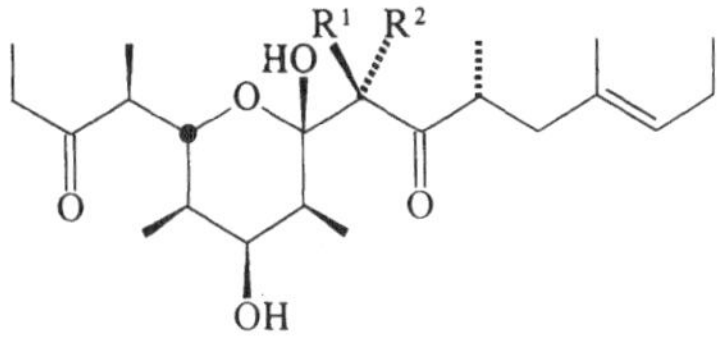

(885) R^1=CH$_3$ R^2=H
(886) R^1=H R^2=CH$_3$

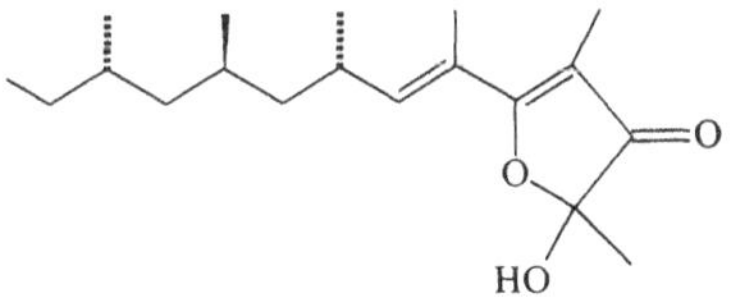

(888)

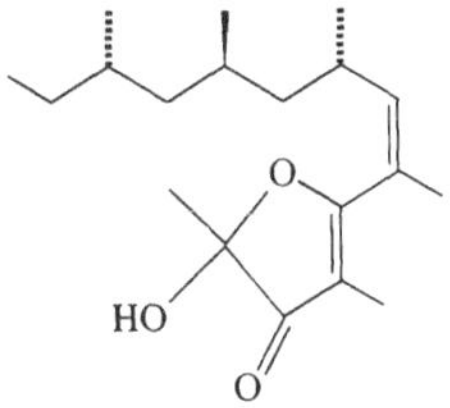

(889)

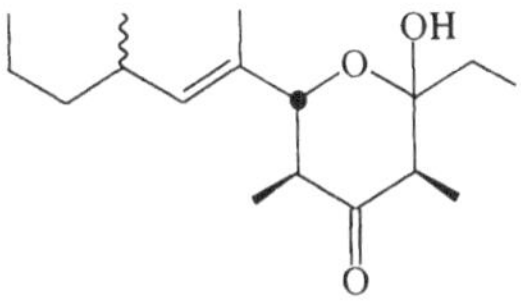

(890)

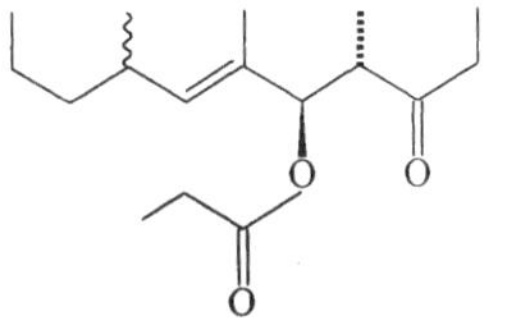

(891)

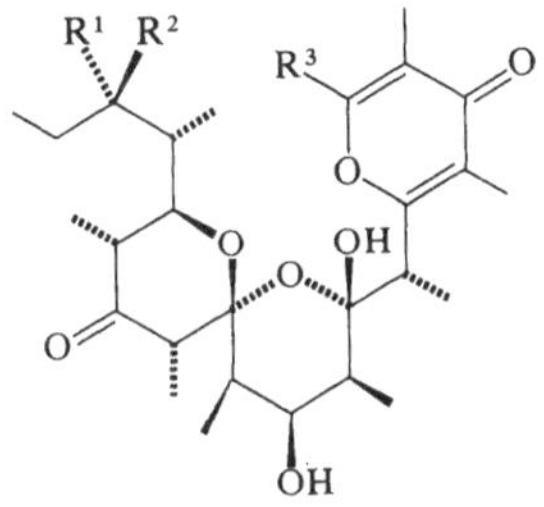

(892) R^1/R^2=O   R^3=CH$_3$
(893) R^1/R^2=O   R^3=C$_2$H$_5$
(894) R^1=OH R^2=H R^3=CH$_3$
(895) R^1=OH R^2=H R^3=C$_2$H$_5$

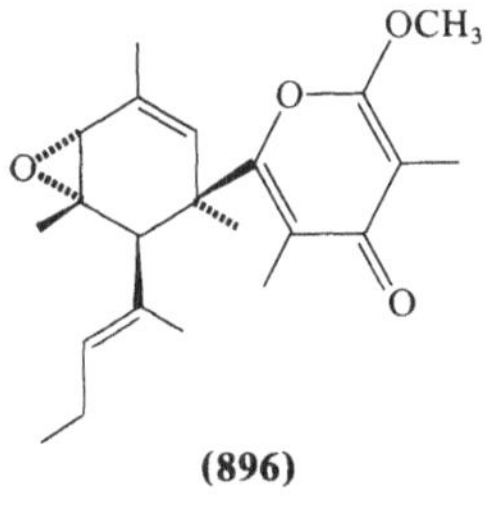

(896)

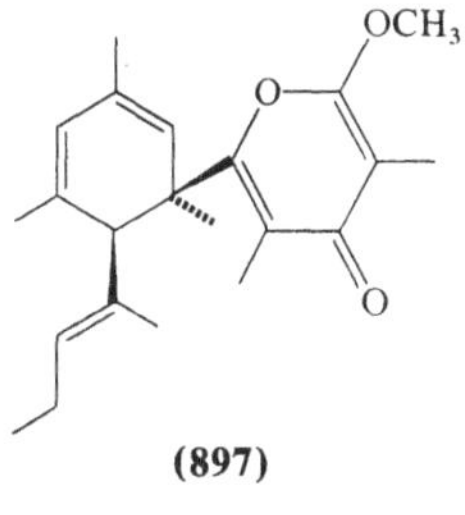

(897)

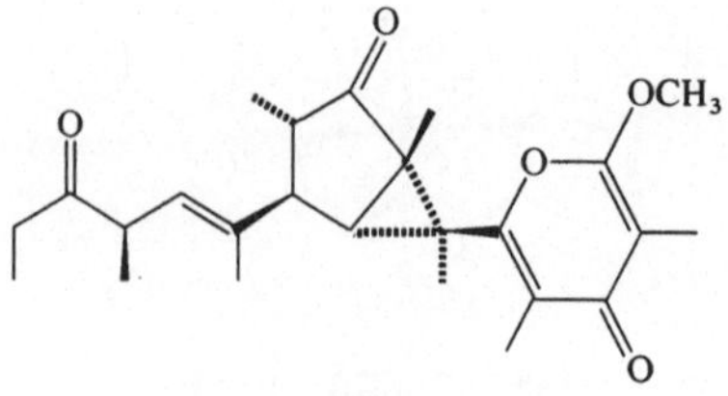

(898)

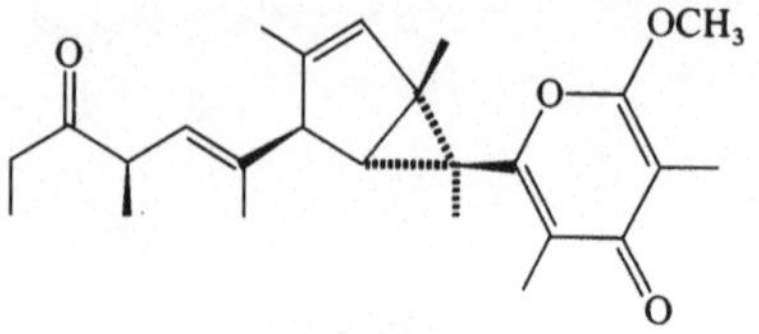

(899)

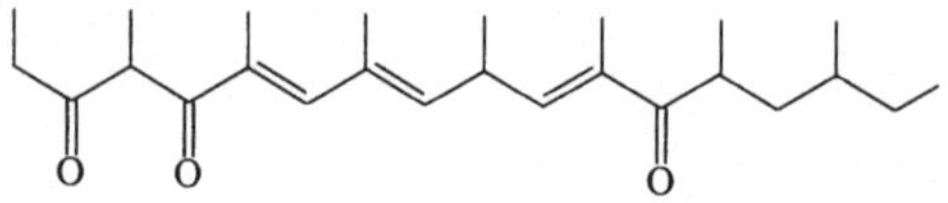

(900)

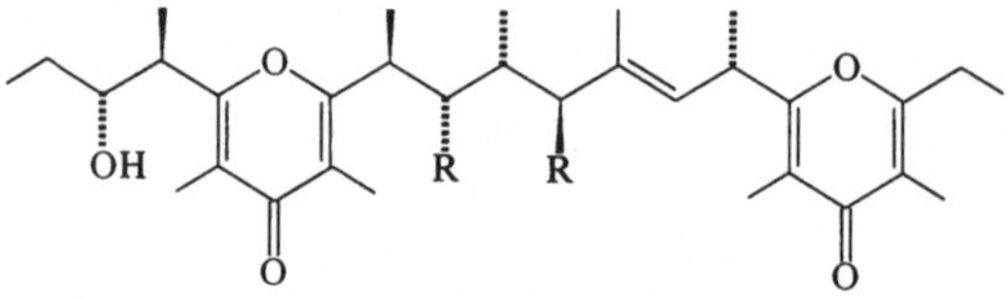

(901) R=OH
(902) R=O—CO—CH$_2$—CH$_3$

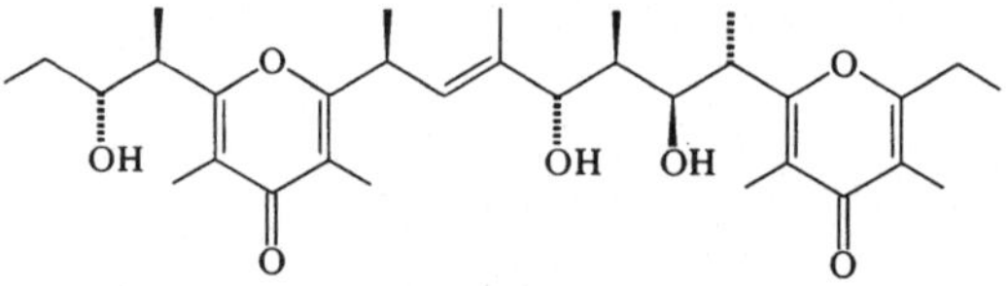

(903)

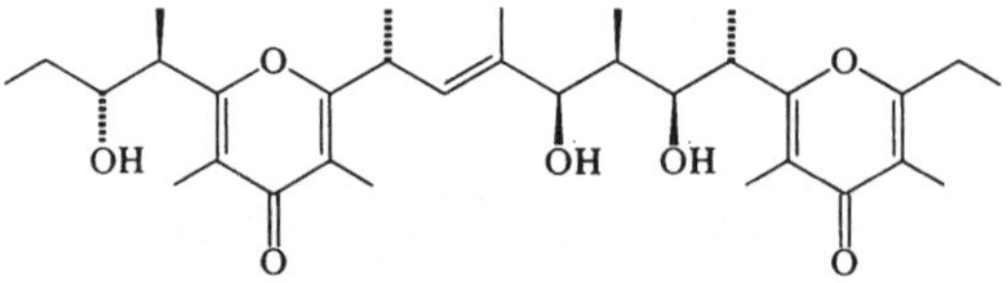

(904)

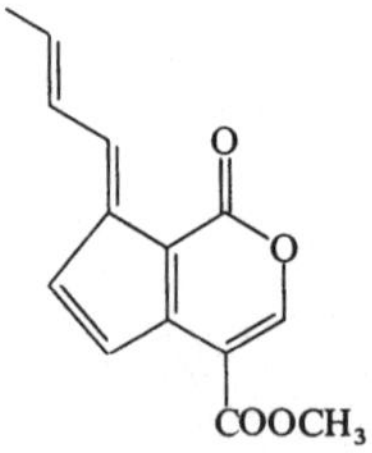

(905)

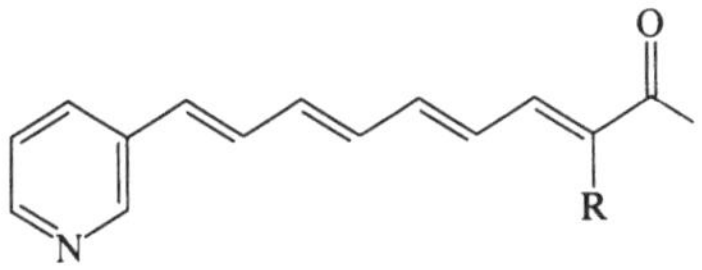

(906) R=H
(909) R=CH₃

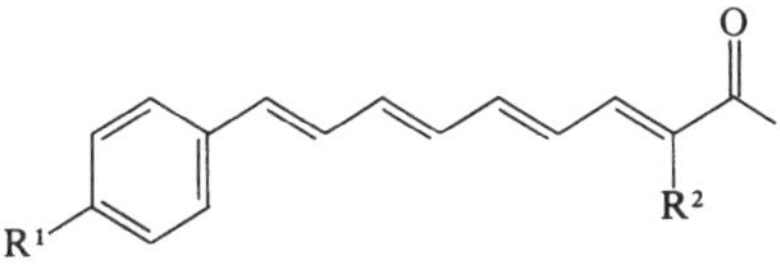

(907) R¹=R²=H
(908) R¹=OH R²=H
(911) R¹=H   R²=CH₃

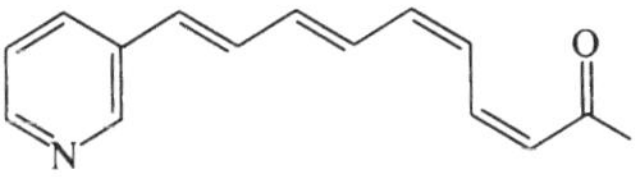

(910)

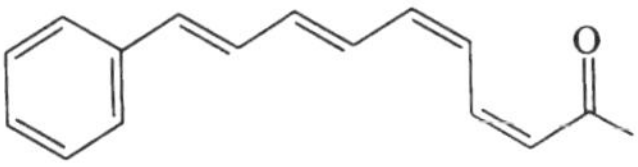

(912)

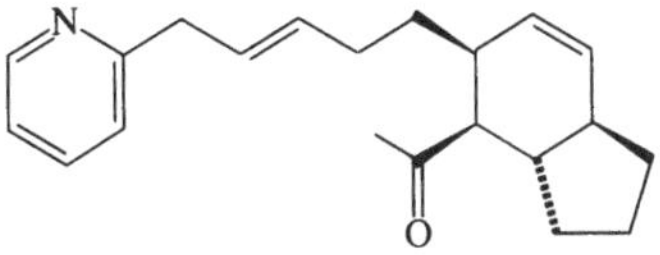

(913)

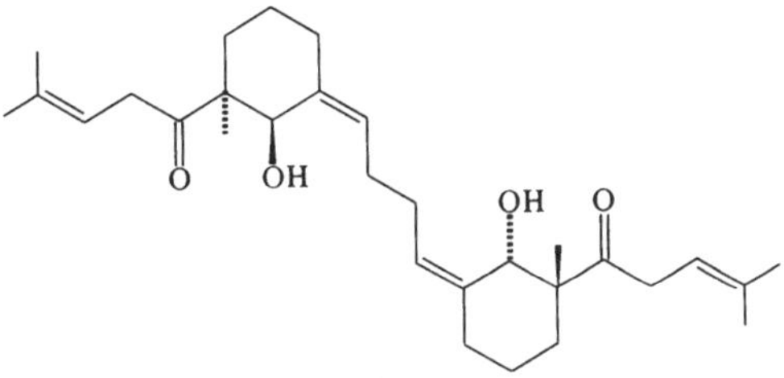

(914)

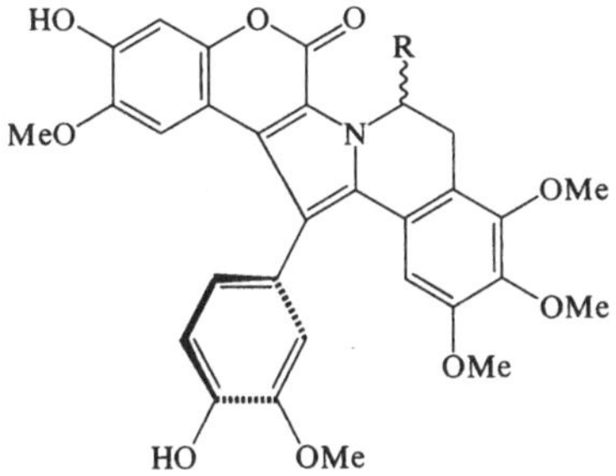

(915) R=OH
(917) R=H

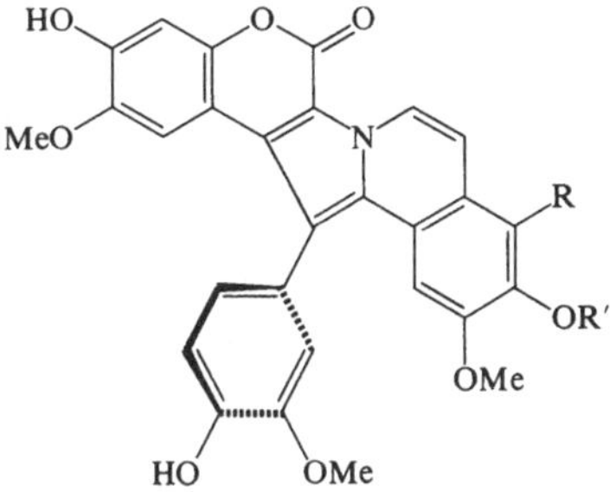

(916) R=OMe R'=Me
(918) R=H   R'=H

V.2. Bivalvia

In an article which includes a description of biologically active compounds from mussels one should deal with saxitoxin and other paralytic shellfish poisons, toxins which are responsible for acute and often fatal poisonings caused by the consumption of certain shellfish. However, these compounds derived from dinoflagelates will not be reviewed here because a thorough discussion by SHIMIZU (*32*) can be found in an earlier volume of this series. Considerable efforts have been devoted to elucidation of the sterol and lipid content in the scallop *Placopecten magellanicus* (*601*), the oyster *Crassostrea virginica* (*602–605*), in *Mytilus edulis* (*606*), *Aulacomya ater* (*607*), and *Chlamys tehuelcha* (*608*). MATSUNO et al. (*609–611*) identified novel marine carotenoids from several species of bivalves. An acid extract of pedal ganglia of the mollusc *Mytilus edulis* was fractionated by high-pressure chromatography to furnish endogenous neuropeptides which had the same primary structures as [Met]- and [Leu]enkephalin (*612, 613*). This study demonstrated that in relatively primitive organisms a variety of opioids can be found, and suggested that the genes responsible for these peptides are extremely "old".

V.3. Cephalopoda

Tetrodotoxin has been isolated from various vertebrate and invertebrate species, among them from the blue-ringed octopus, *Hapalochlaena maculosa*. SHEUMACK et al. (*614*) described the occurrence of a lethal toxin in the eggs of this species. The properties of this toxin were indistinguishable from those of authentic tetrodotoxin.

VI. Echinodermata

The main compounds responsible for the biological activity of Echinoderms are saponins. Although saponins are rather common in plants, they are very rare in the animal kingdom. They have been found only in the classes Holothuroidea and Asteroidea of the phylum Echinodermata. Saponins are composed of a carbohydrate part attached to an aglycone. While the aglycones of asterosaponins are steroids, those of the holothurian compounds are triterpenes. In former times saponins or saponin mixtures were hydrolyzed by acid prior to structure determination of the resulting aglycones. In many cases this

References, pp. 320–363

resulted in isolation not of the natural aglycone, but of an artifact. However, during the past few years advances in separation and spectroscopic techniques such as reversed phase hplc, ^{13}C-NMR, and FAB mass spectrometry have facilitated the task of determinating the structures of the saponins themselves. A thorough review of work on the aglycones as well as of earlier work dealing with the structures and the biological activities of saponins from Echinoderms can be found in an article by BURNELL and APSIMON (615).

VI.1. Saponins from Starfish

Asterosaponins are highly toxic. Dilute solutions obtained from many species of starfish are lethal to fish, annelides, mollusks, arthopods, and vertebrates. It has also been reported that dried starfish meal has been used in Japan to eradicate harmful insects and maggots (6). Asterosaponins are reported to have hemolytic, antineoplastic, cytotoxic (616), antitumor, antibacterial, antiviral, antifungal und antiinflammatory activities, they induce abnormalities in the early development of larvae of sea urchins, cause inhibition of the multiplication of influenza virus, and blockade contractions of a rat phrenic nerve-diaphragm preparation as reviewed in detail by BURNELL and APSIMON (615) and by MINALE et al. (617). These dramatic effects have provided a motivation for structure elucidation not only of the aglycones or their artifacts, but also for structure elucidation of the native saponins themselves, although before 1980 the structures of only few asterosaponins were known. These included asterosaponin A (919) from *Asterias amurensis, Astropecten scoparius, Asterina pectinifera* and *Coscinasterias acutispina*, thornasteroside A (920) from *Acanthaster planci*, glycoside B_2 (921) from *Asterias amurensis*, and a pentaglycoside of thornasterol B (922) from *Acanthaster planci* (615). More recently OKANO et al. (618, 619) isolated socalled OA-1 to -5 from the ovary of *Asterias amurensis*. OA-1 (923) contained a 6-deoxy-xylo-hexos-4-ulopyranosyl residue which was readily hydrated in media containing water. The structure elucidation was based mainly on ^{13}C-NMR studies.

DE SIMONE et al. (620) reported the structure of a new type of saponin, sepositoside A (924), from *Echinaster sepositus*. This compound was unique, apart from the interesting $\Delta^7,3\beta,6\beta$-dioxygenated steroid moiety, because it contained no sulfate group and because the sugar moiety bridged C-3 and C-6 of the steroid nucleus. Sepositoside (924) was isolated after extraction with water and separation on Amberlite XAD-2, silicagel and reverse-phase hplc. Structure elucidation utilized ^{13}C- and ^{1}H-NMR spectroscopy and acid hydrolysis. Very mild acid hydrolysis with 1N HCl at 50° C for about 1 min, provided a ring-

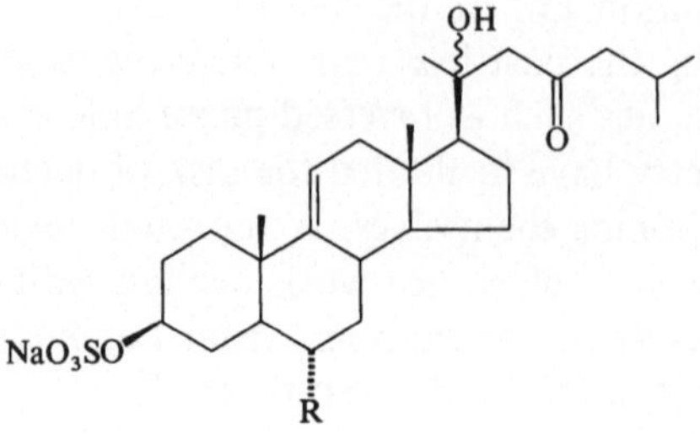

(919) R =

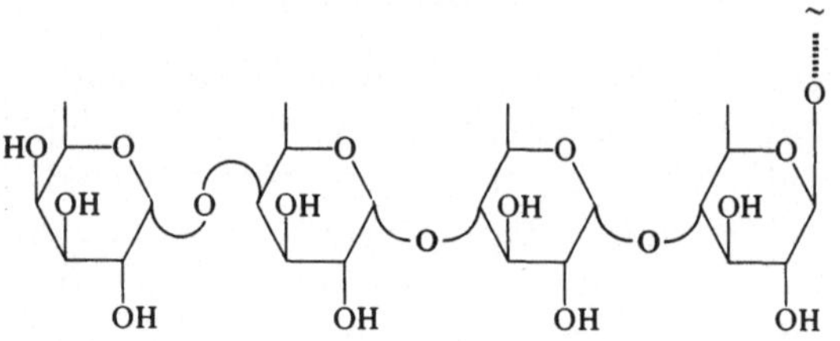

(920) R = 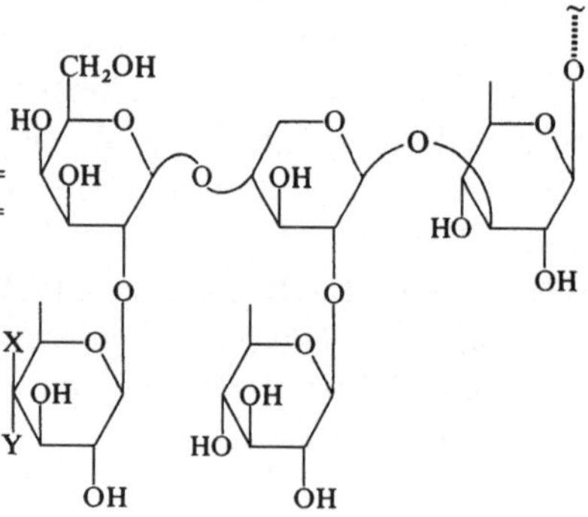 with X=OH and Y=H
(921) R = with X=H and Y=OH

(923) R =

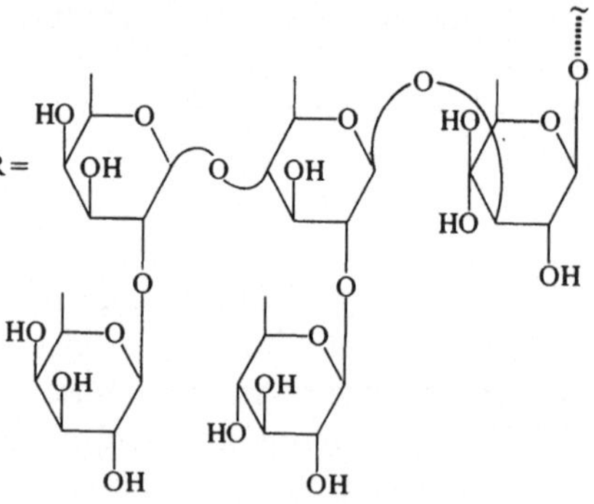

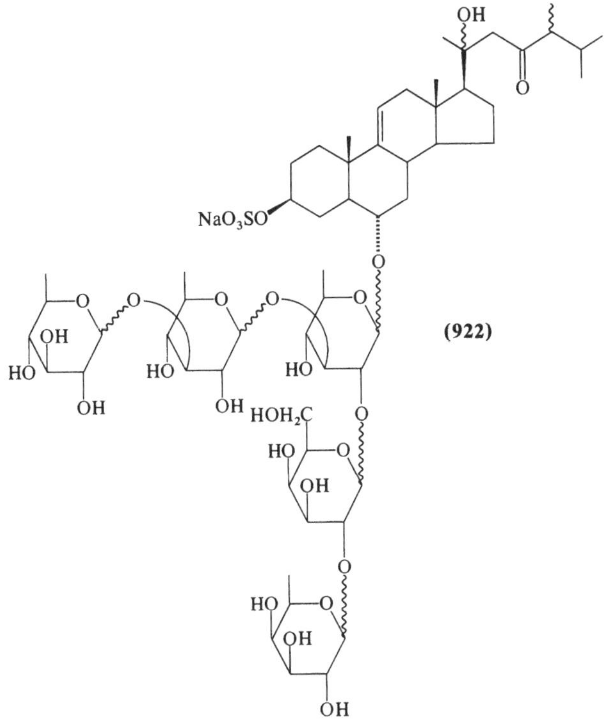

(922)

opened saponin (925). The sequence of the sugar residues in 925 was established by permethylation followed by acid hydrolysis and study of the mass-spectral fragmentation pattern of the permethylated hydrolysate. Sepositoside A (924) was toxic intraperitoneally with an LD_{50} of 43 mg kg^{-1}.

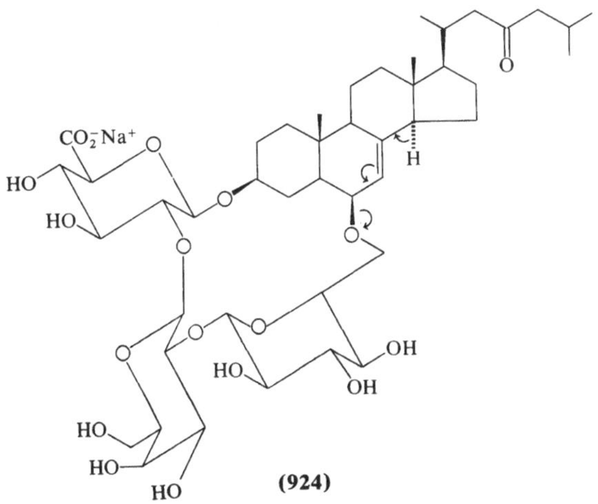

(924)

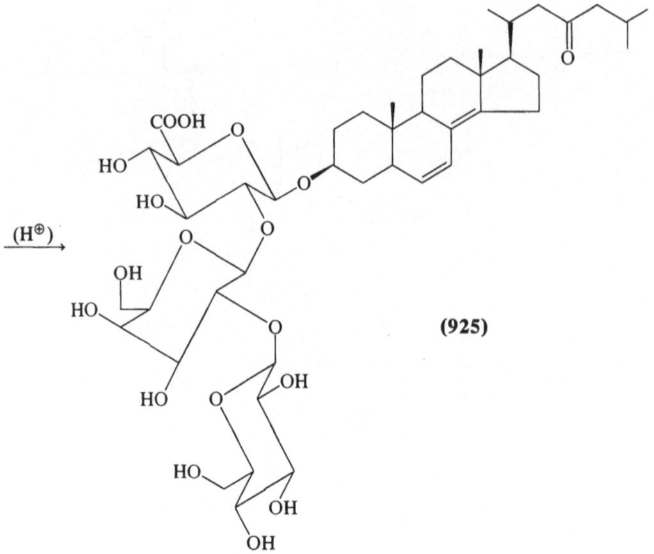

(925)

Three minor saponins (**926–928**) from the same starfish differed only in the nature of the aglycone *(621)*. Like **924**, compounds **926–928** gave the correspondenting opened glycosides on very mild acid treatment.

A similar asterosaponin, luzonicoside (**929**), has been obtained from *Echinaster luzonicus (622)*. It differed from sepositoside A (**924**) in the nature of the sugar moiety which contained D-galactose, L-arabinose and D-glucuronic acid. Like **924**, luzonicoside (**929**) was accompanied by three minor saponins (**930–932**) with epoxide functions in three different steroidal side chains.

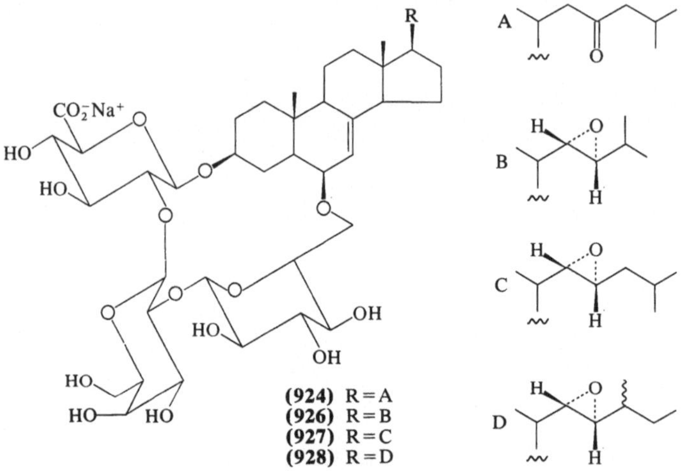

(924) R = A
(926) R = B
(927) R = C
(928) R = D

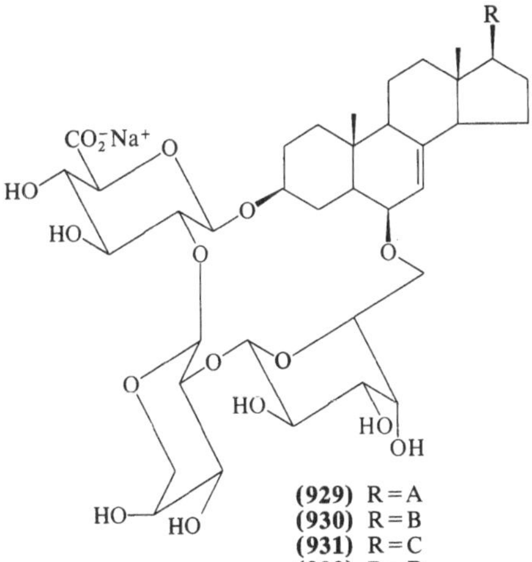

(929) R=A
(930) R=B
(931) R=C
(932) R=D

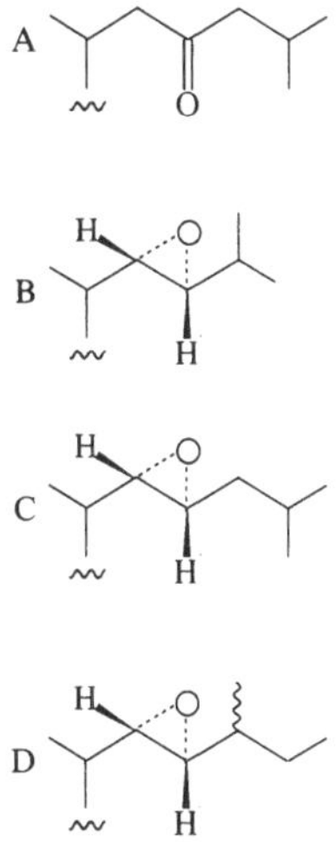

Another group of glycosides from *Protoreaster nodosus* and *Hacelia attenuata* was composed of a polyhydroxylated steroidal aglycone and a carbohydrate residue which was glycosidally attached to C-24 of the steroid (*617*, *623*). The cytotoxic nodososide (**933**) was isolated from *P. nodosus* by methanolic extraction and chromatography on silicagel and $C_{18}\mu$-bondapak; the structure was deduced by 500 MHz ^{1}H NMR- and ^{13}C NMR-spectrometry and by acetylation to an hexaacetate (**934**) and determination of the acid hydrolysis products. Five more glycosides (**935–939**) all with the carbohydrate moiety attached to C-24, could be isolated from *H. attenuata* by MINALE et al. (*617*, *624–626*). More recently, MINALE et al. (*627*) isolated a steroidal glycoside in small amounts from *Acanthaster planci* and *Linckia laevigata*, identical with nodososide (**933**).

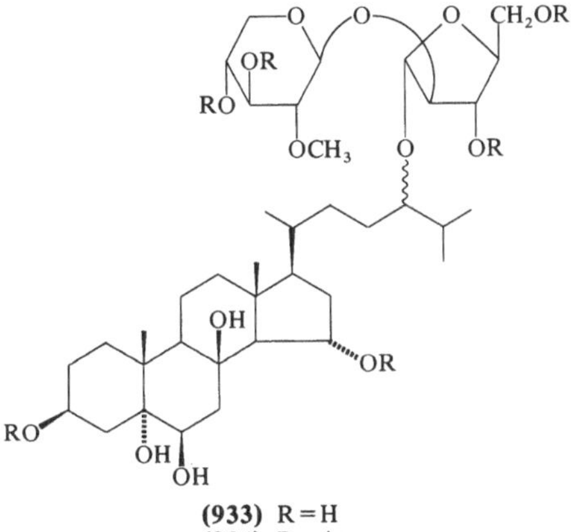

(933) R=H
(934) R=Ac

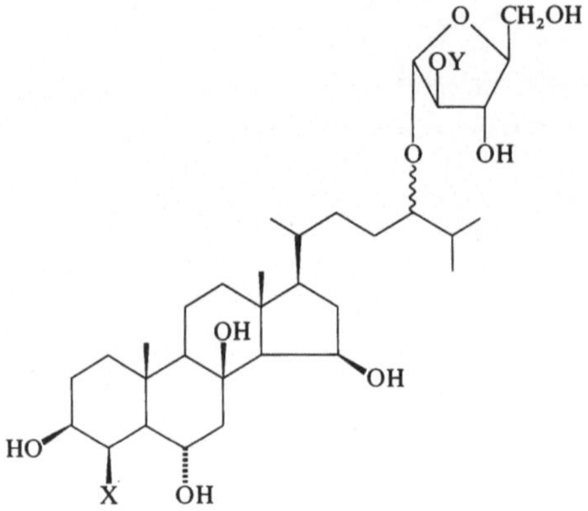

(935) X=H   Y=R
(936) X=H   Y=H
(937) X=OH Y=R
(938) X=OH Y=H

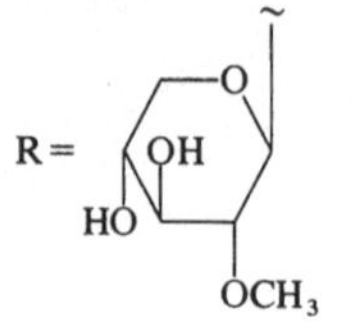

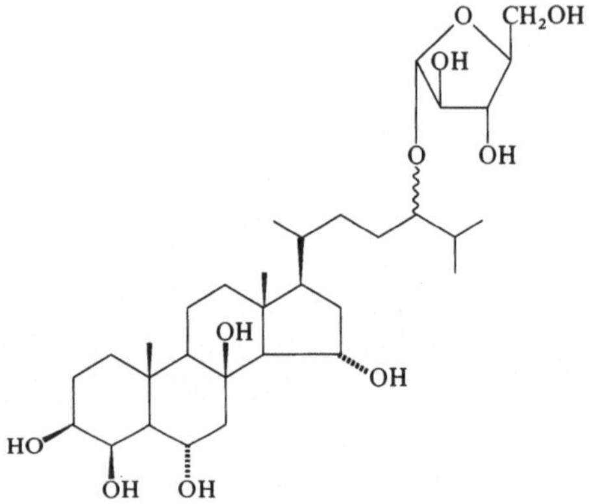

(939)

Thus, the asterosaponins can be subdivided into three groups. The first type, which is of widespread occurrence, has a $\Delta^{9,11}$,3β,6α-dihydroxysteroidal moiety. The oligosaccharide chain is attached to C-6 and a sulfate group is at C-3 of the aglycone. The second group has a Δ^7,3β,6β-dihydroxysteroidal moiety and no sulfate group. The sugar part bridges the C-3 and the C-6 atoms of the aglycone. The last group includes the saponins from *P. nodosus* and *H. attenuata*, mentioned in the previous paragraph, which contain a polyhydroxylated steroid glycosidated at C-24.

Most saponins which have been isolated recently belong to the first group. For example, in studying the starfish *Acanthaster planci* KOMORI et al. (*628, 629*) found in addition to thornasteroside A (**920**) a new saponin, acanthaglycoside A (**940**) and another oligoglycoside sulfate (**941**). The latter was obviously an artifact generated from thornasteroside A (**920**) via a retro-aldol reaction. The structures were established by acid hydrolysis, partial hydrolysis with a glycosidase mixture from *Charonia lampas* and solvolysis with dioxane-pyridine to give a desulfated glycoside. Permethylation of the saponin followed by methanolysis led, together with the results of the field desorption (FD) and FAB mass and ^{1}H- and ^{13}C-NMR spectrometry and the results of the partial hydrolysis to the sequence of the sugar chain.

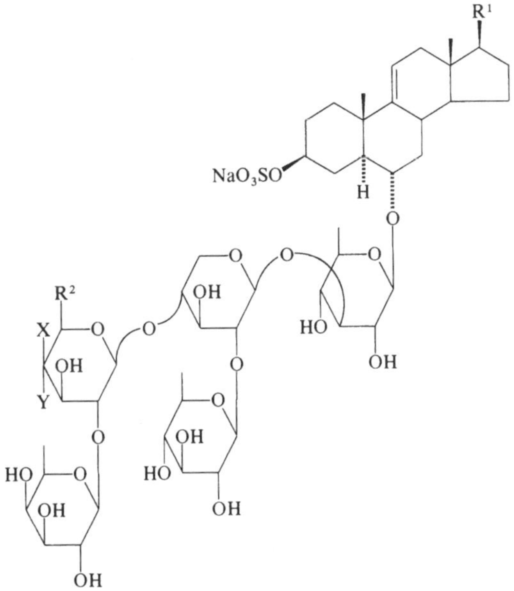

(940) R^1 = [structure]

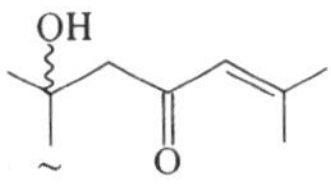

R^2 = CH$_3$ X = H   Y = OH

(941) R^1 = [structure]

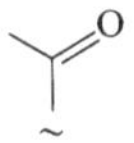

R^2 = CH$_2$OH   X = OH   Y = H

The same methods were utilized to determine the structures of two oligoglycosides (**942**) and (**943**) from *Astropecten latespinosus* after they had been desulfated by solvolysis (*630*), thornasteroside A (**920**) and a new saponin, versicoside A (**944**), from *Asterias amurensis versicolor* (*631*), and four new compounds, the luidiaglycosides A–D (**945–948**), from *Luidia maculata* (*632, 633*). Luidiaglycosides C (**947**) and D (**948**) contain the aglycones marthasterone and dihydromarthasterone which have long been known as hydrolysis products of starfish glycosides, and are the first such saponins whose structure has been completely elucidated. Recently, dihydromarthasterone has been isolated after hydrolysis of the crude extract from *Stichaster striatus*, but the nature of the sugar part has not been reported *(634)*. The starfish *Asterias vulgaris* contained glycosides which also belong to the first group mentioned in the previous paragraph. FINDLAY et al. *(635, 636)* described three saponins **(949–951)** based on the aglycone thornasterol A but the complete structures of the saponins has not yet been determined.

MINALE et al. *(617)* obtained six individual compounds from the saponin mixture of *Marthasterias glacialis*. Using *inter al.* FAB mass spectrometry as a tool for structure elucidation of these polar nonvolatile substances they proposed the structures shown below for three of the six components.

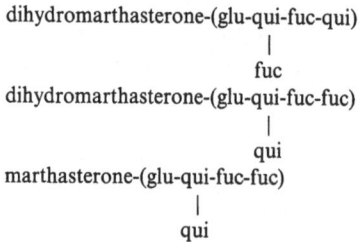

Subsequently the same group described in detail isolation *(637)* and characterization *(638)* of two hexa- and two pentaglycosides from the same source. The first hexaglycoside, marthasteroside A_1 **(952)**, was a derivative of **942** containing the sulfate residue while the second was identical with luidiaglycoside A **(945)** from *Luidia maculata (632)*. The two pentaglycosides **953** and **954** contained the same sugar moiety but different aglycones, with **953** being based on marthasterone and **954** on dihydromarthasterone, and differed from luidiaglycosides C **(947)** and D **(948)**, previously isolated by KREBS et al. *(633)* only in one sugar residue. While the compounds from *L. maculata* had three quinovose units and only one fucose, **953** and **954** included two of each.

MINALE et al. have also studied the saponins of *L. maculata* and have reported isolation of luidiaglycoside A **(945)**, B **(946)**, marthasteroside B **(953)**, C **(954)**, and thornasteroside A **(920)** *(639)*. RICCIO et al. investigated five additional species of starfish, *Ophidiaster ophidianus (640)*, *Hacelia attenuata (640)*, *Linckia laevigata (641)*, *Protoreaster nodosus (642)*, and *Pentaceraster alveolatus (642)*. *O. ophidianus* and *H. attenuata* both contained two saponins with a C_{26} steroidal aglycone, named ophidianoside B **(955)** and C **(956)**. Two further saponins from *O. ophidianus* were thornasteroside A (**920**) and ophidianoside F (**957**). *L. laevigata* furnished, in addition to the previously known saponins **920**, **946**, **952**, and **957** the new laevigatoside (**958**). A new sulfated asterosaponin, protoreasteroside (**959**) from *P. nodosus* and *P. alveolatus*, contained an unprecedented aglycone (20 R,22 S)-5α-cholesta-9(11),24(25)-diene-3β,6α,20,22-tetraol.

The biological activities of the steroidal saponins from the Japanese starfish *Acanthaster planci*, *Luidia maculata*, and *Asterias amurensis versicolor* have been examined by FUSETANI et al. (*643*) who investigated the effects of 17 saponines on fertilized sea urchin and starfish eggs. These compounds inhibited sea urchin embryos from further development in the morula stage, while starfish eggs were much more resistant. Some of these asterosaponins showed hemolytic effects. The authors discussed the relationship between the structure and the biological activity.

	R^1	R^2	R^3	X	Y	
(942)	H	CH_2OH	A	OH	H	
(943)	H	CH_2OH	H	H	OH	
(944)	SO_3Na	CH_2OH	B	OH	H	
(945)	SO_3Na	CH_3	A	H	OH	
(946)	$SO_3^\ominus X^\oplus$	CH_3	H	H	OH	$X^\oplus = Na^\oplus$, or $NH_4^\oplus$
(952)	SO_3Na	CH_2OH	A	OH	H	
(957)	SO_3Na	H	H	H	OH	

(947) X=H   Y=OH
(948) X=H   Y=OH   24,25-dihydro
(953) X=OH   Y=H
(954) X=OH   Y=H   24,25-dihydro

(949) R = (quinovose)₃(fucose)₂
(950) R = (quinovose)₃(sugar X)₁
(951) R = (quinovose)₃(fucose)₁

(955) R=CH₂OH   X=OH   Y=H
(956) R=H   X=H   Y=OH

(958)

(959)

KICHA et al. *(644)* reported the occurrence of a new substance, asterosaponin P_1 **(960)**, in the Pacific starfish *Patiria pectinifera*. This saponin belongs to the group of 24-O-glycosidal steroids, but differed from those encounted previously, because the sugar moiety bears a sulfate residue. The structure of **960** was elucidated by spectroscopic and chemical methods. These latter included solvolysis of **960** to **961** and subsequent conversion by acetylation or methylation to the pentaacetate **962** or hexa-O-methyl derivative **963**. Oxidation with CrO_3-C_5H_5N resulted in the triketone **964**, oxidation with CrO_3-CH_3COOH in the diketone **965**.

(960) $R^1=H$ $R^2=SO_3Na$
(961) $R^1=R^2=H$
(962) $R^1=R^2=Ac$
(963) $R^1=R^2=CH_3$

VOOGT and VAN RHEENEN (*645*) isolated asterosaponins from the gonads, the pyloric caeca, the stomach and the body wall of female specimens of *Asterias rubens* and determined the sugar and sulfate content for each saponin. On the basis of a statistical analysis which showed that the sugar content in the substances of the pyloric cecum was significantly lower than those in the other parts, they hypothesized that the saponins are formed in the pyloric cecum, and that they are modified slightly in each tissue, in particular by addition of fucose units. The authors found that the haemolytic activity increased with the molar ratio of carbohydrate to sulfate.

Recently, BURNELL *et al.* (*645a*) investigated the levels of two steroids, asterone (**966**) and asterogenol (**967**), obtained by hydrolysis of the crude asterosaponin mixture from the starfish *Asterias vulgaris*. Concentrations were highest in winter and spring, then the steroid levels fell to their annual minima in July, after the spawning period.

VI.2. Steroids from Starfish

As the glycosides from starfish are steroidal in nature and as some known polyhydroxylated steroids with moderate cytotoxicity have been isolated from these animals, the literature since 1980 dealing with such compounds will be reviewed here. Aside from some monohydroxysterols from *Echinaster sepositus* (*646*), *Acanthaster planci* (*647*) – this being a source of a new cyclopropane sterol –, *Asterias vulgaris* (*648*), and *Comasterias lurida* (*649*), the reports deal mainly with polyhydroxylated steroids.

MINALE *et al.* investigated the starfish species *Protoreaster nodosus* (*617, 650, 651*), *Hacelia attenuata* (*617, 626, 652*), and *Luidia maculata* (*653*). Structure determination of these compounds, isolated from the methanol or chloroform-methanol extracts was based upon ^{1}H- and

^{13}C-NMR studies and chemical transformations such as acetylation, hydrogenation and oxidation. In the same manner six steroids from *P. nodosus* with 6–8 hydroxylgroups could be identified as 5α-cholestane-3β,6α,8,15α,16β,26-hexol (**968**), 5α-cholestane-3β,6α,7α,8,15α,16β,26-heptol (**969**), and 5α-cholestane-3β,4β,6α,7α,8,15α,16β,26-octol (**970**) (*651*), 5α-cholestane-3β,4β,6α,8,15α,16β,26-heptol (**971**), 24ξ-methyl-5α-cholestane-3β,6α,8,15α,16β,26-hexol (**972**), and (22 E)-24ξ-methyl-5α-cholest-22-en-3β,4β,6α,8,15α,16β,26-heptol (**973**) (*650*). The major polyhydroxylated sterol from *H. attenuata* lacked the 8β-hydroxyl group and was characterized as 5α-cholestane-3β,6β,15α,16β,26-pentol (**974**) (*652*). Three additional very minor compounds from the same species were interesting because they contained side chains of the 24-methyl-27-*nor*-cholestane and 24-methyl-26,27-bis*nor* type and were 24ξ-methyl-26,27-bis*nor*-5α-cholest-22-ene-3β,4β,6β,8,15α,16β,25-heptol (**975**), 24ξ-methyl-27-*nor*-5α-cholestane-3β,6β,8,15α,16β,26-hexol (**976**), and 24ξ-methyl-27-*nor*-5α-cholestane-3β,4β,6β,8,15α,16β,26-heptol (**977**) (*626*). From the third starfish species, *L. maculata*, MINALE *et al.* (*653*) isolated two hexols and one heptol which were identified as 5α-cholestane-3β,5,6β,15α,16β,26-hexol (**978**), 5α-cholestane-3β,6β,7α,15α,16β,26-hexol (**979**), and 5α-cholestane-3β,5,6β,7α,15α,16β,26-heptol (**980**), all of which, like **974**, lacked the 8-hydroxylfunction.

D'AURIA *et al.* (*654*) found two groups of sterol sulfates in the aqueous extract of the starfish *Euretaster insignis*. Hydrolysis of the less polar fraction gave cholest-7-en-3β-ol and C_{26}, C_{27}, C_{28}, and C_{29} 5α-steroidal alcohols. The more polar fraction contained sulfated 3β,21-dihydroxysteroids. After solvolysis to remove the sulfate groups the steroids were identified as (20 R)-24-methylenecholestane-3β,21-diol (**981**), (20 R)-24-methylenecholest-5-ene-3β,21-diol (**982**), (20 R,22 E)-24-methylcholest-22-ene-3β,21-diol (**983**), (20 R)-cholestane-3β,21-diol (**984**), (20 R,22 E)-cholest-22-ene-3β,21-diol (**985**), and (20 R)-cholest-5-ene-3β,21-diol (**986**). This starfish was apparently devoid of saponins.

	R¹	R²	R³	
(**968**)	H	H	H	
(**969**)	OH	H	H	
(**970**)	OH	OH	H	
(**971**)	H	OH	H	
(**972**)	H	H	CH₃	
(**973**)	H	OH	CH₃	Δ²²-trans

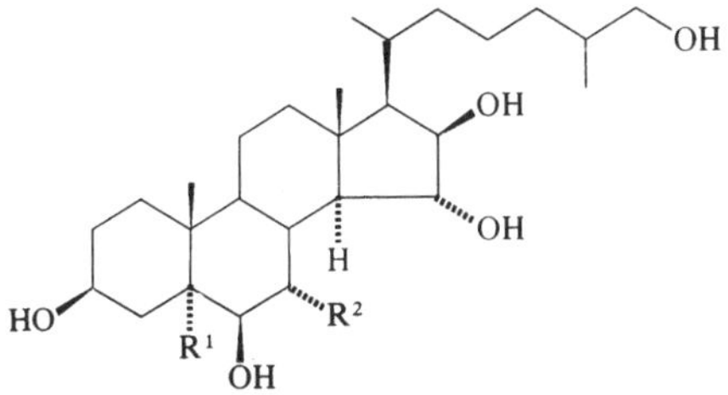

(974) R¹=R²=H
(978) R¹=OH R²=H
(979) R¹=H R²=OH
(980) R¹=R²=OH

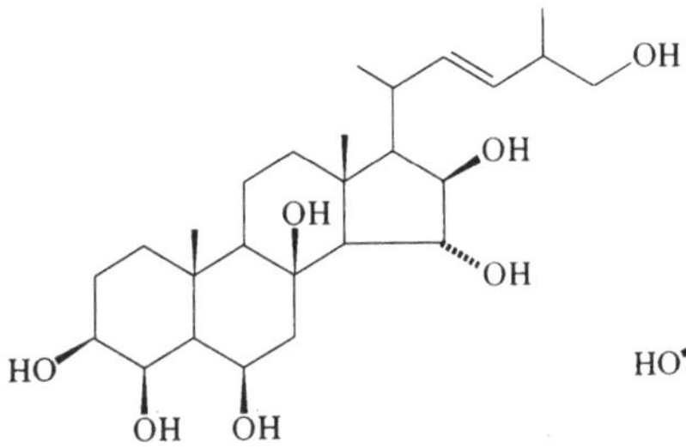

(975)

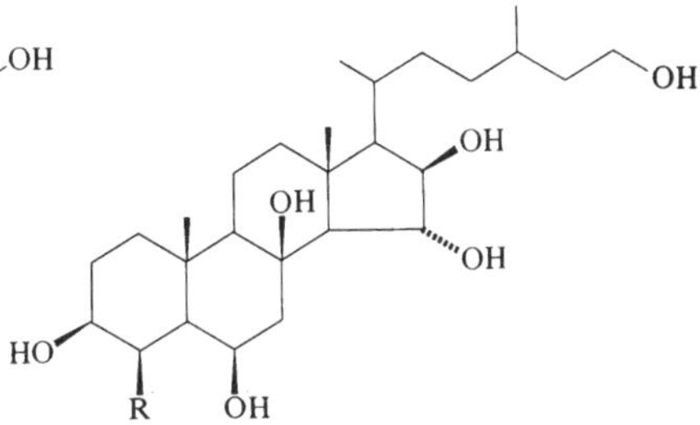

(976) R=H
(977) R=OH

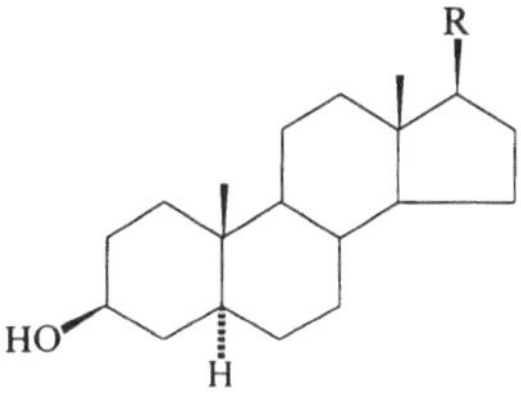

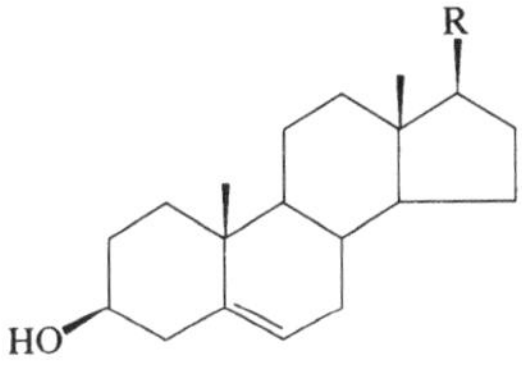

(981) R= 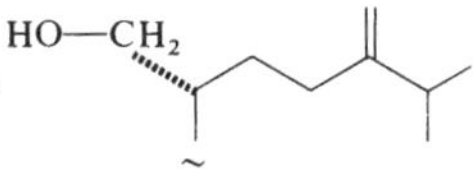
(982) R= 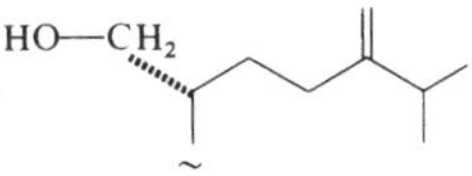
(983) R= 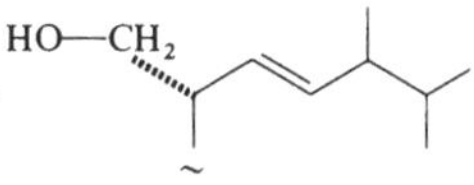
(986) R= 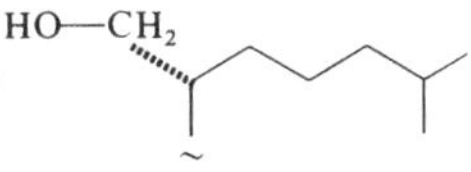
(984) R= 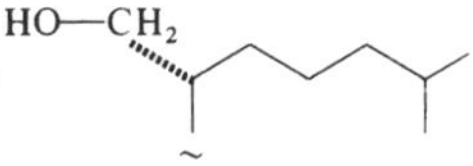
(985) R=

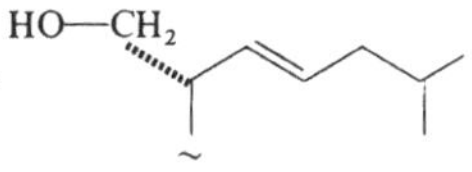

VI.3. Miscellaneous Compounds from Starfish

KOMORI et al. (655) reported the isolation of a cerebroside mixture (**987**) and two nucleosides, thymine desoxyribonucleoside (**988**) and uracil desoxyribonucleoside (**989**), from *Acanthaster planci*. Structure elucidation was based on ^{1}H- and ^{13}C-NMR spectrometry as well as on EI- and FD mass spectrometry. Similar cerebroside mixtures which showed nearly identical IR-spectra could be extracted from *Luidia quinaria* and *Astropecten latespinosus*. The mycosporine-like amino acids (**990–994**) have been obtained from alcoholic extracts of the starfish *Asterina pectinifera* (656).

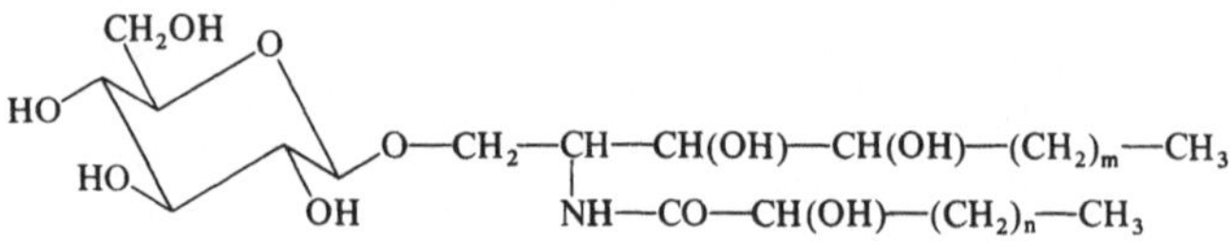

(**987**) n = <u>21</u>, 20, 19, 16
m = <u>12</u>, 13

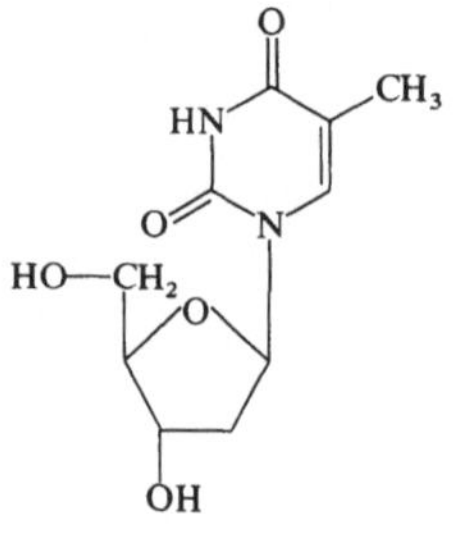

(**988**)

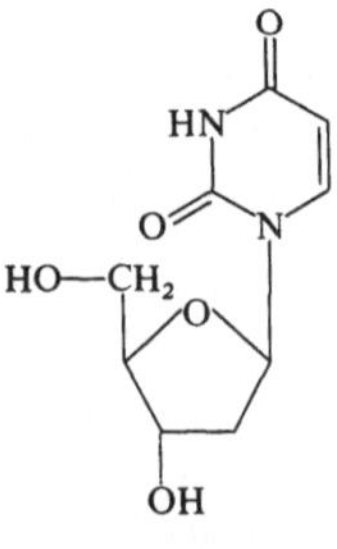

(**989**)

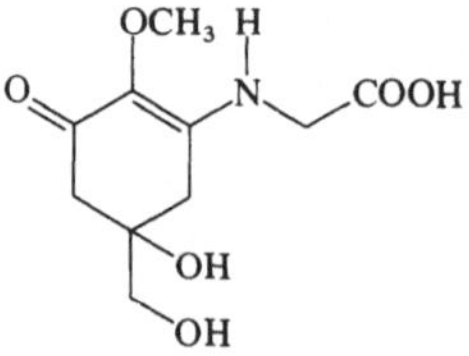

(**990**)

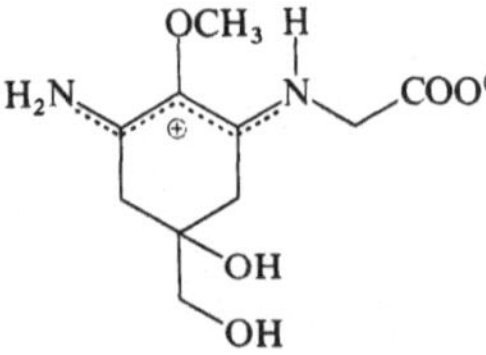

(**991**)

$$\begin{array}{c}\text{(992) R=COOH}\\ \text{(993) R=CH}_3\\ \text{(994) R=H}\end{array}$$

The last substance to be mentioned in this chapter dealing with natural products from starfish is a well known toxin, tetrodotoxin. During examination of the mechanism involved in tetrodotoxin-intoxification of *Charonia sauliae*, NOGUCHI et al. (*657*) found parts of the starfish *Astropecten polyacanthus* in the contents of the trumpet shell's digestive tract. They isolated the starfish toxin and proved by thin layer chromatography, electrophoretic analysis and gas chromatography-mass spectrometry of its trimethylsilyl derivative that it was identical with tetrodotoxin. The authors assumed that the tetrodotoxin found in this starfish is introduced through the food chain. More recently the same group demonstrated that *Astropecten latespinosus* also contained tetrodotoxin (*658*).

VI.4. Saponins from Sea Cucumbers

The poisonous properties of sea cucumbers have been known for many centuries. They have been used to poison fish in tidal pools and there are reports about fatal poisonings after consumption of sea cucumbers. On the other hand, they are considered acceptable food for humans (trepang, haishen) in some Asiatic countries. Research on the toxicity of sea cucumbers began already in the 1920's, but the complete structures of their toxins have been determined only in the last ten years. Early investigations dealt with isolation and characterization of the aglycones, although hydrolysis of the saponins generally resulted in artifacts. Thus, elimination of the C-12 hydroxy group and rearrangement of the double bonds led to a 7,9(11)-dien-system. In addition to their ichthyotoxicity, the extracts of holothurians are toxic to other animals and possess cytostatic, antimicrobial (*659*), hemolytic, antifungal, and neurotoxic properties. They act as shark repellents and induce abnormalities in the development of fertilized sea urchin eggs. The biological activities, structures of the aglycones and early reports on complete structures of the saponins themselves were reviewed by BURNELL and APSIMON (*615*).

Most of the sea cucumbers which have been investigated belong to the family *Holothuriidae*. Holothurin A (**995**), the major lanostane-type triterpene oligoglycoside from the Cuvierian tubules of *Holothuria leucospilota* BRANDT (= *H. vagabunda* SELENKA) has been isolated by KITAGAWA et al. (*660*). The major compound of the body walls of this species is holothurin B (**996**) (*661*) which differs from **995** only in the nature of the sugar part, **995** being a tetraglycoside while **996** contains only two sugar residues. To obtain these compounds, body walls and Cuvierian tubules were extracted with 70% ethanol, followed by partitioning between butanol-water, silica gel chromatography and recrystallization. While acid hydrolysis of both saponins with 3N hydrochloric acid resulted in an artifact, 22,25-oxidoholothurinogenin, treatment with snail enzyme gave in the case of **996** two prosapogenols. Both of these were monoglycosides and one of them lacked the sulfate residue. Solvolysis of holothurin B (**996**) with dioxane-pyridine furnished a desulfated derivative; methylation of **996** or of the desulfated derivative followed by methanolysis permitted determination of the sequence of sugar residues and the position of the sulfate group. IR, CD, ^{1}H-NMR, and ^{13}C-NMR spectra gave further evidence for the structure of holothurin B (**966**). Enzymatic hydrolysis of holothurin A (**995**) gave holothurin B (**996**) and the same two prosapogenols obtained also from **996**. The usual series of reactions, solvolysis and per-methylation followed by methanolysis, together with the spectroscopic data led to formulation of structure **995**. Recently, holothurin A (**995**) has also been isolated from *Holothuria squamifera* (*661a*) and *Bohadschia graeffei* (*661b*).

A tetraglycoside similar to **995**, but differing in the nature of the aglycone side chain, has been isolated by OLEINIKOVA et al. (*662*) from two species of holothurians, *Holothuria floridana* and *H. grisea*. They named it holothurin A_1 (**997**). Echinosides A (**998**) and B (**999**) from *Actinopyga echinites* had the same carbohydrate moiety as **995** and **996**, respectively, and similary differed from each other only in the side chain of the aglycones (*663*). Later, KALININ and STONIK (*664, 665*) and OLEINIKOVA et al. (*666*) described isolation and structure elucidation of holothurin A_2 from *Holothuria edulis*, *H. floridana*, and *Bohadschia graeffei*, and of holothurin B_1 from *H. floridana* (*667–669*). Holothurin A_2 and B_1 were identical with echinosides A (**998**) and B (**999**), respectively, except that the cation associated with the sulfate group in the glycoside from *H. edulis* was an ammonium ion. Another tetraglycoside sulfate belonging to the same group of saponins is 24-dehydroechinoside A (**1000**) which together with holothurin A (**995**) has

Recent Developments in the Field of Marine Natural Products 301

been isolated from *Actinopyga agassizi* (*670*). On the basis of chemical and physicochemical evidence, structures were assigned to four oligoglycosides lacking a sulfate group, bivittoside A (**1001**), B (**1002**), C (**1003**), and D (**1004**) from *Bohadschia bivittata* by KITAGAWA *et al.* (*671*). The aglycones of these four compounds did not have a 17-hydroxyl and **1003** even lacked the one at C-12. The structure of the native genin of bivittoside C (**1003**) has been confirmed by partial synthesis (*672*). The same native aglycones, holost-9(11)-ene-3β,12α-diol and holost-9(11)-ene-3β-ol – these names being assigned in accordance with the convention proposed by HABERMEHL and VOLKWEIN (*673*) already in 1971 –, could be obtained after Smith degradation of the glycosides from *Bohadschia argus* (*674*). ANISIMOV *et al.* (*616*) compared the cytotoxicity of holothurin A (**995**) and B (**996**), with the cytotoxicity of other saponins and found that the glycosides of holothurians were more cytotoxic than other saponins.

	R^1	R^2	R^3	R^4	R^5
(**995**)	(tetrahydrofuran)	OH	OH	SO_3Na	Z
(**996**)	(tetrahydrofuran)	OH	OH	SO_3Na	H
(**997**)	HO–CH=CH–CH₂–C(CH₃)=	OH	OH	SO_3Na	Z

	R¹	R²	R³	R⁴	R⁵
(998)	(branched alkyl)	OH	OH	SO_3Na	Z
(999)	(branched alkyl)	OH	OH	SO_3Na	H
(1000)	(branched alkenyl)	OH	OH	SO_3Na	Z
(1001)	(branched alkyl)	H	OH	H	H
(1002)	(branched alkyl)	H	OH	Z	H
(1003)	(branched alkyl)	H	H	Z	Z
(1004)	(branched alkyl)	H	OH	Z	Z

Glycosides of four species of the family *Stichopodidae* have been investigated, *Stichopus chloronotus*, *S. variegatus*, *Astichopus multifidus*, and *Thelenota ananas*. *S. chloronotus* was examined simultaneously by a Japanese and a Russian group and provided eight saponins. KITAGAWA *et al.* (*675, 676*) isolated stichlorosides A_1 (**1005**), A_2 (**1006**), B_1 (**1007**), B_2 (**1008**), C_1 (**1009**), and C_2 (**1010**) which contained, like all other glycosides known sofar from holothurians of this family, lanost-7-ene type aglycones and no sulfate residue. On enzymatic hydrolysis, **1005** and **1006** liberated stichlorogenol and 25-dehydrostichlorogenol, respectively (*675*). The structure of stichlorogenol was clarified by X-ray analysis and the structure of the saponins themselves were established, as in the case of holothurin A (**995**) and B (**996**), by chemical and spectroscopic means. Catalytic hydrogenation over 5% Pd-C of **1006**, **1008**, and **1010** gave quantitatively **1005**, **1007**, and **1009**, respectively. The Russian group named the substances isolated by them stichoposides and assigned the structure of stichlorogenol to the aglycone derived from them (*677*). Stichoposides A (**1011**) and B (**1012**) were two new diglycosides (*678*), whereas the structures given for stichoposides C (*679*), D (*680*), and E (*681*) were identical with those of **1009**, **1007**, and **1005**, respectively. Stichoposide D (**1007**) has also been found in *Stichopus variegatus*, (*680*). The genuine aglycone of astichoposide C from *Astichopus multifidus* (*677*) had the same structure as dehydrostichlorogenol and the saponin itself (*679*) was identical with stichloroside C_2 (**1010**). Two thelothurins, A (**1013**) and B (**1014**), from *Thelenota ananas* (*682*) had an unusual sugar chain which contained an unidentified substituent. Later, STONIK *et al.* (*683*) isolated from the same sea cucumber thelenotoside A (**1015**) and B (**1016**), two glycosides with structures similar to those from the other species of this family.

	R¹	R²	R³	R⁴
(1005)	⌇⎯⎯(OAc)⎯⎯	X	H	X
(1006)	⌇⎯⎯(OAc)⎯⎯=	X	H	X
(1007)	⌇⎯⎯(OAc)⎯⎯	X	CH₂OH	Y
(1008)	⌇⎯⎯(OAc)⎯⎯=	X	CH₂OH	Y
(1009)	⌇⎯⎯(OAc)⎯⎯	X	CH₃	Y
(1010)	⌇⎯⎯(OAc)⎯⎯=	X	CH₃	Y
(1011)	⌇⎯⎯(OAc)⎯⎯	H	CH₃	H
(1012)	⌇⎯⎯(OAc)⎯⎯	H	CH₂OH	H
(1015)	⌇⎯⎯(OAc)⎯⎯	H	CH₃	Y
(1016)	⎯⎯(OAc)⎯⎯	H	CH₂OH	Y

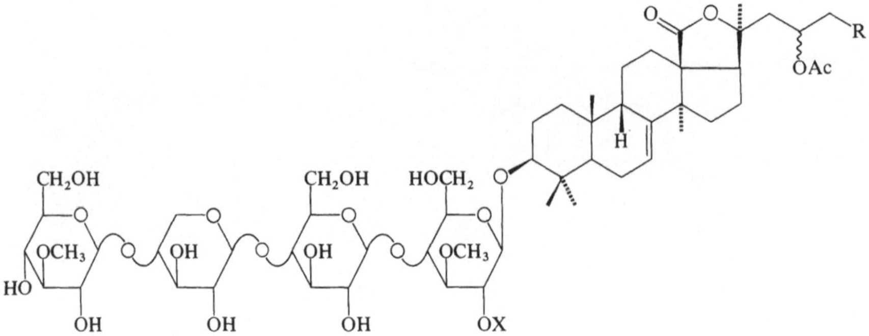

(1013) R = ⌇—⟨

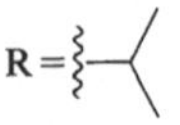

(1014) R = ⌇—⟨

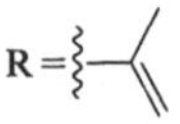

X = unidentified substituent

The complete structures of only two glycosides from the family *Cucumariidae* have been described, cucumarioside G_1 (**1017**) (*684*) from *Cucumaria fraudatrix* and cucumarioside A_2 (**1018**) from *Cucumaria japonica* (*684a*). The native genin of **1017** was 16β-acetoxyholosta-7,24-dien-3β-ol (*677, 685*) while **1018** contains (3β-hydroxyholosta-7,25-dien-16-one) (*677*). Further native genines were isolated after mild hydrolysis of the corresponding glycosides from *C. frondosa* (16α-acetoxyholost-7-ene-3β-ol) (*686*) and a *Parathyona* species (16,23-epoxyholosta-7,24-dien-3β-ol) (*687*). The last-named aglycone was a component of parathyonoside R (**1019**).

GARNEAU *et al.* (*688*) determined the structure of psoluthurin A (**1020**), the major saponin of *Psolus fabricii*, which was unusual because it had two sulfate groups. KALININ *et al.* (*689*) isolated from the same species psolusoside A, a tetraglycoside, which liberated after hydrolysis 3β-hydroxyholosta-9(11),25(26)-dien-16-one, quinovose, xylose, glucose, and 3-O-methylglucose. Hence this saponin is possibly identical with psoluthurin A (**1020**).

Acid hydrolysis of the glycosides from the Far Eastern holothurian *Duasmodactyla kurilensis* has yielded a new aglycone the structure of which is 3β-hydroxy-4,4,14-trimethylpregna-9(11),16-dien-20-one (**1021**) (*690*).

The glycosides from the sea cucumbers *Actinopyga agassizi*, *Holothuria atra*, *Bohadschia argus*, *Cucumaria fraudatrix*, *Astichopus multifidus*, and *Thelenota ananas* inhibited both Na^+-K^+-ATPase and Mg^{2+}-ATPase of rat brain *in vitro* (*691*). Glycosides from holothurians

(1017)

(1018)

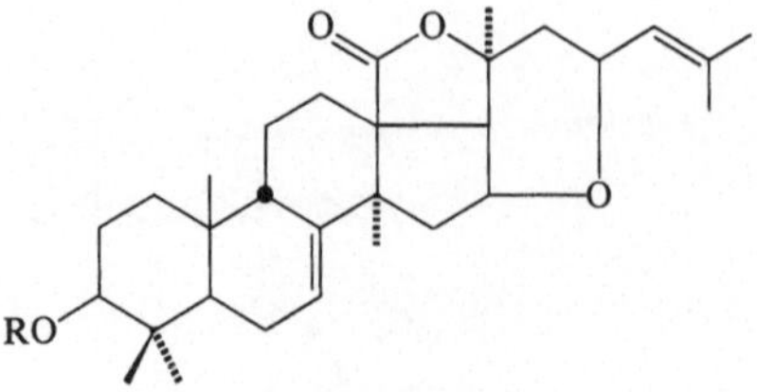

(1019) R = { D-quinovose, D-xylose, 3-O-Me-D-glucose }

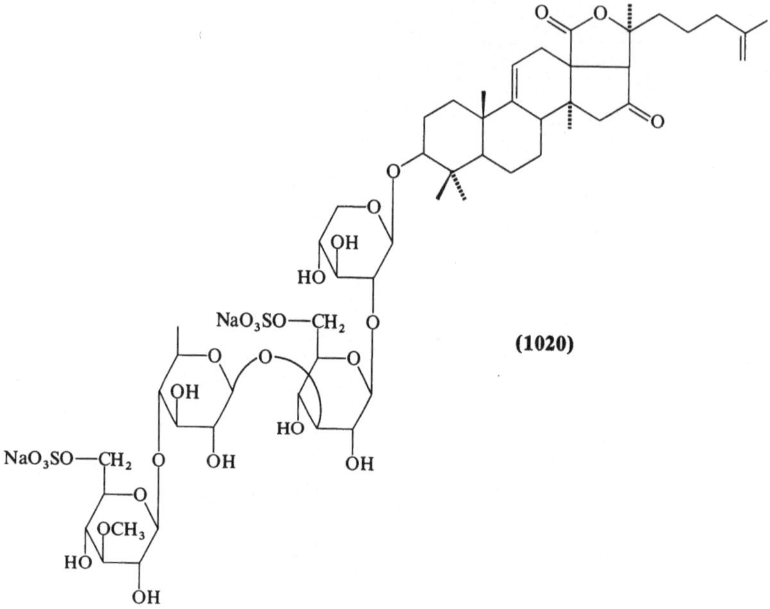

(1020)

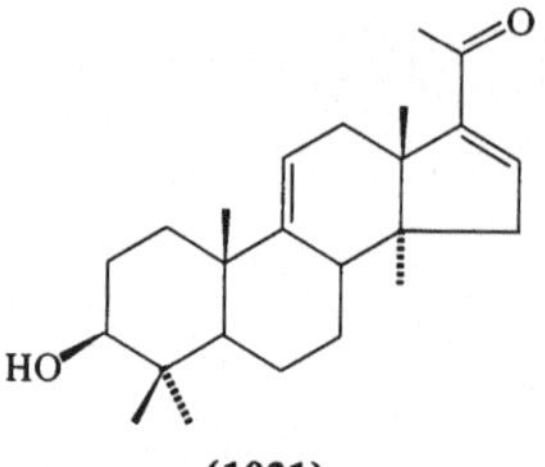

(1021)

showed significantly higher inhibition than those of starfish. Incubation together with cholesterol did not influence ATPase activity.

VI.5. Steroids from Sea Cucumbers

The free sterols of the Pacific holothurian *Stichopus japonicus* have been investigated by KALINOVSKAYA *et al.* (*692*) who found that C_{27}-, C_{28}-, C_{29}-, and C_{30}-sterols with saturated and monounsaturated (Δ^7) nucleus were present. The major sterol was 5α-cholest-7-en-3β-ol. The composition of the free sterol fraction was basically similar to the sterol part of the steroid xylosides in the same species (*693*). Compounds resembling these glycosides have also been identified in *Isostichopus badionotus*, *Synapta lappa* and other species. Injection of ^{14}C-cholesterol into *S. japonicus* gave labelled fractions of xylosides and sulfated sterols (*692*). Sulfated saturated and Δ^5-monounsaturated sterols have been isolated by SMETANINA *et al.* (*694*) from the holothurian *Parathyona* sp., compounds which had not previously been identified in sea cucumbers. Recently, BATRAKOV *et al.* (*694a*) isolated from *C. japonica* a fraction of sterol sulfates, the main components of which were derivatives of cholest-5-en-3β-ol, 24-methylene-, 24-ethyl-, and 24-ethylidenecholest-5-en-3β-ol, 5α-cholestan-3β-ol, and 24-methyl- and 24-methylene-5α-cholesten-3β-ol. Fractionation of extracts of *Cucumaria frondosa* resulted in the isolation and spectroscopic identification of 4α,14α-dimethylcholest-9(11)-en-3β-ol in addition to plastochromanol 8, canthaxanthin, 3-octadecyloxy-1,2-propanediol, and N^5-acetylornithine (*695*).

VI.6. Crinoids and Ophiuroids

Sulfated sterols from brittle star *Macrophiotrix longipeda* have been isolated by ELYAKOV *et al.* (*696*) who compared the sterol composition of the sulfated compounds with that of the free steroids. The latter were Δ^5-, $\Delta^{5,22}$-, and $\Delta^{5,24(28)}$-sterols, with cholesterol as the main component. The composition of the free and sulfated sterols from *M. longipeda* did not qualitatively differ from each other. D'AURIA *et al.* (*697*) isolated ophioxanthin (**1022**), a new marine carotenoid sul-

(**1022**)

fate from *Ophioderma longicaudum*. RIDEOUT *et al.* (*698, 699, 699a*) reported isolation of some polyketide sulfates, highly oxygenated naphthalenes and anthraquinones as chemical defense substances from *Comatula pectinata* and *Lamprometra palmata gyges*.

VI.7. Sea Urchins

The pedicellaria of some species of sea urchins contain toxic substances which are capable of causing injury. Most investigations have been carried out with *Tripneustes gratilla* and *Toxopneustes pileolus*. KIMURA and NAKAGAWA (*700*) demonstrated that an extract from the pedicellariae of the latter species caused histamine release from isolated smooth muscle. The extract produced contraction of the longitudinal muscle of isolated guinea pig ileum in a concentration as low as 3×10^{-8} g/mL. Later, the same group examined the mechanism of histamine release from rat peritoneal mast cells by *T. pileolus* extract (*701*) and found a dose-dependent effect between 0.5 and 4 mg/mL. The authors stated that the extract-induced histamine release seemed to be dependent on aerobic glycolysis. Partial purification with gel chromatography furnished a protein fraction which had a molecular weight of about 20,000 (*702*). Since the contractile activity of the crude extract fluctuated KIMURA *et al.* (*703*) investigated seasonal changes in contractile activity, and found it to increased greatly from October through December, with a secondary peak in May or June, while it was lowest in July. MEBS (*704*) isolated a heat labile toxic protein with a molecular weight of about 25,000 from *Tripneustes gratilla*. An LD_{50} of 0.85 mg toxin protein/kg mouse was found on intraperitoneal injection.

Antineoplastic glycoproteins have been found in the ethanolic extracts of *Lytechinus variegatus* (*705*) and *Strongylocentrotus droebachiensis* (*706*). Strongylostatin 1 (molecular weight 4×10^7) and 2 (molecular weight 65.000) from the second species were found to cause 35–53% and 39–42% life extension, respectively, in the murine P388 lymphocytic leukemia system. The dose of strongylostatin 2 was 4.5 mg/kg and that of strongylostatin 1 5–10 mg/kg. Lytechinastatin from *L. variegatus* was a high molecular weight glycoprotein with antineoplastic activity. KAMERLING *et al.* (*707*) identified the sialic acids in purified sulfated sialoglycoproteins from the egg jelly coat of the sea urchin *Pseudocentrotus depressus*. Analysis by colorimetry, thin layer chromatography and gas-liquid chromatography-mass spectrometry identified the sialic acids as N-glycoloylneuraminic acid and its 9-O-acetyl-derivative.

References, pp. 320–363

PALUMBO et al. (*708*) isolated a new sulphur-containing amino acid from sea urchin eggs (*Paracentrotus lividus*) and identified it as 1-methyl-5-thiol-L-histidine (**1023**).

(**1023**)

VII. Tunicata

Tunicates provide a rich source of biologically active compounds with potentially useful medicinal properties. The most interesting substances identified thus far are cyclic oligopeptides. Methanol extraction of the ascidian *Lissoclinum patella* furnished ulicyclamide (**1024**) and ulithiacyclamide (**1025**) (*709, 710*), as well as three other cyclic peptides (**1026–1028**) (*710*). Their structures have been determined by spectroscopic methods using FAB mass spectrometry (*710*) and by hydrolysis (*711*). Recently, SCHMIDT and GLEICH (*711a*) described a total synthesis of **1024**. Three additional cyclic peptides from the same species, patellamides A (**1029**), B (**1030**), and C (**1031**) have been described by IRELAND et al. (*712*). All eight of these metabolites were tested for antitumor activity against L1210 murine leukemia cells cultured in vitro and exhibited 50% inhibition at low doses of 0.35 to 7.2 µg/mL (for **1024, 1025**, and **1029–1031**) and greater than 10 µg/mL in the case of the peptides **1026–1028**. Patellamide A (**1029**) and ulithiacyclamide (**1025**) also inhibited the human ALL cell line (T cell acute leukemia) CEM with IC_{50} values of 0.028 and 0.010 µg/mL, respectively. Recently, the structures of patellamides B and C have been revised as **1030a** and **1031a**, respectively, by syntheses (*712a, b*) as well as patellamide A (*712c*).

RINEHART et al. (*14, 713, 714*) isolated from a member of the *Trididemnum* genus of the family *Didemnidae* a new class of depsipeptides, including highly active antiviral (against RNA and DNA viruses) and antitumor (against L1210 and P388 leukemia and B16 melanoma) agents. Various extraction methods and chromatography over silica gel gave the didemnins A (**1032**), B (**1033**), and C (**1034**). While didemnin C (**1034**) was available only in traces, didemnin A (**1032**) and B (**1033**) were tested against herpes simplex virus and for antitumor activity. Didemnin B (**1033**) was the most active in inhibiting viral replication *in vitro* and P388 leukemia *in vivo*. Structure elucidation was based mainly upon FD-, HRFD- and HREI-mass spectrometry, ^{1}H-NMR spectroscopy and upon acid hydrolysis *(715)*. The authors also used FAB mass spectrometry for determination of the amino acid sequence *(716)* which was a new method at that time. HAMAMOTO *et al.* *(717)* isolated ascidiacyclamide (**1035**) from an unidentified species of ascidian which has been collected from Rodda Reef, Queensland, Australia. The structure of this new cytotoxic cyclic peptide **1035** was elucidated by NMR spectroscopy.

Recently, BRUENING *et al.* *(717a)* isolated tunichrome B-1 (**1036**), which is derived from three (3,4,5-trihydroxyphenyl)alanine units, from *Ascidia nigra*.

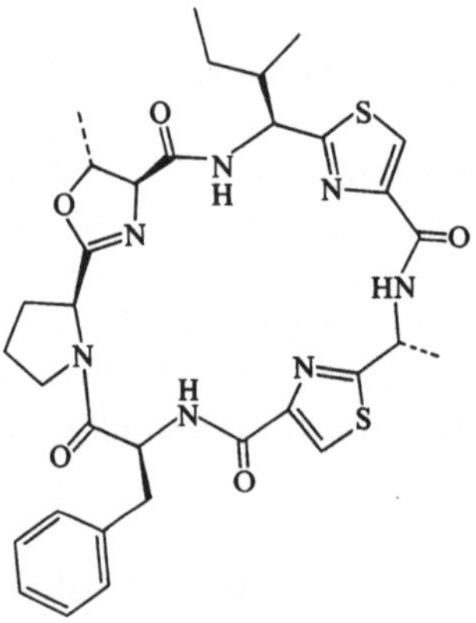

(**1024**)

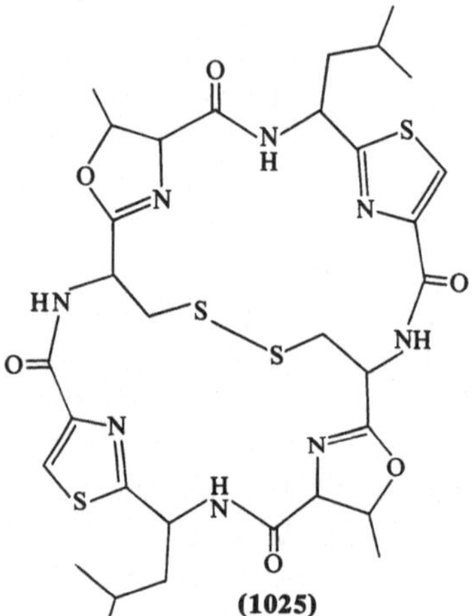

(**1025**)

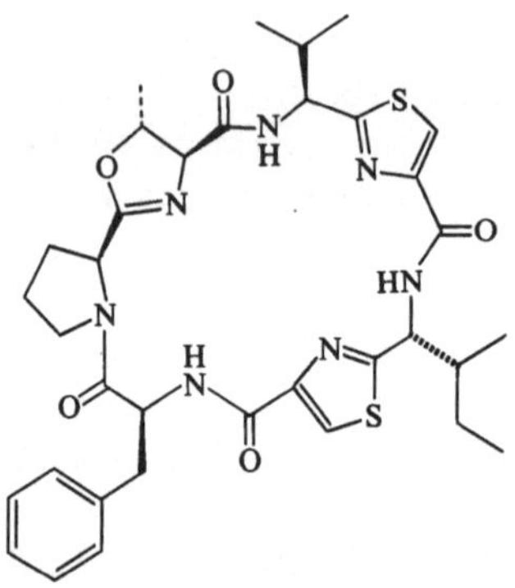

(**1026**)

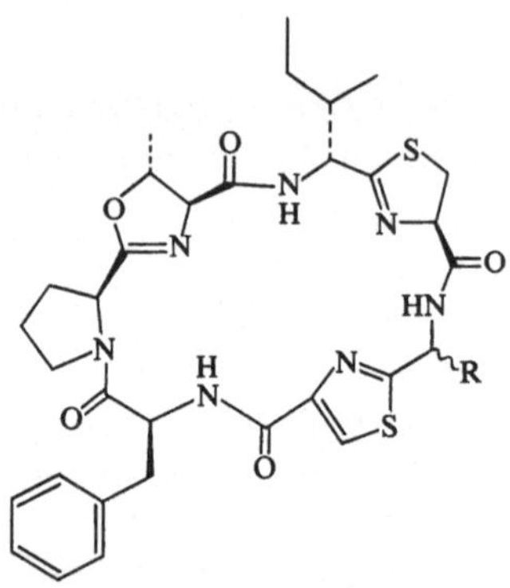

(**1027**) R = ⋯⋯CH$_3$
(**1028**) R = ⎯⎯CH$_3$

References, pp. 320–363

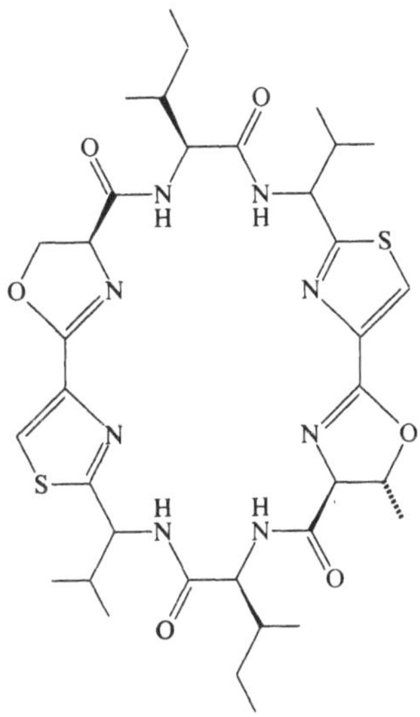

(1029)

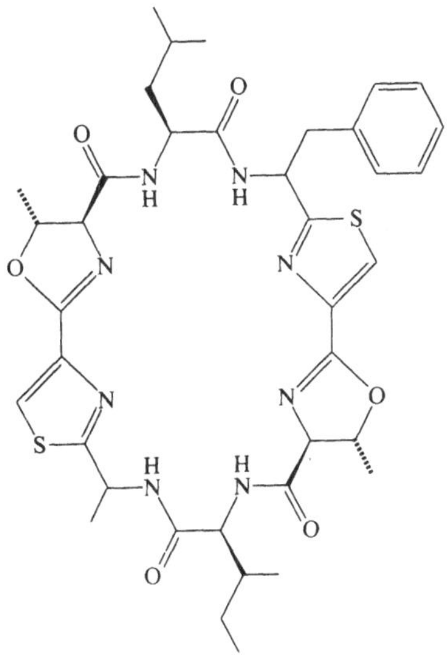

(1030)

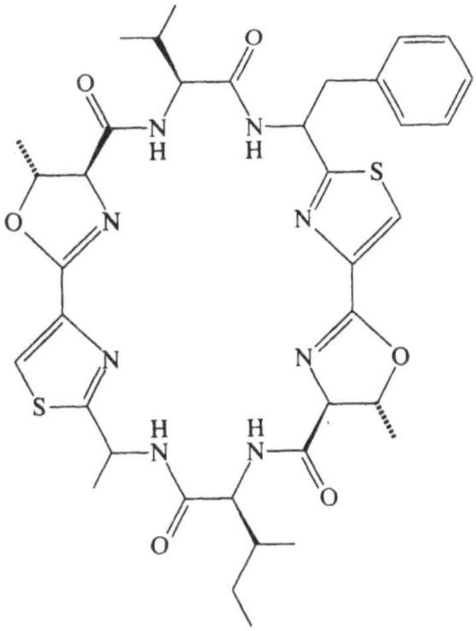

(1031)

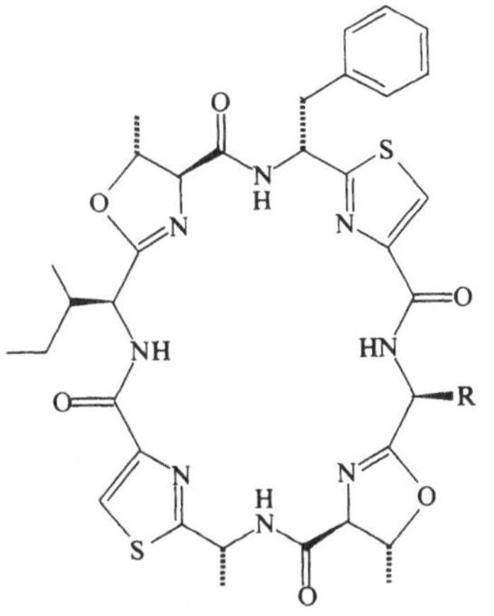

(1030a) R=(CH$_3$)$_2$CHCH$_2$
(1031a) R=(CH$_3$)$_2$CH

Urea derivatives have been reported as constituents of marine ascidians belonging to the genus *Didemnum*. IRELAND et al. (*718*) isolated N,N'-diphenethylurea (**1037**), which was apparently a product of phenylalanine metabolism, from *Didemnum ternatanum* and N,N'-(3,5-diiodo-4-methoxyphenethyl)-urea (**1038**) has been found in addition to 3,5-diiodo-4-methoxyphenethylamine (**1039**) as a metabolite from an unidentified *Didemnum* species (*719*). Compound **1037** was found to be a weak depressant lacking acute toxicity and **1039** showed *in* *vitro* activity against the yeast *Candida albicans*. Furthermore, **1039** was mildly cytotoxic.

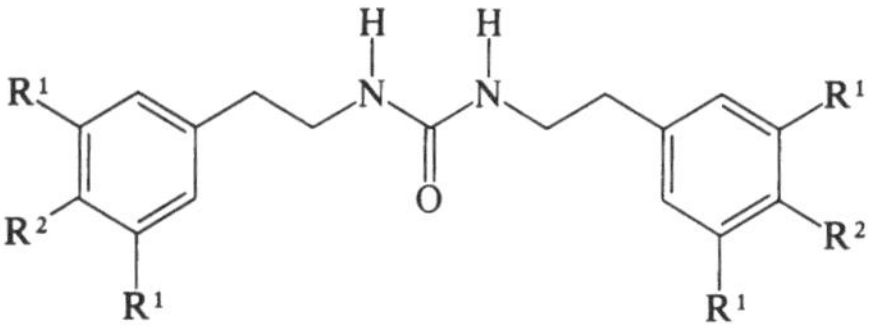

(1037) R¹=R²=H
(1038) R¹=I R²=OCH₃

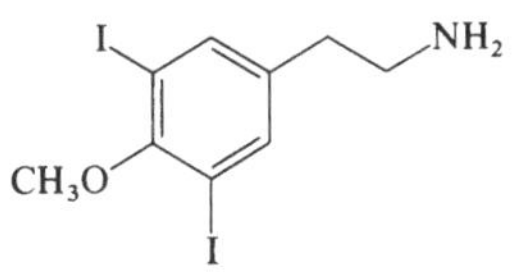

(1039)

Another group of biologically highly active compounds from tunicates consists of derivatives of indole. In 1980 HEITZ *et al.* (*720*) reported structure determination of the cytotoxic dendrodoine from *Dendrodoa grossularia* by X-ray analysis and spectral data as 5-((3-N-dimethylamino)-1,2,4-thiadiazolyl)-3-indanyl-methanone **(1040)**. The molecule has been synthesized (*721*). *Polycitorella mariae* furnished a novel bromoindole derivative, citorellamine **(1041)**, which possesses both cytotoxic and potent antimicrobial activity (*721 a*). Recently MOQUIN and GUYOT (*722*) isolated another indole derivative, grossularine **(1042)**, from *D. grossularia*. The proposed structure was based upon spectroscopic evidence. Several compounds called eudistomins have been reported as constituents of the colonial Caribbean tunicate *Eudistoma olivaceum* (*723*, *724*). Eudistomins C **(1043)**, E **(1044)**, K **(1045)**, and L **(1046)** contained a previously unreported condensed oxathiazepine ring system. These substances which can be considered to be biosynthetically derived from tryptophan and cysteine showed very strong inhibition of *Herpes simplex* virus, type 1 (HSV-1). Other eudistomins, A, D, G, H, I, J, M, N, O, P, and Q **(1047–1057)**, bromo-, hydroxy-, pyrrolyl-, and 1-pyrrolinyl-β-carbolines from the extract of the same species exhibited modest activity against HSV-1 **(1048–1051, 1054, and 1055)**, *Saccharomyces cerevisiae* **(1050, and 1054–1056)**, and *Bacillus subtilis* **(1048, 1051, and 1054–1057)**. Modern spectroscopic methods like high resolution FAB mass- and 2D NMR spectrometry were used for determination of their structures.

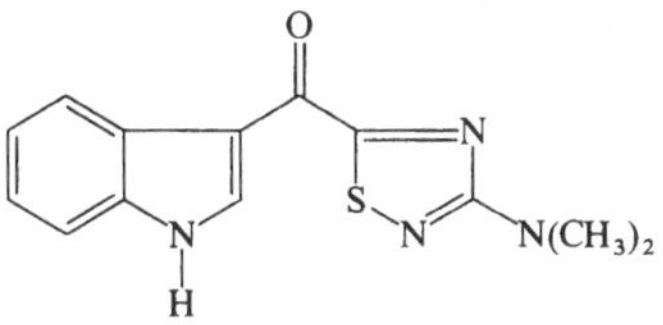

(1040)

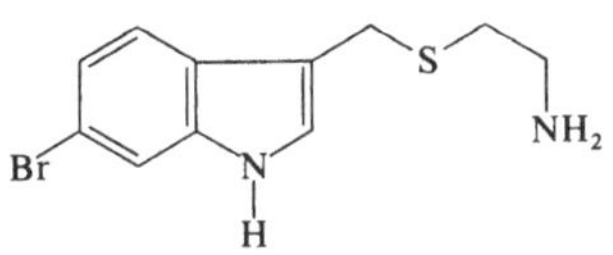

(1041)

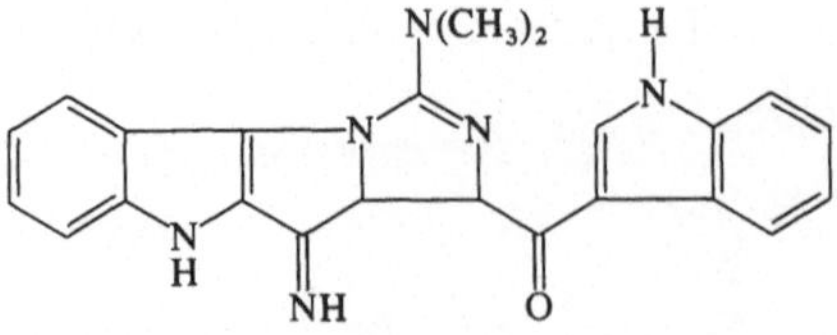

(1042)

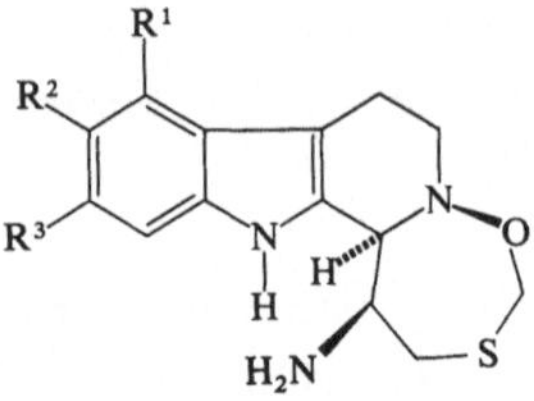

	R¹	R²	R³
(1043)	H	OH	Br
(1044)	Br	OH	H
(1045)	H	H	Br
(1046)	H	Br	H

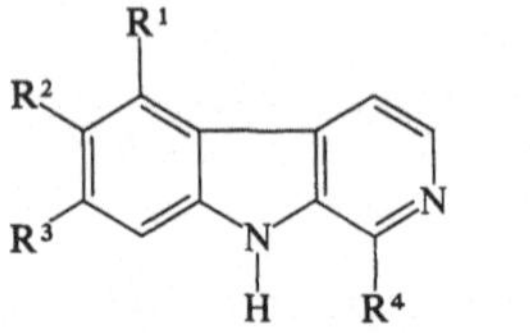

	R¹	R²	R³	R⁴
(1047)	H	OH	Br	a
(1048)	Br	OH	H	H
(1049)	H	H	Br	b
(1050)	H	Br	H	b
(1051)	H	H	H	b
(1052)	H	OH	Br	H
(1053)	H	OH	H	a
(1054)	H	Br	H	H
(1055)	H	H	Br	H
(1056)	H	OH	Br	b
(1057)	H	OH	H	b

a = 2-Pyrrolyl
b = 1-Pyridin-2-yl

From the muscle of the ascidian *Halocynthia roretzi* WATANABE et al. (725) isolated a new betaine, halocynine, whose structure was shown to be (*R*)-2-hydroxy-6-trimethylammoniohexanoate (**1058**). Halocynine was present in considerable amounts in the tissue of the ascidian but the mode of its accumulation and its physiological role in the animal have not been investigated. The crude extracts of a red encrusting tunicate belonging to the genus *Polyandrocarpa* showed potent antimicrobial activity. CARTÉ and FAULKNER (726) examined several specimens of this species and isolated the polyandrocarpidines A–D (**1059–1062**). A deep blue pigment, isolated from a Western Australian ascidian, has been shown to be identical with tetrapyrrole (**1063**) from a mutant strain of a bacterium (727). Screening of the pigment **1063** on the electrically driven isolated guinea pig ileum gave a dose-dependent increase in the contractile force of the atrium. A new mycosporin like amino acid (**1064**) has been isolated from the ascidian *Halocynthia roretzi* in addition to four previously known compounds (**991, 992,**

References, pp. 320–363

1065, and **1066**) *(728)*. WATANABE *et al.* *(729)* have investigated the seasonal variation of extractive nitrogen and free amino acids in the muscle of cultured *H. roretzi*.

Terpene hydroquinones which are well known as constituents of other marine organisms have been isolated from the ascidian *Aplidium constellatum*, *(730)*. These compounds demonstrated activity as anticancer agents. In the case of *A. constellatum*, 2-methyl-2-(4-methylpent-3-enyl)-2*H*-chromen-6-ol could be identified. Five new carotenoids have been obtained from tunicates, halocynthiaxanthin (**1067**) *(731)* and mytiloxanthinone (**1068**) *(732)* from *Halocynthia roretzi*, sidnyaxanthin (**1069**) from *Sidnyum argus* *(733)* and amarouciaxanthin A (**1070**) and B (**1071**) from *Amaroucium pliciferum (733a)*. *Halocynthia aurantium* furnished two known carotenoids, astaxanthin and diatoxanthin *(733b)*.

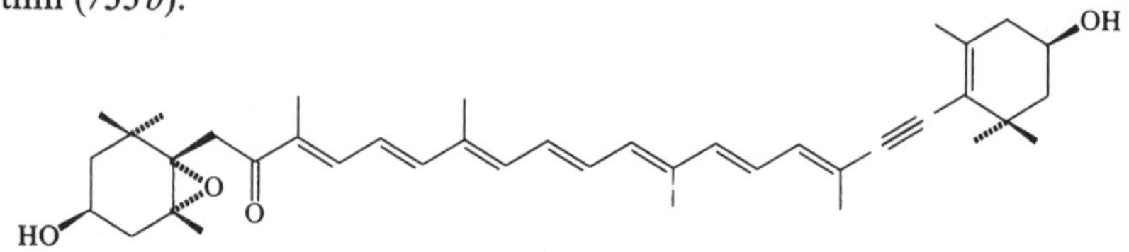

(**1067**)

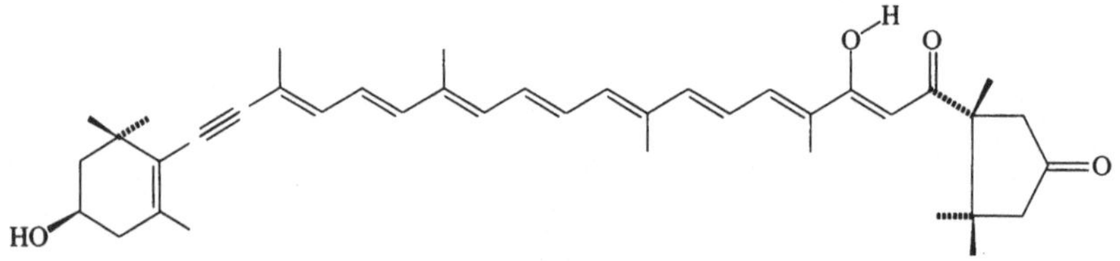

(**1068**)

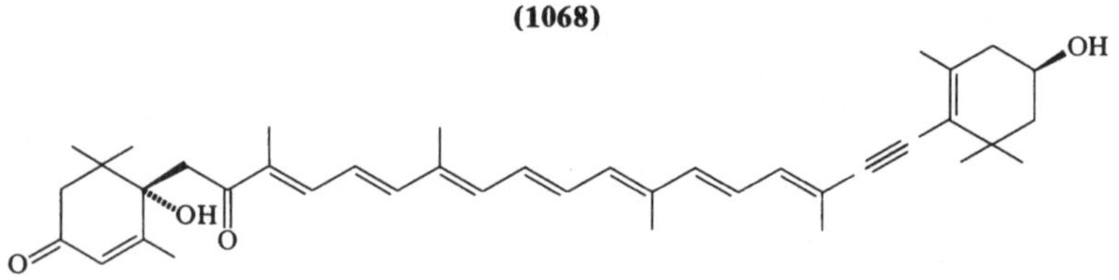

(**1069**)

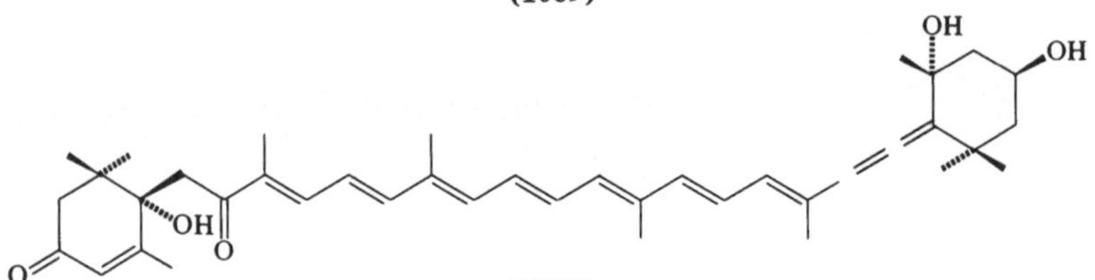

(**1070**)

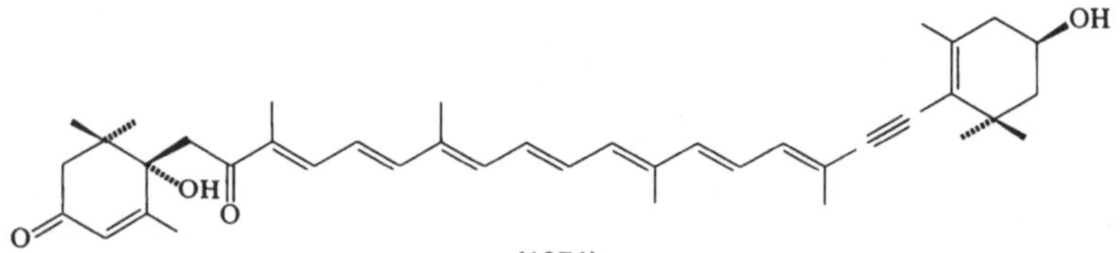

(**1071**)

References, pp. 320–363

As to steroids, a complex mixture of Δ^4-3-keto steroids, 5β-stanols and 4α-methyl sterols was isolated in addition to "regular" sterols from *Ascidia nigra* (*734*) and some steroids, with cholesterol and brassicasterol as major components have been found in *Microcosmus sulcatus* and *Halocynthia papillosa* (*735*). There have also been some reports on the occurrence of steroidal peroxides and hydroperoxides in tunicates. GUYOT and DURGEAT (*736*) isolated four 9(11)-unsaturated sterol peroxides (**12, 14, 16,** and **1072**) from *Phallusia mamillata* and *Ciona intestinalis*. The same species furnished 24-hydroperoxy-24-vinylcholesterol (**1073**) (*737*). From *A. nigra* GUNATILAKA *et al.* (*67*) obtained sixteen 5α,8α-epidioxy Δ^6- and $\Delta^{6,9(11)}$-sterols (**3–16, 454,** and **1072**). In excluding the possibility that these compounds are artifacts the authors expressed the opinion that they are not merely metabolic dead ends but that they probably act as substrates for various enzyme systems.

(**1072**) (**1073**)

Sulfated glycans from different ascidian species and from the sea cucumber *Ludwigothurea grisca* have been distinguished by the pattern of electrophoretic migration and by the molar proportion of amino sugars, hexoses and sulfate (*738*). A crude extract from *Ecteinascidia turbinata* was found to be a potent inhibitor of semiconservative DNA synthesis in human fibroblasts (*739*).

VIII. Miscellaneous Other Sources

Nemertines are carnivorous and feed on annelids, crustaceans, molluscs, and occasionally fish. Certain nemertines paralyze their prey by ejecting a tenacious toxic fluid from the proboscis. *Cerebratulus lacteus* has been found to contain several polypeptides which were believed to be used for defensive purposes, but not for predation. BLUMENTHAL investigated the toxins A-III *(740–742)*, B-II *(743)*, and B-VI *(743–745)* from this species. Toxin A-III **(1074)** containing 95 amino acid residues was strongly basic and highly active in lysing human erythrocytes. It also caused rapid lysis of Ehrlich ascites cells and cardiac muscle cells. Comparison of the sequences of toxins-B-II **(1075)** and B-VI **(1076)** revealed a high degree of homology between these two neurotoxins.

BALLANTINE et al. *(746)* established the structure of bonellin **(1077)**, a physiologically active pigment of the marine echurian worm *Bonellia viridis*. It is possible that bonellin **(1077)** induces masculinity in sexually undifferentiated larvae which settle on the proboscis of the female. When the larvae were removed from the proboscis, intersexes resulted.

Several brominated phenols and indoles have been reported from acorn worms *(747)*. Comparative studies of four species have led to isolation of the compounds listed below.

Species	Isolated compounds
Ptychodera flava	**1079, 1081, 1085–1091**
Glossobalanus sp.	**1079, 1082–1084, 1086, 1090**
Balanoglossus carnosus	**1078–1081, 1092**
Balanoglossus misakiensis	**1078, 1083**

The phenols **1078** and **1080** as well as the indoles **1086** and **1087** were responsible for the characteristic odor of acorn worms. **1078** and **1079** were antiseptic constituents. These compounds are not only disinfectant and antifungal, but are also toxic to molluscs.

The lobster, *Homarus americanus*, furnished an unusual heparan sulfate with a high content of sulfate and D-glucuronic acid *(748)*. Arseno betain **(1093)** has been isolated from the tail muscle of the western rock lobster, *Panulirus cygnus*, as well as trigonelline *(749)*. Several species of crab have been investigated for occurrence and properties of toxins *(750–756)*. A sialic acid- *(757)* and a teichoic acid-binding *(758)* lectin have been obtained from two species of horseshoe carb, *Carcinoscorpius rotunda cauda* and *Limulus polyphemus*, respectively. WATSON and SPAZIANI *(759)* developed a liquid chromatographic method for rapid isolation of ecdysteroids from crustacean tissue using Sep-Pak C-18 chromatographic cartridges and NEWCOMB *(760)* used reversed phase liquid chromatography to isolate most of the peptides of the sinus gland of the land crab *Cardisoma carnifex*.

References, pp. 320–363

1074 NH$_2$-Ile-Ser-Trp-Pro-Ser-Tyr-Pro-Gly-Ser-Glu-Gly-Ile-Arg-Ser-Ser-Asn-Cys-Gln-Lys-Lys-Leu-Asn-Cys-Gly-Thr-Lys-Asn-Ile-Ala-Thr-Lys-Gly-Val-Cys-Lys-Ala-Phe-Cys-Leu-Gly-Arg-Lys-Arg-Phe-Trp-Gln-Lys-Cys-Gly-Lys-Asn-Gly-Ser-Gly-Ser-Lys-Gly-Ser-Lys-Val-Cys-Asn-Ala-Val-Leu-Ala-His-Ala-Val-Glu-Lys-Ala-Gly-Lys-Gly-Leu-Ile-Ala-Val-Thr-Asp-Lys-Ala-Val-Ala-Ala-Ile-Val-Lys-Leu-Ala-Ala-Gly-Ile-Ala-COOH
Difulfide bonds between positions 17–38, 23–34, 48–61

1075 Ala-Ser-Ser-Thr-Trp-Gly-Gly-Ser-Tyr-Hyp-Ala-Cys-Glu-Asn-Asn-Cys-Arg-Lys-Gln-Tyr-Asp-Asp-Cys-Ile-Lys-Cys-Gln-Gly-Lys-Trp-Ala-Gly-Lys-Arg-Gly-Lys-Cys-Ala-Ala-His-Cys-Ala-Val-Gln-Thr-Thr-Ser-Cys-Asn-Asp-Lys-Cys-Lys-Lys-His

1076 Ala-Ser-Ala-Thr-Trp-Gly-Ala-Ala-Tyr-Hyp-Ala-Cys-Glu-Asn-Asn-Cys-Arg-Lys-Lys-Tyr-Asp-Leu-Cys-Ile-Arg-Cys-Gln-Gly-Lys-Trp-Ala-Gly-Lys-Arg-Gly-Lys-Cys-Ala-Ala-His-Cys-Ile-Ile-Gln-Lys-Asn-Asn-Cys-Lys-Gly-Lys-Cys-Lys-Lys-Glu

	R¹	R²	R³	R⁴	R⁵	R⁶		R¹	R²	R³	R⁴	R⁵
(1078)	OH	Br	H	H	H	Br	(1086)	H	H	H	H	Cl
(1079)	OH	Br	H	Br	H	Br	(1087)	H	H	H	H	Br
(1080)	OH	Br	H	Br	H	H	(1088)	H	H	Br	H	Cl
(1081)	OH	Br	Br	OH	Br	H	(1089)	H	Br	H	Br	Br
(1082)	OH	Br	H	OH	H	H	(1090)	H	H	Br	H	Br
(1083)	OH	Br	H	OH	H	Br	(1091)	H	Br	OMe	Br	Br
(1084)	OH	H	H	OH	H	H	(1092)	Br	H	Br	H	Br

References

1. HALSTEAD, B.W.: Current Status of Marine Biotoxicology – An Overview. Clinical Toxicology **18**, 1 (1981).
2. BAKUS, G.J.: Chemical Defense Mechanism on the Great Barrier Reef, Australia. Science **211**, 497 (1981).
3. ENDEAN, R., and A.M. CAMERON: Toxins in Coral Reef Organisms. Toxicon Suppl. **3**, 105 (1983).
4. BAKUS, G.J., T. EVANS, B. MADING, and P. KOUROS: The Use of Natural and Synthetic Toxins as Shark Repellents and Antifouling Agents. Toxicon Suppl. **3**, 25 (1983).
5. a) HALSTEAD, B.W.: Poisonous and Venomous Marine Animals of the World. Vols. 1–3. Washington, D.C.: U.S. Government Printing Office. 1965–1970. b) Rev. ed. Princeton, N.J.: Darwin Press. 1978.
6. HASHIMOTO, Y.: Marine Toxins and Other Bioactive Marine Metabolites. Tokyo: Japan Scientific Societies Press. 1979.
7. a) HABERMEHL, G.: Gift-Tiere und ihre Waffen, 3. Auflage. Berlin-Heidelberg-New York: Springer 1983. b) Venomous Animals and Their Toxins. Berlin-Heidelberg-New York: Springer. 1981.
8. SCHEUER, P.J., ed.: Marine Natural Products. Chemical and Biological Perspectives, Vols. I–V. New York: Academic Press. 1978–1983.
9. OKUDA, R.K., D. KLEIN, R.B. KINNEL, M. LI, and P.J. SCHEUER: Marine Natural Products: The Past Twenty Years and Beyond. Pure and Appl. Chem. **54**, 1907 (1982).
10. FENICAL, W.: Natural Products Chemistry in the Marine Environment. Science **215**, 923 (1982).
11. KASHMAN, Y., A. GROWEISS, S. CARMELY, Z. KINAMONI, D. CZARKIE, and M. ROTEM: Recent Research in Marine Natural Products from the Red Sea. Pure and Appl. Chem. **54**, 1995 (1982).
12. FAULKNER, D.J.: Marine Natural Products: Metabolites of Marine Algae and Herbivorous Marine Molluscs. Nat. Prod. Rep. **1984**, 251.
13. – Marine Natural Products: Metabolites of Marine Invertebrates. Nat. Prod. Rep. **1984**, 551.
14. RINEHART, JR., K.L., P.D. SHAW, L.S. SHIELD, J.B. GLOER, G.C. HARBOUR, M.E.S. KOKER, D. SAMAIN, R.E. SCHWARTZ, A.A. TYMIAK, D.L. WELLER, G.T. CARTER, M.H.G. MUNRO, R.G. HUGHES, JR., H.E. RENIS, E.B. SWYNENBERG, D.A. STRINGFELLOW, J.J. VAVRA, J.H. COATS, G.E. ZURENKO, S.L. KUENTZEL, L.H. LI, G.J. BAKUS, R.C. BRUSCA, L.L. CRAFT, D.N. YOUNG, and J.L. CONNOR: Marine Natural Products as Sources of Antiviral, Antimicrobial, and Antineoplastic agents. Pure and Appl. Chem. **53**, 795 (1981).
15. JACOBS, R.S., S. WHITE, and L. WILSON: Selective Compounds Derived from Marine Organisms: Effects on Cell Division in Fertilized Sea Urchin Eggs. Fed. Proc. **40**, 26 (1981).
16. BERLEPSCH, K.: Drug from Marine Organisms. The Target of the Roche Research Institute of Marine Pharmacology in Australia. Naturwissenschaften **67**, 338 (1980).
17. KAUL, P.N.: Biomedical Potential of the Sea. Pure and Appl. Chem. **54**, 1963 (1982).
18. – Compounds from the Sea with Actions on the Cardiovascular and Central Nervous Systems. Fed. Proc. **40**, 10 (1981).
19. JACOBS, R.S., P. CULVER, R. LANGDON, T. O'BRIEN, and S. WHITE: Some Pharmacological Observations on Marine Natural Products. Tetrahedron **41**, 981 (1985).
20. SHIMIZU, Y.: Bioactive Marine Natural Products, with Emphasis on Handling of Water-Soluble Compounds. J. Nat. Prod. **48**, 223 (1985).
21. DJERASSI, C.: Recent Studies in the Marine Sterol Field. Pure and Appl. Chem. **53**, 873 (1981).

22. GOAD, L.J.: Sterol Biosynthesis and Metabolism in Marine Invertebrates. Pure and Appl. Chem. **51**, 837 (1981).
23. CATALAN, C.A.N., J.E. THOMPSON, W.C.M.C. KOKKE, and C. DJERASSI: Biosynthetic Studies of Marine Lipids – 3. Experimental Demonstration of the Course of Side Chain Extension in Marine Sterols. Tetrahedron **41**, 1073 (1985).
24. WITHERS, N.W., W.C.M.C. KOKKE, W. FENICAL, and C. DJERASSI: Sterol Patterns of Cultured Zooxanthellae Isolated from Marine Invertebrates: Synthesis of Gorgosterol and 23-Desmethylgorgosterol by Aposymbiotic Algae. Proc. Natl. Acad. Sci. U.S.A. **79**, 3764 (1982).
25. BLANC, P.-A., and C. DJERASSI: Isolation and Structure Elucidation of 22(S),23(S)-Methylenecholesterol. Evidence for Direct Bioalkylation of 22-Dehydrocholesterol. J. Am. Chem. Soc. **102**, 7113 (1980). Erratum: J. Am. Chem. Soc. **103**, 7036 (1981).
26. ZIELINSKI, J., W.C.M.C. KOKKE, T.B. T. HA, A.Y.L. SHU, W.L. DUAX, and C. DJERASSI: Isolation, Partial Synthesis, and Structure Determination of Sterols with the Four Possible 23,24-Dimethyl-Substituted Side Chains. J. Org. Chem. **48**, 3471 (1983).
27. SARGENT, J.R., R.R. GATTEN, and R.J. HENDERSON: Marine Wax Esters. Pure and Appl. Chem. **53**, 867 (1981).
28. KOKKE, W.C.M.C., S. EPSTEIN, S.A. LOOK, G.H. RAU, W. FENICAL, and C. DJERASSI: On the Origin of Terpenes in Symbiotic Associations between Marine Invertebrates and Algae (Zooxanthellae). Culture Studies and an Application of $^{13}C/^{12}C$ Isotope Ratio Mass Spectrometry. J. Biol. Chem. **259**, 8168 (1984).
29. CREWS, P., S. NAYLOR, F.J. HANKE, E.R. HOGUE, E. KHO, and R. BRASLAU: Halogen Regiochemistry and Substituent Stereochemistry Determination in Marine Monoterpenes by ^{13}C NMR. J. Org. Chem. **49**, 1371 (1984).
30. COLWELL, R.R.: Biotechnology in the Marine Sciences. Science **222**, 19 (1983).
31. OKAMI, Y.: Potential Use of Marine Microorganisms for Antibiotics and Enzyme Production. Pure and Appl. Chem. **54**, 1951 (1982).
32. SHIMIZU, Y.: Paralytic Shellfish Poisons. Fortschritte d. Chem. organ. Naturstoffe (W. HERZ, H. GRISEBACH, G.W. KIRBY, and C. TAMM, eds.), Vol. **45**, p. 235. Wien-New York: Springer. 1984.
33. BERGQUIST, P.R., and R.J. WELLS: Chemotaxonomy of the Porifera: The Development and Current Status of the Field. In: Marine Natural Products. Chemical and Biological Perspectives (P. SCHEUER, ed.), Vol. V, p. 1. New York-San Francisco-London: Academic Press. 1983.
34. MCCAFFREY, E.J.: Biologically Active Material from a Marine Sponge. Toxicon Suppl. **3**, 277 (1983).
35. SEVCIK, C., and C.A. BARBOZA: The Presynaptic Effect of Fractions Isolated from the Sponge *Tedania ignis*. Toxicon **21**, 191 (1983).
36. MEBS, D.: Lack of Toxins in Alcoholic Extracts of the Sponge *Tedania ignis*. Toxicon **22**, 821 (1984).
37. BOHLIN, L., H.P. GEHRKEN, P.J. SCHEUER, and C. DJERASSI: Minor and Trace Sterols in Marine Invertebrates XVI. 3ξ-Hydroxymethyl-A-*nor*-5α-gorgostane, a Novel Sponge Sterol. Steroids **35**, 295 (1980).
38. KHALIL, M.W., L.J. DURHAM, C. DJERASSI, and D. SICA: Ficisterol (23-Ethyl-24-methyl-27-*nor*cholesta-5,25-dien-3β-ol). A Biosynthetically Unprecedented Sterol from the Marine Sponge *Petrosia ficiformis*. J. Am. Chem. Soc. **102**, 2133 (1980).
39. KARPF, M., and C. DJERASSI: Synthesis of Ficisterol – The First Natural Sterol with 23-Ethyl Substitution. Tetrahedron Lett. **21**, 1603 (1980).
40. BERGQUIST, P.R., W. HOFHEINZ, and G. OESTERHELT: Sterol Composition and the Classification of the Demospongiae. Biochem. Syst. Ecol. **8**, 423 (1980).
41. KHALIL, M.W., C. DJERASSI, and D. SICA: Minor and Trace Sterols in Marine Inver-

tebrates XVII. (24 R)-24,26-Dimethylcholesta-5,26-dien-3 β-ol, a New Sterol from the Sponge *Petrosia ficiformis*. Steroids **35**, 707 (1980).

42. LI, L.N., U. SJÖSTRAND, and C. DJERASSI: Minor and Trace Sterols in Marine Invertebrates. 19. Isolation, Structure Elucidation, and Partial Synthesis of 24-Methylene-25-ethylcholesterol (Mutasterol): First Example of Sterol Side-Chain Bioalkylation at Position 25. J. Am. Chem. Soc. **103**, 115 (1981).

43. BOHLIN, L., U. SJÖSTRAND, C. DJERASSI, and B.W. SULLIVAN: Minor and Trace Sterols in Marine Invertebrates. Part 20. 3 ξ-Hydroxymethyl-A-*nor*-patinosterol and 3 ξ-Hydroxymethyl-A-*nor*-dinosterol. Two New Sterols with Modified Nucleus and Side-chain from the Sponge *Teichaxinella morchella*. J. Chem. Soc., Perkin Trans. I **1981**, 1023.

44. KHO, E., D.K. IMAGAWA, M. ROHMER, Y. KASHMAN, and C. DJERASSI: Sterols in Marine Invertebrates. 22. Isolation and Structure Elucidation of Conicasterol and Theonellasterol, Two New 4-Methylene Sterols from the Red Sea Sponges *Theonella conica* and *Theonella swinhoei*. J. Org. Chem. **46**, 1836 (1981).

45. LI, L.N., and C. DJERASSI: Minor and Trace Sterols in Marine Invertebrates. 23. Xestospongesterol and Isoxestospongesterol – First Examples of Quadruple Biomethylation of the Sterol Side Chain. J. Am. Chem. Soc. **103**, 3606 (1981).

46. ZIELINSKI, J., H. LI, T.S. MILKOVA, S. POPOV, N.L. MAREKOV, and C. DJERASSI: Isolation of 22-Methylenecholesterol. Further Evidence for Direct Bioalkylation of 22-Dehydrocholesterol. Tetrahedron Lett. **22**, 2345 (1981).

47. LI, L.N., U. SJÖSTRAND, and C. DJERASSI: Minor and Trace Sterols in Marine Invertebrates. 27. Isolation, Structure Elucidation, and Partial Synthesis of 25-Methylxestosterol, a New Sterol Arising from Quadruple Biomethylation in the Side Chain. J. Org. Chem. **46**, 3867 (1981).

48. SJÖSTRAND, U., J.M. KORNPROBST, and C. DJERASSI: Minor and Trace Sterols from Marine Invertebrates. 29. (22 E)-Ergosta-5,22,25-trien-3 β-ol and (22 E,24 R)-24,26-Dimethylcholesta-5,22,25(27)-trien-3 β-ol. Two New Marine Sterols from the Sponge *Pseudaxinella lunacharta*. Steroids **38**, 355 (1981).

49. LI, L.N., and C. DJERASSI: Minor and Trace Sterols in Marine Invertebrates. 30. Isolation Structure Elucidation and Partial Synthesis of 26-Methylstrongylosterol and 28-Methylxestosterol – Two Marine Sterols Arising by a Novel Quadruple Biomethylation Sequence. Tetrahedron Lett. **22**, 4639 (1981).

50. LI, L.N., H. LI, R.W. LANG, T. ITOH, D. SICA, and C. DJERASSI: Minor and Trace Sterols in Marine Invertebrates. 31. Isolation and Structure Elucidation of 23 H-Isocalysterol, a Naturally Occurring Cyclopropene. Some Comparative Observations on the Course of Hydrogenolytic Ring Opening of Steroidal Cyclopropenes and Cyclopropanes. J. Am. Chem. Soc. **104**, 6726 (1982).

51. ZIELINSKI, J., T. MILKOVA, S. POPOV, N. MAREKOV, and C. DJERASSI: Minor and Trace Sterols in Marine Invertebrates. 34. Isolation and Structure Elucidation of 24-Methyl-5α-cholesta-7,9(11),24(28)-trien-3 β-ol, the First Naturally Occurring $\Delta^{7,9(11)}$ Unsaturated Marine Sterol. Steroids **39**, 675 (1982).

52. EGGERSDORFER, M.L., W.C.M.C. KOKKE, C.W. CRANDELL, J.E. HOCHLOWSKI, and C. DJERASSI: Sterols in Marine Invertebrates. 32. Isolation of 3 β-(Hydroxymethyl)-A-*nor*-5α-cholest-15-ene, the First Naturally Occurring Sterol with a 15–16 Double Bond. J. Org. Chem. **47**, 5304 (1982).

53. BOHLIN, L., U. SJÖSTRAND, G. SODANO, and C. DJERASSI: Sterols in Marine Invertebrates. 33. Structures of Five New 3 β-(Hydroxymethyl)-A-*nor* Steranes: Indirect Evidence for Transformation of Dietary Precursors in Sponges. J. Org. Chem. **47**, 5309 (1982).

54. ITOH, T., D. SICA, and C. DJERASSI: Minor and Trace Sterols in Marine Invertebrates. 36. (24 S)-24 H-Isocalysterol: A New Steroidal Cyclopropene from the Marine Sponge *Calyx niceaensis*. J. Org. Chem. **48**, 890 (1983).

55. DI GIACOMO, G., A. DINI, B. FALCO, A. MARINO, and D. SICA: Sterols from the Sponge *Agelas oroides*. Comp. Biochem. Physiol. **74B**, 499 (1983).
56. CATALAN, C.A.N., V. LAKSHMI, F.J. SCHMITZ, and C. DJERASSI: Minor and Trace Sterols in Marine Invertebrates. 39. 24ξ,25ξ-24,26-Cyclocholest-5-en-3β-ol, a Novel Cyclopropyl Sterol. Steroids **40**, 455 (1982).
57. CRIST, B.V., X. LI, P.R. BERGQUIST, and C. DJERASSI: Sterols of Marine Invertebrates. 44. Isolation, Structure Elucidation, Partial Synthesis, and Determination of Absolute Configuration of Pulchrasterol. The First Example of Double Bioalkylation of the Sterol Side Chain at Position 26. J. Org. Chem. **48**, 4472 (1983).
58. CRIST, B.V., and C. DJERASSI: Minor and Trace Sterols in Marine Invertebrates. 47. A Re-investigation of the 19-*Nor* Stanols Isolated from the Sponge *Axinella polypoides*. Steroids **42**, 331 (1983).
58a. PROUDFOOT, J.R., X. LI, and C. DJERASSI: Minor and Trace Sterols from Marine Invertebrates. 50. Stereostructure and Synthesis of Nicasterol, a Novel Cyclopropane-Containing Sponge Sterol. J. Org. Chem. **50**, 2026 (1985).
59. LI, X., and C. DJERASSI: Minor and Trace Sterols in Marine Invertebrates. 40. Structure and Synthesis of Axinyssasterol, 25-Methylfucosterol and 24-Ethyl-24-methylcholesterol – Novel Sponge Sterols with Highly Branched Side Chains. Tetrahedron Lett. **24**, 665 (1983).
60. ITOH, T., D. SICA, and C. DJERASSI: Minor and Trace Sterols in Marine Invertebrates. Part 35. Isolation and Structure Elucidation of Seventyfour Sterols from the Sponge *Axinella cannabina*. J. Chem. Soc., Perkin Trans. I **1983**, 147.
61. SHUBINA, L.K., T.N. MAKARIEVA, V.M. BOGUSLAVSKII, and V.A. STONIK: Steroid Compounds of Marine Sponges. 1. Sterols of *Esperiopsis digitata*. Khim. Prir. Soedin. **1983**, 740.
62. KOCH, P., C. DJERASSI, V. LAKSHMI, and F.J. SCHMITZ: 240. Identification de stérols à chaîne latérale oxygénée dans une éponge de l'espèce *Hyrtios*. Helv. Chim. Acta **66**, 2431 (1983).
63. CAFIERI, F., P. CIMINIELLO, M.V. D'AURIA, and C. SANTACROCE: Sterol Composition of the Marine Sponge *Crambe crambe*. Biochem. Syst. Ecol. **12**, 203 (1984).
64. MAKARIEVA, T.N., L.K. SHUBINA, A.I. KALINOVSKY, V.A. STONIK, and G.B. ELYAKOV: Steroids in Porifera. II. Steroid Derivatives from two Sponges of the Family *Halichondriidae*. Sokotrasterol sulfate, a Marine Steroid with a New Pattern of Side Chain Alkylation. Steroids **42**, 267 (1983).
64a. SHUBINA, L.K., T.N. MAKARIEVA, and V.A. STONIK: Steroid Compounds of Marine Sponges. III. 24-Ethyl-25-methylcholesta-5,22-dien-3β-ol – A New Marine Sterol from the Sponge *Halichondria* sp. Khim.Prir. Soedin. **1984**, 464.
65. KITAGAWA, I., M. KOBAYASHI, K. KITANAKA, M. KIDO, and Y. KYOGOKU: Marine Natural Products. XII. On the Chemical Constituents of the Okinawan Marine Sponge *Hymeniacidon aldis*. Chem. Pharm. Bull. (Japan) **31**, 2321 (1983).
66. DE STEFANO, A., and G. SODANO: Metabolism in Porifera. XII. Further Informations on the Biosynthesis of 3β-Hydroxymethyl-A-*nor*-steranes in the Sponge *Axinella verrucosa*. Experientia **36**, 630 (1980).
67. GUNATILAKA, A.A.L., Y. GOPICHAND, F.J. SCHMITZ, and C. DJERASSI: Minor and Trace Sterols in Marine Invertebrates. 26. Isolation and Structure Elucidation of Nine New 5α,8α-Epidioxy Sterols from Four Marine Organisms. J. Org. Chem. **46**, 3860 (1981).
68. FUSETANI, N., S. MATSUNAGA, and S. KONOSU: Bioactive Marine Metabolites. II. Halistanol sulfate, an Antimicrobial Novel Steroid Sulfate from the Marine Sponge *Halichondria* CF. *moorei* Bergquist. Tetrahedron Lett. **22**, 1985 (1981).
69. NAKATSU, T., R.P. WALKER, J.E. THOMPSON, and D.J. FAULKNER: Biologically Active Sterol Sulfates from the Marine Sponge *Toxadocia zumi*. Experientia **39**, 759 (1983).
70. GUNASEKERA, S.P., and F.J. SCHMITZ: Marine Natural Products: 9α,11α-Epoxycho-

lest-7-ene-3β,5α,6β,19-tetrol 6-Acetate from a Sponge, *Dysidea* sp. J. Org. Chem. **48**, 885 (1983).
71. ROSSER, R.M., and D.J. FAULKNER: Two Steroidal Alkaloids from a Marine Sponge, *Plakina* sp. J. Org. Chem. **49**, 5157 (1984).
72. CARIELLO, L., M. DE NICOLA GIUDICI, and L. ZANETTI: Developmental Aberrations in Sea-Urchin Eggs Induced by Avarol and Two Cogeners. The Main Sesquiterpenoid Hydroquinones from the Marine Sponge, *Dysidea avara*. Comp. Biochem. Physiol. **65C**, 37 (1980).
73. CARIELLO, L., L. ZANETTI, V. CUOMO, and F. VANZANELLA: Antimicrobial Activity of Avarol, a Sesquiterpenoid Hydroquinone from the Marine Sponge, *Dysidea avara*. Comp. Biochem. Physiol. **71B**, 281 (1982).
74. SCHULTE, G., P.J. SCHEUER, and O.J. MCCONNELL: 227. Two Furanosesquiterpene Marine Metabolites with Antifeedant Properties. Helv. Chim. Acta **63**, 2159 (1980).
75. GUELLA, G., A. GUERRIERO, P. TRALDI, and F. PIETRA: Penlanfuran, a New Furanoid Sesquiterpene from the Marine Sponge *Dysidea fragilis* (Mont.) of Brittany. A Striking Difference with the Same Hawaiian Species. Tetrahedron Lett. **24**, 3897 (1983).
76. GUELLA, G., A. GUERRIERO, and F. PIETRA: 4. Sesquiterpenoids of the Sponge *Dysidea fragilis* of the North-Brittany Sea. Helv. Chim. Acta **68**, 39 (1985).
76a. GUELLA, G., I. MANCINI, A. GUERRIERO, and F. PIETRA: 140. New Furanosesquiterpenoids from Mediterranean Sponges. Helv. Chim. Acta **68**, 1276 (1985).
77. WALKER, R.P., R.M. ROSSER, D.J. FAULKNER, L.S. BASS, H. CUN-HENG, and J. CLARDY: Two New Metabolites of the Sponge *Dysidea amblia* and Revision of the Structure of Ambliol B. J. Org. Chem. **49**, 5160 (1984).
78. HOCHLOWSKI, J.E., R.P. WALKER, C. IRELAND, and D.J. FAULKNER: Metabolites of Four Nudibranchs of the Genus *Hypselodoris*. J. Org. Chem. **47**, 88 (1982).
79. DUNLOP, R.W., R. KAZLAUSKAS, G. MARCH, P.T. MURPHY, and R.J. WELLS: New Furano-Sesquiterpenes from the Sponge *Dysidea herbacea*. Aust. J. Chem. **35**, 95 (1982).
80. GRODE, S.H., and J.H. CARDELLINA II: Sesquiterpenes from the Sponge *Dysidea etheria* and the Nudibranch *Hypselodoris zebra*. J. Nat. Prod. **47**, 76 (1984).
81. SULLIVAN, B., D.J. FAULKNER, and L. WEBB: Siphonodictidine, a Metabolite of the Burrowing Sponge *Siphonodictyon* sp. That Inhibits Coral Growth, Science **221**, 1175 (1983).
82. SCHULTE, G., P.J. SCHEUER, and O.J. MCCONNELL: Upial, a Sesquiterpenoid Bicyclo[3.3.1]nonane Aldehyde Lactone from the Marine Sponge *Dysidea fragilis*. J. Org. Chem. **45**, 552 (1980).
82a. TASCHNER, M.J., and A. SHAHRIPOUR: Total Synthesis of (−)-Upial. J. Am. Chem. Soc. **107**, 5570 (1985).
83. SULLIVAN, B., P. DJURA, D.E. MCINTYRE, and D.J. FAULKNER: Antimicrobial Constituents of the Sponge *Siphonodictyon coralliphagum*. Tetrahedron **37**, 979 (1981).
84. SCHMITZ, F.J., V. LAKSHMI, D.R. POWELL, and D. VAN DER HELM: Arenarol and Arenarone: Sesquiterpenoids with Rearranged Drimane Skeletons from the Marine Sponge *Dysidea arenaria*. J. Org. Chem. **49**, 241 (1984).
84a. CARTÉ, B., C.B. ROSE, and D.J. FAULKNER: 5-*epi*-Ilimaquinone, a Metabolite of the Sponge *Fenestraspongia* Sp. J. Org. Chem. **50**, 2785 (1985).
85. AMADE, P., L. CHEVOLOT, H.P. PERZANOWSKI, and P.J. SCHEUER: 161. A Dimer of Puupehenone. Helv. Chim. Acta **66**, 1672 (1983). Erratum: Helv. Chim. Acta **67**, 927 (1984).
86. DJURA, P., D.B. STIERLE, B. SULLIVAN, D.J. FAULKNER, E. ARNOLD, and J. CLARDY: Some Metabolites of the Marine Sponges *Smenospongia aurea* and *Smenospongia* (= *Polyfibrospongia*) *echina*. J. Org. Chem. **45**, 1435 (1980).
87. CAPON, R.J., E.L. GHISALBERTI, and P.R. JEFFERIES: New Aromatic Sesquiterpenes from a *Halichondria* sp. Aust. J. Chem. **35**, 2583 (1982).

88. CIMINIELLO, P., E. FATTORUSSO, S. MAGNO, and L. MAYOL: New Nitrogenous Sesquiterpenes from the Marine Sponge *Axinella cannabina*. J. Org. Chem. **49**, 3949 (1984).
89. – – – – Sesquiterpenoids Based on the epi-Maaliane Skeleton from the Marine Sponge *Axinella cannabina*. J. Nat. Prod. **48**, 64 (1985).
90. NAKAMURA, H., J. KOBAYASHI, Y. OHIZUMI, and Y. HIRATA: Novel Bisabolene-Type Sesquiterpenoids with a Conjugated Diene Isolated from the Okinawan Sea Sponge *Theonella* cf. *swinhoei*. Tetrahedron Lett. **25**, 5401 (1984).
91. NAKAMURA, H., H. WU, J. KOBAYASHI, Y. OHIZUMI, Y. HIRATA, T. HIGASHIJIMA, and T. MIYAZAWA: Agelasidine-A, a Novel Sesquiterpene Possessing Antispasmodic Activity from the Okinawa Sea Sponge *Agelas* sp. Tetrahedron Lett. **24**, 4105 (1983).
92. CAPON, R.J., and D.J. FAULKNER: Antimicrobial Metabolites from a Pacific Sponge, *Agelas* sp. J. Am. Chem. Soc. **106**, 1819 (1984).
93. WALKER, R.P., and D.J. FAULKNER: Diterpenes from the Sponge *Dysidea amblia*. J. Org. Chem. **46**, 1098 (1981).
94. SULLIVAN, B., and D.J. FAULKNER: Metabolites of the Marine Sponge *Dendrilla* sp. J. Org. Chem. **49**, 3204 (1984).
95. CAPELLE, N., J.C. BRAEKMAN, D. DALOZE, and B. TURSCH: Chemical Studies of Marine Invertebrates. XLIV. Three New Spongian Diterpenes from *Spongia officinalis*. Bull. Soc. Chim. Belg. **89**, 399 (1980).
96. CIMINO, G., R. MORRONE, and G. SODANO: New Diterpenes from *Spongia officinalis*. Tetrahedron Lett. **23**, 4139 (1982).
97. GONZALEZ, A.G., D.M. ESTRADA, J.D. MARTIN, V.S. MARTIN, C. PEREZ, and R. PEREZ: New Antimicrobial Diterpenes from the Sponge *Spongia officinalis*. Tetrahedron **40**, 4109 (1984).
98. NAKANO, T., and M.I. HERNÁNDEZ: Stereoselective Total Syntheses of (±)-Isoagatholactone and (±)-12α-Hydroxyspongia-13(16),14-diene, Two Marine Sponge Metabolites. J. Chem. Soc., Perkin Trans. I **1983**, 135.
98a. SCHMITZ, F.J., J.S. CHANG, M.B. HOSSAIN, and D. VAN DER HELM: Marine Natural Products: Spongiane Derivatives from the Sponge *Igernella notabilis*. J. Org. Chem. **50**, 2862 (1985).
99. SCHMITZ, F.J., D.J. VANDERAH, K.H. HOLLENBEAK, C.E.L. ENWALL, Y. GOPICHAND, P.K. SENGUPTA, M.B. HOSSAIN, and D. VAN DER HELM: Metabolites from the Marine Sponge *Tedania ignis*. A New Atisanediol and Several Known Diketopiperazines. J. Org. Chem. **48**, 3941 (1983).
100. MAYOL, L., V. PICCIALLI, and D. SICA: Gracilin A, an Unique *Nor*-Diterpene Metabolite from the Marine Sponge *Spongionella gracilis*. Tetrahedron Lett. **26**, 1357 (1985).
101. – – – Application of 2D-NMR Spectroscopy in the Structural Determination of Gracilin B, a Bis-*nor*-diterpene from the Sponge *Spongionella gracilis*. Tetrahedron Lett. **26**, 1253 (1985).
102. HOCHLOWSKI, J.E., D.J. FAULKNER, G.K. MATSUMOTO, and J. CLARDY: Norrisolide, a Novel Diterpene from the Dorid Nudibranch *Chromodoris norrisi*. J. Org. Chem. **48**, 1141 (1983).
103. KARUSO, P., B.W. SKELTON, W.C. TAYLOR, and A.H. WHITE: The Constituents of Marine Sponges. I. The Isolation from *Aplysilla sulphurea* (Dendroceratida) of (1R^*,1'S^*,1''R^*,3R^*)-1-Acetoxy-4-ethyl-5-(1,3,3-trimethylcyclohexyl)-1,3-dihydroisobenzofuran-1'(4),3-carbolactone and the Determination of its Crystal Structure. Aust. J. Chem. **37**, 1081 (1984).
104. NAKAMURA, H., H. WU, Y. OHIZUMI, and Y. HIRATA: Agelasine-A, -B, -C, and -D, Novel Bicyclic Diterpenoids with a 9-Methyladeninium Unit Possessing Inhibitory Effects on Na,K-ATPase from the Okinawan Sea Sponge *Agelas* sp. Tetrahedron Lett. **25**, 2989 (1984).
105. WU, H., H. NAKAMURA, J. KOBAYASHI, Y. OHIZUMI, and Y. HIRATA: Agelasine-E

and -F, Novel Monocyclic Diterpenoids with a 9-Methyladeninium Unit Possessing Inhibitory Effects on Na,K-ATPase Isolated from the Okinawan Sea Sponge *Agelas nakamurai* Hoshino. Tetrahedron Lett. **25**, 3719 (1984).
106. NAKATSU, T., D.J. FAULKNER, G.K. MATSUMOTO, and J. CLARDY: Structure of the Diterpene Portion of a Novel Base from the Sponge *Agelas mauritiana*. Tetrahedron Lett. **25**, 935 (1984).
107. CHANG, C.W.J., A. PATRA, D.M. ROLL, P.J. SCHEUER, G.K. MATSUMOTO, and J. CLARDY: Kalihinol-A, a Highly Functionalized Diisocyano Diterpenoid Antibiotic from a Sponge. J. Am. Chem. Soc. **106**, 4644 (1984).
108. PATRA, A., C.W.J. CHANG, P.J. SCHEUER, G.D. VAN DUYNE, G.K. MATSUMOTO, and J. CLARDY: An Unprecedented Triisocyano Diterpenoid Antibiotic from a Sponge. J. Am. Chem. Soc. **106**, 7981 (1984).
109. KAZLAUSKAS, R., P.T. MURPHY, R.J. WELLS, and F.J. BLOUNT: New Diterpene Isocyanides from a Sponge. Tetrahedron Lett. **21**, 315 (1980).
110. MANES, L.V., G.J. BAKUS, and P. CREWS: Bioactive Marine Sponge *Nor*diterpene and *Nor*sesterterpene Peroxides. Tetrahedron Lett. **25**, 931 (1984).
110a. CAPON, R.J., and J.K. MACLEOD: Structural and Stereochemical Studies on Marine *Nor*terpene Cyclic Peroxides. Tetrahedron **41**, 3391 (1985).
110b. NAKAMURA, H., H. WU, J. KOBAYASHI, M. KOBAYASHI, Y. OHIZUMI, and Y. HIRATA: Agelasidines. Novel Hypotaurocyamine Derivatives from the Okinawan Sea Sponge *Agelas nakamurai* Hoshino. J. Org. Chem. **50**, 2494 (1985).
110c. CREWS, P., and S. NAYLOR: Sesterterpenes: An Emerging Group of Metabolites from Marine and Terrestrial Organisms. Fortschr. Chem. Organ. Naturstoffe (W. HERZ, H. GRISEBACH, G.W. KIRBY, and CH. TAMM, eds.) **48**, p. 203. Wien-New York: Springer. 1985.
111. GONZÁLEZ, A.G., M.L. RODRÍGUEZ, and A.S.M. BARRIENTOS: On the Stereochemistry and Biogenesis of C_{21} Linear Furanoterpenes in *Ircinia* sp. J. Nat. Prod. **46**, 256 (1983).
112. KAZLAUSKAS, R., P.T. MURPHY, and R.J. WELLS: Furodendin, a C_{22} Degraded Terpene from the Sponge *Phyllospongia dendyi*. Experientia **36**, 814 (1980).
113. KASHMAN, Y., and M. ZVIELY: Furospongolide, a New C_{21} Furanoterpene from a Marine Organism. Experientia **36**, 1279 (1980).
114. CAPON, R.J., E.L. GHISALBERTI, and P.R. JEFFERIES: A New Furanoterpene from a *Spongia* sp. Experientia **38**, 1444 (1982).
115. WALKER, R.P., J.E. THOMPSON, and D.J. FAULKNER: Sesterterpenes from *Spongia idia*. J. Org. Chem. **45**, 4976 (1980).
116. FUSETANI, N., Y. KATO, S. MATSUNAGA, and K. HASHIMOTO: Bioactive Marine Metabolites. V. Two New Furanosesterterpenes, Inhibitors of Cell Division of the Fertilized Starfish Eggs, from the Marine Sponge *Cacospongia scalaris*. Tetrahedron Lett. **25**, 4941 (1984).
117. DE SILVA, E.D., and P.J. SCHEUER: Manoalide, an Antibiotic Sesterterpenoid from the Marine Sponge *Luffariella variabilis* (Polejaeff). Tetrahedron Lett. **21**, 1611 (1980).
118. – – Three New Sesterterpenoid Antibiotics from the Marine Sponge *Luffariella variabilis* (Polejaeff). Tetrahedron Lett. **22**, 3147 (1981).
119. BLANKEMEIER, L.A., and R.S. JACOBS: Biochemical Analysis of the Inhibition of Purified Phospholipase A_2 by the Marine Natural Product, Manoalide. Fed. Proc. **42**, 374 (1983).
120. DE FREITAS, J.C., and R.S. JACOBS: Antagonism of the Neurotoxic Action of the β-Bungarotoxin on the Rat Phrenic Nerve-Hemidiaphragm by the Marine Natural Product, Manoalide. Fed. Proc. **42**, 374 (1983).
121. BRAEKMAN, J.C., D. DALOZE, R. BERTAU, and P. MACEDO DE ABREU: Cavernosine,

a Novel Ichthyotoxic Terpenoid Lactone from the Sponge *Fasciospongia cavernosa*. Bull. Soc. Chim. Belg. **91**, 791 (1982).
122. SULLIVAN, B., and D.J. FAULKNER: An Antimicrobial Sesterterpene from a Palauan Sponge. Tetrahedron Lett. **23**, 907 (1982).
123. YASUDA, F., and H. TADA: Desacetylscalaradial, a Cytotoxic Metabolite from the Sponge *Cacospongia scalaris*. Experientia **37**, 110 (1981).
124. KAZLAUSKAS, R., P.T. MURPHY, R.J. WELLS, and J.J. DALY: Terpenoid Constituents from Two *Phyllospongia* spp. Aust. J. Chem. **33**, 1783 (1980).
125. KIKUCHI, H., Y. TSUKITANI, I. SHIMIZU, M. KOBAYASHI, and I. KITAGAWA: Foliaspongin, an Antiinflammatory Bishomosesterterpene from the Marine Sponge *Phyllospongia foliascens* (Pallas). Chem. Pharm. Bull. (Japan) **29**, 1492 (1981).
126. ----- Marine Natural Products. XI. An Antiinflammatory Scalarane-type Bishomosesterterpene, Foliaspongin, from the Okinawan Marine Sponge *Phyllospongia foliascens* (Pallas). Chem. Pharm. Bull. (Japan) **31**, 552 (1983).
127. KASHMAN, Y., and M. ZVIELY: New Alkylated Scalarins from the Sponge *Dysidea herbacea*. Tetrahedron Lett. **40**, 3879 (1979).
128. KAZLAUSKAS, R., P.T. MURPHY, and R.J. WELLS: Five New C_{26} Tetracyclic Terpenes from a Sponge (*Lendenfeldia* sp.). Aust. J. Chem. **35**, 51 (1982).
129. CROFT, K.D., E.L. GHISALBERTI, B.W. SKELTON, and A.H. WHITE: Structural Study of a New Dialkylated Scalarane from a *Carteriospongia* sp. J. Chem. Soc., Perkin Trans. I **1983**, 155.
130. CIMINO, G., S. DE ROSA, and S. DE STEFANO: Scalarobutenolide, a New Sesterterpenoid from the Marine Sponge *Spongia nitens*. Experientia **37**, 214 (1981).
131. GREGSON, R.P., and D. OUVRIER: Wistarin, a Tetracyclic Furanosesterterpene from the Marine Sponge *Ircinia wistarii*. J. Nat. Prod. **45**, 412 (1982).
132. ALBERICCI, M., J.C. BRAEKMAN, D. DALOZE, and B. TURSCH: Chemical Studies of Marine Invertebrates – XLV. The Chemistry of Three *N*orsesterterpene Peroxides from the Sponge *Sigmosceptrella laevis*. Tetrahedron **38**, 1881 (1982).
133. SOKOLOFF, S., S. HALEVY, V. USIELI, A. COLORNI, and S. SAREL: Prianicin A and B, *Nor*-sesterterpenoid Peroxide Antibiotics from Red Sea Sponges. Experientia **38**, 337 (1982).
134. KASHMAN, Y., and M. ROTEM: Muqubilin, a New C_{24}-Isopenoid from a Marine Sponge. Tetrahedron Lett. **19**, 1707 (1979).
135. MANES, L.V., S. NAYLOR, P. CREWS, and G.J. BAKUS: Suvanine, a Novel Sesterterpene from an *Ircinia* Marine Sponge. J. Org. Chem. **50**, 284 (1985).
136. SHMUELI, U., S. CARMELY, A. GROWEISS, and Y. KASHMAN: Sipholenol and Sipholenone, Two New Triterpenes from the Marine Sponge *Siphonochalina siphonella* (Levi). Tetrahedron Lett. **22**, 709 (1981).
137. CARMELY, S., and Y. KASHMAN: The Sipholanes: A Novel Group of Triterpenes from the Marine Sponge *Siphonochalina siphonella*. J. Org. Chem. **48**, 3517 (1983).
138. CARMELY, S., Y. LOYA, and Y. KASHMAN: Siphonellinol, a New Triterpene from the Marine Sponge *Siphonochalina siphonella*. Tetrahedron Lett. **24**, 3673 (1983).
139. RAVI, B.N., R.J. WELLS, and K.D. CROFT: Malabaricane Triterpenes from a Fijian Collection of the Sponge *Jaspis stellifera*. J. Org. Chem. **46**, 1998 (1981).
140. RAVI, B.N., and R.J. WELLS: Malabaricane Triterpenes from a Great Barrier Reef Collection of the Sponge *Jaspis stellifera*. Aust. J. Chem. **35**, 39 (1982).
141. MCCABE, T., J. CLARDY, L. MINALE, C. PIZZA, F. ZOLLO, and R. RICCIO: A Triterpenoid Pigment with the Isomalabaricane Skeleton from the Marine Sponge *Stelletta* sp. Tetrahedron Lett. **23**, 3307 (1982).
142. LIAAEN-JENSEN, S., B. RENSTRØM, T. RAMDAHL, M. HALLENSTVET, and P. BERGQUIST: Carotenoids of Marine Sponges. Biochem. Syst. Ecol. **10**, 167 (1982).

143. LITCHFIELD, C., and S. LIAAEN-JENSEN: Carotenoids of the Marine Sponge *Microciona prolifera*. Comp. Biochem. Physiol. **66B**, 359 (1980).
144. HERTZBERG, S., T. RAMDAHL, J.E. JOHANSEN, and S. LIAAEN-JENSEN: Carotenoid Sulfates. 2. Structural Elucidation of Bastaxanthin. Acta Chem. Scand. **B 37**, 267 (1983).
145. Djura, P., and D.J. FAULKNER: Metabolites of the Marine Sponge *Dercitus* sp. J. Org. Chem. **45**, 735 (1980).
146. TAYLOR, K.M., J.A. BAIRD-LAMBERT, P.A. DAVIS, and I. SPENCE: Methylaplysinopsin and Other Marine Natural Products Affecting Neurotransmission. Fed. Proc. **40**, 15 (1981).
147. TYMIAK, A.A., K.L. RINEHART, JR., and G.J. BAKUS: Constituents of Morphologically Similar Sponges *Aplysina* and *Smenospongia* Species. Tetrahedron **41**, 1039 (1985).
148. DELLAR, G., P. DJURA, and M.V. SARGENT: Structure and Synthesis of a New Bromoindole from a Marine Sponge. J. Chem. Soc., Perkin Trans. I **1981**, 1679.
149. MAKARIEVÁ, T.N., V.A. STONIK, P. ALCOLADO, and Y.B. ELYAKOV: Comparative Study of the Halogenated Tyrosine Derivatives from Demospongiae (Porifera). Comp. Biochem. Physiol. **68B**, 481 (1981).
150. GORSHKOV, B.A., I.A. GORSHKOVA, T.N. MAKARIEVA, and V.A. STONIK: Inhibiting Effect of Cytotoxic Bromine-Containing Compounds from Sponges (Aplysinidae) on Na^+-K^+-ATPase Activity. Toxicon **20**, 1092 (1982).
151. GORSHKOV, B.A., I.A. GORSHKOVA, and T.N. MAKARIEVA: Inhibitory Characteristics of 3,5-Dibromo-1-acetoxy-4-oxo-2,5-cyclohexadien-1-acetonitrile, a Semisynthetic Derivative of Aeroplysinin-1 from Sponges (Aplysinidae), on Na^+-K^+-ATPase. Toxicon **22**, 441 (1984).
152. D'AMBROSIO, M., A. GUERRIERO, and F. PIETRA: 168. Novel, Racemic or Nearly-Racemic Antibacterial Bromo- and Chloroquinols and γ-Lactams of the Verongiaquinol and the Cavernicolin Type from the Marine Sponge *Aplysina* (= *Verongia*) *cavernicola*. Helv. Chim. Acta **67**, 1484 (1984).
153. D'AMBROSIO, M., A. GUERRIERO, R. DE CLAUSER, G. DE STANCHINA, and F. PIETRA: Dichloroverongiaquinol, a New Marine Antibacterial Compound from *Aplysina cavernicola*. Isolation and Synthesis. Experientia **39**, 1091 (1983).
154. D'AMBROSIO, M., A. GUERRIERO, P. TRALDI, and F. PIETRA: Cavernicolin-1 and Cavernicolin-2, Two Epimeric Dibromolactams from the Mediterranean Sponge *Aplysina* (*Verongia*) *cavernicola*. Tetrahedron Lett. **23**, 4403 (1982).
155. GUERRIERO, A., M. D'AMBROSIO, P. TRALDI, and F. PIETRA: On the First Marine Natural Product Having Low Enantiomeric Purity. Naturwissenschaften **71**, 425 (1984).
156. TYMIAK, A.A., and K.L. RINEHART, JR.: Biosynthesis of Dibromotyrosine-Derived Antimicrobial Compounds by the Marine Sponge *Aplysina fistularis* (*Verongia aurea*). J. Am. Chem. Soc. **103**, 6763 (1981).
157. ROTEM, M., S. CARMELY, Y. KASHMAN, and Y. LOYA: Two New Antibiotics from the Red Sea Sponge *Psammaplysilla purpurea*. Total ^{13}C-NMR Line Assignment of Psammaplysins A and B and Aerothionin. Tetrahedron **39**, 667 (1983).
158. MCMILLAN, J.A., I.C. PAUL, Y.M. GOO, K.L. RINEHART, JR., W.C. KRUEGER, and L.M. PSCHIGODA: An X-Ray Study of Aerothionin from *Aplysina fistularis* (Pallas). Tetrahedron Lett. **22**, 39 (1981).
159. CIMINO, G., S. DE ROSA, S. DE STEFANO, R. SELF, and G. SODANO: The Bromo-Compounds of the True Sponge *Verongia aerophoba*. Tetrahedron Lett. **24**, 3029 (1983).
160. KAZLAUSKAS, R., R.O. LIDGARD, P.T. MURPHY, and R.J. WELLS: Brominated Tyrosine-Derived Metabolites from the Sponge *Ianthella basta*. Tetrahedron Lett. **21**, 2277 (1980).

161. KAZLAUSKAS, R., R.O. LIDGARD, P.T. MURPHY, R.J. WELLS, and J.F. BLOUNT: Brominated Tyrosine-Derived Metabolites from the Sponge *Ianthella basta*. Aust. J. Chem. **34**, 765 (1981).
162. ROLL, D.M., C.W.J. CHANG, P.J. SCHEUER, G.A. GRAY, J.N. SHOOLERY, G.K. MATSUMOTO, G.D. VAN DUYNE, and J. CLARDY: Structure of the Psammaplysins. J. Am. Chem. Soc. **107**, 2916 (1985).
162a. NAKAMURA, H., H. WU, J. KOBAYASHI, Y. NAKAMURA, Y. OHIZUMI, and Y. HIRATA: Purealin, a Novel Enzyme Activator from the Okinawan Marine Sponge *Psammaplysilla purea*. Tetrahedron Lett. **26**, 4517 (1985).
163. CHARLES, C., J.C. BRAEKMAN, D. DALOZE, and B. TURSCH: Chemical Studies of Marine Invertebrates – XLII. The Relative and Absolute Configuration of Dysidenin. Tetrahedron **36**, 2133 (1980).
163a. CHARLES, C., J.C. BRAEKMAN, D. DALOZE, B. TURSCH, and R. KARLSSON: Chemical Studies of Marine Invertebrates. XXXII. Isodysidenin, a Further Hexachlorinated Metabolite from the Sponge *Dysidea herbacea*. Tetrahedron Lett. **17**, 1519 (1978).
164. BISKUPIAK, J.E., and C.M. IRELAND: Revised Absolute Configuration of Dysidenin and Isodysidenin. Tetrahedron Lett. **25**, 2935 (1984).
165. ERICKSON, K.L., and R.J. WELLS: New Polychlorinated Metabolites from a Barrier Reef Collection of the Sponge *Dysidea herbacea*. Aust. J. Chem. **35**, 31 (1982).
166. DE LASZLO, S.E., and P.G. WILLIARD: Total Synthesis of (+)-Demethyldysidenin and (−)-Demethylisodysidenin, Hexachlorinated Amino Acids from the Marine Sponge *Dysidea herbacea*. Assignment of Absolute Stereochemistry. J. Am. Chem. Soc. **107**, 199 (1985).
167. WILLIARD, P.G., and S.E. DE LASZLO: Total Synthesis of (±)-Dysidin, a Marine Metabolite Containing an N-Acyl-O-methyltetramic Acid. J. Org. Chem. **49**, 3489 (1984).
168. STONARD, R.J., and R.J. ANDERSEN: Celenamides A and B, Linear Peptide Alkaloids from the Sponge *Cliona celata*. J. Org. Chem. **45**, 3687 (1980).
169. – – Linear Peptide Alkaloids from the Sponge *Cliona celata* (Grant). Celenamides C and D. Can. J. Chem. **58**, 2121 (1980).
170. SCHMIDT, U., and J. WILD: Totalsynthese von Hexaacetylcelenamid A. Angew. Chem. **96**, 996 (1984).
171. MATSUNAGA, S., N. FUSETANI, and S. KONOSU: Bioactive Marine Metabolites, IV. Isolation and the Amino Acid Composition of Discodermin A, an Antimicrobial Peptide, from the Marine Sponge *Discodermia kiiensis*. J. Nat. Prod. **48**, 236 (1985).
172. – – – Bioactive Marine Metabolites VI. Structure Elucidation of Discodermin A, an Antimicrobial Peptide from the Marine Sponge *Discodermia kiiensis*. Tetrahedron Lett. **25**, 5165 (1984).
173. – – – Bioactive Marine Metabolites VII. Structures of Discodermins B, C, and D, Antimicrobial Peptides from the Marine Sponge *Discodermia kiiensis*. Tetrahedron Lett. **26**, 855 (1985).
174. PETTIT, G.R., J.A. RIDEOUT, and J.A. HASLER: Isolation of Geodiastatins 1 and 2 from the Marine Sponge *Geodia mesotriaena*. J. Nat. Prod. **44**, 588 (1981).
175. MÜLLER, W.E.G., A. BERND, R.K. ZAHN, B. KURELEC, K. DAWES, I. MÜLLER, and G. UHLENBRUCK: Xenograft Rejection in Marine Sponges. Isolation and Purification of an Inhibitory Aggregation Factor from *Geodia cydonium*. Eur. J. Biochem. **116**, 573 (1981).
176. QUINN, R.J., R.P. GREGSON, A.F. COOK, and R.T. BARTLETT: Isolation and Synthesis of 1-Methylisoguanosine, a Potent Pharmacologically Active Constituent from the Marine Sponge *Tedania digitata*. Tetrahedron Lett. **21**, 567 (1980).
177. COOK, A.F., R.T. BARTLETT, R.P. GREGSON, and R.J. QUINN: 1-Methylisoguanosine, a Pharmacologically Active Agent from a Marine Sponge. J. Org. Chem. **45**, 4020 (1980).

178. BAIRD-LAMBERT, J., J.F. MARWOOD, L.P. DAVIES, and K.M. TAYLOR: 1-Methylisoguanosine: An Orally Active Marine Natural Product with Skeletal Muscle and Cardiovascular Effects. Life Sci. **26**, 1069 (1980).
179. DAVIES, L.P., K.M. TAYLOR, R.P. GREGSON, and R.J. QUINN: Stimulation of Guinea-Pig Brain Adenylate Cyclase by Adenosine Analogues with Potent Pharmacological Activity *in vivo*. Life Sci. **26**, 1079 (1980).
180. DAVIES, L.P., A.F. COOK, M. POONIAN, and K.M. TAYLOR: Displacement of [^{3}H] Diazepam Binding in Rat Brain by Dipyridamole and by 1-Methylisoguanosine, a Marine Natural Product with Muscle Relaxant Activity. Life Sci. **26**, 1089 (1980).
180a. KATO, Y., N. FUSETANI, S. MATSUNAGA, and K. HASHIMOTO: Bioactive Marine Metabolites. IX. Mycalisines A and B, Novel Nucleosides which Inhibit Cell Division of Fertilized Starfish Eggs, from the Marine Sponge *Mycale* Sp. Tetrahedron Lett. **26**, 3483 (1985).
181. WEBER, J.F., F.A. FUHRMAN, G.J. FUHRMAN, and H.S. MOSHER: Isolation of Allantoin and Adenosine from the Marine Sponge *Tethya aurantia*. Comp. Biochem. Physiol. **70B**, 799 (1981).
182. KAZLAUSKAS, R., P.T. MURPHY, R.J. WELLS, J.A. BAIRD-LAMBERT, and D.D. JAMIESON: Halogenated Pyrrolo[2,3-d]pyrimidine Nucleosides from Marine Organisms. Aust. J. Chem. **36**, 165 (1983).
183. STIERLE, D.B., and D.J. FAULKNER: Metabolites of the Marine Sponge *Laxosuberites* sp. J. Org. Chem. **45**, 4980 (1980).
184. NAKAMURA, H., Y. OHIZUMI, J. KOBAYASHI, and Y. HIRATA: Keramadine, a Novel Antagonist of Serotonergic Receptors Isolated from the Okinawan Sea Sponge *Agelas* sp. Tetrahedron Lett. **25**, 2475 (1984).
185. WALKER, R.P., D.J. FAULKNER, D. VAN ENGEN, and J. CLARDY: Sceptrin, an Antimicrobial Agent from the Sponge *Agelas sceptrum*. J. Am. Chem. Soc. **103**, 6772 (1981).
186. SHARMA, G.M., J.S. BUYER, and M.W. POMERANTZ: Characterization of a Yellow Compound Isolated from the Marine Sponge *Phakellia flabellata*. J. Chem. Soc., Chem. Comm. **1980**, 435.
187. CIMINO, G., S. DE ROSA, S. DE STEFANO, L. MAZZARELLA, R. PULITI, and G. SODANO: Isolation and X-Ray Crystal Structure of a Novel Bromo-Compound from Two Marine Sponges. Tetrahedron Lett. **23**, 767 (1982).
188. SCHMITZ, F.J., S.P. GUNASEKERA, V. LAKSHMI, and L.M.V. TILLEKERATNE: Marine Natural Products: Pyrrololactams from Several Sponges. J. Nat. Prod. **48**, 47 (1985).
189. HARBOUR, G.C., A.A. TYMIAK, K.L. RINEHART, JR., P.D. SHAW, R.G. HUGHES, JR., S.A. MIZSAK, J.H. COATS, G.E. ZURENKO, L.H. LI, and S.L. KUENTZEL: Ptilocaulin and Isoptilocaulin, Antimicrobial and Cytotoxic Cyclic Guanidines from the Caribbean Sponge *Ptilocaulis* aff. *P. spiculifer* (Lamarck, 1814). J. Am. Chem. Soc. **103**, 5604 (1981).
189a. SNIDER, B.B., and W.C. FAITH: The Total Synthesis of ($\pm$)-Ptilocaulin. Tetrahedron Lett. **24**, 861 (1983).
189b. – – Total Synthesis of ($\pm$)- and ($-$)-Ptilocaulin. J. Am. Chem. Soc. **106**, 1443 (1984).
189c. ROUSH, W.R., and A.E. WALTS: Total Synthesis of ($-$)-Ptilocaulin. J. Am. Chem. Soc. **106**, 721 (1984).
189d. WALTS, A.E., and W.R. ROUSH: A Stereorational Total Synthesis of ($-$)-Ptilocaulin. Tetrahedron **41**, 3463 (1985).
190. CIMINO, G., S. DE ROSA, S. DE STEFANO, A. SPINELLA, and G. SODANO: The Zoochrome of the Sponge *Verongia aerophoba* ("Uranidine"). Tetrahedron Lett. **25**, 2925 (1984).
191. CARDELLINA II, J.H., and J. MEINWALD: Leucettidine, a Novel Pteridine from the Calcareous Sponge *Leucetta microraphis*. J. Org. Chem. **46**, 4782 (1981).

191a. CIMINO, G., A. DE GIULIO, S. DE ROSA, S. DE STEFANO, R. PULITI, C.A. MATTIA, and L. MAZZARELLA: Isolation and X-ray Crystal Structure of a Novel 8-Oxopurine Compound from a Marine Sponge. J. Nat. Prod. **48**, 523 (1985).
192. NAKAMURA, H., J. KOBAYASHI, Y. OHIZUMI, and Y. HIRATA: Isolation and Structure of Aaptamine, a Novel Heteroaromatic Substance Possessing α-Blocking Activity from the Sea Sponge *Aaptos aaptos*. Tetrahedron Lett. **23**, 5555 (1982).
193. SCHMITZ, F.J., S.K. AGARWAL, S.P. GUNASEKERA, P.G. SCHMIDT, and J.N. SHOOLERY: Amphimedine, New Aromatic Alkaloid from a Pacific Sponge, *Amphimedon* sp. Carbon Connectivity Determination from Natural Abundance ^{13}C-^{13}C Coupling Constants. J. Am. Chem. Soc. **105**, 4835 (1983).
194. BRAEKMAN, J.C., D. DALOZE, P. MACEDO DE ABREU, C. PICCINNI-LEOPARDI, G. GERMAIN, and M. VAN MEERSSCHE: A Novel Type of Bis-Quinolizidine Alkaloid from the Sponge *Petrosia seriata*. Tetrahedron Lett. **23**, 4277 (1982).
195. BRAEKMAN, J.C., D. DALOZE, N. DEFAY, and D. ZIMMERMANN: Petrosin-A and -B, Two New Bis-Quinolizidine Alkaloids from the Sponge *Petrosia seriata*. Bull. Soc. Chim. Belg. **93**, 941 (1984).
196. NAKAGAWA, M., M. ENDO, N. TANAKA, and L. GEN-PEI: Structures of Xestospongin A, B, C and D, Novel Vasodilative Compounds from Marine Sponge, *Xestospongia exigua*. Tetrahedron Lett. **25**, 3227 (1984).
197. FRINCKE, J.M., and D.J. FAULKNER: Antimicrobial Metabolites of the Sponge *Reniera* sp. J. Am. Chem. Soc. **104**, 265 (1982).
198. KASHMAN, Y., A. GROWEISS, and U. SHMUELI: Latrunculin, a New 2-Thiazolidinone Macrolide from the Marine Sponge *Latrunculia magnifica*. Tetrahedron Lett. **21**, 3629 (1980).
199. GROWEISS, A., U. SHMUELI, and Y. KASHMAN: Marine Toxins of *Latrunculia magnifica*. J. Org. Chem. **48**, 3512 (1983).
199a. KASHMAN, Y., A. GROWEISS, R. LIDOR, D. BLASBERGER, and S. CARMELY: Latrunculins: NMR Study, Two New Toxins and a Synthetic Approach. Tetrahedron **41**, 1905 (1985).
200. SPECTOR, I., N.R. SHOCHET, Y. KASHMAN, and A. GROWEISS: Latrunculins: Novel Marine Toxins That Disrupt Microfilament Organization in Cultured Cells. Science **219**, 493 (1983).
201. CARMELY, S., and Y. KASHMAN: Structure of Swinholide-A, a New Macrolide from the Marine Sponge *Theonella swinhoei*. Tetrahedron Lett. **26**, 511 (1985).
202. SCHMITZ, F.J., S.P. GUNASEKERA, G. YALAMANCHILI, M.B. HOSSAIN, and D. VAN DER HELM: Tedanolide: A Potent Cytotoxic Macrolide from the Caribbean Sponge *Tedania ignis*. J. Am. Chem. Soc. **106**, 7251 (1984).
202a. UEMURA, D., K. TAKAHASHI, T. YAMAMOTO, C. KATAYAMA, J. TANAKA, Y. OKUMURA, and Y. HIRATA: Norhalichondrin A: An Antitumor Polyether Macrolide from a Marine Sponge. J. Am. Chem. Soc. **107**, 4796 (1985).
203. CARTÉ, B., and D.J. FAULKNER: Polybrominated Diphenyl Ethers from *Dysidea herbacea*, *Dysidea chlorea* and *Phyllospongia foliascens*. Tetrahedron **37**, 2335 (1981).
204. NORTON, R.S., K.D. CROFT, and R.J. WELLS: Polybrominated Oxydiphenol Derivatives from the Sponge *Dysidea herbacea*. Structure Determination by Analysis of ^{13}C Spin-Lattice Relaxation Data for Quaternary Carbons and ^{13}C-^{1}H Coupling Constants. Tetrahedron **37**, 2341 (1981).
205. NORTON, R.S., and R.J. WELLS: Use of ^{13}C Spin-Lattice Relaxation Measurements to Determine the Structure of a Tetrabromo Diphenyl Ether from the Sponge *Dysidea herbacea*. Tetrahedron Lett. **21**, 3801 (1980).
206. CAPON, R., E.L. GHISALBERTI, P.R. JEFFERIES, B.W. SKELTON, and A.H. WHITE: Structural Studies of Halogenated Diphenyl Ethers from a Marine Sponge. J. Chem. Soc., Perkin Trans. I **1981**, 2464.

207. WRATTEN, S.J., and J. MEINWALD: Antimicrobial Metabolites of the Marine Sponge *Axinella polycapella*. Experientia **37**, 13 (1981).
208. BERGQUIST, P.R., M.P. LAWSON, A. LAVIS, and R.C. CAMBIE: Fatty Acid Composition and the Classification of the Porifera. Biochem. Syst. Ecol. **12**, 63 (1984).
209. LAWSON, M.P., P.R. BERGQUIST, and R.C. CAMBIE: Fatty Acid Composition and the Classification of the Porifera. Biochem. Syst. Ecol. **12**, 375 (1984).
210. LITCHFIELD, C., J. TYSZKIEWICS, and V. DATO: 5,9,23-Triacontatrienoic Acid, Principal Fatty Acid of the Marine Sponge *Chondrilla nucula*. Lipids **15**, 200 (1980).
211. WALKUP, R.D., G.C. JAMIESON, M.R. RATCLIFF, and C. DJERASSI: Phospholipid Studies of Marine Organisms: 2. Phospholipids, Phospholipid-Bound Fatty Acids and Free Sterols of the Sponge *Aplysina fistularis* (Pallas) Forma *fulva* (Pallas) (= *Verongia thiona*). Isolation and Structure Elucidation of Unprecedented Branched Fatty Acids. Lipids **16**, 631 (1981).
212. AYANOGLU, E., R.D. WALKUP, D. SICA, and C. DJERASSI: Phospholipid Studies of Marine Organisms: III. New Phospholipid Fatty Acids from *Petrosia ficiformis*. Lipids **17**, 617 (1982).
213. AYANOGLU, E., J.M. KORNPROBST, A. ABOUD-BICHARA, and C. DJERASSI: Phospholipid Studies of Marine Organisms 4. (2R,21Z)-2-Methoxy-21-octacosenoic Acid, the First Naturally Occurring α-Methoxy Acid from a Phospholipid. Tetrahedron Lett. **24**, 1111 (1983).
214. AYANOGLU, E., S. POPOV, J.M. KORNPROBST, A. ABOUD-BICHARA, and C. DJERASSI: Phospholipid Studies of Marine Organisms: V. New α-Methoxy Acids from *Higginsia tethyoides*. Lipids **18**, 830 (1983).
215. LANKELMA, J., E. AYANOGLU, and C. DJERASSI: Double-Bond Location in Long-Chain Polyunsaturated Fatty Acids by Chemical Ionization-Mass Spectrometry. Lipids **18**, 853 (1983).
216. DASGUPTA, A., E. AYANOGLU, and C. DJERASSI: Phospholipid Studies of Marine Organisms: New Branched Fatty Acids from *Strongylophora durissima*. Lipids **19**, 768 (1984).
217. CIMINO, G., A. CRISPINO, S. DE ROSA, S. DE STEFANO, and G. SODANO: Polyacetylenes from the Sponge *Petrosia ficiformis* Found in Dark Caves. Experientia **37**, 924 (1981).
218. WIJEKOON, W.M.D., E. AYANOGLU, and C. DJERASSI: Phospholipid Studies of Marine Organisms 9. New Brominated Demospongic Acids from the Phospholipids of Two *Petrosia* Species. Tetrahedron Lett. **25**, 3285 (1984).
219. QUINN, R.J., and D.J. TUCKER: A Brominated Bisacetylenic Acid from the Marine Sponge *Xestospongia testudinaria*. Tetrahedron Lett. **26**, 1671 (1985).
220. SCHMITZ, F.J., R.S. PRASAD, Y. GOPICHAND, M.B. HOSSAIN, D. VAN DER HELM, and P. SCHMIDT: Acanthifolicin, a New Episulfide-Containing Polyether Carboxylic Acid from Extracts of the Marine Sponge *Pandaros acanthifolium*. J. Am. Chem. Soc. **103**, 2467 (1981).
221. TACHIBANA, K., P.J. SCHEUER, Y. TSUKITANI, H. KIKUCHI, D. VAN ENGEN, J. CLARDY, Y. GOPICHAND, and F.J. SCHMITZ: Okadaic Acid, a Cytotoxic Polyether from Two Marine Sponges of the Genus *Halichondria*. J. Am. Chem. Soc. **103**, 2469 (1981).
222. SHIBATA, S., Y. ISHIDA, H. KITANO, Y. OHIZUMI, J. HABON, Y. TSUKITANI, and H. KIKUCHI: Contractile Effects of Okadaic Acid, a Novel Ionophore-Like Substance from Black Sponge, on Isolated Smooth Muscles Under the Condition of Ca Deficiency. J. Pharmacol. Exp. Ther. **223**, 135 (1982).
223. STIERLE, D.B., and D.J. FAULKNER: Metabolites of Three Marine Sponges of the Genus *Plakortis*. J. Org. Chem. **45**, 3396 (1980).
224. PHILLIPSON, D.W., and K.L. RINEHART, JR.: Antifungal Peroxide-Containing Acids from Two Caribbean Sponges. J. Am. Chem. Soc. **105**, 7735 (1983).
225. CIMINO, G., A. DE GIULIO, S. DE ROSA, S. DE STEFANO, and G. SODANO: Further

High Molecular Weight Polyacetylenes from the Sponge *Petrosia ficiformis*. J. Nat. Prod. **48**, 22 (1985).
226. FUSETANI, N., Y. KATO, S. MATSUNAGA, and K. HASHIMOTO: Bioactive Marine Metabolites III. A Novel Polyacetylene Alcohol, Inhibitor of Cell Division in Fertilized Sea Urchin Eggs, from the Marine Sponge *Tetrosia* sp. Tetrahedron Lett. **24**, 2771 (1983).
227. CASTIELLO, D., G. CIMINO, S. DE ROSA, S. DE STEFANO, and G. SODANO: High Molecular Weight Polyacetylenes from the Nudibranch *Peltodoris atromaculata* and the Sponge *Petrosia ficiformis*. Tetrahedron Lett. **21**, 5047 (1980).
228. CARDELLINA II, J.H., C.J. GRADEN, B.J. GREER, and J.R. KERN: 17Z-Tetracosenyl 1-Glycerol Ether from the Sponges *Cinachyra alloclada* and *Ulosa ruetzleri*. Lipids **18**, 107 (1983).
229. MYERS, B.L., and P. CREWS: Chiral Ether Glycerides from a Marine Sponge. J. Org. Chem. **48**, 3583 (1983).
230. DO, M.N., and K.L. ERICKSON: Branched Chain Mono-Glycerol Ethers from a Taiwanese Marine Sponge of the Genus *Aaptos*. Tetrahedron Lett. **24**, 5699 (1983).
231. ROLL, D.M., P.J. SCHEUER, G.K. MATSUMOTO, and J. CLARDY: Halenaquinone, a Pentacyclic Polyketide from a Marine Sponge. J. Am. Chem. Soc. **105**, 6177 (1983).
231a. KOBAYASHI, M., N. SHIMIZU, I. KITAGAWA, Y. KYOGOKU, N. HARADA, and H. UDA: Absolute Stereostructures of Halenaquinol and Halenaquinol Sulfate, Pentacyclic Hydroquinones from the Okinawan Marine Sponge *Xestospongia sapra*, as Determined by Theoretical Calculation of CD Spectra. Tetrahedron Lett. **26**, 3833 (1985).
232. CAPON, R.J., E.L. GHISALBERTI, and P.R. JEFFERIES: New Tetrahydropyrans from a Marine Sponge. Tetrahedron **38**, 1699 (1982).
233. HAGADONE, M.R., P.J. SCHEUER, and A. HOLM: On the Origin of the Isocyano Function in Marine Sponges. J. Am. Chem. Soc. **106**, 2447 (1984).
234. FAULKNER, D.J., and B.N. RAVI: Metabolites of the Marine Sponge *Plakortis zygompha*. Tetrahedron Lett. **21**, 23 (1980).
235. BÉRESS, L.: Biologically Active Compounds from Coelenterates. Pure and Appl. Chem. **54**, 1981 (1982).
236. NEEMAN, I., G.J. CALTON, and J.W. BURNETT: Purification and Characterization of the Endonuclease Present in *Physalia physalis* Venom. Comp. Biochem. Physiol. **67B**, 155 (1980).
237. LAL, D.M., G.J. CALTON, I. NEEMAN, and J.W. BURNETT: Characterization of *Physalia physalis* (Portuguese Man-O'War) Nematocyst Venom Collagenase. Comp. Biochem. Physiol. **70B**, 635 (1981).
238. NEEMAN, I., G.J. CALTON, and J.W. BURNETT: Purification of an Endonuclease Present in *Chrysaora quinquecirrha* Venom (41077). Proceedings of the Society for Experimental Biology and Medicine **166**, 374 (1981).
239. LAL, D.M., G.J. CALTON, I. NEEMAN, and J.W. BURNETT: Characterization of *Chrysaora quinquecirrha* (Sea Nettle) Nematocyst Venom Collagenase. Comp. Biochem. Physiol. **69B**, 529 (1981).
240. CALTON, G.J., and J.W. BURNETT: Partial Purification and Characterization of the Acid Protease of Sea Nettle (*Chrysaora quinquecirrha*) Nematocyst Venom. Comp. Biochem. Physiol. **72B**, 93 (1982).
241. – – Partial Purification and Characterization of the Alkaline Protease of Sea Nettle (*Chrysaora quinquecirrha*) Nematocyst Venom. Comp. Biochem. Physiol. **74C**, 361 (1983).
242. TAMKUN, M.M., and D.A. HESSINGER: Isolation and Partial Characterization of a Hemolytic and Toxic Protein from the Nematocyst Venom of the Portuguese Man-of-War, *Physalia physalis*. Biochim. Biophys. Acta **667**, 87 (1981).

243. WATROUS, J., and K. THOMPSON: *Chrysaora quinquecirrha* (Sea Nettle) Toxin: A Comparison Between a Commercial and Our Own Preparation. Toxicon **19**, 319 (1981).
244. COBBS, C.S., P. GOLD, G.J. CALTON, and J.W. BURNETT: Sea Nettle (*Chrysaora quinquecirrha*) Nematocyst Venom Hemagglutinins. Comp. Biochem. Physiol. **74C**, 225 (1983).
245. KELMAN, S.N., G.J. CALTON, and J.W. BURNETT: Isolation and Partial Characterization of a Lethal Sea Nettle (*Chrysaora quinquecirrha*) Mesenteric Toxin. Toxicon **22**, 139 (1984).
246. HARTMAN, K.R., G.J. CALTON, and J.W. BURNETT: The Utilization of the Bradykinin Radioimmunoassay for the Study of a Kinin-Like Factor in Jellyfish Toxin. Comp. Biochem. Physiol. **66C**, 163 (1980).
247. – – – A Comparison of the Kinin-Like Factor in the Sea Nettle Fishing and Mesenteric Tentacles. Comp. Biochem. Physiol. **68C**, 235 (1981).
248. GAUR, P.K., G.J. CALTON, and J.W. BURNETT: Enzyme-Linked Immunosorbent Assay to Detect Anti-Sea Nettle Venom Antibodies. Experientia **37**, 1005 (1981).
249. GAUR, P.K., R.L. ANTHONY, G.J. CALTON, and J.W. BURNETT: Isolation of Hybridomas Secreting Monoclonal Antibodies Against *Physalia physalis* (Portuguese Man-O'War) Nematocyst Venom. Toxicon **20**, 419 (1982).
250. COBBS, C.S., P.K. GAUR, A.J. RUSSO, J.E. WARNICK, G.J. CALTON, and J.W. BURNETT: Immunosorbent Chromatography of Sea Nettle (*Chrysaora quinquecirrha*) Venom and Characterization of Toxins. Toxicon **21**, 385 (1983).
251. OLSON, C.E., D.G. CARGO, G.J. CALTON, and J.W. BURNETT: Immunochromatography and Cardiotoxicity of Sea Nettle (*Chrysaora quinquecirrha*) Polyps and Cysts. Toxicon **23**, 127 (1985).
252. WARNICK, J.E., D. WEINREICH, and J.W. BURNETT: Sea Nettle (*Chrysaora quinquecirrha*) Toxin on Electrogenic and Chemosensitive Properties of Nerve and Muscle. Toxicon **19**, 361 (1981).
253. NEEMAN, I., G.J. CALTON, and J.W. BURNETT: Cytotoxicity and Dermonecrosis of Sea Nettle (*Chrysaora quinquecirrha*) Venom. Toxicon **18**, 55 (1980).
254. – – – An Ultrastructural Study of the Cytotoxic Effect of the Venoms from the Sea Nettle (*Chrysaora quinquecirrha*) and Portuguese Man-of-War (*Physalia physalis*) on Cultured Chinese Hamster Ovary K-1 Cells. Toxicon **18**, 495 (1980).
255. SHRYOCK, J.C., and C.P. BIANCHI: Sea Nettle (*Chrysaora quinquecirrha*) Nematocyst Venom: Mechanism of Action on Muscle. Toxicon **21**, 81 (1983).
256. COBBS, C.S., R.E. DRZYMALA, A.E. SHAMOO, G.J. CALTON, and J.W. BURNETT: Sea Nettle (*Chrysaora quinquecirrha*) Lethal Factor: Effect on Black Lipid Membranes. Toxicon **21**, 558 (1983).
257. BURNETT, J.W., G.J. CALTON, C.S. COBBS, and S.N. KELMAN: Sea Nettle and Portuguese Man-O'War Nematocyst Venoms: Studies with Monoclonal Antibodies and Affinity Chromatography. Toxicon Suppl. **3**, 49 (1983).
258. FLOWERS, A.L., and D.A. HESSINGER: Mast Cell Histamine Release Induced by Portuguese Man-of-War (*Physalia*) Venom. Biochem. Biophys. Res. Commun. **103**, 1083 (1981).
259. BEHRENS-BAUMANN, W.: Zur Bedeutung der Quallen und deren Giftwirkung auf den Menschen. Dermatosen **29**, 121 (1981).
260. ENDEAN, R., and J. RIFKIN: Envenomation Involving Nematocysts of the Box Jellyfish *Chironex fleckeri*. Toxicon Suppl. **3**, 115 (1983).
261. OLSON, C.E., E.E. POCKL, G.J. CALTON, and J.W. BURNETT: Immunochromatographic Purification of a Nematocyst Toxin from the Cnidarian *Chironex fleckeri* (Sea Wasp). Toxicon **22**, 733 (1984).
262. CIMINO, G., S. DE ROSA, S. DE STEFANO, and G. SODANO: Cholest-4-en-4,16β,18,22R-

tetrol-3-one 16,18-diacetate, a Novel Polyhydroxylated Steroid from the Hydroid *Eudendrium* sp. Tetrahedron Lett. **21**, 3303 (1980).
262a. FATTORUSSO, E., V. LANZOTTI, S. MAGNO, and E. NOVELLINO: Cholest-5-ene-2α,3α,7β,15β,18-pentol 2,7,15,18-Tetraacetate, a Novel Highly Hydroxylated Sterol from the Marine Hydroid *Eudendrium glomeratum*. J. Org. Chem. **50**, 2868 (1985).
262b. ---- Two New Polyoxygenated Sterols from the Marine Hydroid *Eudendrium glomeratum*. J. Nat. Prod. **48**, 784 (1985).
263. DE NAPOLI, L., E. FATTORUSSO, S. MAGNO, and L. MAYOL: Acyclic Polyhalogenated Monoterpenes from Four Marine Hydroids. Biochem. Syst. Ecol. **12**, 321 (1984).
264. FAHY, E., R.J. ANDERSEN, H. CUN-HENG, and J. CLARDY: Garveatin A, an Antimicrobial 1(4H)-Anthracenone Derivative from the Hydroid *Garveia annulata*. J. Org. Chem. **50**, 1149 (1985).
265. LUBBOCK, R.: Why are Clownfishes not Stung by Sea Anemones? Proc. R. Soc. Lond. B **207**, 35 (1980).
266. MARETIC, Z., and F.E. RUSSELL: Some Epidemiological and Clinical Aspects of Stings by the Sea Anemone *Anemonia sulcata*. Toxicon **20**, 360 (1982).
267. SCHEFFLER, J.-J., A. TSUGITA, G. LINDEN, H. SCHWEITZ, and M. LAZDUNSKI: The Amino Acid Sequence of Toxin V from *Anemonia sulcata*. Biochem. Biophys. Res. Commun. **107**, 272 (1982).
268. BLUMENTHAL, K.M., and W.R. KEM: Primary Structure of *Stoichactis helianthus* Cytolysin III. J. Biol. Chem. **258**, 5574 (1983).
269. KREBS, H.C., G. HABERMEHL, and E. WACHTER: Biologic Active Substances from the Sea Anemone *Metridium senile*. Third International Symposium on Marine Natural Products, Brussels, Belgium, 1980.
270. MEBS, D., and E. GEBAUER: Isolation of Proteinase Inhibitory, Toxic and Hemolytic Polypeptides from a Sea Anemone, *Stoichactis* sp. Toxicon **18**, 97 (1980).
271. SAMEJIMA, Y., and D. MEBS: Structural Studies on a Proteinase Inhibitor from the Sea Anemone *Stoichactis* sp. Toxicon **20**, 335 (1982).
272. SCHWEITZ, H., J.-P. VINCENT, J. BARHANIN, C. FRELIN, G. LINDEN, M. HUGUES, and M. LAZDUNSKI: Purification and Pharmacological Properties of Eight Sea Anemone Toxins from *Anemonia sulcata, Anthopleura xanthogrammica, Stoichactis giganteus*, and *Actinodendron plumosum*. Biochemistry **20**, 5245 (1981).
273. ------- Purification and Pharmacological Properties of Eight Sea Anemone Toxins from *Anemonia sulcata, Anthopleura xanthogrammica, Stoichactis giganteus*, and *Actinodendron plumosum*. Toxicon **20**, 77 (1982).
274. NORTON, T.R.: Cardiotonic Polypeptides from *Anthopleura xanthogrammica* (Brandt) and *A. elegantissima* (Brandt). Fed. Proc. **40**, 21 (1981).
275. ALDEEN, S.I., R.C. ELLIOTT, and M. SHEARDOWN: The Partial Purification and Bioassay of a Toxin Present in Extracts of the Sea Anemone, *Tealia felina* (L.). Br. J. Pharmac. **72**, 211 (1981).
276. BERNHEIMER, A.W., L.S. AVIGAD, and C.Y. LAI: Purification and Properties of a Toxin from the Sea Anemone *Condylactis gigantea*. Arch. Biochem. Biophys. **214**, 840 (1982).
277. MÁCEK, P., L. SENČIČ, and D. LEBEZ: Isolation and Partial Characterization of Three Lethal and Hemolytic Toxins from the Sea Anemone *Actinia cari*. Toxicon **20**, 181 (1982).
278. BERNHEIMER, A.W., and L.S. AVIGAD: Toxins of the Sea Anemone *Epiactis prolifera*. Arch. Biochem. Biophys. **217**, 174 (1982).
279. FERLAN, I., and K.W. JACKSON: Partial Amino Acid Sequence of Equinatoxin. Toxicon Suppl. **3**, 141 (1983).
280. NABIULLIN, A.A., S.E. ODINOKOV, E.P. KOZLOVSKAYA, and G.B. ELYAKOV: Second-

ary Structure of Sea Anemone Toxins. Circular Dichroism, Infrared Spectroscopy and Chou-Fasman Calculations. Febs Lett. **141**, 124 (1982).
281. MEBS, D., M. LIEBRICH, A. REUL, and Y. SAMEJIMA: Hemolysins and Proteinase Inhibitors from Sea Anemones of the Gulf of Aqaba. Toxicon **21**, 257 (1983).
282. MEBS, D., M. LIEBRICH, and A. REUL: Biologically Active Polypeptides from Sea Anemones of the Red Sea. Toxicon Suppl. **3**, 289 (1983).
283. BÉRESS, L., and J. ZWICK: Purification of Two Crab-Paralysing Polypeptides from the Sea Anemone *Bolocera tuediae*. Mar. Chem. **8**, 333 (1980).
284. BERNHEIMER, A.W., L.S. AVIGAD, G. BRANCH, E. DOWDLE, and C.Y. LAI: Purification and Properties of a Toxin from the South African Sea Anemone, *Pseudactinia varia*. Toxicon **22**, 183 (1984).
285. BERNHEIMER, A.W., and L.S. AVIGAD: New Cytolysins in Sea Anemones from the West Coast of the United States. Toxicon **19**, 529 (1981).
286. NORTON, R.S., J. ZWICK, and L. BÉRESS: Natural-Abundance ^{13}C Nuclear-Magnetic-Resonance Study of Toxin II from *Anemonia sulcata*. Eur. J. Biochem. **113**, 75 (1980).
287. ROMEY, G., J.F. RENAUD, M. FOSSET, and M. LAZDUNSKI: Pharmacological Properties of the Interaction of a Sea Anemone Polypeptide Toxin with Cardiac Cells in Culture. J. Pharmacol. Exp. Ther. **213**, 607 (1980).
288. ELLIOTT, R.C., and M.J. SHEARDOWN: The Action of an Extract of the Sea Anemone *Tealia felina* on Neuromuscular Transmission Studied *in vitro*. Br. J. Pharmac. **69**, 293 P (1980).
289. ALSEN, C., and A. KAMPEN: Induction of Spontaneous Contractions of Left Guinea Pig Atria by ATX II (*Anemonia sulcata*). Naunyn Schmiedebergs Arch. Pharmacol. **311** (Suppl. R), R 39 (1980).
290. STENGELIN, S., and F. HUCHO: Radioactive Labelling of Toxin I from *Anemonia sulcata* and Binding to Crayfish Nerve *in vitro*. Hoppe-Seyler's Z. Physiol. Chem. **361**, 577 (1980).
291. NEUMCKE, B., W. SCHWARZ, and R. STÄMPFLI: Modification of Sodium Inactivation in Myelinated Nerve by *Anemonia* Toxin II and Iodate. Analysis of Current Fluctuations and Current Relaxations. Biochim. Biophys. Acta **600**, 456 (1980).
292. OHIZUMI, Y., and S. SHIBATA: Possible Mechanism of the Dual Action of the New Polypeptide (Anthopleurin-B) from Sea Anemone in the Isolated Ileum and Taenia Caeci of the Guinea-Pig. Br. J. Pharmac. **72**, 239 (1981).
293. KUDO, Y., and S. SHIBATA: The Potent Excitatory Effect of a Novel Polypeptide, Anthopleurin-B, Isolated from a Sea Anemone (*Anthopleura xanthogrammica*) on the Frog Spinal Cord. J. Pharmacol. Exp. Ther. **214**, 443 (1980).
294. HASHIMOTO, K., R. OCHI, K. HASHIMOTO, J. INUI, and Y. MIURA: The Ionic Mechanism of Prolongation of Action Potential Duration of Cardiac Ventricular Muscle by Anthopleurin-A and Its Relationship to the Inotropic Effect. J. Pharmacol. Exp. Ther. **215**, 479 (1980).
295. VINCENT, J.P., M. BALERNA, J. BARHANIN, M. FOSSET, and M. LAZDUNSKI: Binding of Sea Anemone Toxin to Receptor Sites Associated with Gating System of Sodium Channel in Synaptic Nerve Endings *in vitro*. Proc. natl. Acad. Sci. U.S.A. **77**, 1646 (1980).
296. BAILEY, L.E., S. SHIBATA, D.G. SERIGUCHI, and P.E. DRESEL: Inhibition of the Positive Inotropic Effect of Anthopleurin-A (AP-A) by Dantrolene. Life Sci. **26**, 1061 (1980).
297. NORTON, T.R., Y. OHIZUMI, and S. SHIBATA: Excitatory Effect of a New Polypeptide (Anthopleurin-B) from Sea Anemone on the Guinea-Pig Vas Deferens. Br. J. Pharmac. **74**, 23 (1981).
298. LEMEIGNAN, M., J. MOLGO, and F. TAZIEFF-DEPIERRE: Effects of Sea *Anemonia sulcata* Toxin II on Normal and Chronically Denervated Mammalian Neuromuscular Junctions. Br. J. Pharmac. **72**, 546 P (1981).

299. ALSEN, C., J.B. HARRIS, and I. TESSERAUX: Mechanical and Electrophysiological Effects of Sea Anemone (*Anemonia sulcata*) Toxins on Rat Innervated and Denervated Skeletal Muscle. Br. J. Pharmac. **74**, 61 (1981).
300. STENGELIN, S., W. RATHMAYER, G. WUNDERER, L. BÉRESS, and F. HUCHO: Radioactive Labeling of Toxin II from *Anemonia sulcata*. Analytical Biochemistry **113**, 277 (1981).
301. NORTON, R.S., T.R. NORTON, R.W. SLEIGH, and D.G. BISHOP: Interaction of the Polypeptide Cardiac Stimulant Anthopleurin-A with H^+, Ca^{2+}, and Membrane Lipids. Arch. Biochem. Biophys. **213**, 87 (1982).
302. HUCHO, F., and L. LAUFFER: Interaction of Anemone Toxin I with Crayfish Nerve Membranes. Toxicon **20**, 70 (1982).
303. LAUFER, J., and M. PELHATE: Action of Anthopleurin-A on the Isolated Cockroach Axon. Toxicon **20**, 72 (1982).
304. SCHMIDTMAYER, J., M. STOYE-HERZOG, and W. ULBRICHT: Rate of Action of *Anemonia sulcata* Toxin II on Sodium Channels in Myelinated Nerve Fibres. Pflügers Arch. **394**, 313 (1982).
305. ALSEN, C., T. PETERS, and E. SCHEUFLER: Studies on the Mechanism of the Positive Inotropic Effect of ATX II (*Anemonia sulcata*) on the Isolated Guinea Pig Atria. J. Cardiovasc. Pharmacol. **4**, 63 (1982).
306. ALSEN, C.: Biological Significance of Peptides from *Anemonia sulcata*. Fed. Proc. **42**, 101 (1983).
307. FUJITA, S., A. WARASHINA, and M. SATAKE: Binding Characteristics of a Sea Anemone Toxin from *Parasicyonis actinostoloides* with Crayfish Leg Nerves. Comp. Biochem. Physiol. **76C**, 25 (1983).
308. BÉRESS, L., R. RITTER, and U. RAVENS: The Influence of the Rate of Electrical Stimulation on the Effects of the *Anemonia sulcata* Toxin ATX II in Guinea Pig Papillary Muscle. Eur. J. Pharmac. **79**, 265 (1982).
309. HARTUNG, K., and W. RATHMAYER: Effects of Three Anemone Toxins on Sodium Currents in a Crayfish Neurone. Naunyn-Schmiedebergs Arch. Pharmacol. **322**, R 66 (1983).
310. MIYAKE, M., and S. SHIBATA: A Novel Mode of Neurotoxin Action. A Polypeptide Toxin Isolated from *Anemonia sulcata* Shifts the Voltage Dependence of the Maximal Rate of Rise of Na^+ Action Potentials in a Mouse Neuronal Clone. Mol. Pharmac. **20**, 453 (1981).
311. RAVENS, U., and E. SCHÖLLHORN: The Effects of a Toxin (ATX II) from the Sea Anemone *Anemonia sulcata* on the Electrical and Mechanical Activity of Denervated Hemidiaphragm of the Rat. Toxicon **21**, 131 (1983).
312. ULBRICHT, W., and J. SCHMIDTMAYER: Modification of Sodium Channels in Myelinated Nerve by *Anemonia sulcata* Toxin II. J. Physiol., Paris **77**, 1103 (1981).
313. WARASHINA, A., and S. FUJITA: Effect of Sea Anemone Toxin on the Sodium Inactivation Process in Crayfish Axons. J. Gen. Physiol. **81**, 305 (1983).
314. LAFRANCONI, W.M., I. FERLAN, F.E. RUSSELL and R.J. HUXTABLE: The Action of Equinatoxin, a Peptide from the Venom of the Sea Anemone, *Actinia equina*, on the Isolated Lung. Toxicon **22**, 347 (1984).
315. ERXLEBEN, C., and W. RATHMAYER: Effects of the Sea Anemone *Anemonia sulcata* Toxin II on Skeletal Muscle and on Neuromuscular Transmission. Toxicon **22**, 387 (1984).
316. VARANDA, W., and A. FINKELSTEIN: Ion and Nonelectrolyte Permeability Properties at Channels Formed in Planar Lipid Bilayer Membranes by the Cytolytic Toxin from the Sea Anemone *Stoichactis helianthus*. J. Membrane Biol. **55**, 203 (1980).
317. SCHMIDTMAYER, J., M. STOYE-HERZOG, and W. ULBRICHT: Kinetic and Equilibrium Effects of Different Concentrations of *Anemonia sulcata* Toxin II (ATX). Pflügers Arch. ges. Physiol. **293**, R 32 (1982).

318. RACK, M., H. MEVES, L. BÉRESS, and H.H. GRÜNHAGEN: Preparation and Properties of Fluorescence Labeled Neuro- and Cardiotoxin II from the Sea Anemone (*Anemonia sulcata*). Toxicon **21**, 231 (1983).
319. MAČEK, P., and D. LEBEZ: Kinetics of Hemolysis Induced by Equinatoxin, a Cytolytic Toxin from the Sea Anemone *Actinia equina*. Effect of Some Ions and pH. Toxicon **19**, 233 (1981).
320. BARHANIN, J., M. HUGUES, H. SCHWEITZ, J.-P. VINCENT, and M. LAZDUNSKI: Structure-Function Relationships of Sea Anemone Toxin II from *Anemonia sulcata*. J. Biol. Chem. **256**, 5764 (1981).
321. – – – – – Structure-Function Relationships of Sea Anemone Toxin II from *Anemonia sulcata*. Toxicon **20**, 59 (1982).
322. SEDMAK, B., J. ŠKRK, V. KODELJA, and D. LEBEZ: The Influence of Cytolysin from the Sea Anemone *Condylactis aurantiaca* on the Survival of Mice Bearing Ehrlich Ascites Tumor. Toxicon **20**, 207 (1982).
323. BURCHAM, J.M., G.E. DEARLOVE, and S.H. BISHOP: Phosphonomonoesterase Activity in *Metridium senile* L. Comp. Biochem. Physiol. **67B**, 147 (1980).
323a. FINDLAY, J.A., and A.D. PATIL: A Novel Sterol Peroxide from the Sea Anemone *Metridium senile*. Steroids **44**, 261 (1984).
324. MILKOVA, T.S., S.S. POPOV, N.L. MAREKOV, and S.N. ANDREEV: Sterols from Black Sea Invertebrates – I. Sterols from Scyphozoa and Anthozoa (Coelenterata). Comp. Biochem. Physiol. **67B**, 633 (1980).
325. MILKOVA, T., S. POPOV, N. MAREKOV, I. STOILOV, S. ANDREEV, and G. KOVACHEV: Sterols from Black Sea Coelenterata and Mollusca. Bull. Soc. Chim. Belg. **89**, 1081 (1980).
326. MOORE, R.E.: The Structure of Palytoxin. Fortschr. Chem. Organ. Naturstoffe (W. HERZ, H. GRISEBACH, G.W. KIRBY, and CH. TAMM, eds.), Vol. **48**, p. 81. Wien-New York: Springer 1985.
326a. BÉRESS, L., J. ZWICK, H.J. KOLKENBROCK, P.N. KAUL, and O. WASSERMANN: A Method for the Isolation of the Caribbean Palytoxin (C-PTX) from the Coelenterate (Zooanthid) *Palythoa caribaeorum*. Toxicon **21**, 285 (1983).
327. MACFARLANE, R.D., D. UEMURA, K. UEDA, and Y. HIRATA : ^{252}Cf Plasma Desorption Mass Spectrometry of Palytoxin. J. Am. Chem. Soc. **102**, 875 (1980).
328. UEMURA, D., K. UEDA, Y. HIRATA, C. KATAYAMA, and J. TANAKA: Structural Studies on Palytoxin, a Potent Coelenterate Toxin. Tetrahedron Lett. **21**, 4857 (1980).
329. – – – – – Structures of Two Oxidation Products Obtained from Palytoxin. Tetrahedron Lett. **21**, 4861 (1980).
330. MOORE, R.E., F.X. WOOLARD, and G. BARTOLINI: Periodate Oxidation of N-(p-Bromobenzoyl)palytoxin. J. Am. Chem. Soc. **102**, 7370 (1980).
331. MOORE, R.E., and G. BARTOLINI: Structure of Palytoxin. J. Am. Chem. Soc. **103**, 2491 (1981).
332. UEMURA, D., K. UEDA, Y. HIRATA, H. NAOKI, and T. IWASHITA: Further Studies on Palytoxin. I. Tetrahedron Lett. **22**, 1909 (1981).
333. KLEIN, L.L., W.W. MCWHORTER, JR., S.S. KO, K.-P. PFAFF, Y. KISHI, D. UEMURA, and Y. HIRATA: Stereochemistry of Palytoxin. 1. C85–C115 Segment. J. Am. Chem. Soc. **104**, 7362 (1982).
334. KO, S.S., J.M. FINAN, M. YONAGA, Y. KISHI, D. UEMURA, and Y. HIRATA: Stereochemistry of Palytoxin. 2. C1–C6, C47–C74, and C77–C83 Segments. J. Am. Chem. Soc. **104**, 7364 (1982).
335. FUJIOKA, H., W.J. CHRIST, J.K. CHA, J. LEDER, Y. KISHI, D. UEMURA, and Y. HIRATA: Stereochemistry of Palytoxin. 3. C7–C51 Segment. J. Am. Chem. Soc. **104**, 7367 (1982).

336. UEMURA, D., K. UEDA, Y. HIRATA, H. NAOKI, and T. IWASHITA: Further Studies on Palytoxin, II. Structure of Palytoxin. Tetrahedron Lett. **22**, 2781 (1981).
337. MOORE, R.E., G. BARTOLINI, J. BARCHI, A.A. BOTHNER-BY, J. DADOK, and J. FORD: Absolute Stereochemistry of Palytoxin. J. Am. Chem. Soc. **104**, 3776 (1982).
338. CHA, J.K., W.J. CHRIST, J.M. FINAN, H. FUJIOKA, Y. KISHI, L.L. KLEIN, S.S. KO, J. LEDER, W.W. MCWHORTER, JR., K.-P. PFAFF, M. YONAGA, D. UEMURA, and Y. HIRATA: Stereochemistry of Palytoxin. 4. Complete Structure. J. Am. Chem. Soc. **104**, 7369 (1982).
339. UEMURA, D., Y. HIRATA, T. IWASHITA, and H. NAOKI: Studies on Palytoxins. Tetrahedron **41**, 1007 (1985).
340. STILL, W.C., and I. GALYNKER: Stereospecific Synthesis of the C30–C43 Segment of Palytoxin by Macrocyclically Controlled Remote Asymmetric Induction. J. Am. Chem. Soc. **104**, 1774 (1982).
341. KO, S.S., L.L. KLEIN, K.-P. PFAFF, and Y. KISHI: Synthetic Studies on Palytoxin. Stereocontrolled, Practical Synthesis of the C.101–C.115 Segment. Tetrahedron Letters **23**, 4415 (1982).
341a. LEDER, J., H. FUJIOKA, and Y. KISHI: Synthetic Studies on Palytoxin. Stereocontrolled Practical Synthesis of the C.23–C.37 Segment. Tetrahedron Lett. **24**, 1463 (1983).
342. KUDO, Y., and S. SHIBATA: The Potent Depolarizing Action of Palytoxin Isolated from *Palythoa tubercurosa* on the Isolated Spinal Cord of the Frog. Br. J. Pharmac. **71**, 575 (1980).
343. OHIZUMI, Y., and S. SHIBATA: Mechanism of the Excitatory Action of Palytoxin and N-Acetylpalytoxin in the Isolated Guinea-Pig Vas Deferens. J. Pharmacol. exp. Ther. **214**, 209 (1980).
344. HABERMANN, E., G.S. CHHATWAL, and H.J. HESSLER: Palytoxin Raises the Nonspecific Permeability of Erythrocytes in an Ouabain-Sensitive Manner. Naunyn-Schmiedeberg's Arch. Pharmacol. **317**, P 374 (1981).
345. HABERMANN, E., G. AHNERT-HILGER, G.S. CHHATWAL, and L. BÉRESS: Delayed Haemolytic Action of Palytoxin. General Characteristics. Biochim. Biophys. Acta **649**, 481 (1981).
346. HABERMANN, E., and G.S. CHHATWAL: Ouabain Inhibits the Increase Due to Palytoxin of Cation Permeability of Erythrocytes. Naunyn-Schmiedeberg's Arch. Pharmacol. **319**, 101 (1982).
347. CHHATWAL, G.S., G. AHNERT-HILGER, L. BÉRESS, and E. HABERMANN: Palytoxin Induces As Well As Inhibits the Release of Histamine from Rat Mast Cells. Int. Archs. Allergy appl. Immun. **68**, 97 (1982).
348. ITO, K., N. URAKAWA, and H. KOIKE: Cardiovascular Toxicity of Palytoxin in Anesthetized Dogs. Arch. int. Pharmacodyn. **258**, 146 (1982).
349. PICHON, Y.: Effects of Palytoxin on Sodium and Potassium Permeabilities in Unmyelinated Axons. Toxicon **20**, 41 (1982).
350. CHHATWAL, G.S., G. AHNERT-HILGER, L. BÉRESS, and E. HABERMANN: Palytoxin: Its Action on Erythrocytes and Rat Mast Cells. Toxicon **20**, 62 (1982).
351. DRAIJER, F., P. SUVANTO, I. TESSERAUX, and L. BÉRESS: Mechanical Studies About the Action of Palytoxin (*Palythoa caribaeorum*) on Innervated and Denervated Rat Diaphragms. Toxicon **20**, 65 (1982).
352. ALSEN, C., G. AGENA, and L. BÉRESS: The Action of Palytoxin (*Palythoa caribaeorum*) on Isolated Atria of Guinea Pig Hearts. Toxicon **20**, 57 (1982).
353. ISHIDA, Y., K. TAKAGI, M. TAKAHASHI, N. SATAKE, and S. SHIBATA: Palytoxin Isolated from Marine Coelenterates. The Inhibitory Action on (Na,K)-ATPase. J. Biol. Chem. **258**, 7900 (1983).
354. PETTIT, G.R., Y. FUJII, J.A. HASLER, J.M. SCHMIDT, and C. MICHEL: Antineoplastic

Agents. 78. Isolation of Palystatins 1-3 from the Indian Ocean *Palythoa liscia.* J. Nat. Prod. **45**, 263 (1982).
355. PETTIT, G.R., Y. FUJII, J.A. HASLER, and J.M. SCHMIDT: Isolation and Characterization of Palystatins A–D. J. Nat. Prod. **45**, 272 (1982).
356. PETTIT, G.R., and Y. FUJII: Antineoplastic Agents. 81. The Glycerol Ethers of *Palythoa liscia.* J. Nat. Prod. **45**, 640 (1982).
357. KELECOM, A., and A.M. SOLÉ-CAVA: Comparative Study of Zoanthid Sterols, the Genus *Palythoa* (Hexacorallia, Zoanthidea). Comp. Biochem. Physiol. **72B**, 677 (1982).
358. KELECOM, A.: Studies of Brazilian Marine Invertebrates. VIII. Zoanthosterol, a New Sterol from the Zoanthid *Zoanthus sociatus* (Hexacorallia, Zoanthidae). Bull. Soc. Chim. Belg. **90**, 971 (1981).
359. STURARO, A., A. GUERRIERO, R. DE CLAUSER, and F. PIETRA: A New, Unexpected Marine Source of a Molting Hormone. Isolation of Ecdysterone in Large Amounts from the Zoanthid *Gerardia savaglia.* Experientia **38**, 1184 (1982).
360. KOMODA, Y., M. SHIMIZU, S. KANEKO, M. YAMAMOTO, and M. ISHIKAWA: Chemistry of Paragracine, a Biologically Active Marine Base from *Parazoanthus gracilis* (Lwowsky). Chem. Pharm. Bull. (Japan) **30**, 502 (1982).
361. RAO, C.B., A.S.R. ANJANEYULU, N.S. SARMA, Y. VENKATESWARLU, R.M. ROSSER, D.J. FAULKNER, M.H.M. CHEN, and J. CLARDY: Zoanthamine: A Novel Alkaloid from a Marine Zoanthid. J. Am. Chem. Soc. **106**, 7983 (1984).
361a. RAO, C.B., A.S.R. ANJANEYULU, N.S. SARMA, Y. VENKATESWARLU, R.M. ROSSER, and D.J. FAULKNER: Alkaloids from a Marine Zoanthid. J. Org. Chem. **50**, 3757 (1985)
362. KOKKE, W.C.M.C., W. FENICAL, L. BOHLIN, and C. DJERASSI: Sterol Synthesis by Cultured Zooxanthellae; Implications Concerning Sterol Metabolism in the Host-Symbiont Association in Caribbean Gorgonians. Comp. Biochem. Physiol. **68B**, 281 (1981).
363. KELECOM, A., A.M. SOLÉ CAVA, and G.J. KANNENGIESSER: Occurrence of 23,24ξ-Dimethylcholesta-5,22-dien-3β-ol in the Brazilian Gorgonian *Phyllogorgia dilatata* (Octocorallia, Gorgonacea) and in its Associated Zooxanthella. Bull. Soc. Chim. Belg. **89**, 1013 (1980).
364. KOKKE, W.C.M.C., L. BOHLIN, W. FENICAL, and C. DJERASSI: Novel Dinoflagellate 4α-Methylated Sterols from Four Caribbean Gorgonians. Phytochemistry **21**, 881 (1982).
365. KOBAYASHI, M., T. ISHIZAKA, and H. MITSUHASHI: Marine Sterols X. Minor Constituents of the Sterols of the Soft Coral *Sarcophyton glaucum.* Steroids **40**, 209 (1982).
366. BONINI, C., R.B. KINNEL, M. LI, P.J. SCHEUER, and C. DJERASSI: Minor and Trace Sterols in Marine Invertebrates 38: Isolation, Structure Elucidation and Partial Synthesis of Papakusterol, a New Biosynthetically Unusual Marine Sterol with a Cyclopropyl-Containing Side Chain. Tetrahedron Lett. **24**, 277 (1983).
367. KOBAYASHI, M., and H. MITSUHASHI: Marine Sterols. XII. Glaucasterol, a Novel C_{27} Sterol with a Unique Side Chain, from the Soft Coral *Sarcophyton glaucum.* Steroids **40**, 665 (1982).
368. KOBAYASHI, M., T. ISHIZAKA, and H. MITSUHASHI: Isolation of 5α,6-Dihydroglaucasterol, a New Marine C_{27} Sterol with a 24,26-Cyclized Side Chain, from the Soft Coral *Sarcophyton glaucum.* Chem. Pharm. Bull. (Japan) **31**, 1803 (1983).
369. POPOV, S., R.M.K. CARLSON, and C. DJERASSI: Occurrence and Seasonal Variation of 19-*Nor*cholest-4-en-3-one and 3β-Monohydroxy Sterols in the Californian Gorgonian, *Muricea californica.* Steroids **41**, 537 (1983).
370. KASHMAN, Y., and S. CARMELY: Four Novel C_{28} Sterols from *Lobophytum depressum.* Tetrahedron Lett. **21**, 4939 (1980).

371. CARMELY, S., and Y. KASHMAN: Isolation and Structure Elucidation of Lobophytosterol, Depresosterol and Three Other Closely Related Sterols. Tetrahedron 37, 2397 (1981).
372. YAMADA, Y., S. SUZUKI, K. IGUCHI, H. KIKUCHI, Y. TSUKITANI, H. HORIAI, and H. NAKANISHI: Studies on Marine Natural Products. II. New Polyhydroxylated Sterols from the Soft Coral *Lobophytum pauciflorum* (Ehrenberg). Chem. Pharm. Bull. (Japan) 28, 473 (1980).
373. KOBAYASHI, M., T. HAYASHI, K. HAYASHI, M. TANABE, T. NAKAGAWA, and H. MITSUHASHI: Marine Sterols. XI. Polyhydroxysterols of the Soft Coral *Sarcophyton glaucum*: Isolation and Synthesis of 5α-Cholestane-1β,3β,5,6β-tetrol. Chem. Pharm. Bull. (Japan) 31, 1848 (1983).
374. KOBAYASHI, M., and H. MITSUHASHI: Marine Sterols. XIV. Isolation of (24S)-24-Methyl-5α-cholestane-3β,5,6β,25ξ,26-pentol from the Soft Coral *Sarcophyton glaucum*. Chem. Pharm. Bull. (Japan) 31, 4127 (1983).
375. KSEBATI, M.B., and F.J. SCHMITZ: 24ξ-Methyl-5α-cholestane-3β,5,6β,22R,24-pentol 6-Acetate: New Polyhydroxylated Sterol from the Soft Coral *Asterospicularia randalli*. Steroids 43, 639 (1984).
376. KOBAYASHI, M., and H. MITSUHASHI: Marine Sterols XIII. Isolation and Synthesis of 1β,3β,5,6β-Tetrahydroxy-5α-androstan-17-one from the Soft Coral *Sarcophyton glaucum*. Steroids 40, 673 (1982).
377. JAGODZINSKA, B.M., J.S. TRIMMER, W. FENICAL, and C. DJERASSI: Sterols in Marine Invertebrates. 49. Isolation and Structure Elucidation of Eight New Polyhydroxylated Sterols from the Soft Coral *Sinularia dissecta*. J. Org. Chem. 50, 1435 (1985).
378. KOBAYASHI, M., N.K. LEE, B.W. SON, K. YANAGI, Y. KYOGOKU, and I. KITAGAWA: Stoloniferone-A, -B, -C, and -D, Four New Cytotoxic Steroids from the Okinawan Soft Coral *Clavularia viridis*. Tetrahedron Lett. 25, 5925 (1984).
379. CIMINO, G., S. DE ROSA, S. DE STEFANO, G. SCOGNAMIGLIO, and G. SODANO: Cholest-4,14-dien-15,20ξ-diol-3,16-dione, a Novel Polyoxygenated Marine Steroid which Easily Loses the Side Chain. Tetrahedron Lett. 22, 3013 (1981).
380. BENVEGNÙ, R., G. CIMINO, S. DE ROSA, and S. DE STEFANO: Guggulsterol-Like Steroids from the Mediterranean Gorgonian *Leptogorgia sarmentosa*. Experientia 38, 1443 (1982).
381. KINGSTON, J.F., and A.G. FALLIS: Marine Natural Products: Highly Functionalized Steroids (12β-Hydroxy-24-*nor*cholesta-1,4,22-trien-3-one and 12β-Acetoxy-24-*nor*cholesta-1,4,22-trien-3-one) from the Sea Raspberry, *Gersemia rubiformis*. Can. J. Chem. 60, 820 (1982).
382. SJÖSTRAND, U., L. BOHLIN, L. FISHER, M. COLIN, and C. DJERASSI: Minor and Trace Sterols from Marine Invertebrates. 28. A Novel Polyhydroxylated Sterol from the Soft Coral *Anthelia glauca*. Steroids 38, 347 (1981).
383. TANAKA, J., T. HIGA, K. TACHIBANA, and T. IWASHITA: Gorgost-5-ene-3β,7α,11α,12β-tetraol 12-Monoacetate, a New Marine Sterol from the Gorgonian *Isis hippuris*. Chem. Lett. 1982, 1295.
384. HIGA, T., J. TANAKA, and K. TACHIBANA: 18-Oxygenated Polyfunctional Steroids from the Gorgonian *Isis hippuris*. Tetrahedron Lett. 22, 2777 (1981).
385. HIGA, T., J. TANAKA, Y. TSUKITANI, and H. KIKUCHI: Hippuristanols, Cytotoxic Polyoxygenated Steroids from the Gorgonian *Isis hippuris*. Chem. Lett. 1981, 1647.
386. STONARD, R.J., J.C. PETROVICH, and R.J. ANDERSEN: New C_{26} Sterol Peroxide from the Opisthobranch Mollusk *Adalaria* sp. and the Sea Pen *Virgularia* sp. Steroids 36, 81 (1980).
387. KAZLAUSKAS, R., P.T. MURPHY, B.N. RAVI, R.L. SANDERS, and R.J. WELLS: Spermidine Derivatives and 9,11-Secosteroids from a Soft Coral (*Sinularia* sp.). Aust. J. Chem. 35, 69 (1982).

388. BONINI, C., C.B. COOPER, R. KAZLAUSKAS, R.J. WELLS, and C. DJERASSI: Minor and Trace Sterols in Marine Invertebrates. 41. Structure and Stereochemistry of Naturally Occurring 9,11-Seco Sterols. J. Org. Chem. **48**, 2108 (1983).
389. KOBAYASHI, M., Y. KIYOTA, S. ORITO, Y. KYOGOKU, and I. KITAGAWA: Five New Steroidal Glycosides, Pregnedioside-A, -B, and their Three Monoacetates, from an Okinawan Soft Coral of *Alcyonium* sp. Tetrahedron Lett. **25**, 3731 (1984).
390. BANDURRAGA, M.M., and W. FENICAL: Isolation of the Muricins. Evidence of a Chemical Adaptation Against Fouling in the Marine Octocoral *Muricea fruticosa* (Gorgonacea). Tetrahedron **41**, 1057 (1985).
391. COLL, J.C., and P.W. SAMMARCO: Terpenoid Toxins of Soft Corals (Cnidaria, Octocorallia): Their Nature, Toxicity, and Ecological Significance. Toxicon Suppl. **3**, 69 (1983).
392. COLL, J.C., B.F. BOWDEN, D.M. TAPIOLAS, R.H. WILLIS, P. DJURA, M. STREAMER, and L. TROTT: Studies of Australian Soft Corals – XXXV. The Terpenoid Chemistry of Soft Corals and Its Implications. Tetrahedron **41**, 1085 (1985).
393. WEBB, L., and J.C. COLL: Effects of Alcyonarian Coral Terpenes on Scleractinian Coral Photosynthesis and Respiration. Toxicon Suppl. **3**, 485 (1983).
394. BOWDEN, B.F., J.C. COLL, and D.M. TAPIOLAS: Studies of Australian Soft Corals. XXX. A Novel Tris*nor*sesquiterpene from a *Cespitularia* Species and the Isolation of Guaiazulene from a Small Blue *Alcyonium* Species. Aust. J. Chem. **36**, 211 (1983).
395. KOBAYASHI, M., B.W. SON, M. KIDO, Y. KYOGOKU, and I. KITAGAWA: Clavukerin A, a New Tri*nor*-guaiane Sesquiterpene from the Okinawan Soft Coral *Clavularia koellikeri*. Chem. Pharm. Bull. (Japan) **31**, 2160 (1983).
396. KOBAYASHI, M., B.W. SON, Y KYOGOKU, and I. KITAGAWA: Clavukerin C, a New Tri*nor*-guaiane Sesquiterpene Having a Hydroperoxy Function, from the Okinawan Soft Coral *Clavularia koellikeri*. Chem. Pharm. Bull. (Japan) **32**, 1667 (1984).
397. IZAC, R.R., W. FENICAL, and J.M. WRIGHT: Inflatene, an Ichthyotoxic C_{12} Hydrocarbon from the Stoloniferan Soft Coral *Clavularia inflata* var. *luzoniana*. Tetrahedron Lett. **25**, 1325 (1984).
398. KASHMAN, Y., M. BODNER, J.S. FINER-MOORE, and J. CLARDY: $\Delta^{9(15)}$-Africanene, a New Sesquiterpene Hydrocarbon from the Soft Coral *Sinularia erecta*. Experientia **36**, 891 (1980).
399. BRAEKMAN, J.C., D. DALOZE, B. TURSCH, S.E. HULL, J.P. DECLERCQ, G. GERMAIN, and M. VAN MEERSSCHE: Chemical Studies of Marine Invertebrates. XXXVIII. $\Delta^{9(15)}$-Africanene, a New Sesquiterpene Hydrocarbon from *Sinularia polydactyla* (Coelenterata, Octocorallia, Alcyonaceae). Experientia **36**, 893 (1980).
400. BOWDEN, B.F., J.C. COLL, and S.J. MITCHELL: Studies of Australian Soft Corals. XXI. A New Sesquiterpene from *Nephthea chabrolii* and an Investigation of the Common Clam *Tridacna maxima*. Aust. J. Chem. **33**, 1833 (1980).
401. FUSETANI, N., S. MATSUNAGA, and S. KONOSU: Bioactive Marine Metabolites. I. Isolation of Guaiazulene from the Gorgonian *Euplexaura erecta*. Experientia **37**, 680 (1981).
402. LI, M.K.W., and P.J. SCHEUER: Halogenated Blue Pigments of a Deep Sea Gorgonian. Tetrahedron Lett. **25**, 587 (1984).
403. – – N,N-Dimethylamino-3-guaiazulenylmethane from a Deep Sea Gorgonian. Tetrahedron Lett. **25**, 4707 (1984).
404. IMRE, S., R.H. THOMSON, and B. YALHI: Linderazulene, a New Naturally Occurring Pigment from the Gorgonian *Paramuricea chamaeleon*. Experientia **37**, 442 (1981).
405. LI, M.K.W., and P.J. SCHEUER: A Guaianolide Pigment from a Deep Sea Gorgonian. Tetrahedron Lett. **25**, 2109 (1984).
406. ALPERTUNGA, B., S. IMRE, H.J. COWE, P.J. COX, and R.H. THOMSON: A Photo Artefact from Linderazulene. Tetrahedron Lett. **24**, 4461 (1983).
407. GOPICHAND, Y., F.J. SCHMITZ, and P.G. SCHMIDT: Marine Natural Products: Two

New Acyclic Sesquiterpene Hydrocarbons from the Gorgonian *Plexaurella grisea*. J. Org. Chem. **45**, 2523 (1980).
408. BOWDEN, B.F., J.C. COLL, E.D. DE SILVA, M.S.L. DE COSTA, P.J. DJURA, M. MAHENDRAN, and D.M. TAPIOLAS: Studies of Australian Soft Corals. XXXI. Novel Furanosesquiterpenes from Several Sinularian Soft Corals (Coelenterata, Octocorallia, Alcyonacea). Aust. J. Chem. **36**, 371 (1983).
409. CIMINO, G., S. DE ROSA, S. DE STEFANO, and G. SODANO: A New Furanosesquiterpene from the Mediterranean Alcyonacean *Alcyonum palmatum*. J. Nat. Prod. **47**, 877 (1984).
410. BOWDEN, B.F., J.C. BRAEKMAN, J.C. COLL, and S.J. MITCHELL: Studies of Australian Soft Corals. XX. A New Sesquiterpene Furan from Soft Corals of the Family Xeniidae and an Examination of *Clavularia inflata* from North Queensland Waters. Aust. J. Chem. **33**, 927 (1980).
411. IZAC, R.R., M.M. BANDURRAGA, J.M. WASYLYK, F.W. DUNN, and W. FENICAL: Germacrene Derivatives from Diverse Marine Soft-Corals (Octocorallia). Tetrahedron **38**, 301 (1982).
412. BOWDEN, B.F., J.C. COLL, S.J. MITCHELL, J.L.E. NEMORIN, and S. STERNHELL: Studies of Australian Soft Corals. XXIII. The Co-occurrence of Bicyclogermacrene and Lemnacarnol Derivatives in *Parerythropodium fulvum*. Tetrahedron Lett. **21**, 3105 (1980).
413. KOBAYASHI, M., T. YASUZAWA, Y. KYOGOKU, M. KIDO, and I. KITAGAWA: Three New *ent*-Valerenane Sesquiterpenes from an Okinawan Soft Coral. Chem. Pharm. Bull. (Japan) **30**, 3431 (1982).
414. IZAC, R.R., S.E. POET, W. FENICAL, D. VAN ENGEN, and J. CLARDY: The Structure of Pacifigorgiol, an Ichthyotoxic Sesquiterpenoid from the Pacific Gorgonian Coral *Pacifigorgia* cf. *adamsii*. Tetrahedron Lett. **23**, 3743 (1982).
415. MARTIN, M., and J. CLARDY: The Synthesis of Pacifigorgiol. Pure & Appl. Chem. **54**, 1915 (1982).
416. BOWDEN, B.F., J.C. COLL, and S.J. MITCHELL: Studies of Australian Soft Corals. XVI. Two New Sesquiterpenes from *Lemnalia humesi*. Aust. J. Chem. **33**, 681 (1980).
417. DO, M.N., and K.L. ERICKSON: An Aristolane Sesquiterpenoid from the Sea Pen *Scytalium splendens*. J. Org. Chem. **48**, 4410 (1983).
418. BOWDEN, B.F., J.C. COLL, and S.J. MITCHELL: Studies of Australian Soft Corals. XIX. Two New Sesquiterpenes with the Nardosinane Skeleton from a *Paralemnalia* Species. Aust. J. Chem. **33**, 885 (1980).
419. BOWDEN, B.F., J.C. COLL, S.J. MITCHELL, B.W. SKELTON, and A.H. WHITE: Studies of Australian Soft Corals. XXII. The Structures of Two Novel Sesquiterpenes and a *Nor* Sesquiterpene from *Lemnalia africana*, Confirmed by a Single-Crystal X-Ray Study. Aust. J. Chem. **33**, 2737 (1980).
420. IZAC, R.R., P. SCHNEIDER, M. SWAIN, and W. FENICAL: New *Nor*sesquiterpenoids of Apparent Nardosinane Origin from the Pacific Soft-Coral *Paralemnalia thyrsoides*. Tetrahedron Lett. **23**, 817 (1982).
421. IZAC, R.R., W. FENICAL, B. TAGLE, and J. CLARDY: Neolemnane and Eremophilane Sesquiterpenoids from the Pacific Soft Coral *Lemnalia africana*. Tetrahedron **37**, 2569 (1981).
422. BRAEKMAN, J.C., D. DALOZE, A. DUPONT, B. TURSCH, J.P. DECLERCQ, G. GERMAIN, and M. VAN MEERSSCHE: Chemical Studies of Marine Invertebrates-XLIII. Novel Sesquiterpenes from *Clavularia inflata* and *Clavularia koellikeri* (Coelenterata, Octocorallia, Stolonifera). Tetrahedron **37**, 179 (1981).
423. KIKUCHI, H., Y. TSUKITANI, Y. YAMADA, K. IGUCHI, S.A. DREXLER, and J. CLARDY: Lemnalol, a New Sesquiterpenoid from the Soft Coral *Lemnalia tenuis* Verseveldt. Tetrahedron Lett. **23**, 1063 (1982).
424. KIKUCHI, H., T. MANDA, K. KOBAYASHI, Y. YAMADA, and K. IGUCHI: Anti-Tumor

Activity of Lemnalol Isolated from the Soft Coral *Lemnalia tenuis* Verseveldt. Chem. Pharm. Bull. (Japan) **31**, 1086 (1983).
425. KAISIN, M., J.C. BRAEKMAN, D. DALOZE, and B. TURSCH: Novel Acetoxycapnellenes from the Alcyonacean *Capnella imbricata*. Tetrahedron **41**, 1067 (1985).
425a. SHEIKH, Y.M., G. SINGY, M. KAISIN, H. EGGERT, C. DJERASSI, B. TURSCH, D. DALOZE, and J.C. BRAEKMAN: Terpenoids – LXXI. Chemical Studies of Marine Invertebrates – XIV. Four Representatives of a Novel Sesquiterpene Class – The Capnellane Skeleton. Tetrahedron **32**, 1171 (1976).
426. GROWEISS, A., W. FENICAL, H. CUN-HENG, J. CLARDY, W. ZHONGDE, Y. ZHONGNIAN, and L. KANGHOU: Subergorgic Acid, a Novel Tricyclopentanoid Cardiotoxin from the Pacific Gorgonian Coral *Subergorgia suberosa*. Tetrahedron Lett. **26**, 2379 (1985).
427. FENICAL, W., R.K. OKUDA, M.M. BANDURRAGA, P. CULVER, and R.S. JACOBS: Lophotoxin: A Novel Neuromuscular Toxin from Pacific Sea Whips of the Genus *Lophogorgia*. Science **212**, 1512 (1981).
428. BANDURRAGA, M.M., B. MCKITTRICK, W. FENICAL, E. ARNOLD, and J. CLARDY: Diketone Cembrenolides from the Pacific Gorgonian *Lophogorgia alba*. Tetrahedron **38**, 305 (1982).
429. CULVER, P., and R.S. JACOBS: Lophotoxin: A Neuromuscular Acting Toxin from the Sea Whip (*Lophogorgia rigida*). Toxicon **19**, 825 (1981).
430. KAZLAUSKAS, R., J.A. BAIRD-LAMBERT, P.T. MURPHY, and R.J. WELLS: Two New Cembrane Diterpenes from a Soft Coral (*Sarcophyton* Species). Aust. J. Chem. **35**, 61 (1982).
431. KOBAYASHI, J., Y. OHIZUMI, H. NAKAMURA, T. YAMAKADO, T. MATSUZAKI, and Y. HIRATA: Ca-antagonistic Substance from Soft Coral of the Genus *Sarcophyton*. Experientia **39**, 67 (1983).
432. BOWDEN, B.F., J.C. COLL, and S.J. MITCHELL: Studies of Australian Soft Corals. XVIII. Further Cembranoid Diterpenes from Soft Corals of the Genus *Sarcophyton*. Aust. J. Chem. **33**, 879 (1980).
433. UCHIO, Y., J. TOYOTA, H. NOZAKI, M. NAKAYAMA, Y. NISHIZONO, and T. HASE: Lobohedleolide and (7Z)-Lobohedleolide, New Cembranolides from the Soft Coral *Lobophytum hedleyi* Whitelegge. Tetrahedron Lett. **22**, 4089 (1981).
434. NORTON, R.S., and R. KAZLAUSKAS: ^{13}C NMR-Study of Flexibilide, an Anti-Inflammatory Agent from a Soft Coral. Experientia **36**, 276 (1980).
435. MORI, K., S. SUZUKI, K. IGUCHI, and Y. YAMADA: 8,11-Epoxy Bridged Cembranolide Diterpene from the Soft Coral *Sinularia flexibilis*. Chem. Lett. **1983**, 1515.
435a. UCHIO, Y., S. EGUCHI, J. KURAMOTO, M. NAKAYAMA, and T. HASE: Denticulatolide, an Ichthyotoxic Peroxide-Containing Cembranolide from the Soft Coral *Lobophytum denticulatum*. Tetrahedron Lett. **26**, 4487 (1985).
436. KASHMAN, Y., S. CARMELY, and A. GROWEISS: Further Cembranoid Derivatives from the Red Sea Soft Corals *Alcyonium flaccidum* and *Lobophytum crassum*. J. Org. Chem. **46**, 3592 (1981).
437. POET, S.E., and B.N. RAVI: Three New Diterpenes from a Soft Coral *Nephthea* Species. Aust. J. Chem. **35**, 77 (1982).
438. BLACKMAN, A.J., B.F. BOWDEN, J.C. COLL, B. FRICK, M. MAHENDRAN, and S.J. MITCHELL: Studies of Australian Soft Corals. XXIX[.] Several New Cembranoid Diterpenes from *Nephthea brassica* and Related Diterpenes from a *Sarcophyton* Species. Aust. J. Chem. **35**, 1873 (1982).
439. KINAMONI, Z., A. GROWEISS, S. CARMELY, Y. KASHMAN, and Y LOYA: Several New Cembranoid Diterpenes from Three Soft Corals of the Red Sea. Tetrahedron **39**, 1643 (1983).
440. BOWDEN, B.F., J.C. COLL, and D.M. TAPIOLAS: Studies of Australian Soft Corals. XXXIII. New Cembranoid Diterpenes from a *Lobophytum* Species. Aust. J. Chem. **36**, 2289 (1983).

441. NAKAGAWA, T., M. KOBAYASHI, K. HAYASHI, and H. MITSUHASHI: Marine Terpenes and Terpenoids. II. Structures of Three Cembrane-type Diterpenes, Sarcophytol-C, Sarcophytol-D, and Sarcophytol-E, from the Soft Coral, *Sarcophyton glaucum* Q. et G. Chem. Pharm. Bull. (Japan) **29**, 82 (1981).
442. UCHIO, Y., H. NABEYA, M. NAKAYAMA, S. HAYASHI, and T. HASE: Cembrenene and Mayol, Two New Cembranoid Diterpenes from the Soft Coral *Sinularia mayi*. Tetrahedron Lett. **22**, 1689 (1981).
443. BOWDEN, B.F., J.C. COLL, S.J. MITCHELL, and R. KAZLAUSKAS: Studies of Australian Soft Corals. XXIV. Two Cembranoid Diterpenes from the Soft Coral *Sinularia facile*. Aust. J. Chem. **34**, 1551 (1981).
444. CARMELY, S., A. GROWEISS, and Y. KASHMAN: Decaryiol, a New Cembrane Diterpene from the Marine Soft Coral *Sarcophyton decaryi*. J. Org. Chem. **46**, 4279 (1981).
445. UCHIO, Y., M. NITTA, H. NOZAKI, M. NAKAYAMA, T. IWAGAWA, and T. HASE: 10-Oxo-, 10-Hydroxy-, and 10-Methoxycembrenes from the Soft Coral *Sarcophyta elegans*. Chem. Lett. **1983**, 1719.
446. UCHIO, Y., S. EGUCHI, M. NAKAYAMA, and T. HASE: The Isolation of Two Simple γ-Lactonic Cembranolides from the Soft Coral *Sinularia mayi*. Chem. Lett. **1982**, 277.
447. YAMADA, Y., S. SUZUKI, K. IGUCHI, H. KIKUCHI, Y. TSUKITANI, H. HORIAI, and F. SHIBAYAMA: Studies on Marine Natural Products. IV. The Stereochemistry of 13-Membered Carbocyclic Cembranolide Diterpenes from the Soft Coral *Lobophytum pauciflorum* (Ehrenberg). Tetrahedron Lett. **21**, 3911 (1980).
448. FRINCKE, J.M., D.E. MCINTYRE, and D.J. FAULKNER: Deoxosarcophine from a Soft Coral, *Sarcophyton* sp. Tetrahedron Lett. **21**, 735 (1980).
449. YAMADA, Y., S. SUZUKI, K. IGUCHI, H. KIKUCHI, Y. TSUKITANI, and H. HORIAI: Studies on Marine Natural Products. III. Two New Cembranolides from the Soft Coral *Lobophytum pauciflorum* (Ehrenberg). Chem. Pharm. Bull. (Japan) **28**, 2035 (1980).
450. BURNS, K.P., R. KAZLAUSKAS, P.T. MURPHY, R.J. WELLS, and P. SCHÖNHOLZER: Two Cembranes from *Cespitularia* Species (Soft Coral). Aust. J. Chem. **35**, 85 (1982).
451. GOPICHAND, Y., L.S. CIERESZKO, F.J. SCHMITZ, D. SWITZNER, A. RAHMAN, M.B. HOSSAIN, and D. VAN DER HELM: Further Studies of the Terpenoid Content in the Gorgonian *Eunicea succinea*: 12,13-Bisepieupalmerin, a New Cembranolide. J. Nat. Prod. **47**, 607 (1984).
452. TOTH, J.A., B.J. BURRESON, P.J. SCHEUER, J. FINER-MOORE, and J. CLARDY: Emblide, a New Polyfunctional Cembranolide from the Soft Coral *Sarcophyton glaucum*. Tetrahedron **36**, 1307 (1980).
453. BOWDEN, B.F., J.C. COLL, and R.H. WILLIS: Studies of Australian Soft Corals. XXVII. Two Novel Diterpenes from *Sarcophyton glaucum*. Aust. J. Chem. **35**, 621 (1982).
454. UCHIO, Y., M. NITTA, M. NAKAYAMA, T. IWAGAWA, and T. HASE: Ketoemblide and Sarcophytolide, Two New Cembranolides with ε-Lactone Function from the Soft Coral *Sarcophyta elegans*. Chem. Lett. **1983**, 613.
455. ESTRELLA, D.J., and R.S. JACOBS: Deoxosarcophine (DXS), a Potentiator of Skeletal Muscle Contraction. Fed. Proc. **43**, 586 (1984).
456. BANDURRAGA, M.M., W. FENICAL, S.F. DONOVAN, and J. CLARDY: Pseudopterolide, an Irregular Diterpenoid with Unusual Cytotoxic Properties from the Caribbean Sea Whip *Pseudopterogorgia acerosa* (Pallas) (Gorgonacea). J. Am. Chem. Soc. **104**, 6463 (1982).
457. LOOK, S.A., W. FENICAL, Z. QI-TAI, and J. CLARDY: Calyculones, New Cubitane Diterpenoids from the Caribbean Gorgonian Octocoral *Eunicea calyculata*. J. Org. Chem. **49**, 1417 (1984).

458. LOOK, S.A., and W. FENICAL: New Bicyclic Diterpenoids from the Caribbean Gorgonian Octocoral *Eunicea calyculata*. J. Org. Chem. **47**, 4129 (1982).
459. KOBAYASHI, M., B.W. SON, T. FUJIWARA, Y. KYOGOKU, and I. KITAGAWA: Neodolabelline, a Methyl Migrated Dolabellane-Type Diterpene from the Okinawan Soft Coral *Clavularia koellikeri*. Tetrahedron Lett. **25**, 5543 (1984).
460. LOOK, S.A., W. FENICAL, D. VAN ENGEN, and J. CLARDY: Erythrolides: Unique Marine Diterpenoids Interrelated by a Naturally Occurring Di-π-methane Rearrangement. J. Am. Chem. Soc. **106**, 5026 (1984).
461. GRODE, S.H., T.R. JAMES, JR., J.H. CARDELLINA II, and K.D. ONAN: Molecular Structures of the Briantheins, New Insecticidal Diterpenes from *Briareum polyanthes*. J. Org. Chem. **48**, 5203 (1983).
462. GRODE, S.H., T.R. JAMES, and J.H. CARDELLINA II: Brianthein Z, a New Polyfunctional Diterpene from the Gorgonian *Briareum polyanthes*. Tetrahedron Lett. **24**, 691 (1983).
463. CARDELLINA II, J.H., T.R. JAMES, JR., M.H.M. CHEN, and J. CLARDY: Structure of Brianthein W, from the Soft Coral *Briareum polyanthes*. J. Org. Chem. **49**, 3398 (1984).
464. CLASTRES, A., A. AHOND, C. POUPAT, P. POTIER, and S.K. KAN: Invertébrés Marins Du Lagon Néo-Calédonien, II. Étude Structurale De Trois Nouveaux Diterpènes Isolés Du Pennatulaire *Pteroides laboutei*. J. Nat. Prod. **47**, 155 (1984).
465. CLASTRES, A., P. LABOUTE, A. AHOND, C. POUPAT, and P. POTIER: Invertébrés Marins Du Lagon Néo-Calédonien, III. Étude Structurale De Trois Nouveaux Diterpènes Isolés Du Pennatulaire, *Cavernulina grandiflora*. J. Nat. Prod. **47**, 162 (1984).
466. RAVI, B.N., J.F. MARWOOD, and R.J. WELLS: Three New Diterpenes from the Sea Pen *Scytalium tentaculatum*. Aust. J. Chem. **33**, 2307 (1980).
467. KASHMAN, Y.: A New Diterpenoid Related to Eunicellin and Cladiellin from a *Muricella* sp. Tetrahedron Lett. **21**, 879 (1980).
468. HOCHLOWSKI, J.E., and D.J. FAULKNER: A Diterpene Related to Cladiellin from a Pacific Soft Coral. Tetrahedron Lett. **21**, 4055 (1980).
469. STIERLE, D.B., B. CARTÉ, D.J. FAULKNER, B. TAGLE, and J. CLARDY: The Asbestinins, a Novel Class of Diterpenes from the Gorgonian *Briareum asbestinum*. J. Am. Chem. Soc. **102**, 5088 (1980).
470. SELOVER, S.J., P. CREWS, B. TAGLE, and J. CLARDY: New Diterpenes from the Common Caribbean Gorgonian *Briareum asbestinum* (Pallas). J. Org. Chem. **46**, 964 (1981).
471. KASHMAN, Y., and A. GROWEISS: New Diterpenoids from the Soft Corals *Xenia macrospiculata* and *Xenia obscuronata*. J. Org. Chem. **45**, 3814 (1980).
471a. VANDERAH, D.J., P.A. STEUDLER, L.S. CIERESZKO, F.J. SCHMITZ, J.D. EKSTRAND, and D. VAN DER HELM: Marine Natural Products. Xenicin: a Diterpenoid Possessing a Nine-Membered Ring from the Soft Coral *Xenia elongata*. J. Am. Chem. Soc. **99**, 5780 (1977).
472. BOWDEN, B.F., J.C. COLL, E. DITZEL, S.J. MITCHELL, and W.T. ROBINSON: Studies of the Australian Soft Corals. XXVIII. The Structure Determination of Two New Diterpenes from the Genus *Xenia* (Alcyonacea). Aust. J. Chem. **35**, 997 (1982).
473. GROWEISS, A., and Y. KASHMAN: Eight New Xenia Diterpenoids from Three Soft Corals of the Red Sea. Tetrahedron **39**, 3385 (1983).
474. AHOND, A., B.F. BOWDEN, J.C. COLL, J.-D. FOURNERON, and S.J. MITCHELL: Studies of Australian Soft Corals. XXV. Several Caryophyllene-Based Diterpenes from a *Nephthea* Species. Aust. J. Chem. **34**, 2657 (1981).
475. COVAL, S.J., P.J. SCHEUER, G.K. MATSUMOTO, and J. CLARDY: Two New Xenicin Diterpenoids from the Octocoral *Anthelia edmondsoni*. Tetrahedron **40**, 3823 (1984).

476. SCHWARTZ, R.E., P.J. SCHEUER, V. ZABEL, and W.H. WATSON: The Coraxeniolides, Constituents of Pink Coral, *Corallium* sp. Tetrahedron **37**, 2725 (1981).
477. BURNS, K.P., G. ENGLERT, R. KAZLAUSKAS, P.T. MURPHY, P. SCHÖNHOLZER, and R.J. WELLS: The Structure of a Diterpene with a New Carbocyclic Ring Skeleton from the Soft Coral *Efflatounaria* sp. nov. Aust. J. Chem. **36**, 171 (1983).
478. KOBAYASHI, M., T. YASUZAWA, Y. KOBAYASHI, Y. KYOGOKU, and I. KITAGAWA: Alcyonolide, a Novel Diterpenoid from a Soft Coral. Tetrahedron Lett. **22**, 4445 (1981).
479. BOWDEN, B.F., J.C. COLL, V.A. PATRICK, D.M. TAPIOLAS, and A.H. WHITE: Studies of Australian Soft Corals. XXXII. The Structure Determination of Degraded Xenicin-Type Diterpenes from Several *Efflatounaria* Species. Aust. J. Chem. **36**, 2279 (1983).
480. BAIRD-LAMBERT, J., R.W. DUNLOP, and D.D. JAMIESON: Anticonvulsant Activity of a Novel Diterpene Isolated from a Soft Coral of the Genus *Lobophytum*. Arzneim. Forsch./Drud Res. **30 (I)**, 964 (1980).
481. RAVI, B.N., and R.J. WELLS: Lipid and Terpenoid Metabolites of the Gorgonian *Plexaura flava*. Aust. J. Chem. **35**, 105 (1982).
482. BOWDEN, B.F., and J.C. COLL: Studies of Australian Soft Corals. XXVI. Tetraprenylbenzoquinone Derivatives from a *Nephthea* Species of Soft Coral (Octocorallia, Alcyonacea). Aust. J. Chem. **34**, 2677 (1981).
483. CARMELY, S., Y. KASHMAN, Y. LOYA, and Y. BENAYAHU: New Prostaglandin (PGF) Derivatives from the Soft Coral *Lobophyton depressum*. Tetrahedron Lett. **21**, 875 (1980).
483a. COREY, E.J., P.T. LANSBURY, JR., and Y. YAMADA: Identification of a New Eicosanoid from *in Vitro* Biosynthetic Experiments with *Clavularia viridis*. Implications for the Biosynthesis of Clavulones. Tetrahedron Lett. **26**, 4171 (1985).
484. KOBAYASHI, M., T. YASUZAWA, M. YOSHIHARA, H. AKUTSU, Y. KYOGOKU, and I. KITAGAWA: Four New Prostanoids: Claviridenone-A, -B, -C, and -D from the Okinawan Soft Coral *Clavularia viridis*. Tetrahedron Lett. **23**, 5331 (1982).
485. KOBAYASHI, M., T. YASUZAWA, M. YOSHIHARA, B.W. SON, Y. KYOGOKU, and I. KITAGAWA: Absolute Stereostructures of Claviridenone-A, -B, -C, and -D, Four Prostanoids from the Okinawan Soft Coral *Clavularia viridis*. Chem. Pharm. Bull. (Japan) **31**, 1440 (1983).
486. KITAGAWA, I., M. KOBAYASHI, T. YASUZAWA, B.W. SON, M. YOSHIHARA, and Y. KYOGOKU: New Prostanoids from Soft Coral. Tetrahedron **41**, 995 (1985).
487. KIKUCHI, H., Y. TSUKITANI, K. IGUCHI, and Y. YAMADA: Clavulones, New Type of Prostanoids from the Stolonifer *Clavularia viridis* Quoy and Gaimard. Tetrahedron Lett. **23**, 5171 (1982).
488. ---- Absolute Stereochemistry of New Prostanoids Clavulone I, II and III, from *Clavularia viridis* Quoy and Gaimard. Tetrahedron Lett. **24**, 1549 (1983).
489. IGUCHI, K., Y. YAMADA, H. KIKUCHI, and Y. TSUKITANI: Novel C-20-Oxygenated Prostanoids, 20-Acetoxyclavulones, from the Stolonifer *Clavularia viridis* Quoy and Gaimard. Tetrahedron Lett. **24**, 4433 (1983).
490. BAKER, B.J., R.K. OKUDA, P.T.K. YU, and P.J. SCHEUER: Punaglandins: Halogenated Antitumor Eicosanoids from the Octocoral *Telesto riisei*. J. Am. Chem. Soc. **107**, 2976 (1985).
490a. IGUCHI, K., S. KANETA, K. MORI, Y. YAMADA, A. HONDA, and Y. MORI: Chlorovulones, New Halogenated Marine Prostanoids with an Antitumor Activity from the Stolonifer *Clavularia viridis* Quoy and Gaimard. Tetrahedron Lett. **26**, 5787 (1985).
491. COREY, E.J., and M.M. MEHROTRA: Total Synthesis of ($\pm$)-Clavulones. J. Am. Chem. Soc. **106**, 3384 (1984).

492. NAGAOKA, H., T. MIYAKOSHI, and Y. YAMADA: Total Synthesis of Marine Prostanoids Clavulones. Tetrahedron Lett. **25**, 3621 (1984).
493. HASHIMOTO, S., Y. ARAI, and N. HAMANAKA: Synthesis of Clavulones (Claviridenones). Tetrahedron Lett. **26**, 2679 (1985).
494. KIKUCHI, H., Y. TSUKITANI, H. NAKANISHI, I. SHIMIZU, S. SAITOH, K. IGUCHI, and Y. YAMADA: New Butenolides from the Gorgonian *Euplexaura flava* (Nutting). Chem. Lett. **1982**, 233.
495. − − − − − − − Studies on Marine Natural Products. VIII. New Butenolides from the Gorgonian *Euplexaura flava* (Nutting). Chem. Pharm. Bull. (Japan) **31**, 1172 (1983).
496. ENDO, M., M. NAKAGAWA, Y. HAMAMOTO, and T. NAKANISHI: Clavularins, a New Class of Cytotoxic Compounds Isolated from the Soft Coral, *Clavularia koellikeri*. J. Chem. Soc., Chem. Commun. **1983**, 322. Erratum: J. Chem. Soc., Chem. Commun. **1983**, 980.
497. CHANTRAPROMMA, K., J.S. MCMANIS, and B. GANEM: Synthesis of Cytotoxic Spermidine Metabolites from the Soft Coral *Sinularia brongersmai*. Tetrahedron Lett. **21**, 2605 (1980).
498. KAZLAUSKAS, R., J.F. MARWOOD, and R.J. WELLS: 2-Phenylethylamides of a Novel Lipid Acid; Atrial Stimulants from the Soft Coral *Sinularia flexibilis*. Aust. J. Chem. **33**, 1799 (1980).
499. BOWDEN, B.F., P.S. CLEZY, J.C. COLL, B.N. RAVI, and D.M. TAPIOLAS: Studies of Australian Soft Corals. XXXIV. A New Substituted Pyrrole from a Soft Coral-Sponge Association. Aust. J. Chem. **37**, 227 (1984).
500. MURAMATSU, I., M. FUJIWARA, S. IKUSHIMA, and K. ASHIDA: Effects of Goniopora Toxin on Guinea-Pig Blood Vessels. Naunyn-Schmiedeberg's Arch. Pharmacol. **312**, 193 (1980).
501. FUJIWARA, M., S.-C. HONG, and I. MURAMATSU: Effects of Goniopora Toxin on Non-Adrenergic, Non-Cholinergic Response and Purine Nucleotide Release in Guinea-Pig Taenia Coli. J. Physiol. **326**, 515 (1982).
502. IKUSHIMA, S., I. MURAMATSU, and M. FUJIWARA: Nicotine-Induced Response in Guinea-Pig Aorta Enhanced by Goniopora Toxin. J. Pharmacol. exp. Ther. **223**, 790 (1982).
503. LESNIAK, A.P., and E.H. LIU: Characterization of a Hemagglutinin from *Leptogorgia virgulata*. Comp. Biochem. Physiol. **71B**, 305 (1982).
504. CARLÉ, J.S., and C. CHRISTOPHERSEN: Dogger Bank Itch. The Allergen Is (2-Hydroxyethyl)dimethylsulfonium Ion. J. Am. Chem. Soc. **102**, 5107 (1980).
505. − − Dogger Bank Itch. 2. An Allergic Contact Dermatitis. Bull. Soc. Chim. Belg. **89**, 1087 (1980).
506. − − Dogger Bank Itch. 4. An Eczema-Causing Sulfoxonium Ion from the Marine Animal, *Alcyonidium gelatinosum* (Bryozoa). Toxicon **20**, 307 (1982).
507. − − Marine Alkaloids. 2. Bromo Alkaloids from the Marine Bryozoan *Flustra foliacea*. Isolation and Structure Elucidation. J. Org. Chem. **45**, 1586 (1980).
508. − − Marine Alkaloids. 3. Bromo-Substituted Alkaloids from the Marine Bryozoan *Flustra foliacea*, Flustramine C and Flustraminol A and B. J. Org. Chem. **46**, 3440 (1981).
509. WRIGHT, J.L.C.: A New Antibiotic from the Marine Bryozoan *Flustra foliaceae*. J. Nat. Prod. **47**, 893 (1984).
510. WULFF, P., J.S. CARLÉ, and C. CHRISTOPHERSEN: Marine Alkaloids. Part 4. A Formamide, Flustrabromine, from the Marine Bryozoan *Flustra foliacea*. J. Chem. Soc., Perkin Trans. I **1981**, 2895.
511. − − − Marine Alkaloids. 6. The First Naturally Occurring Bromo-Substituted Quinoline from *Flustra foliacea*. Comp. Biochem. Physiol. **71B**, 525 (1982).

512. SATO, A., and W. FENICAL: Gramine-Derived Bromo-Alkaloids from the Marine Bryozoan *Zoobotryon verticillatum*. Tetrahedron Lett. **24**, 481 (1983).
513. AYER, S.W., R.J. ANDERSEN, H. CUN-HENG, and J. CLARDY: Phidolopin, a New Purine Derivative from the Bryozoan *Phidolopora pacifica*. J. Org. Chem. **49**, 3869 (1984).
514. HIROTA, K., K. KUBO, Y KITADE, and Y. MAKI: Synthesis of Phidolopin, 7-(4-Hydroxy-3-nitrobenzyl)-1,3-dimethylxanthine from the Bryozoan *Phidolopora pacifica*. Tetrahedron Lett. **26**, 2355 (1985).
515. PETTIT, G.R., C.L. HERALD, D.L. DOUBEK, D.L. HERALD, E. ARNOLD, and J. CLARDY: Isolation and Structure of Bryostatin 1. J. Am. Chem. Soc. **104**, 6846 (1982).
516. PETTIT, G.R., C.L. HERALD, Y. KAMANO, D. GUST, and R. AOYAGI: The Structure of Bryostatin 2 from the Marine Bryozoan *Bugula neritina*. J. Nat. Prod. **46**, 528 (1983).
517. PETTIT, G.R., C.L. HERALD, and Y. KAMANO: Structure of the *Bugula neritina* (Marine Bryozoa) Antineoplastic Component Bryostatin 3. J. Org. Chem. **48**, 5354 (1983).
518. PETTIT, G.R., Y. KAMANO, C.L. HERALD, and M. TOZAWA: Structure of Bryostatin 4. An Important Antineoplastic Constituent of Geographically Diverse *Bugula neritina* (Bryozoa). J. Am. Chem. Soc. **106**, 6768 (1984).
519. PETTIT, G.R., Y. KAMANO, R. AOYAGI, C.L. HERALD, D.L. DOUBEK, J.M. SCHMIDT, and J.J. RUDLOE: Antineoplastic Agents 100. The Marine Bryozoan *Amathia convoluta*. Tetrahedron **41**, 985 (1985).
520. CIMINO, G., S. DE STEFANO, S. DE ROSA, G. SODANO, and G. VILLANI: Novel Metabolites from Some Predator-Prey Pairs. Bull. Soc. Chim. Belg. **89**, 1069 (1980).
521. FUHRMAN, F.A., G.J. FUHRMAN, Y.H. KIM, L.A. PAVELKA, and H.S. MOSHER: Doridosine: A New Hypotensive N-Methylpurine Riboside from the Nudibranch *Anisodoris nobilis*. Science **207**, 193 (1980).
522. KIM, Y.H., R.J. NACHMAN, L. PAVELKA, H.S. MOSHER, F.A. FUHRMAN, and G.J. FUHRMAN: Doridosine, 1-Methylisoguanosine, from *Anisodoris nobilis*; Structure, Pharmacological Properties, and Synthesis. J. Nat. Prod. **44**, 206 (1981).
523. FUHRMAN, F.A., G.J. FUHRMAN, R.J. NACHMAN, and H.S. MOSHER: Isoguanosine: Isolation from an Animal. Science **212**, 557 (1981).
524. THOMPSON, J.E., R.P. WALKER, S.J. WRATTEN, and D.J. FAULKNER: A Chemical Defense Mechanism for the Nudibranch *Cadlina luteomarginata*. Tetrahedron **38**, 1865 (1982).
525. HELLOU, J., R.J. ANDERSEN, and J.E. THOMPSON: Terpenoids from the Dorid Nudibranch *Cadlina luteomarginata*. Tetrahedron **38**, 1875 (1982).
526. CIMINO, G., S. DE ROSA, S. DE STEFANO, and G. SODANO: The Chemical Defense of Four Mediterranean Nudibranchs. Comp. Biochem. Physiol. **73B**, 471 (1982).
527. DE SILVA, E.D., and P.J. SCHEUER: Furanoditerpenoids from the Dorid Nudibranch *Casella atromarginata*. Heterocycles **17**, 167 (1982).
528. CIMINO, G., S. DE ROSA, S. DE STEFANO, R. MORRONE, and G. SODANO: The Chemical Defense of Nudibranch Molluscs. Structure, Biosynthetic Origin, and Defensive Properties of Terpenoids from the Dorid Nudibranch *Dendrodoris grandiflora*. Tetrahedron **41**, 1093 (1985).
529. SCHULTE, G.R., and P.J. SCHEUER: Defense Allomones of Some Marine Mollusks. Tetrahedron **38**, 1857 (1982).
530. AYER, S.W., and R.J. ANDERSEN: Steroidal Antifeedants from the Dorid Nudibranch *Aldisa sanguinea cooperi*. Tetrahedron Lett. **23**, 1039 (1982).
531. — — Degraded Monoterpenes from the Opisthobranch Mollusc *Melibe leonina*. Experientia **39**, 255 (1983).

532. ANDERSEN, R.J., and F.W. SUM: Farnesic Acid Glycerides from the Nudibranch *Archidoris odhneri*: Tetrahedron Lett. **21**, 797 (1980).
533. GUSTAFSON, K., and R.J. ANDERSEN: Chemical Studies of British Columbia Nudibranchs. Tetrahedron **41**, 1101 (1985).
534. HOCHLOWSKI, J.E., and D.J. FAULKNER: Chemical Constituents of the Nudibranch *Chromodoris marislae*. Tetrahedron Lett. **22**, 271 (1981).
535. CIMINO, G., S. DE ROSA, S. DE STEFANO, and G. SODANO: Novel Sesquiterpenoid Esters from the Nudibranch *Dendrodoris limbata*. Tetrahedron Lett. **22**, 1271 (1981).
536. D'ISCHIA, M., G. PROTA, and G. SODANO: Reaction of Polygodial with Primary Amines: An Alternative Explanation to the Antifeedant Activity. Tetrahedron Lett. **23**, 3295 (1982).
537. CIMINO, G., S. DE ROSA, S. DE STEFANO, G. SODANO, and G. VILLANI: Dorid Nudibranch Elaborates Its Own Chemical Defense. Science **219**, 1237 (1983).
538. OKUDA, R.K., P.J. SCHEUER, J.E. HOCHLOWSKI, R.P. WALKER, and D.J. FAULKNER: Sesquiterpenoid Constituents of Eight Porostome Nudibranchs. J. Org. Chem. **48**, 1866 (1983).
539. AYER, S.W., J. HELLOU, M. TISCHLER, and R.J. ANDERSEN: Nanaimoal, a Sesquiterpenoid Aldehyde from the Dorid Nudibranch *Acanthodoris nanaimoensis*. Tetrahedron Lett. **25**, 141 (1984).
540. AYER, S.W., R.J. ANDERSEN, H. CUN-HENG, and J. CLARDY: Acanthodoral and Isoacanthodoral, Two Sesquiterpenoids with New Carbon Skeletons from the Dorid Nudibranch *Acanthodoris nanaimoensis*. J. Org. Chem. **49**, 2653 (1984).
541. GUSTAFSON, K., R.J. ANDERSEN, M.H.M. CHEN, J. CLARDY, and J.E. HOCHLOWSKI: Terpenoic Acid Glycerides from the Dorid Nudibranch *Archidoris montereyensis*. Tetrahedron Lett. **25**, 11 (1984).
541a. RICKETS, E.F., and J. CALVIN: Between Pacific Tides (Fourth Edition), p. 118, Stanford University Press, Stanford, Calif. (1968).
542. GUSTAFSON, K., R.J. ANDERSEN, H. CUN-HENG, and J. CLARDY: Marginatafuran, a Furanoditerpene with a New Carbon Skeleton from the Dorid Nudibranch *Cadlina luteomarginata*. Tetrahedron Lett. **26**, 2521 (1985).
543. HELLOU, J., R.J. ANDERSEN, S. RAFII, E. ARNOLD, and J. CLARDY: Luteone, a Twenty Three Carbon Terpenoid from the Dorid Nudibranch *Cadlina luteomarginata*. Tetrahedron Lett. **22**, 4173 (1981).
544. HOCHLOWSKI, J.E., D.J. FAULKNER, L.S. BASS, and J. CLARDY: Metabolites of the Dorid Nudibranch *Chromodoris sedna*. J. Org. Chem. **48**, 1738 (1983).
545. WALKER, R.P., and D.J. FAULKNER: Chlorinated Acetylenes from the Nudibranch *Diaulula sandiegensis*. J. Org. Chem. **46**, 1475 (1981).
546. CARTÉ, B., and D.J. FAULKNER: Defensive Metabolites from Three Nembrothid Nudibranchs. J. Org. Chem. **48**, 2314 (1983).
547. GUSTAFSON, K., and R.J. ANDERSEN: Triophamine, a Unique Diacylguanidine from the Dorid Nudibranch *Triopha catalinae* (Cooper). J. Org. Chem. **47**, 2167 (1982).
548. PIERS, E., J.M. CHONG, K. GUSTAFSON, and R.J. ANDERSEN: A Total Synthesis of ($\pm$)-Triophamine. Can. J. Chem. **62**, 1 (1984).
548a. YAMAMURA, S., and Y. HIRATA: Structures of Aplysin and Aplysinol, Naturally Occurring Bromo-Compounds. Tetrahedron **19**, 1485 (1963).
549. HOYE, T.R., A.J. CARUSO, J.F. DELLARIA, JR., and M.J. KURTH: Two Syntheses of *dl* -Aplysistatin. J. Am. Chem. Soc. **104**, 6704 (1982).
550. SCHMITZ, F.J., Y. GOPICHAND, D.P. MICHAUD, R.S. PRASAD, S. REMALEY, M.B. HOSSAIN, A. RAHMAN, P.K. SENGUPTA, and D. VAN DER HELM: Recent Developments in Research on Metabolites from Caribbean Marine Invertebrates. Pure and Appl. Chem. **51**, 853 (1981).
551. SCHMITZ, F.J., D.P. MICHAUD, and K.H. HOLLENBEAK: Marine Natural Products:

Dihydroxydeodactol Monoacetate, a Halogenated Sesquiterpene Ether from the Sea Hare *Aplysia dactylomela*. J. Org. Chem. **45**, 1525 (1980).
552. GOPICHAND, Y., F.J. SCHMITZ, J. SHELLY, A. RAHMAN, and D. VAN DER HELM: Marine Natural Products: Halogenated Acetylenic Ethers from the Sea Hare *Aplysia dactylomela*. J. Org. Chem. **46**, 5192 (1981).
553. GONZÁLEZ, A.G., J.D. MARTIN, M. NORTE, R. PÉREZ, V. WEYLER, A. PERALES, and J. FAYOS: New Halogenated Constituents of the Digestive gland of the Sea Hare *Aplysia dactylomela*. Tetrahedron Lett. **24**, 847 (1983).
554. SCHMITZ, F.J., D.P. MICHAUD, and P.G. SCHMIDT: Marine Natural Products: Parguerol, Deoxyparguerol, and Isoparguerol. New Brominated Diterpenes with Modified Pimarane Skeletons from the Sea Hare *Aplysia dactylomela*. J. Am. Chem. Soc. **104**, 6415 (1982). Erratum: J. Am. Chem. Soc. **106**, 3385 (1984).
555. GONZÁLEZ, A.G., J.D. MARTIN, M. NORTE, R. PÉREZ, V. WEYLER, S. RAFII, and J. CLARDY: A New Diterpene from *Aplysia dactylomela*. Tetrahedron Lett. **24**, 1075 (1983).
556. PETTIT, G.R., C.L. HERALD, R.H. ODE, P. BROWN, D.J. GUST, and C. MICHEL: The Isolation of Loliolide from an Indian Ocean Opisthobranch Mollusc. J. Nat. Prod. **43**, 752 (1980).
557. MIDLAND, S.L., R.M. WING, and J.J. SIMS: New Crenulides from the Sea Hare *Aplysia vaccaria*. J. Org. Chem. **48**, 1906 (1983).
558. SUN, H.H., F.J. MCENROE, and W. FENICAL: Acetoxycrenulide, a New Bicyclic Cyclopropane-Containing Diterpenoid from the Brown Seaweed *Dictyota crenulata*. J. Org. Chem. **48**, 1903 (1983).
559. SCHULTE, G.R., M.C.H. CHUNG, and P.J. SCHEUER: Two Bicyclic C_{15} Enynes from the Sea Hare *Aplysia oculifera*. J. Org. Chem. **46**, 3870 (1981).
560. GOPICHAND, Y., and F.J. SCHMITZ: Bursatellin: A New Diol Dinitrile from the Sea Hare *Bursatella leachii pleii*. J. Org. Chem. **45**, 5383 (1980).
561. PETTIT, G.R., Y. KAMANO, Y. FUJII, C.L. HERALD, M. INOUE, P. BROWN, D. GUST, K. KITAHARA, J.M. SCHMIDT, D.L. DOUBEK, and C. MICHEL: Marine Animal Biosynthetic Constituents for Cancer Chemotherapy. J. Nat. Prod. **44**, 482 (1981).
562. PETTIT, G.R., Y. KAMANO, P. BROWN, D. GUST, M. INOUE, and C.L. HERALD: Structure of the Cyclic Peptide Dolastatin 3 from *Dolabella auricularia*. J. Am. Chem. Soc. **104**, 905 (1982).
562a. HAMADA, Y., K. KOHDA, and T. SHIOIRI: Proposed Structure of the Cyclic Peptide Dolastatin 3, a Powerful Cell Growth Inhibitor, Should be Revised! Tetrahedron Lett. **25**, 5303 (1984).
562b. SCHMIDT, U., and R. UTZ: Synthesestudien zur Ermittlung von Struktur und Konfiguration des Dolastatins 3. Angew. Chem. **96**, 723 (1984).
563. YOSHIBA, S., and S. SAKURAI: Comparative Studies on the Toxicities of Venoms of Five Species of *Conus* and the Effect of Ligation as a Treatment for *Conus*-stings. Jpn. J. med. Sci. Biol. **35**, 139 (1982).
564. KOBAYASHI, J., H. NAKAMURA, Y. HIRATA, and Y. OHIZUMI: Effect of Venoms from *Conidae* on Skeletal, Cardiac, and Smooth Muscles. Toxicon **20**, 823 (1982).
565. GRAY, W.R., A. LUQUE, B.M. OLIVERA, J. BARRETT, and L.J. CRUZ: Peptide Toxins from *Conus geographus* Venom. J. Biol. Chem. **256**, 4734 (1981).
566. NISHIUCHI, Y., and S. SAKAKIBARA: Primary and Secondary Structure of Conotoxin GI, a Neurotoxic Tridecapeptide from a Marine Snail. Febs Lett. **148**, 260 (1982).
567. STONE, B.L., and W.R. GRAY: Occurrence of Hydroxyproline in a Toxin from the Marine Snail *Conus geographus*. Arch. Biochem. Biophys. **216**, 765 (1982).
568. NAKAMURA, H., J. KOBAYASHI, Y. OHIZUMI, and Y. HIRATA: Isolation and Amino Acid Compositions of Geographutoxin I and II from the Marine Snail *Conus geographus* Linné. Experientia **39**, 590 (1983).

569. SATO, S., H. NAKAMURA, Y. OHIZUMI, J. KOBAYASHI, and Y. HIRATA: The Amino Acid Sequences of Homologous Hydroxyproline-Containing Myotoxins from the Marine Snail *Conus geographus* Venom. Febs Lett. **155**, 277 (1983).
570. CLARK, C., B.M. OLIVERA, and L.J. CRUZ: A Toxin from the Venom of the Marine Snail *Conus geographus* which Acts on the Vertebrate Central Nervous System. Toxicon **19**, 691 (1981).
571. MCINTOSH, M., L.J. CRUZ, M.W. HUNKAPILLER, W.R. GRAY, and B.M. OLIVERA: Isolation and Structure of a Peptide Toxin from the Marine Snail *Conus magus*. Arch. Biochem. Biophys. **218**, 329 (1982).
572. KOBAYASHI, J., H. NAKAMURA, and Y. OHIZUMI: Excitatory and Inhibitory Effects of a Myotoxin from *Conus magus* Venom on the Mouse Diaphragm, the Guinea-pig Atria, Taenia Caeci, Ileum and Vas Deferens. Eur. J. Pharmacol. **86**, 283 (1983).
573. STRICHARTZ, G.R., G.K. WANG, J. SCHMIDT, R. HAHIN, and B.I. SHAPIRO: Modification of Ionic Currents in Frog Nerve by Crude Venom and Isolated Peptides of the Mollusc *Conus striatus*. Fed. Proc. **39**, 2065 (1980).
574. KOBAYASHI, J., H. NAKAMURA, and Y. OHIZUMI: Biphasic Mechanical Responses of the Guinea-pig Isolated Ileum to the Venom of the Marine Snail *Conus striatus*. Br. J. Pharmac. **73**, 583 (1981).
575. KOBAYASHI, J., H. NAKAMURA, Y. HIRATA, and Y. OHIZUMI: Isolation of a Cardiotonic Glycoprotein, Striatoxin, from the Venom of the Marine Snail *Conus striatus*. Biochem. Biophys. Res. Commun. **105**, 1389 (1982).
576. JIMENEZ, E.C., B.M. OLIVERA, and L.J. CRUZ: Localization of Enzymes and Possible Toxin Precursors in Granules from *Conus striatus* Venom. Toxicon Suppl. **3**, 199 (1983).
577. KOBAYASHI, J., Y. OHIZUMI, H. NAKAMURA, and Y. HIRATA: Pharmacological Study on the Venom of the Marine Snail *Conus textile*. Toxicon **19**, 757 (1981).
578. NAKAMURA, H., J. KOBAYASHI, Y. OHIZUMI, and Y. HIRATA: The Occurrence of Arachidonic Acid in the Venom Duct of the Marine Snail *Conus textile*. Experientia **38**, 897 (1982).
579. KOBAYASHI, J., H. NAKAMURA, Y. HIRATA, and Y. OHIZUMI: Tessulatoxin, the Vasoactive Protein from the Venom of the Marine Snail *Conus tessulatus*. Comp. Biochem. Physiol. **74B**, 381 (1983).
580. – – – – Isolation of Eburnetoxin, a Vasoactive Substance from the *Conus eburneus* Venom. Life Sciences **31**, 1085 (1982).
581. YOSHIBA, S., and S. SAKURAI: Analysis on the Toxicity of the Venom of a Cone Shell Bekko-Imogai *Chelyconus fulmen* (Reeve, 1843). Japan. J. Med. Sci. Biol. **33**, 38 (1980).
582. OLIVERA, B.M., J.M. MCINTOSH, C. CLARK, D. MIDDLEMAS, W.R. GRAY, and L.J. CRUZ: A Sleep-Inducing Peptide from *Conus geographus* Venom. Toxicon **23**, 277 (1985).
583. TYMIAK, A.A., and K.L. RINEHART, JR.: Structures of Kelletinins I and II, Antibacterial Metabolites of the Marine Mollusk *Kelletia kelletii*. J. Am. Chem. Soc. **105**, 7396 (1983).
584. KOSUGE, T., K. TSUJI, K. HIRAI, K. YAMAGUCHI, T. OKAMOTO, and Y. IITAKA: Isolation and Structure Determination of a New Marine Toxin, Neosurugatoxin, from the Japanese Ivory Shell, *Babylonia japonica*. Tetrahedron Lett. **22**, 3417 (1981).
585. KOSUGE, T., K. TSUJI, and K. HIRAI: Isolation of Neosurugatoxin from the Japanese Ivory Shell, *Babylonia japonica*. Chem. Pharm. Bull. (Japan) **30**, 3255 (1982).
586. KOSUGE, T., K. TSUJI, K. HIRAI, K. YAMAGUCHI, T. OKAMOTO, and Y. IITAKA: Isolation and Structural Determination of a New Marine Toxin, Neosurugatoxin, from the Japanese Ivory Shell *Babylonia japonica*. Tennen Yuki Kagobutsu Toronkai Koen Yoshishu, 24th **1981**, 127.

587. NOGUCHI, T., J. MARUYAMA, Y. UEDA, K. HASHIMOTO, and T. HARADA: Occurrence of Tetrodotoxin in the Japanese Ivory Shell *Babylonia japonica*. Bull. Japan. Soc. Sci. Fish. **47**, 909 (1981).
588. YASUMOTO, T., Y. OSHIMA, M. HOSAKA, and S. MIYAKOSHI: Occurrence of Tetrodotoxin in the Ivory Shell *Babylonia japonica* from Wakasa Bay. Bull. Japan. Soc. Sci. Fish. **47**, 929 (1981).
589. NARITA, H., T. NOGUCHI, J. MARUYAMA, Y. UEDA, K. HASHIMOTO, Y. WATANABE, and K. HIDA: Occurrence of Tetrodotoxin in a Trumpet Shell, "Boshubora" *Charonia sauliae*. Bull. Japan. Soc. Sci. Fish. **47**, 935 (1981).
590. NOGUCHI, T., J. MARUYAMA, H. NARITA, and K. HASHIMOTO: Occurrence of Tetrodotoxin in the Gastropod Mollusk *Tutufa lissostoma* (Frog Shell). Toxicon **22**, 219 (1984).
591. HOCHLOWSKI, J.E., and D.J. FAULKNER: Antibiotics from the Marine Pulmonate *Siphonaria diemenensis*. Tetrahedron Lett. **24**, 1917 (1983).
592. BISKUPIAK, J.E., and C.M. IRELAND: Pectinatone, a New Antibiotic from the Mollusc *Siphonaria pectinata*. Tetrahedron Lett. **24**, 3055 (1983).
593. HOCHLOWSKI, J.E., D.J. FAULKNER, G.K. MATSUMOTO, and J. CLARDY: The Denticulatins, Two Polypropionate Metabolites from the Pulmonate *Siphonaria denticulata*. J. Am. Chem. Soc. **105**, 7413 (1983).
594. CAPON, R.J., and D.J. FAULKNER: Metabolites of the Pulmonate *Siphonaria lessoni*. J. Org. Chem. **49**, 2506 (1984).
595. HOCHLOWSKI, J.E., and D.J. FAULKNER: Metabolites of the Marine Pulmonate *Siphonaria australis*. J. Org. Chem. **49**, 3838 (1984).
596. HOCHLOWSKI, J.E., J.C. COLL, D.J. FAULKNER, J.E. BISKUPIAK, C.M. IRELAND, Z. QI-TAI, H. CUN-HENG, and J. CLARDY: Novel Metabolites of Four *Siphonaria* Species. J. Am. Chem. Soc. **106**, 6748 (1984).
597. IRELAND, C., and J. FAULKNER: The Metabolites of the Marine Molluscs *Tridachiella diomedea* and *Tridachia crispata*. Tetrahedron **37**, Supplement No. 1, 233 (1981).
597a. CIMINO, G., G. SODANO, A. SPINELLA, and E. TRIVELLONE: Aglajne-1, a Polypropionate Metabolite from the Opisthobranch Mollusk *Aglaja depicta*. Determination of Carbon-Carbon Connectivity Via Long-Range ^{1}H-^{13}C Couplings. Tetrahedron Lett. **26**, 3389 (1985).
598. IRELAND, C.M., J.E. BISKUPIAK, G.J. HITE, M. RAPPOSCH, P.J. SCHEUER, and J.R. RUBLE: Ilikonapyrone Esters, Likely Defense Allomones from the Mollusc *Onchidium verruculatum*. J. Org. Chem. **49**, 559 (1984).
598a. BISKUPIAK, J.E., and C.M. IRELAND: Cytotoxic Metabolites from the Mollusc *Peronia peronii*. Tetrahedron Lett. **26**, 4307 (1985).
599. SANDUJA, R., A.J. WEINHEIMER, K.L. EULER, and M. ALAM: Unusual Occurrence of Fulvoplumierin, an Antibacterial Pigment, in the Marine Mollusk *Nerita albicilla*. J. Nat. Prod. **48**, 335 (1985).
600. SLEEPER, H.L., V.J. PAUL, and W. FENICAL: Alarm Pheromones from the Marine Opisthobranch *Navanax inermis*. J. Chem. Ecol. **6**, 57 (1980).
600a. COVAL, S.J., and P.J. SCHEUER: An Intriguing C_{16}-Alkadienone-Substituted 2-Pyridine from a Marine Mollusc. J. Org. Chem. **50**, 3024 (1985).
600b. ALBIZATI, K.F., J.R. PAWLIK, and D.J. FAULKNER: Limatulone, a Potent Defensive Metabolite of the Intertidal Limpet *Collisella limatula*. J. Org. Chem. **50**, 3428 (1985).
600c. ANDERSEN, R.J., D.J. FAULKNER, H. CUN-HENG, G.D. VAN DUYNE, and J. CLARDY: Metabolites of the Marine Prosobranch Mollusc *Lamellaria* sp. J. Am. Chem. Soc. **107**, 5492 (1985).
601. KHALIL, M.W., D.R. IDLER, and G.W. PATTERSON: Sterols of Scallop. III. Characterization of Some C-24 Epimeric Sterols by High Resolution (220 MHz) Nuclear Magnetic Resonance Spectroscopy. Lipids **15**, 69 (1980).

602. SWIFT, M.L., D. WHITE, and M.B. GHASSEMIEH: Distribution of Neutral Lipids in the Tissues of the Oyster *Crassostrea virginica*. Lipids **15**, 129 (1980).
603. TESHIMA, S., G.W. PATTERSON, and S.R. DUTKY: Sterols of the Oyster, *Crassostrea virginica*. Lipids **15**, 1004 (1980).
604. TESHIMA, S., and G.W. PATTERSON: Identification of 4α-Methylsterols in the Oyster *Crassostrea virginica*. Comp. Biochem. Physiol. **69 B**, 175 (1981).
605. BERENBERG, C.J., and G.W. PATTERSON: The Relationship between Dietary Phytosterols and the Sterols of Wild and Cultivated Oysters. Lipids **16**, 276 (1981).
606. BOUTRY, J.L., and M. BARBIER: New Results About the C_{26} Sterols Also Concerning Their Possible Origin. Bull. Soc. Chim. Belg. **89**, 1075 (1980).
607. ROMERO, M.S., and A.M. SELDES: Sterol Composition of the Mollusk *Aulacomya ater*. J. Nat. Prod. **46**, 588 (1983).
608. – – Sterols of the Clam *Chlamys tehuelcha*. Experientia **39**, 258 (1983).
609. MATSUNO, T., and S. SAKAGUCHI: A Novel Marine Carotenoid, Mactraxanthin from the Japanese Edible Surf Clam. Tetrahedron Lett. **24**, 911 (1983).
610. MATSUNO, T., and T. MAOKA: Carotenoids of Shellfishes. I. Isolation of a New Carotenoid, 3,4,3'-Trihydroxy-7',8'-didehydro-β-carotene from Sea Mussels. Nippon Suisan Gakkaishi **47**, 377 (1981).
611. – – Carotenoids of Shellfish. III. Isolation of Diatoxanthin, Pectenoxanthin, Pectenolone, and a New Carotenoid 3,4,3'-Trihydroxy-7',8'-didehydro-β-carotene from Arkshell and Related Three Species of Bivalves. Nippon Suisan Gakkaishi **47**, 495 (1981).
612. LEUNG, M., and G.B. STEFANO: Isolation of Molluscan Opioid Peptides. Life Sciences **33**, Sup. I, 77 (1983).
613. – – Isolation and Identification of Enkephalins in Pedal Ganglia of *Mytilus edulis* (Mollusca). Proc. Natl. Acad. Sci. U.S.A. **81**, 955 (1984).
614. SHEUMACK, D.D., M.E.H. HOWDEN, and I. SPENCE: Occurrence of a Tetrodotoxin-Like Compound in the Eggs of the Venomous Blue-ringed Octopus (*Hapalochlaena maculosa*). Toxicon **22**, 811 (1984).
615. BURNELL, D.J., and J.W. APSIMON: Echinoderm Saponins. In: Marine Natural Products. Chemical and Biological Perspectives (P.J. SCHEUER, ed.), vol. V, p. 287. New York-San Francisco-London: Academic Press. 1983.
616. ANISIMOV, M.M., N.G. PROKOFIEVA, L.Y. KOROTKIKH, I.I. KAPUSTINA, and V.A. STONIK: Comparative Study of Cytotoxic Activity of Triterpene Glycosides from Marine Organisms. Toxicon **18**, 221 (1980).
617. MINALE, L., C. PIZZA, R. RICCIO, and F. ZOLLO: Steroidal Glycosides from Starfishes. Pure & Appl. Chem. **54**, 1935 (1982).
618. OKANO, K., and S. IKEGAMI: Separation of Ovarian Asterosaponins in the Starfish *Asterias amurensis*. Agric. Biol. Chem. **45**, 801 (1981).
619. OKANO, K., T. NAKAMURA, Y. KAMIYA, and S. IKEGAMI: Structure of Ovarian Asterosaponin-1 in the Starfish *Asterias amurensis*. Agric. Biol. Chem. **45**, 805 (1981).
620. DE SIMONE, F., A. DINI, E. FINAMORE, L. MINALE, C. PIZZA, R. RICCIO, and F. ZOLLO: Starfish Saponins. Part 5. Structure of Sepositoside A, a Novel Steroidal Cyclic Glycoside from the Starfish *Echinaster sepositus*. J. Chem. Soc., Perkin Trans. I **1981**, 1855.
621. RICCIO, R., E. DE SIMONE, A. DINI, L. MINALE, C. PIZZA, F. SENATORE, and F. ZOLLO: Starfish Saponins VI – Unique 22,23-Epoxysteroidal Cyclic Glycosides, Minor Constituents from *Echinaster sepositus*. Tetrahedron Lett. **22**, 1557 (1981). Erratum: Tetrahedron Lett. **22**, 2242 (1981).
622. RICCIO, R., A. DINI, L. MINALE, C. PIZZA, F. ZOLLO, and T. SEVENET: Starfish Saponins VII. Structure of Luzonicoside, a Further Steroidal Cyclic Glycoside from the Pacific Starfish *Echinaster luzonicus*. Experientia **38**, 68 (1982).
623. RICCIO, R., L. MINALE, C. PIZZA, F. ZOLLO, and J. PUSSET: Starfish Saponins. Part 8.

Structure of Nodososide, a Novel Type of Steroidal Glycoside from the Starfish *Protoreaster nodosus*. Tetrahedron Lett. **23**, 2899 (1982).
624. MINALE, L., C. PIZZA, R. RICCIO, and F. ZOLLO: Starfish Saponins. Part 9. A Novel 24-O-Glycosidated Steroid from the Starfish *Hacelia attenuata*. Experientia **39**, 567 (1983).
625. — — — — Starfish Saponins. Part 10. Further 24-O-Glycosidated Steroids from the Starfish *Hacelia attenuata*. Experientia **39**, 569 (1983).
626. MINALE, L., C. PIZZA, F. ZOLLO, and R. RICCIO: Trace Polyhydroxylated Steroids from Starfish *Hacelia attenuata*. J. Nat. Prod. **46**, 736 (1983).
627. MINALE, L., C. PIZZA, R. RICCIO, F. ZOLLO, J. PUSSET, and P. LABOUTE: Starfish Saponins, XIII. Occurrence of Nodososide in the Starfish *Acanthaster planci* and *Linckia laevigata*. J. Nat. Prod. **47**, 558 (1984).
628. KOMORI, T., J. MATSUO, Y. ITAKURA, K. SAKAMOTO, Y. ITO, S. TAGUCHI, and T. KAWASAKI: Biologically Active Glycosides from Asteroidea, II. Steroid Oligoglycosides from the Starfish *Acanthaster planci* L., 1. Isolation and Structure of the Oligoglycoside Sulfates. Liebigs Ann. Chem. **1983**, 24.
629. KOMORI, T., H. NANRI, Y. ITAKURA, K. SAKAMOTO, S. TAGUCHI, R. HIGUCHI, T. KAWASAKI, and T. HIGUCHI: Biologically Active Glycosides from Asteroidea, III. Steroid Oligoglycosides from the Starfish *Acanthaster planci* L., 2. Structure of Two Newly Characterized Genuine Sapogenins and an Oligoglycoside Sulfate. Liebigs Ann. Chem. **1983**, 37.
630. ITAKURA, Y., T. KOMORI, and T. KAWASAKI: Biologically Active Gylcosides from Asteroidea, IV. Steroid Oligoglycosides from the Starfish *Astropecten latespinosus* Meissner. Liebigs Ann. Chem. **1983**, 56.
631. — — — — Biologically Active Glycosides from Asteroidea, V. Steroid Oligoglycosides from the Starfish *Asterias amurensis* (cf.). *versicolor* Sladen, 1. Structural Elucidation of a New Oligoglycoside Sulfate. Liebigs Ann. Chem. **1983**, 2079.
632. KOMORI, T., H.C. KREBS, Y. ITAKURA, R. HIGUCHI, K. SAKAMOTO, S. TAGUCHI, and T. KAWASAKI: Biologisch aktive Glycoside aus Asteroidea, VI. Steroid-Oligoglycoside aus dem Seestern *Luidia maculata* Müller et Troschel, 1. Die Strukturen eines neuen Aglyconsulfats und von zwei neuen Oligoglycosidsulfaten. Liebigs Ann. Chem. **1983**, 2092.
633. KREBS, H.C., T. KOMORI, and T. KAWASAKI: Biologisch aktive Glycoside aus Asteroidea, VII. Steroid-Oligoglycoside aus dem Seestern *Luidia maculata* Müller et Troschel, 2. Die Strukturen von zwei neuen Oligoglycosidsulfaten. Liebigs Ann. Chem. **1984**, 296.
634. NEIRA, C., M. HOENEISEN, M. SILVA, and P.G. SAMMES: Marine Organic Chemistry, III. Isolation of 3β,6α,Dihydroxy-5α-cholest-9(11)-en-23-one from the Starfish *Stichaster striatus*. J. Nat. Prod. **47**, 182 (1984).
635. FINDLAY, J.A., and V.K. AGARWAL: Aglycones from the Saponin of the Starfish *Asterias vulgaris*. J. Nat. Prod. **46**, 876 (1983).
636. FINDLAY, J.A., V.K. AGARWAL, and Y.E. MOHARIR: On the Saponins of the Starfish *Asterias vulgaris*. J. Nat. Prod. **47**, 113 (1984).
637. DINI, A., F.A. MELLON, L. MINALE, C. PIZZA, R. RICCIO, R. SELF, and F. ZOLLO: Starfish Saponins - XI. Isolation and Partial Characterization of the Saponins from the Starfish *Marthasterias glacialis*. Comp. Biochem. Physiol. **76B**, 839 (1983).
638. BRUNO, I., L. MINALE, C. PIZZA, F. ZOLLO, R. RICCIO, and F.M. MELLON: Starfish Saponins. Part 14. Structures of the Steroidal Glycoside Sulphates from the Starfish *Marthasterias glacialis*. J. Chem. Soc., Perkin Trans. I **1984**, 1875.
639. MINALE, L., R. RICCIO, O. SQUILLACE-GRECO, J. PUSSET, and J.L. MENOU: Starfish Saponins - XVI. Composition of the Steroidal Glycoside Sulphates from the Starfish *Luidia maculata*. Comp. Biochem. Physiol. **80B**, 113 (1985).
640. RICCIO, R., C. PIZZA, O. SQUILLACE-GRECO, and L. MINALE: Starfish Saponins.

Part 17. Steroidal Glycoside Sulphates from the Starfish *Ophidiaster ophidianus* (Lamarck) and *Hacelia attenuata* (Gray). J. Chem. Soc., Perkin Trans. I **1985**, 655.

641. RICCIO, R., O. SQUILLACE-GRECO, L. MINALE, J. PUSSET, and J.L. MENOU: Starfish Saponins, Part 18. Steroidal Glycoside Sulfates from the Starfish *Linckia laevigata*. J. Nat. Prod. **48**, 97 (1985).

642. RICCIO, R., F. ZOLLO, E. FINAMORE, L. MINALE, D. LAURENT, G. BARGIBANT, and J. PUSSET: Starfish Saponins, 19. A Novel Steroidal Glycoside Sulfate from the Starfishes *Protoreaster nodosus* and *Pentaceraster alveolatus*. J. Nat. Prod. **48**, 266 (1985).

643. FUSETANI, N., Y. KATO, K. HASHIMOTO, T. KOMORI, Y. ITAKURA, and T. KAWASAKI: Biological Activities of Asterosaponins with Special Reference to Structure-Activity Relationships. J. Nat. Prod. **47**, 997 (1984).

644. KICHA, A.A., A.I. KALINOVSKY, E.V. LEVINA, V.A. STONIK, and G.B. ELYAKOV: Asterosaponin P_1 from the Starfish *Patiria pectinifera*. Tetrahedron Lett. **24**, 3893 (1983).

645. VOOGT, P.A., and J.W.A. VAN RHEENEN: Carbohydrate Content and Composition of Asterosaponins from Different Organs of the Sea Star *Asterias rubens*: Relation to Their Haemolytic Activity and Implications for Their Biosynthesis. Comp. Biochem. Physiol. **72B**, 683 (1982).

645a. BURNELL, D.J., J.W. APSIMON, and M.W. GILGAN: Variations in the Levels of Asterone and Asterogenol, Two Steroids from the Saponins of the Starfish, *Asterias vulgaris* (Verrill). Steroids **44**, 67 (1984).

646. DE SIMONE, F., A. DINI, L. MINALE, R. RICCIO, and F. ZOLLO: The Sterols of the Asteroid *Echinaster sepositus*. Comp. Biochem. Physiol. **66B**, 351 (1980).

647. SATO, S., N. IKEKAWA, A. KANAZAWA, and T. ANDO: Identification of 23-Demethylacanthasterol in an Asteroid, *Acanthaster planci* and Its Synthesis. Steroids **36**, 65 (1980).

648. BURNELL, D.J., J.W. APSIMON, and M.W. GILGAN: Seasonal and Geographic Variations of the Sterols from the Starfish *Asterias vulgaris* (Verrill). Steroids **39**, 357 (1982).

649. SELDES, A.M., and E.G. GROS: Main Sterols from the Starfish *Comasterias lurida*. Comp. Biochem. Physiol. **80B**, 337 (1985).

650. MINALE, L., C. PIZZA, R. RICCIO, C. SORRENTINO, F. ZOLLO, J. PUSSET, and G. BARGIBANT: Minor Polyhydroxylated Sterols from the Starfish *Protoreaster nodosus*. J. Nat. Prod. **47**, 790 (1984).

651. RICCIO, R., L. MINALE, S. PAGONIS, C. PIZZA, F. ZOLLO, and J. PUSSET: A Novel Group of Highly Hydroxylated Steroids from the Starfish *Protoreaster nodosus*. Tetrahedron **38**, 3615 (1982).

652. MINALE, L., C. PIZZA, F. ZOLLO, and R. RICCIO: 5α-Cholestane-3β,6β,15α,16β,26-pentol: a Polyhydroxylated Sterol from the Starfish *Hacelia attenuata*. Tetrahedron Lett. **23**, 1841 (1982).

653. MINALE, L., C. PIZZA, R. RICCIO, O. SQUILLACE GRECO, F. ZOLLO, J. PUSSET, and J.L. MENOU: New Polyhydroxylated Sterols from the Starfish *Luidia maculata*. J. Nat. Prod. **47**, 784 (1984).

654. D'AURIA, M.V., E. FINAMORE, L. MINALE, C. PIZZA, R. RICCIO, F. ZOLLO, M. PUSSET, and P. TIRARD: Steroids from the Starfish *Euretaster insignis*: a Novel Group of Sulphated 3β,21-Dihydroxysteroids. J. Chem. Soc., Perkin Trans. I **1984**, 2277.

655. KOMORI, T., Y. SANECHIKA, Y. ITO, J. MATSUO, T. NOHARA, T. KAWASAKI, and H.-R. SCHULTEN: Biologisch aktive Glycoside aus Asteroidea, I. Strukturen eines neuen Cerebrosidgemischs und von Nucleosiden aus dem Seestern *Acanthaster planci*. Liebigs Ann. Chem. **1980**, 653.

656. NAKAMURA, H., J. KOBAYASHI, and Y. HIRATA: Isolation and Structure of a 330 nm UV-absorbing Substance, Asterina-330 from the Starfish *Asterina pectinifera*. Chem. Lett. **1981**, 1413.

657. NOGUCHI, T., H. NARITA, J. MARUYAMA, and K. HASHIMOTO: Tetrodotoxin in the Starfish *Astropecten polyacanthus*, in Association with Toxification of a Trumpet Shell, "Boshubora" *Charonia sauliae*. Bull. Japan. Soc. Sci. Fish. **48**, 1173 (1982).
658. MARUYAMA, J., T. NOGUCHI, J.K. JEON, T. HARADA, and K. HASHIMOTO: Occurrence of Tetrodotoxin in the Starfish *Astropecten latespinosus*. Experientia **40**, 1395 (1984).
659. KUZNETSOVA, T.A., M.M. ANISIMOV, A.M. POPOV, S.I. BARANOVA, S.S. AFIYATULLOV, I.I. KAPUSTINA, A.S. ANTONOV, and G.B. ELYAKOV: A Comparative Study *in vitro* of Physiological Activity of Triterpene Glycosides of Marine Invertebrates of Echinoderm Type. Comp. Biochem. Physiol. **73C**, 41 (1982).
660. KITAGAWA, I., T. NISHINO, M. KOBAYASHI, and Y. KYOGOKU: Marine Natural Products. VIII. Bioactive Triterpene-Oligoglycosides from the Sea Cucumber *Holothuria leucospilota* Brandt (2). Structure of Holothurin A. Chem. Pharm. Bull. (Japan) **29**, 1951 (1981).
661. KITAGAWA, I., T. NISHINO, M. KOBAYASHI, T. MATSUNO, H. AKUTSU, and Y. KYOGOKU: Marine Natural Products. VII. Bioactive Triterpene Oligoglycosides from the Sea Cucumber *Holothuria leucospilota* Brandt (1). Structure of Holothurin B. Chem. Pharm. Bull. (Japan) **29**, 1942 (1981).
661a. IVANOVA, N.S., O.F. SMETANINA, and T.A. KUZNETSOVA: Glycosides of Marine Invertebrates. XXVI. Holothurin A from the Pacific Ocean Holothurian *Holothuria squamifera*. Isolation of the Native Aglycone. Khim. Prir. Soedin. **1984**, 448.
661b. IVANOVA, N.S., and T.A. KUZNETSOVA: Holothurin A – The Main Triterpene Glycoside of the Cuvierian Organs of the Holothurian *Bohadschia graeffei*. The Structures of the Native Aglycone and of Progenins. Khim. Prir. Soedin. **1985**, 123.
662. OLEINIKOVA, G.K., T.A. KUZNETSOVA, N.S. IVANOVA, A.I. KALINOVSKII, N.V. ROVNYKH, and G.B. ELYAKOV: Glycosides of Marine Invertebrates. XV. A New Triterpene Glycoside – Holothurin A_1 – from Caribbean Holothurians of the Family Holothuriidae. Khim. Prir. Soedin. **1982**, 464.
663. KITAGAWA, I., T. INAMOTO, M. FUCHIDA, S. OKADA, M. KOBAYASHI, T. NISHINO, and Y. KYOGOKU: Structures of Echinoside A and B, Two Antifungal Oligoglycosides from the Sea Cucumber *Actinopyga echinites* (Jaeger). Chem. Pharm. Bull. (Japan) **23**, 1651 (1980).
664. KALININ, V.I., and V.A. STONIK: Glycosides of Marine Invertebrates. Structure of Holothurin A_2 from the Holothurian *Holothuria edulis*. Khim. Prir. Soedin. **1982**, 215.
665. – – Glycosides of the Holothurian *Bohadschia graeffei*. Khim. Prir. Soedin. **1982**, 789.
666. OLEINIKOVA, G.K., T.A. KUZNETSOVA, N.V. ROVNYKH, A.I. KALINOVSKII, and G.B. ELYAKOV: Glycosides of Marine Invertebrates. XVIII. Holothurin A_2 from the Caribbean Holothurian *Holothuria floridana*. Khim. Prir. Soedin. **1982**, 527.
667. ELYAKOV, G.B., N.I. KALINOVSKAYA, A.I. KALINOVSKII, V.A. STONIK, and T.A. KUZNETSOVA: Glycosides of Marine Invertebrates. XIII. New Holothurinogenins of Holothurin B_1 from *Holothuria floridana*. Khim. Prir. Soedin. **1982**, 323.
668. KUZNETSOVA, T.A., N.I. KALINOVSKAYA, A.I. KALINOVSKII, G.K. OLEINIKOVA, N.V. ROVNYKH, and G.B. ELYAKOV: Glycosides of Marine Invertebrates. XIV. Structure of Holothurin B_1 from the Holothurian *Holothuria floridana*. Khim. Prir. Soedin. **1982**, 482.
669. OLEINIKOVA, G.K., and T.A. KUZNETSOVA: Two-Stage Smith Degradation of Holothurin B_1 from the Holothurian *Holothuria floridana*. Khim. Prir. Soedin. **1983**, 534.
670. KITAGAWA, I., M. KOBAYASHI, and Y. KYOGOKU: Marine Natural Products. IX. Structural Elucidation of Triterpenoidal Oligoglycosides from the Bahamean Sea Cucumber *Actinopyga agassizi* Selenka. Chem. Pharm. Bull. (Japan) **30**, 2045 (1982).
671. KITAGAWA, I., M. KOBAYASHI, M. HORI, and Y. KYOGOKU: Structures of Four

New Triterpenoidal Oligoglycosides, Bivittoside A, B, C, and D, from the Sea Cucumber *Bohadschia bivittata* Mitsukuri. Chem. Pharm. Bull. (Japan) **29**, 282 (1981).
672. HABERMEHL, G.G., and J.H. KIRSCH: Synthesis of Bivittoside C Genine. Tetrahedron Lett. **24**, 2981 (1983).
673. HABERMEHL, G., and G. VOLKWEIN: Aglycones of the Toxins from the Cuvierian Organs of *Holothuria forskali* and a New Nomenclature for the Aglycones from Holothurioideae. Toxicon **9**, 319 (1971).
674. STONIK, V.A., V.F. SHARYPOV, T.A. KUZNETSOVA, and A.I. KALINOVSKII: Native Aglycones of Triterpene Glycosides of the Holothurian *Bohadschia argus*. Khim. Prir. Soedin. **1982**, 790.
675. KITAGAWA, I., M. KOBAYASHI, T. INAMOTO, T. YASUZAWA, Y. KYOGOKU, and M. KIDO: Stichlorogenol and Dehydrostichlorogenol, Genuine Aglycones of Stichlorosides A_1, B_1, C_1, and A_2, B_2, C_2, from the Sea Cucumber *Stichopus chloronotus* (Brandt). Chem. Pharm. Bull. (Japan) **29**, 1189 (1981).
676. KITAGAWA, I., M. KOBAYASHI, T. INAMOTO, T. YASUZAWA, and Y. KYOGOKU: The Structures of Six Antifungal Oligoglycosides, Stichlorosides A_1, A_2, B_1, B_2, C_1, and C_2, from the Sea Cucumber *Stichopus chloronotus* (Brandt). Chem. Pharm. Bull. (Japan) **29**, 2387 (1981).
677. ELYAKOV, G.B., V.A. STONIK, S.S. AFIYATULLOV, A.I. KALINOVSKII, V.F. SHARYPOV, and L.Y. KOROTKIKH: Native Genins from Glycosides of Holothurians. Dokl. Akad. Nauk SSSR **259**, 1367 (1981).
678. SHARYPOV, V.F., A.D. CHUMAK, V.A. STONIK, and G.B. ELYAKOV: Glycosides of Marine Invertebrates. X. The Structure of Stichoposides A and B from the Holothurian *Stichopus chloronotus*. Khim. Prir. Soedin. **1981**, 181.
679. STONIK, V.A., I.I. MAL'TSEV, A.I. KALINOVSKII, C. CONDE, and G.B. ELYAKOV: Glycosides of Marine Invertebrates. XI. Two New Triterpene Glycosides from Holothurians of the Family *Stichopodidae*. Khim. Prir. Soedin. **1982**, 194.
680. STONIK, V.A. I.I. MAL'TSEV, A.I. KALINOVSKII, and G.B. ELYAKOV: Glycosides of Marine Invertebrates. XII. Structure of a New Triterpene Oligoglycoside from Holothurians of the Family *Stichopodidae*. Khim. Prir. Soedin. **1982**, 200.
681. MAL'TSEV, I.I., V.A. STONIK, and A.I. KALINOVSKII: Stichoposide E – A New Triterpene Glycoside from Holothurians of the Family *Stichopodidae*. Khim. Prir. Soedin. **1983**, 308.
682. KELECOM, A., B. TURSCH, and M. VANHAELEN: Chemical Studies of Marine Invertebrates XIX. Glycosidic Chain Structure of Thelothurins A and B, Two New Saponins from the Indo-Pacific Sea Cucumber *Thelenota ananas* Jaeger (Echinodermata). Bull. Soc. Chim. Belg. **85**, 277 (1976).
683. STONIK, V.A., I.I. MAL'TSEV, and G.B. ELYAKOV: The Structure of Thelenotosides A and B from the Holothurian *Thelenota ananas*. Khim. Prir. Soedin. **1982**, 624.
684. AFIYATULLOV, S.S., V.A. STONIK, and G.B. ELYAKOV: Glycosides of Marine Invertebrates. Cucumarioside G_1 from the Holothurian *Cucumaria fraudatrix*. Khim. Prir. Soedin. **1983**, 654.
684a. AVILOV, S.A., L.Y. TISHCHENKO, and V.A. STONIK: Structure of Cucumarioside A_2 – 2 – A Triterpene Glycoside from the Holothurian *Cucumaria japonica*. Khim. Prir. Soedin. **1984**, 799.
685. AFIYATULLOV, S.S., V.A. STONIK, A.I. KALINOVSKII, and G.B. ELYAKOV: Glycosides of Marine Invertebrates. XVI. Cucumariogenin from Glycosides of the Holothurian *Cucumaria fraudatrix*. Khim. Prir. Soedin. **1983**, 59.
686. FINDLAY, J.A., and A. DALJEET: Frondogenin, a New Aglycone from the Sea Cucumber *Cucumaria frondosa*. J. Nat. Prod. **47**, 320 (1984).
687. SMETANINA, O.F., A.I. KALINOVSKII, T.A. KUZNETSOVA, V.A. STONIK, and G.B. ELYAKOV: Glycosides of Marine Invertebrates. XVII. New Genins of Glycosides

from the Caribbean Holothurian *Parathyona* sp. (Holothurioidae, Cucumariidae). Khim. Prir. Soedin. **1983**, 64.
688. GARNEAU, F.-X., J.-L. SIMARD, O. HARVEY, J.W. APSIMON, and M. GIRARD: The Structure of Psoluthurin A, the Major Triterpene Glycoside of the Sea Cucumber *Psolus fabricii*. Can. J. Chem. **61**, 1465 (1983).
689. KALININ, V.I., V.R. STEPANOV, and V.A. STONIK: Psolusoside A – A New Triterpene Glycoside from the Holothurian *Psolus fabricii*. Khim. Prir. Soedin. **1983**, 789.
690. KALINOVSKII, A.I., S.A. AVILOV, V.R. STEPANOV, and V.A. STONIK: Glycosides of Marine Invertebrates. XXIII. Kurilogenin – A New Genin from the Glycosides of the Holothurian *Duasmodactyla kurilensis*. Khim. Prir. Soedin. **1983**, 724.
691. GORSHKOV, B.A., I.A. GORSHOKOVA, V.A. STONIK, and G.B. ELYAKOV: Effect of Marine Glycosides on Adenosinetriphosphatase Activity. Toxicon **20**, 655 (1982).
692. KALINOVSKAYA, N.I., T.A. KUZNETSOVA, and G.B. ELYAKOV: Sterol Composition of Pacific Holothurian *Stichopus japonicus selenka*. Comp. Biochem. Physiol. **74 B**, 597 (1983).
693. ELYAKOV, G.B., N.I. KALINOVSKAYA, V.A. STONIK, and T.A. KUZNETSOVA: Glycosides of Marine Invertebrates VI. Steroid Glycosides from Holothurian *Stichopus japonicus*. Comp. Biochem. Physiol. **65 B**, 309 (1980).
694. SMETANINA, O.F., E. NAIT, T.A. KUZNETSOVA, and G.B. ELYAKOV: Sulfated Derivatives from Marine Organisms. II. Sulfated Steroyl Alcohols from the Holothurian *Parathyona* sp. (Holothurioidea, Cucumariidae). Khim. Prir. Soedin. **1981**, 585.
694a. BATRAKOV, S.G., V.B. MURATOV, O.G. SAKANDELIDZE, O.S. RESHETOVA, and B.V. ROZYNOV: Sterol Sulfates from the Far Eastern Holothurian *Cucumaria japonica*. Khim. Prir. Soedin. **1984**, 470.
695. FINDLAY, J.A., A. DALJEET, J. MATSOUKAS, and Y.E. MOHARIR: Constituents of the Sea Cucumber *Cucumaria frondosa*. J. Nat. Prod. **47**, 560 (1984).
696. ELYAKOV, G.B., S.N. FEDOROV, A.D. CHUMAK, V.V. ISAKOV, and V.A. STONIK: Sulfated Derivatives from Marine Invertebrates – 1. Sulfated Sterols from Some Species of Echinoderms. Comp. Biochem. Physiol. **71 B**, 325 (1982).
697. D'AURIA, M.V., R. RICCIO, and L. MINALE: Ophioxanthin, a New Marine Carotenoid Sulfate from the Ophiuroid *Ophioderma longicaudum*. Tetrahedron Lett. **26**, 1871 (1985).
698. RIDEOUT, J.A., N.B. SMITH, and M.D. SUTHERLAND: Chemical Defense of Crinoids by Polyketide Sulfates. Experientia **35**, 1273 (1979).
699. RIDEOUT, J.A., and M.D. SUTHERLAND: Pigments of Marine Animals. XIV. Polyketide Sulfates from the Crinoid *Comatula pectinata*. Aust. J. Chem. **34**, 2385 (1981).
699a. – – Pigments of Marine Animals. XV. Bianthrones and Realated Polyketides from *Lamprometra palmata gyges* and Other Species of Crinoids. Aust. J. Chem. **38**, 793 (1985).
700. KIMURA, A., and H. NAKAGAWA: Action of an Extract from the Sea Urchin *Toxopneustes pileolus* on Isolated Smooth Muscle. Toxicon **18**, 689 (1980).
701. NAKAGAWA, H., A. KIMURA, M. TAKEI, and K. ENDO: Histamine Release from Rat Mast Cells Induced by an Extract from the Sea Urchin *Toxopneustes pileolus*. Toxicon **20**, 1095 (1982).
702. NAKAGAWA, H., and A. KIMURA: Partial Purification and Characterization of a Toxic Substance from Pedicellariae of the Sea Urchin *Toxopneustes pileolus*. Japan. J. Pharmacol. **32**, 966 (1982).
703. KIMURA, A., H. NAKAGAWA, H. HAYASHI, and K. ENDO: Seasonal Changes in Contractile Activity of a Toxic Substance from the Pedicellaria of the Sea Urchin *Toxopneustes pileolus*. Toxicon **22**, 353 (1984).
704. MEBS, D.: A Toxin from the Sea Urchin *Tripneustes gratilla*. Toxicon **22**, 306 (1984).

705. PETTIT, G.R., J.A. RIDEOUT, J.A. HASLER, D.L. DOUBEK, and P.R. REUCROFT: Isolation and Characterization of Lytechinastatin. J. Nat. Prod. **44**, 713 (1981).
706. PETTIT, G.R., J.A. HASLER, K.D. PAULL, and C.L. HERALD: Antineoplastic Agents. 76. The Sea Urchin *Strongylocentrotus droebachiensis*. J. Nat. Prod. **44**, 701 (1981).
707. KAMERLING, J.P., R. SCHAUER, J.F.G. VLIEGENTHART, and K. HOTTA: Identification of the Sialic Acids from the Egg Jelly Coat of the Sea Urchin *Pseudocentrotus depressus* (Okayama). Hoppe-Seyler's Z. Physiol. Chem. **361**, 1511 (1980).
708. PALUMBO, A., M. D'ISCHIA, G. MISURACA, and G. PROTA: Isolation and Structure of a New Sulphur-Containing Aminoacid from Sea Urchin Eggs. Tetrahedron Lett. **23**, 3207 (1982).
709. IRELAND, C., and P.J. SCHEUER: Ulicyclamide and Ulithiacyclamide, Two New Small Peptides from a Marine Tunicate. J. Am. Chem. **102**, 5688 (1980).
710. WASYLYK, J.M., J.E. BISKUPIAK, C.E. COSTELLO, and C.M. IRELAND: Cyclic Peptide Structures from the Tunicate *Lissoclinum patella* by FAB Mass Spectrometry. J. Org. Chem. **48**, 4445 (1983).
711. BISKUPIAK, J.E., and C.M. IRELAND: Absolute Configuration of Thiazole Amino Acids in Peptides. J. Org. Chem. **48**, 2302 (1983).
711a. SCHMIDT, U., and P. GLEICH: Totalsynthese von Ulicyclamid. Angew. Chem. **97**, 606 (1985).
712. IRELAND, C.M., A.R. DURSO, JR., R.A. NEWMAN, and M.P. HACKER: Antineoplastic Cyclic Peptides from the Marine Tunicate *Lissoclinum patella*. J. Org. Chem. **47**, 1807 (1982).
712a. HAMADA, Y., M. SHIBATA, and T. SHIOIRI: New Methods and Reagents in Organic Synthesis. 55. Total Syntheses of Patellamides B and C, Cytotoxic Cyclic Peptides from a Tunicate. 1. Their Proposed Structures Should be Corrected. Tetrahedron Lett. **26**, 5155 (1985).
712b. – – – New Methods and Reagents in Organic Synthesis. 56. Total Syntheses of Patellamides B and C, Cytotoxic Cyclic Peptides from a Tunicate. 2. Their Real Structures have been Determined by Their Syntheses. Tetrahedron Lett. **26**, 5159 (1985).
712c. – – – New Methods and Reagents in Organic Synthesis. 58. A Synthesis of Patellamide A, a Cytotoxic Cyclic Peptide from a Tunicate. Revision of Its Proposed Structure. Tetrahedron Lett. **26**, 6501 (1985).
713. RINEHART, JR., K.L., J.B. GLOER, R.G. HUGHES, JR., H.E. RENIS, J.P. MCGOVREN, E.B. SWYNENBERG, D.A. STRINGFELLOW, S.L. KUENTZEL, and L.H. LI: Didemnins: Antiviral and Antitumor Depsipeptides from a Caribbean Tunicate. Science **212**, 933 (1981).
714. RINEHART, JR, K.L., J.B. GLOER, G.R. WILSON, R.G. HUGHES, JR., L.H. LI, H.E. RENIS, and J.P. MCGOVREN: Antiviral and Antitumor Compounds from Tunicates. Fed. Proc. **42**, 87 (1983).
715. RINEHART, JR., K.L., J.B. GLOER, J.C. COOK, JR., S.A. MIZSAK, and T.A. SCAHILL: Structures of the Didemnins, Antiviral and Cytotoxic Depsipeptides from a Caribbean Tunicate. J. Am. Chem. Soc. **103**, 1857 (1981).
716. RINEHART, JR., K.L.: Fast Atom Bombardment Mass Spectrometry. Science **218**, 254 (1982).
717. HAMAMOTO, Y., M. ENDO, M. NAKAGAWA, T. NAKANISHI, and K. MIZUKAWA: A New Cyclic Peptide, Ascidiacyclamide, Isolated from Ascidian. J. Chem. Soc., Chem. Commun. **1983**, 323.
717a. BRUENING, R.C., E.M. OLTZ, J. FURUKAWA, K. NAKANISHI, and K. KUSTIN: Isolation and Structure of Tunichrome B-1, a Reducing Blood Pigment from the Tunicate *Ascidia nigra* L. J. Am. Chem. Soc. **107**, 5298 (1985).
718. IRELAND, C.M., A.R. DURSO, JR., and P.J. SCHEUER: N,N'-Diphenethylurea, a Metabolite from the Marine Ascidian *Didemnum ternatanum*. J. Nat. Prod. **44**, 360 (1981).

719. SESIN, D.F., and C.M. IRELAND: Iodinated Phenethylamine Products from a Didemnid Tunicate. Tetrahedron Lett. **25**, 403 (1984).
720. HEITZ, S., M. DURGEAT, M. GUYOT, C. BRASSY, and B. BACHET: Nouveau Dérivé Indolique du Thiadiazole-1,2,4, Isolé D'Un Tunicier (*Dendrodoa grossularia*). Tetrahedron Lett. **21**, 1457 (1980).
721. HOGAN, I.T., and M. SAINSBURY: The Synthesis of Dendrodoine, 5-[3-(N,N-Dimethylamino)-1,2,4-thiadiazolyl]-3-indolylmethanone, a Metabolite of the Marine Tunicate *Dendroda grossular*. Tetrahedron **40**, 681 (1984).
721a. ROLL, D.M., and C.M. IRELAND: Citorellamine, a New Bromoindole Derivative from *Polycitorella mariae*. Tetrahedron Lett. **26**, 4303 (1985).
722. MOQUIN, C., and M. GUYOT: Grossularine, a Novel Indole Derivative from the Marine Tunicate, *Dendrodoa grossularia*. Tetrahedron Lett. **25**, 5047 (1984).
723. RINEHART, JR., K.L., J. KOBAYASHI, G.C. HARBOUR, R.G. HUGHES, JR., S.A. MIZSAK, and T.A. SCAHILL: Eudistomins C, E, K, and L, Potent Antiviral Compounds Containing a Novel Oxathiazepine Ring from the Caribbean Tunicate *Eudistoma olivaceum*. J. Am. Chem. Soc. **106**, 1524 (1984).
724. KOBAYASHI, J., G.C. HARBOUR, J. GILMORE, and K.L. RINEHART, JR.: Eudistomins A, D, G, H, I, J, M, N, O, P, and Q, Bromo-, Hydroxy-, Pyrrolyl-, and 1-Pyrrolinyl-β-carbolines from the Antiviral Caribbean Tunicate *Eudistoma olivaceum*. J. Am. Chem. Soc. **106**, 1526 (1984).
725. WATANABE, K., S. MATSUNAGA, and S. KONOSU: Studies on the Extractive Components of Ascidians II. Halocynine, a Novel Betaine Isolated from the Muscle of Ascidian *Halocynthia roretzi*. Tetrahedron Lett. **25**, 2003 (1984).
726. CARTÉ, B., and D.J. FAULKNER: Revised Structures for the Polyandrocarpidines. Tetrahedron Lett. **23**, 3863 (1982).
727. KAZLAUSKAS, R., J.F. MARWOOD, P.T. MURPHY, and R.J. WELLS: A Blue Pigment from a Compound Ascidian. Aust. J. Chem. **35**, 215 (1982).
728. KOBAYASHI, J., H. NAKAMURA, and Y. HIRATA: Isolation and Structure of a UV-Absorbing Substance 337 from the Ascidian *Halocynthia roretzi*. Tetrahedron Lett. **22**, 3001 (1981).
729. WATANABE, K., H. MAEZAWA, H. NAKAMURA, and S. KONOSU: Seasonal Variation of Extractive Nitrogen and Free Amino Acids in the Muscle of the Ascidian *Halocynthia roretzi*. Bull. Japan. Soc. Sci. Fish. **49**, 1755 (1983).
730. TARGETT, N.M., and W.S. KEERAN: A Terpenehydroquinone from the Marine Ascidian *Aplidium constellatum*. J. Nat. Prod. **47**, 556 (1984).
731. MATSUNO, T., and M. OOKUBO: A New Carotenoid, Halocynthiaxanthin from the Sea Squirt, *Halocynthia roretzi*. Tetrahedron Lett. **22**, 4659 (1981).
732. — —: A New Marine Carotenoid, Mytiloxanthinone from the Sea Squirt, *Halocynthia roretzi*. Chem. Lett. **1982**, 1605.
733. BELAUD, C., and M. GUYOT: Sidnyaxanthin, a New Carotenoid from the Tunicate, *Sidnyum argus*. Tetrahedron Lett. **25**, 3087 (1984).
733a. MATSUNO, T., M. OOKUBO, and T. KOMORI: Carotenoids of Tunicates, III. The Structural Elucidation of Two New Marine Carotenoids, Amarouciaxanthin A and B. J. Nat. Prod. **48**, 606 (1985).
733b. REBACHUK, N.M., O.B. MAKSIMOV, L.S. BOGUSLAVSKAYA, and S.A. FEDOREEV: Carotenoids of the Ascidian *Halocynthia aurantium*. Khim. Prir. Soedin. **1984**, 431.
734. HA, T.B.T, W.C.M.C. KOKKE, and C. DJERASSI: Minor Sterols of Marine Invertebrates 37. Isolation of Novel Coprostanols and 4α-Methyl Sterols from the Tunicate *Ascidia nigra*. Steroids **40**, 433 (1982).
735. KLJAJIĆ, Z., N. DOGOVIĆ, and M.J. GAŠIĆ: Sterols in Adriatic Sea Ascidians. Comp. Biochem. Physiol. **75B**, 519 (1983).
736. GUYOT, M., and M. DURGEAT: Occurrence of 9(11)-Unsaturated Sterol Peroxides in Tunicates. Tetrahedron Lett. **22**, 1391 (1981).

737. GUYOT, M., D. DAVOUST, and C. BELAUD: Hydroperoxy-24 Vinyl-24 Cholesterol, Nouvel Hydroperoxide Naturel Isolé De Deux Tuniciers: *Phallusia mamillata* et *Ciona intestinalis*. Tetrahedron Lett. **23**, 1905 (1982).
738. ALBANO, R.M., and P.A.S. MOURÃO: Presence of Sulfated Glycans in Ascidian Tunic and in the Body Wall of a Sea Cucumber. Biochim. Biophys. Acta **760**, 192 (1983).
739. DUNN, W.C., W.L. CARRIER, and J.D. REGAN: Effects of an Extract from the Sea Squirt *Ecteinascidia turbinata* on DNA Synthesis and Excision Repair in Human Fibroblasts. Toxicon **20**, 703 (1982).
740. BLUMENTHAL, K.M., and W.R. KEM: Structure and Action of Heteronemertine Polypeptide Toxins. Primary Structure of *Cerebratulus lacteus* Toxin A-III. J. Biol. Chem. **255**, 8266 (1980).
741. BLUMENTHAL, K.M.: Structure and Action of Heteronemertine Polypeptide Toxins. Disulfide Bonds of *Cerebratulus lacteus* Toxin A-III. J. Biol. Chem. **255**, 8273 (1980).
742. — Structure and Action of Heteronemertine Polypeptide Toxins. Membrane Penetration by *Cerebratulus lacteus* Toxin A-III. Biochem. **21**, 4229 (1982).
743. BLUMENTHAL, K.M., P.S. KEIM, R.L. HEINRIKSON, and W.R. KEM: Structure and Action of Heteronemertine Polypeptide Toxins. Amino Acid Sequence of *Cerebratulus lacteus* Toxin B-II and Revised Structure of Toxin B-IV. J. Biol. Chem. **256**, 9063 (1981).
744. BLUMENTHAL, K.M., and W.R. KEM: Structure and Action of Heteronemertine Polypeptide Toxins: Inactivation of *Cerebratulus lacteus* Toxin B-IV by Tyrosine Nitration. Arch. Biochem. Biophys. **203**, 816 (1980).
745. BLUMENTHAL, K.M.: Structure and Action of Heteronemertine Polypeptide Toxins: Inactivation of *Cerebratulus lacteus* Toxin B-IV Concomitant with Tryptophan Alkylation. Arch. Biochem. Biophys. **203**, 822 (1980).
746. BALLANTINE, J.A., A.F. PSAILA, A. PELTER, P. MURRAY-RUST, V. FERRITO, P. SCHEMBRI, and V. JACCARINI: The Structure of Bonellin and its Derivatives. Unique Physiologically Active Chlorins from the Marine Echurian *Bonellia viridis*. J. Chem. Soc., Perkin Trans. I **1980**, 1080.
747. HIGA, T., T. FUJIYAMA, and P.J. SCHEUER: Halogenated Phenol and Indole Constituents of Acorn Worms. Comp. Biochem. Physiol. **65B**, 525 (1980).
748. HOVINGH, P., and A. LINKER: An Unusual Heparan Sulfate Isolated from Lobsters (*Homarus americanus*). J. Biol. Chem. **257**, 9840 (1982).
749. CANNON, J.R., J.S. EDMONDS, K.A. FRANCESCONI, C.L. RASTON, J.B. SAUNDERS, B.W. SKELTON, and A.H. WHITE: Isolation, Crystal Structure and Synthesis of Arsenobetain, a Constituent of the Western Rock Lobster, the Dusky Shark, and Some Samples of Human Urine. Aust. J. Chem. **34**, 787 (1981).
750. FUSETANI, N., H. ENDO, K. HASHIMOTO, and K. TAKAHASHI: Occurrence of Potent Toxins in the Horseshoe Crab *Carcinoscorpius rotundicauda*. Toxicon **20**, 662 (1982).
751. FUSETANI, N., H. ENDO, K. HASHIMOTO, and M. KODAMA: Occurrence and Properties of Toxins in the Horseshoe Crab *Carcinoscorpius rotundicauda*. Toxicon Suppl. **3**, 165 (1983).
752. ENDEAN, R., R. LEWIS, P. GYR, and J. WILLIAMSON: Toxic Material from the Crab *Atergatis floridus*. Toxicon Suppl. **3**, 111 (1983).
753. DE FREITAS, J.C., and R.S. JACOBS: Biotoxins in Brachyuran Decapod Crustaceans. Toxicon Suppl. **3**, 157 (1983).
754. NOGUCHI, T., A. UZU, K. DAIGO (KOYAMA), Y. SHIDA, and K. HASHIMOTO: A Tetrodotoxin-Like Substance as a Minor Toxin in the Xanthid Crab *Atergatis floridus*. Toxicon **22**, 425 (1984).
755. RAJ, U., H. HAQ, Y. OSHIMA, and T. YASUMOTO: The Occurrence of Paralytic Shellfish Toxins in Two Species of Xanthid Crab from Suva Barrier Reef, Fiji Islands. Toxicon **21**, 547 (1983).

756. FUSETANI, N., K. HASHIMOTO, I. MIZUKAMI, H. KAMIYA, and S. YONABARU: Lethality in Mice of the Coconut Crab *Birgus latro*. Toxicon **18**, 694 (1980).
757. BISHAYEE, S., and D.T. DORAI: Isolation and Characterization of a Sialic Acid-Binding Lectin (Carcinoscorpin) from Indian Horseshoe Crab *Carcinoscorpius rotunda cauda*. Biochim. Biophys. Acta **623**, 89 (1980).
758. BRANDIN, E.R., and T.G. PISTOLE: Polyphemin: A Teichoic Acid-Binding Lectin from the Horseshoe Crab, *Limulus polyphemus*. Biochem. Biophys. Res. Commun. **113**, 611 (1983).
759. WATSON, R.D., and E. SPAZIANI: Rapid Isolation of Ecdysteroids from Crustacean Tissues and Culture Media Using Sep-Pak C18 Cartridges. J. Liquid Chromatography **5**, 525 (1982).
760. NEWCOMB, R.W.: Peptides in the Sinus Gland of *Cardisoma carnifex*: Isolation and Amino Acid Analysis. J. Comp. Physiol. **153**, 207 (1983).

(Received September 9, 1985)

Author Index

Page numbers printed in *italics* refer to References

Abe, M. *78*
Abell, C. *63*
Abound-Bichara, A. *332*
Adesida, G.A. *143*
Adesogan, E.K. *143*
Afiyatullov, S.S. *357, 358*
Afzal, S.M. *63*
Agarwal, S.C. *77*
Agarwal, S.K. *331*
Agarwal, V.K. *355*
Agena, G. *339*
Ahmed, S. *72*
Ahnert-Hilger, G. *339*
Ahond, A. 239, *346*
Akita, T. *63*
Akutsu, H. 248, *347*
Alam, M. *353*
Albano, R.M. *362*
Albericci, M. *327*
Albizati, K.F. *353*
Alcolado, P. *328*
Aldeen, S.I. *335*
Aldridge, D.C. *63, 77*
Alpertunga, B. 223, *342*
Alsen, C. *336, 337, 339*
Amade, P. *324*
Amrhein, W. *143*
Andersen, R.J. 268, *329, 335, 341, 349, 350, 353*
Anderson, J.R. *63*
Ando, T. *356*
Andrè, R. *143*
Andreev, S.N. 212, *338*
Andres, W.W. *72, 74*
Anisimov, M.M. 301, *354, 357*
Anjaneyulu, A.S.R. *340*
Anoud, S.R. *78*
Anthony, R.L. *334*
Antonov, A.S. *357*
Antosz, F.J. *63, 72*

Antus, S. *63*
Aoki, H. *70, 74*
Aoyagi, R. *349*
ApSimon, J.W. 283, 295, 300, 304, *354, 356, 359*
Aragozzini, F. *66*
Arahina, H. *64*
Arai, Y. *63, 348*
Arakawa, H. *63, 64*
Aray, K. *63*
Arazashvili, A.I. *64*
Arima, K. *73*
Arnold, E. 158, 254, *324, 344, 349, 350*
Arnone, A. *143*
Asano, J. *64*
Ashida, K. *348*
Assante, G. *64*
Aszalos, A. *143*
Atanosova, I. *76*
Atasaki, S. *71*
Aue, R. *64*
Avigad, L.S. 212, *335, 336*
Avilov, S.A. *358, 359*
Ayanoglu, E. *332*
Ayer, S.W. *349, 350*

Bachet, B. 313, *361*
Baerheim-Svendsen, A. *147, 149*
Bailey, L.E. *336*
Bailleul, F. *143*
Baird-Lambert, J.A. 231, *344, 347, 328, 330*
Baker, B.J. 249, *347*
Baker, T.C. *64, 73*
Bakus, G.J. 165, 173, *320, 326, 378*
Balerna, M. *336*
Ballantine, J.A. 318, *362*
Ballio, A. *64*
Baltus, W. *145*
Balusubramaniam, S. *65*

Bandurraga, M.M. 218, *341, 343, 344, 345*
Banerjee, B.K. *66*
Banks, S. *143*
Banville, J. *143*
Bapat, C.P. *72*
Baranova, S.I. *357*
Barber, J. *64*
Barbezat, B.O. *143*
Barbier, M. *67, 354*
Barboza, C.A. *321*
Barcellona, S. *64*
Barchi, J. *339*
Bargibant, G. *356*
Barhanin, J. *335*
Barras, S.J. *69*
Barre, F.P. *143*
Barrero, A.F. *74*
Barrett, J. *351*
Barrientos, A.S.M. 171, *326*
Barroso, J.T. *145*
Barrowcliff, M. *143*
Barry, R.D. *64*
Bartlett, R.T. *329*
Bartolini, G. *338, 339*
Bass, L.S. *324, 350*
Batrakov, S.G. *359*
Batu, G. *64*
Bauch, H.-J. *143*
Beachen, J.F. *144*, 209
Beesley, T.E. *78*
Behrens-Baumann, W. *334*
Belaud, C. *361*
Belgaonkar, V.H. *64*
Bellinger, G.C.A. *64*
Benayahu, Y. 248, *347*
Bender, M. *143*
Benjamin, C.R. *76*
Benvegnù, R. *341*
Beppu, T. *73*
Berenberg, C.J. *354*
Bèress, L. 213, *333, 336, 337, 338, 339*
Berg, A. van den *143*
Berg, W. *143*
Berger, S. *144*
Berger, Y. *144*
Bergquist, P.R. 153, *321, 323, 327, 331*
Berlepsch, K. *320*
Bernauer, K. *64, 75*
Bernd, A. 193, *329*
Bernheimer, A.W. *335, 336*
Bertau, R. *326*

Beynon, J.H. *144*
Bhakta, C. *66*
Bhide, B.H. *64, 73*
Bi, N. *148*
Bianchi, C.P. 209, *334*
Billedeau, R. *77*
Billek, G. *64*
Birch, A.J. *65*
Birkinshaw, J.H. *65*
Bishayee, S. *363*
Bishop, D.G. 212, *337, 338*
Biskuprik, J.E. 276, 309, *329, 353, 360*
Blackburn, G.M. *65*
Blackmann, A.J. *344*
Blanc, P.-A. *321*
Blank, F. *65, 70, 73*
Blankmeier, L.A. *326*
Blasberger, D. *331*
Blount, F.J. *325, 329*
Blum, M.S. *65, 71*
Blumenthal, K.M. 318, *335, 362*
Bobineau, L.M. *71*
Bodner, M. *342*
Boguslavskaya, L.S. *361*
Boguslavskii, V.M. *323*
Bohlin, L. *321, 322, 340, 341*
Bohlmann, F. *65, 68*
Bohonos, N. *74*
Boit, H.G. *65*
Bonini, C. *340, 342*
Borisov, M.I. *144, 149*
Bountry, J.L. *354*
Bouquet, A. *146*
Bousquet, J.F. *67*
Bowden, B.F. 222, 223, 251, *342, 343, 344, 345, 346, 347, 348*
Bowie, J.H. *144*
Bradsher, C.K. *147*
Bradshow, A.P.W. *68*
Brahmbhatt, D.I. *64*
Branck, G. *336*
Brand, J.M. *65*
Brandin, E.R. *363*
Braslau, R. *321*
Brassard, P. *143, 144, 148*
Brassy, C. 313, *361*
Braz Filho, R. *65, 67*
Breakman, J.C. 163, 190, 195, 225, *325, 326, 327, 329, 331, 342, 343, 344*
Brew, E.J.C. *144*
Brian, P.W. *65*

Author Index

Briggs, L.H. *144*
Brisson, C. *144*
Brophy, J.J. *65*
Brown, P. 270, 272, *351*
Bruening, R.C. 310, *360*
Bruneton, J. *146*
Bruno, I. *355*
Brusca, R.C. *320*
Burcham, J.M. *338*
Burnell, D.J. 283, 295, 300, *354, 356*
Burnet, A.R. *144*
Burnett, J.W. 209, *333, 334*
Burns, K.P. *345, 347*
Burreson, B.J. *345*
Buyer, J.S. *330*

Cabot, C. *76*
Cafieri, F. *323*
Calcagno, M.-P. *145*
Calle, M.V. *74*
Calton, G.J. 209, *333, 334*
Calvin, J. *350*
Camarda, L. *64, 65*
Cambie, R.C. *144, 331*
Cameron, A.M. *320*
Cameron, D.W. *145*
Campbell, W.E. *64*
Cannon, J.R. *362*
Capelle, N. *163, 325*
Capon, R.J. 159, 165, 173, 206, *324, 325, 326, 331, 333, 353*
Cardellina, J.H. 158, *324, 330, 333, 346*
Cardona, R.J. *145*
Carey, S.T. *72*
Cargo, D.G. 209, *334*
Cariello, L. *324*
Carlson, R.M.K. *340*
Carlè, J.S. *348*
Carmely, S. 216, 248, *320, 327, 328, 331, 340, 341, 344, 345, 347*
Carpenter, R.G. *65*
Carrier, W.L. *362*
Carruthers, W.R. *65*
Carté, B. 202, 314, *324, 331, 346, 350, 361*
Carter, G.T. *320*
Carter, R.H. *65, 69*
Caruso, A.J. *350*
Castiello, D. 205, *333*
Castonguay, A. *144*
Catalan, C.A.N. *320, 323*
Cattel, L. *65*
Cavè, A. *146*

Cavill, G.W.K. *65*
Cha, J.K. *338, 339*
Champion, B. *148*
Chang, C.W.J. 185, *326, 329*
Chang, P. *145*
Chantrapromma, K. *348*
Chaplen, P. *65*
Charles, C. 190, *329*
Charubala, R. *66*
Chatterjea, J.N. *66*
Chaudhury, G.R. *66*
Chen, M.H.M. *340, 346, 350*
Chen, Y.-K. *69*
Cheval, K.K. *66*
Chevolot, L. *324*
Chhatwal, G.S. *339*
Chien, M.M. *66*
Chin, C.C. *78*
Chong, J.M. *350*
Christ, W.J. *338, 339*
Christaens, L. *74*
Christophersen, C. *348*
Chu, J.-H. *70*
Chung, M.C.H. *351*
Chumak, A.D. *358, 359*
Ciegler, A. *66*
Ciereszko, L.S. *345, 346*
Ciminiello, P. *323, 325*
Cimino, G. 186, 195, 205, 210, 258, 261, 262, *325, 327, 328, 330, 331, 332, 333, 334, 341, 343, 349, 350, 353*
Clardy, J. 66, 158, 166, 185, 224, 239, 254, 276, *324, 325, 326, 327, 329, 330, 332, 333, 335, 340, 342, 343, 344, 345, 346, 349, 350, 351, 353*
Clark, C. 274, 275, *352*
Clastres, A. 239, *346*
Claydon, N. *66*
Clezy, P.S. 251, *348*
Clutterbuck, P.W. *66*
Coats, J.H. *320, 330*
Cobbs, C.S. *334*
Cole, R.J. *66*
Colin, M. *341*
Coll, J.C. 222, 223, 251, 276, *342, 343, 344, 345, 346, 347, 348, 353*
Colombo, L. *66*
Colorini, A. 173, *327*
Colwell, R.R. 153, *321*
Comer, F.W. *72*
Condon, P. *66*
Conner, J.L. *320*

Cook, A.F. *329, 330*
Cook, J.C. Jr. *360*
Cooke, R.G. *144*
Coombe, R.G. *66*
Cooper, C.B. *341*
Cordova, R. *67*
Corey, E.J. *347*
Costello, C.E. 309, *360*
Coval, S.J. *346, 353*
Covello, M. *145*
Cowe, H.J. 223, *342*
Cox, P.J. 223, *342*
Cox, R.H. *76*
Coxon, D.T. *67*
Craff, L.L. *320*
Crandell, C.W. *322*
Craw, M.R. *144*
Crews, P. 165, 173, *321, 326, 327, 333, 346*
Crispino, A. *332*
Crist, B.V. *323*
Croft, K.D. 173, 180, 202, *327, 331*
Crost, B.V. *323*
Crossley, M.J. *145*
Crumley, F.G. *76*
Cruz, L.J. 274, 275, *351, 352*
Culver, P. *320, 344*
Cun-Heng, H. 225, 276, *324, 335, 344, 349, 350, 353*
Cuomo, V. *324*
Curtis, R.F. *67*
Cutler, H.G. *76*
Czarkie, D. *320*

Dacre, J.C. *144*
Dadok, J. *339*
Daigo, K. *362*
Daljeet, A. *358, 359*
Daloze, D. 163, 190, 195, 225, *325, 326, 327, 329, 331, 341, 343, 344*
Daly, J.J. *172, 327*
D'Ambrosio, M. 185, *328*
Dasgupta, A. *332*
Dato, V. *332*
D'Auria, M.V. 296, 307, *323, 356, 359*
Davies, L.P. *330*
Davies, N.D. *66*
Davies, P.A. *328*
Davies, W.P. *67*
Davis, E.E. *76*
Davoust, D. *361*
Dawes, K. 193, *329*

Day, W.C. *70*
De Alvarenga, M.A. *67*
Dean, B.M. *67*
Dearlove, G.E. 212, *338*
de Clauser, R. 185, *328, 340*
Declercq, J.P. *342, 343*
de Costa, M.S.L. 223, *343*
Detay, N. *331*
De Freitas, J.C. *326, 362*
de Giulio, A. 195, *331, 332*
De Jesus, A.E. *67*
De Laszlo, S.E. 190, *329*
Delaveau, P. *143, 145, 148*
Dellar, G. *328*
Dellaria, J.F. Jr. *350*
Demagos, G.P. *145*
De Megalhaes, G.C. *67*
De Moraes, A.A. *67*
De Morales, M.P.L. *65*
De Napoli, L. *335*
De Nicola Giudici, M. *324*
De Oliveira, A.B. *67*
De Oliveira, G.G. *67*
De P. Dias, J.P. *67*
de Rosa, S. 186, *195, 205, 210, 258, 261, 262, 327, 328, 330, 331, 332, 333, 334, 341, 343, 349, 350*
Desai, H.K. *68*
De Silva, E.D. 223, 265, *326, 343, 349*
De Simone, F. 285, *354, 356*
de Stanchina, G. 185, *328*
De Stefano, A. 154, *323, 327*
De Stefano, S. 185, 195, 205, 210, 258, 261, 262, *328, 341, 343, 349, 350, 351, 352, 353, 354*
Deus, B. *143*
Devadas, B. *145*
Devys, M. *67*
de Walter, A. *147*
Dick, J.W.E. *67*
Diener, U.L. *66, 72*
Dieter, K. *65*
Di Giacomo, G. *323*
Dijkman, A. von *73*
Dini, A. 285, *323, 354, 356*
D'Ischia, M. 309, *350, 360*
Ditzel, E. *346*
Djerassi, C. 217, *320, 321, 322, 323, 332, 340, 341, 342, 344, 361*
Djura, P.J. 158, 223, *324, 327, 328, 343*
Do, M.N. *333, 343*
Dodsworth, D.J. *145*

Dogović, N. *361*
Donovan, S.F. *345*
Dorai, D.T. *363*
Dosseh, Ch. *145*
Doubek, D.L. 254, 272, *349, 351, 360*
Dowdle, E. *336*
Draudt, H.N. *66*
Dreifuss, P.A. *76*
Dreele, R.B. von *77*
Dresel, P.E. *336*
Drexler, S.A. *343*
Drube, C.G. *77*
Drzymala, R.E. *334*
Duax, W.L. *321*
Dudman, N.P.B. *144*
Duffield, R.M. *65*
Dunlop, R.W. 158, *324, 347*
Dunn, A.W. *67*
Dunn, F.W. *343*
Dunn, W.C. *362*
Dupont, A. *343*
Durgeat, M. 313, 317, *361*
Durham, L.J. *321*
Durley, R.C. *67*
Durr, K. *74*
Durso, A.R. Jr. *360*
Durso, J.A.R. 312, *360*
Dutky, S.R. *354*
Duyne, G.D. van 185, *326, 353*

Eaton, M.A.W. *67*
Eckert, R. *75*
Edgar, K.J. *147*
Edmonds, J.S. *362*
Edwards, R.L. *63, 67*
Eggersdorfer, M.L. *322*
Eggert, H. *344*
Eguchi, S. 231, *344, 345*
Ehmke, H. *65*
Ehrmann, E.U. 245
Ekong, D.E.U. *65*
Ekstrand, J.D. *346*
Elezaby, M.S. *145*
Ellestad, G.A. *67, 68*
Elliott, R.C. *336*
El-Shagi, H. *148, 149*
Elyakov, G.B. 155, 294, 300, 302, 307, *323, 328, 339, 356, 357, 358*
Elyakov, Y.B. *328*
Endean, R. *320, 334, 362*
Endo, H. *362*
Endo, K. *359*

Endo, M. 310, *331, 348, 360*
Englert, G. *347*
Enwall, C.E.L. 164, *325*
Eppley, R.M. *76*
Epstein, S. *321*
Erickson, K.L. 190, *329, 333, 343*
Ersoy, L. *146*
Erxleben, C. *337*
Estrada, D.M. *325*
Estrella, D.J. *345*
Eswaran, V. *145*
Euler, K.L. *353*
Evans, S.L. *71*
Evans, T. *320*

Fahy, E. *335*
Faith, W.C. *330*
Falco, B. *323*
Fales, H.M. *65*
Fallis, A.G. 217, *341*
Farkas, L. *149*
Farrell, I.W. *68*
Fattorusso, E. *325, 335*
Faulkner, D.J. 155, 158, 159, 165, 202, 203, 262, 267, 276, 314, *320, 323, 324, 325, 326, 327, 328, 330, 331, 332, 333, 340, 345, 346, 349, 350, 353, 361*
Fayos, J. *351*
Fedoreev, S.A. *361*
Fedorov, S.N. *307, 359*
Fenical, W. 217, 218, 224, 225, 239, 277, *320, 321, 340, 341, 342, 343, 344, 345, 346, 349, 351, 353*
Fennell, D.I. *66, 75*
Ferlan, I. *335, 337*
Ferreira, N.P. *68, 76*
Ferrito, V. 318, *362*
Fieser, L.F. *145*
Filipescu, N.J. *147*
Finamore, E. 285, 296, *354, 356*
Finan, J.M. *338, 339*
Findlay, J.A. *68, 338, 355, 358, 359*
Finer-Moore, J.S. *342, 345*
Fingl, E. *145*
Finkelstein, A. *337*
Fisher, L. *341*
Flowers, A.L. 209, *334*
Ford, J. *339*
Forgione, P. *145*
Formanek, I. *145*
Fosset, M. *336*

Fourie, L. *77*
Fourneron, J.-D. *346*
Fournet, A. *146*
Francesconi, K.A. *362*
Freeman, G.G. *68*
Freire, R. *146*
Frick, B. *344*
Frincke, J.M. *331*, *345*
Fronza, G. *143*
Fuchida, M. *357*
Fuhrman, F.A. *330*, *349*
Fuhrman, G.J. *330*, *349*
Fujii, Y. *272*, *339*, *340*, *351*
Fujimori, T. *70*
Fujioka, H. *338*, *339*
Fujise, S. *68*
Fujita, S. *337*
Fujiwara, M. *348*
Fujiyama, T. *362*
Fukuda, S. *75*
Fukui, H. *146*
Fukuyama, K. *70*
Furukawa, J. 310, *360*
Furukawa, S. *377*
Furutani, Y. *68*
Fusetani, N. 171, 193, 291, *323*, *326*, *329*, *330*, *333*, *342*, *356*, *362*
Fuska, J. *71*

Gager, F. *65*
Galt, S. *63*
Galynker, I. *339*
Gandhi, R.N. *68*
Garneau, F.-X. 304, *359*
Garson, M.J. *63*, *64*, *65*, *68*
Gašič, M.J. *361*
Gatenbeck, S. *68*
Gatten, R.R. *321*
Gauem, B. *348*
Gaur, P.K. *334*
Gebauer, E. *335*
Gehrke, P.I. *143*
Gehrken, H.D. *321*
Gennori, C. *66*
Gen-Pei, L. *331*
Germain, G. 195, *331*, *342*, *343*
Ghassemieh, M.B. *353*
Ghisalberti, E.L. 173, 206, *324*, *326*, *327*, *331*, *333*
Giles, D. *63*, *68*
Giles, R.G.F. *64*
Gilgan, M.W. 295, *356*

Gilmore, J. *361*
Girad, M. 304, *359*
Gleich, P. 309, *360*
Glen, A.T. *67*
Gloer, J.B. 309, *320*, *360*
Gloppe, K.E. *71*
Goad, L.J. *321*
Gold, P. *334*
Gonzales, A.G. *145*, 171, 270, *325*, *326*, *351*
Goo, Y.M. *328*
Gopichand, Y. 164, 217, 223, 269, 271, *323*, *325*, *332*, *342*, *345*, *350*, *351*
Gorshkov, B.A. *328*, *359*
Gorshkova, I.A. *328*, *359*
Gottlieb, D.R. *65*, *67*
Govindachari, T.R. *68*
Graden, C.J. *333*
Grandmaison, J.-L. *143*
Grandmaison, T.I. *74*
Gray, G.A. 185, *329*
Gray, W.R. 275, *351*, *352*
Greer, B.J. *333*
Greger, H. *68*
Gregson, R.P. *327*, *329*, *330*
Grimshaw, J. *68*, *69*
Grindenberg, J. *68*
Grode, S.H. *158*, *324*, *346*
Gros, E.G. *356*
Grove, J.F. *63*, *66*, *68*
Groweiss, A. 200, 215, *327*, *331*, *344*, *345*, *346*
Groweiss, S. *320*
Grünhagen, H.H. *338*
Guella, G. 157, *324*
Guerriero, A. *324*, *328*, *340*
Guggisberg, A. *66*
Gunasekera, S.P. *323*, *330*, *331*
Gunatilaka, A.A.L. 217, *323*
Gupta, N. *147*
Gupta, V.P. *64*
Gust, D. 270, *349*, *351*
Gustafson, K. 268, *350*
Guyot, M. *69*, 313, 317, *361*
Gyr, P. *362*

Ha, T.B.T. *321*, *361*
Habermann, E. *339*
Habermehl, G.G. *301*, *320*, *335*, *358*
Habon, J. 203, *332*
Hacker, M.P. *360*
Hagadone, M.R. 206, *333*, *353*

Hahin, R. *352*
Haimova, M. *76*
Halevy, S. 173, *327*
Hall, I.H. *145*
Hallenstvet, M. *327*
Halsall, T.G. *68*
Halstead, B.W. *320*
Hamada, Y. *351, 360*
Hamamoto, Y. 310, *348, 360*
Hamanaka, N. *348*
Hamid, A. *72*
Hanke, F.J. *321*
Hansen, B.M. *75*
Hanson, J.R. *68*
Haq, H. *362*
Harada, N. *333*
Harada, T. 276, *353, 357*
Harbour, G.C. *320, 330, 361*
Hardegger, E. *69*
Harding, V.K. *69*
Hargreaves, R.T. *67*
Harries, J.B. *337*
Harries, P.C. *67*
Hartman, K.R. 209, *334*
Hartung, K. *337*
Harvan, D. *71, 74*
Harvey, O. 304, *359*
Harwig, J. *69, 75, 77*
Harwood, L.M. *69*
Hase, T. 231, *344, 345*
Hashigaki, K. *78*
Hashimoto, K. 171, 276, 291, 299, *326, 330, 333, 336, 353, 356, 357, 362*
Hashimoto, S. *347*
Hashimoto, Y. *320*
Haskell, T.H. *72*
Hasler, J.A. *329, 339, 340, 360*
Hassall, C.H. *67, 69, 146*
Hattori, S. *69*
Hawksworth, W.A. *69*
Haworth, R.D. *68, 69*
Hay, J.E. *65, 69*
Haynes, L.J. *65, 69*
Hayashi, H. *75,* 308, *359*
Hayashi, K. *341, 345*
Hayashi, T. *341*
Heale, J.B. *69*
Heide, L. *146*
Heinrikson, R.L. *362*
Heitz, S. *313, 361*
Heller, K. *69*
Hellou, J. *349, 350*

Hemingway, R.W. *69, 71*
Hemming, H.G. *65*
Henderson, G.B. *69*
Henderson, R.J. *321*
Henriksen, L.M. 89, *146*
Herald, C.L. 254, 270, 272, *349, 351, 360*
Herald, D.L. Jr. *72*
Hermodsson, S. *68*
Hernàndez, M.I. *325*
Herrmann, M. *143*
Hertzberg, S. *328*
Hesse, A. *143*
Hesse, M. *66*
Hessinger, D.A. 209, *333*
Hessler, H.J. *339*
Hida, K. *353*
Higa, T. *341, 362*
Higuchi, R. 288, *355*
Higuchi, S. *72*
Higuchi, T. 288, *355*
Hill, R.A. *69*
Hirai, K. *352*
Hirata, Y. 159, 165, 200, 274, 275, *325, 326, 329, 330, 331, 338, 339, 344, 350, 351, 352, 356, 361*
Hirayama, T. *145*
Hirose, Y. *146*
Hirota, K. *349*
Hirotsu, K. *66*
Hite, G.J. *353*
Hochlowski, J.E. 158, 262, 275, *322, 324, 325, 346, 350, 353*
Hocquemiller, R. *146*
Hoeneisen, M. *355*
Höfle, G. 87, *145, 146*
Hofle, G. *69*
Hofheinz, W. *321*
Hogan, I.T. *360*
Hogue, E.R. *321*
Holker, J.S.E. *69*
Hollenbeak, K.H. 165, *325, 350*
Holm, A. 206, *333*
Holst, H. *78*
Holzapfel, C.W. *76*
Homma, K. *70*
Honda, A. *347*
Honda, E. *78*
Hong, S.-C. *348*
Hori, M. *357*
Horiai, H. *341, 345*
Hosaka, M. 276, *353*
Hoshen, M. *66*

Hossain, M.B. 165, 269, *325*, *331*, *332*, *345*, *350*
Hotta, K. *308*, *360*
Houlchein, W.S. *70*
Hovingh, P. *362*
Howard, D.F. *65*
Howden, M.E.H. 282, *354*
Hoye, T.R. *350*
Hsu, C.F.J. *72*
Hsu, H.-Y. *70*
Hucho, F. *336*, *337*
Hughes, R.G. Jr. *330*, *369*
Hugues, M. *335*, *338*
Hui, W.H. *146*
Hung, S.-H. *70*
Hunkapiller, D.W. *352*
Hutchinson, D.W. *67*
Hutchinson, R.D. *70*

Ibrahim, R.K. *70*
Idler, D.R. *353*
Iguchi, K. 249, *341*, *343*, *344*, *347*, *348*
Iitaki, Y. *73*, *75*, *352*
Ikeda, T. *76*
Ikegami, S. *354*
Ikekawa, N. *356*
Ikushima, S. *348*
Imagawa, D.K. *322*
Imai, J. *73*
Imre, S. *146*, 223, *342*
Inamoto, T. *357*, *358*
Inoue, H. *89*, *146*
Inoue, K. *89*, *146*
Inoue, M. 272, *351*
Inoue, T. *77*, *78*
Inouye, S. *70*
Inoye, S. *70*
Inui, J. *336*
Ireland, C.M. 158, 276, 309, 312, 313, *324*, *329*, *353*, *360*, *361*
Isakov, V.V. 307, *359*
Isakova, T.I. *144*
Ishida, Y 203, *332*, *339*
Ishikawa, M. *340*
Ishizaki, T. *340*
Issa, R. *145*
Itakura, Y. 288, 291, *355*, *356*
Itatani, Y. *63*
Ito, H. *72*
Ito, K. *339*
Ito, Y. *63*, 288, 298, *355*, *356*
Itoh, J. *70*

Itoh, T. *322*, *323*
Itokawa, H. *87*, *146*
Ivanova, N.S. 300, *357*
Iwagawa, T. *345*
Iwasaki, S. *70*
Iwashita, T. *338*, *339*, *341*
Izac, R.R. 224, *343*

Jaccarini, V. 318, *362*
Jackson, K.W. *335*
Jacobs, J.J. *66*, 152
Jacobs, R.S. *320*, *326*, *344*, *345*, *362*
Jacquemin, A. *143*
Jagodzinska, B.M. 217, *341*
Jalil, S.V. *75*
James, T.R. Jr. *346*
Jamieson, D.D. *330*, *347*
Jamieson, G.C. *332*
Jarchaw, O.H. *75*
Jarrah, M.Y. *70*
Jefferies, P.R. 206, *324*, *326*, *331*, *333*
Jeffs, P.W. *69*
Jeon, J.K. *357*
Jimenez, E.C. 274, *352*
Johannsen, J.E. *328*
Johnstone, R.A.W. *67*
Jones, D.W. *69*
Jones, T.H. *65*
Joshi, B.S. *70*
Joshi, K.C. *70*
Jui, J. *70*, *72*
Just, G. *65*, *70*, *73*

Kaiser, P. *70*
Kaisin, M. 225, *344*
Kale, N. *67*
Kalidhar, S.B. *70*
Kalinin, V.I. 300, 304, *357*, *359*
Kalinovskaya, N.I. 307, *357*, *359*
Kalinovskii, A.I. 155, 294, 300, *323*, *356*, *357*, *358*, *359*
Kamano, Y. 272, *349*, *351*
Kamat, V.N. *70*
Kameda, K. *70*
Kamerling, J.P. 308, *360*
Kamikawa, T. *63*, *71*, *77*
Kamiya, H. *362*
Kamiya, Y. *354*
Kampen, A. *336*
Kan, S.K. 239, *346*
Kanazawa, A. *356*

Kaneko, H. *70*
Kaneko, S. *340*
Kaneko, Y. *71*
Kaneta, S. *347*
Kanghou, L. 225, *344*
Kannengiesser, G.J. *340*
Kapil, R.S. *74*
Kapustina, I.I. 301, *354, 357*
Karlsson, R. *329*
Karpf, M. *321*
Karuso, P. *325*
Kashman, Y. 173, 200, 216, 248, *320, 322, 326, 327, 328, 331, 340, 341, 342, 344, 345, 346, 347*
Katayama, C. *331, 338*
Kato, Y. 171, 294, *326, 330, 333, 356*
Katsube, Y. *70*
Kaul, P.N. 213, *320, 338*
Kawai, K. *63, 72, 73*
Kawasaki, T. 288, 290, 291, 298, *355, 356*
Kazlauskas, R. *71*, 158, 172, 231, 251, *324, 326, 327, 328, 329, 330, 341, 342, 344, 345, 347, 348, 360*
Keeran, W.S. *361*
Keim, P.S. *362*
Kelecom, A. *340, 358*
Kelman, S.W. 209, *334*
Kem, W.R. *335, 362*
Kemertelidze, E.P. *64*
Kern, J.R. *333*
Khalil, M.W. *321, 353*
Khan, A.H. *71*
Kho, E. *321, 322*
Kiagawa, H. *71*
Kicha, A.A. 294, *356*
Kido, M. 195, *323, 342, 343, 358*
Kikuchi, H. 203, 249, *327, 332, 341, 343, 345, 347, 348*
Kim, Y.H. *349*
Kimura, A. 308, *359*
Kimura, Y. *70*
Kinamoni, Z. *320, 344*
Kindl, H. *64*
King, T.J. *67*
Kingston, J.F. 217, *341*
Kinnel, R.B. *320, 340*
Kirksey, J.W. *66*
Kirsch, J.H. 301, *358*
Kishi, T. *73*
Kishi, Y. *338, 339*
Kitade, Y. *349*
Kitagawa, I. 195, 240, 248, *323, 327*

Kitagawa, T. *71, 323, 327, 333, 341, 342, 343, 344, 346, 347, 358*
Kitahara, K. *272, 351*
Kitanaka, K. 195, *323*
Kitano, H. 203, *332*
Kiyota, Y. *342*
Kjøsen, H. *146*
Klaar, M. *71*
Klein, D. *320*
Klein, L.L. *338, 339*
Kljajic, Z. *361*
Knapp, J.E. *66*
Ko, S.S. *338, 339*
Kobayashi, A. *76*
Kobayashi, J. 159, 165, 274, 275, *325, 326, 327, 329, 330, 331, 333, 341, 342, 343, 344, 346, 347, 351, 352, 356, 358, 361*
Kobayashi, K. *343*
Kobayashi, M. 165, 195, 217, 240, 248, *323, 326, 327, 340, 341, 347*
Kobayashi, Y. *240*, 347
Koch, P. *323*
Kodamo, M. *362*
Kodamo, Y. *70*
Kodelja, V. 212, *338*
Kohda, K. *351*
Koike, H. *339*
Koker, M.E.S. *320*
Kokke, W.C.M.C. *321, 322, 340, 361*
Kollmann, A. *67*
Kolkenbrock, H.J. 213, *338*
Kolkmann, R. 81, *149*
Kolle, F. *71*
Komoda, Y. *340*
Komori, T. 288, 290, 291, *355, 356, 361*
Konosu, S. 193, 314, *323, 329, 342, 361*
Kopanski, L. *71*
Kornprobst, J.M. *322, 332*
Korotkikh, L.Y. 301, *354, 358*
Kosuge, T. *351*
Kouros, P. *320*
Kovachev, G. 212, 277, *338*
Koyama, T. *78*
Kozlovskaya, E.P. *335*
Kozu, Y. *78*
Kraft, R. *143*
Krebs, H.C. *335, 355*
Krepinsky, J. *68*
Krivoshchekova, O.E. *71*
Krueger, W.C. *328*
Ksebati, M.B. *341*

Kubo, K. *349*
Kubo, M. *77*
Kubota, T. *63*, *71*, *77*
Kuc, J. *66*
Kudo, Y. *336*, *339*
Kuentzel, S.L. 309, *330*, *360*
Kugler, F. *69*
Kuhr, I. *71*
Kunstmann, M.P. *68*
Kuramoto, J. 231, *344*
Kurelec, B. 293, *329*
Kurth, M.J. *350*
Kustin, K. 310, *360*
Kusuda, J. *146*
Kuznetsova, T.A. 300, 307, *357*, *358*, *359*
Kyogoku, Y. 195, 240, 248, *323*, *333*, *341*, *342*, *343*, *346*, *347*, *358*

Laatsh, H. *78*
Laboute, P. 287, *346*, *355*
Lafranconi, W.H. *337*
Lai, C.Y. *335*, *336*
Lakshmi, V. *323*, *324*, *330*
Lal, D.M. *333*
Lang, G. *143*
Lang, R.W. *322*
Langdon, R. *320*
Lankelma, J. *332*
Lansbury, P.T. *347*
Lanzotti, V. *335*
La Rotonda, M.I. *145*
Lauffer, L. *337*
Laurence, J.W. *77*
Laurent, D. *356*
Lavis, A. *332*
Lawson, M.P. *332*
Lazdunski, M. *335*, *336*, *338*
Lebez, D. 212, *335*, *338*
Le Blanc, G.D. *71*
Leder, J. *338*, *339*
Lee, H.H. *146*
Lee, K.H. *145*
Lee, N.K. *341*
Leeper, F.J. *63*
Leistner, E. 81, *143*, *146*, *149*
Leistner, L. *66*
Lemeignon, M. *336*
Le Quesne, P.W. *144*
Lesniak, A.P. 253, *348*
Lessinger, L. *67*
Leung, M. *354*
Levi, J.D. *67*

Levett, G. *67*
Levina, E.V. 294, *356*
Lewis, B.G. *67*
Lewis, R. *362*
Li, G.-W. *147*
Li, H. *322*
Li, L.H. 309, *330*, *360*
Li, L.N. *322*
Li, M.K.W. *342*
Li, X. *323*
Liaaen-Jensen, S. *327*, *328*
Liau, M.-C. *70*
Lidgard, R.O. *328*, *329*
Lidor, R. *331*
Liebrich, M. *336*
Lillehoj, E.B. *71*, *78*
Lin, J.-Y. *71*
Linden, G. *335*
Linker, A. *362*
Litchfield, C. *328*, *332*
Liu, E.H. 253, *348*
Lloyd, H.A. *71*
Locci, R. *64*
Loeffler, W. *78*
Look, S.A. 239, *321*, *345*, *346*
Lopez, Dorta H. *145*
Lovell, F.M. *67*
Loya, Y. 248, *327*, *328*, *344*, *347*
Lubbock, R. *335*
Luque, A. *351*
Lutomski, J. *147*

McCabe, T. *327*
McCaffrey, E.J. *321*
Macconell, J.G. *65*
McConnell, O.J. 157, *324*
McCorkindale, N.J. *71*
McCrae, W. *72*
Macedo de Abreu, P. 195, *326*, *331*
Maček, P. *335*, *338*
McEnroe, F.J. *351*
Macfarlane, R.D. *338*
McGahren, W.J. *67*, *71*
McGovren, J.P. 309, *360*
McGraw, G.W. *69*, *71*
McIntyre, D.E. 158, *324*, *345*
McIntosh, M. 275, *352*
McKittrick, B. *344*
Macleod, J.K. 165, 173, *326*
McManis, J.S. *348*
MacMillan, J. *67*
McMillan, J.A. *328*

McPhail, A.T. *71*
McWhorter, W.W. Jr. *339*
Mading, B. *320*
Maebayashi, Y. *71, 78*
Maezawa, H. 315, *361*
Magalhaes, A.F. *67*
Magalhaes, E.G. *67*
Magalhaes, G.C. *67*
Magalhaes, M.T. *65, 67*
Makarieva, T.N. 155, *323, 328*
Maki, Y. *349*
Maksimov, O.B. *71, 361*
Mallabaer, A. *71, 72*
Mali, R.S. *64, 73*
Mal'tsev I.I. 302, *358*
Manchanda, A.H. *65*
Mancini, I. 157, *324*
Manda, T. *343*
Manes, L.V. 165, 173, *326, 327*
Maoka, T. 282, *354*
March, G. 158, *324*
Marekov, N.L. 212, 277, *322, 338*
Maretic, Z. 211, *335*
Marks, I.N. *143*
Marino, A. *323*
Marques, R. *67*
Marron, P. *149*
Marsaioli, A.J. *67*
Martin, J.D. 270, *325, 351*
Martin, M. *343*
Martin, V.S. *325*
Marumo, S. *73*
Maruyama, J. 276, 299, *353, 357*
Marwood, J.F. 239, *330, 346, 348, 361*
Masayuki, M. *76*
Masui, Y. *64*
Matlin, S.A. *148*
Matsoukas, J.M. *68, 359*
Matsudo, J. 298, *356*
Matsueda, S. *68*
Matsui, M. *72, 75*
Matsui, T. *72*
Matsumoto, G.K. 165, 185, *325, 326, 329, 333, 346*
Matsumoto, T. *76, 148*
Matsunaga, S. 171, 193, 314, *323, 326, 329, 330, 333, 342, 361*
Matsuno, T. 261, 282, *354*
Matsushita, H. *70*
Matsuzaki, T. *344*
Mattia, C.A. 195, *331*
Mauli, R. *64*

Mayol, L. 164, *325, 335*
Mazzarella, L. 195, *330, 331*
Mazzola, E.P. *76*
Mebs, D. *321, 335, 359*
Medarde, M. *74*
Medina, J.M. *145*
Mehrotra, M.M. *347*
Meinwald, J. *330, 331*
Mellon, F.A. *355*
Menou, J.L. 290, 295, 296, *355, 356*
Merendi, C. *66*
Merlini, L. *64, 65*
Mesquita, A.A.L. *67*
Meves, H. *338*
Miah, M.A. *75*
Michaud, D.P. 270, *350, 351*
Michel, C. 270, 272, *339, 351*
Middlemas, D. 275, *352*
Midland, S.L. 270, *351*
Miguel del Carrol, J.M. *74*
Mihai, G.G. *147*
Mihara, K. *146*
Mikami, Y. *148*
Milburn, M.S. *71*
Miller, R.W. *71*
Milkova, T.S. 212, 277, *322, 338*
Minale, L. 283, 285, 287, 290, 295, 296, 307, *327, 354, 355, 356, 359*
Mintzlaff, H.-J. *66*
Mir, I. *72*
Mirado, P. *68*
Miranda, C.A.S. *65*
Mishchenko, N.P. *71*
Mishra, G. *147*
Mislivec, P.B. *76*
Misuraca, G. 309, *360*
Mitchell, S.J. *342, 343, 344, 345, 346*
Mitscher, L.A. *71, 72*
Mitsuhashi, H. 217, *340, 341, 345*
Miura, Y. *336*
Miyake, M. *337*
Miyaki, K. *71*
Miyakoshi, S. 276, *353*
Miyakoshi, T. *348*
Miyaki, K. *78*
Miyato, S. *70*
Mizsak, S.A. *330, 360, 361*
Mizuba, S.S. *70, 72*
Mizukami, I. *362*
Mizukawa, K. 310, *360*
Mo, L. *65*
Moffat, J.S. *65*

Moharir, Y.E. *355, 359*
Molgo, J. *336*
Molho, D. *69*
Mondelli, R. *143*
Money, T. *72*
Moore, J.H. *66, 72*
Moore, R.E. *338, 339*
Moorthy, N.K. *147*
Moquin, C. *313, 361*
Morgan, B.A. *146*
Mori, K. *72, 75, 344, 347*
Mori, Y. *347*
Morrone, R. *325, 349*
Mosher, H.S. *330, 349*
Moncrief, J.W. *78*
Mourão, P.A.S. *362*
Mueller, H. *72*
Müller, I. *193, 329*
Müller, W.E.G. *193, 329*
Mukerji, I. *66*
Mukherjee, S.K. *66*
Mukrji, J. *66*
Mulder, J. *71*
Mulder-Krieger, Th. *147, 149*
Munakata, K. *74*
Munakata, T. *72*
Munk, M.E. *63, 72, 73*
Munro, M.H.G. *320*
Muramatsu, I. *348*
Murase, M. *76*
Muratov, V.B. *307, 359*
Muro, H. *70*
Murphy, P.T. *71, 158, 172, 231, 251, 324, 326, 327, 328, 329, 330, 341, 344, 345, 347, 361*
Murray-Rust, P. *318, 362*
Murti, U.V.S. *147*
Musilek, V. *75*
Myers, B.L. *333*

Nabeya, H. *345*
Nabiullin, A.A. *335*
Nachman, R.J. *349*
Nadelson, J. *70*
Nadkarni, D.R. *72*
Naganawe, H. *68*
Nagaoka, H. *348*
Nair, E. *307, 359*
Nair, M.S.R. *72*
Nakagawa, H. *308, 359*
Nakagawa, M. *310, 331, 348, 360*
Nakagawa, T. *217, 341, 345*

Nakai, H. *78*
Nakajima, S. *72, 73*
Nakamura, H. *159, 165, 274, 275, 325, 326, 329, 330, 331, 344, 351, 352, 356, 361*
Nakamura, T. *354*
Nakamura, Y. *329*
Nakanishi, H. *341, 348*
Nakanishi, K. *310, 360*
Nakanishi, T. *310, 348, 360*
Nakano, T. *72, 325*
Nakatsu, T. *165, 323, 326*
Nakayama, M. *231, 344, 345*
Namiki, M. *70*
Naoi, Y. *72*
Naoki, H. *338, 339*
Narasimhachari, N. *72*
Narasimhan, N.S. *64, 72, 73*
Narayanan, V. *145*
Narayanaswami, V. *148*
Narita, H. *299, 357*
Nasini, G. *64, 65*
Nauri, H. *288, 353*
Nayeshiro, H. *89, 146*
Naylor, S. *321, 326, 327*
Neelakantan, S. *145*
Neeman, I. *333, 334*
Neilson, A.H. *65*
Neira, C. *355*
Nelson, D.B. *63, 72, 73*
Nemorin, J.L.E. *343*
Neumcke, B. *336*
Newcomb, R.W. *318, 363*
Newman, R.A. *360*
Ng, A.S. *65, 73*
Niazi, H.M. *73*
Nicholl, S.G.A. *144*
Nikinov, G.K. *64*
Nishi, A. *72*
Nishibe, S. *63*
Nishida, R. *64, 73*
Nishikawa, W. *73*
Nishino, T. *357*
Nishioka, I. *78*
Nishiuchi, Y. *273, 351*
Nishzawa, N. *70*
Nishizone, Y. *344*
Nitta, K. *73, 78*
Nitta, M. *345*
Noguchi, M. *70, 76*
Noguchi, T. *276, 299, 353, 357, 362*
Nohara, T. *298, 356*

Nohomura, S. *146*
Norte, M. 270, *351*
Northolt, M.D. *73*
Norton, R.S. 202, *331, 336, 337, 344*
Norton, T.R. *335, 336, 337*
Novellino, E. *335*
Novis, B.H. *143*
Nozaki, H. *344, 345*
Nozawa, K. *73*
Nozawa, Y. *63*
Nozoe, S. *70*
Nukina, M. *73*

Obi, Y. *148*
O'Brien, T. *320*
Ochi, K. *76*
Ochi, R. *336*
Ode, R.H. 270, *351*
Odinokov, S.E. *335*
Oesterhelt, G. *321*
Ogata, Y. *78*
Ogihara, Y. *63, 73, 74, 75*
Ohizumi, Y. 159, 165, 203, 274, 275, *325, 326, 329, 330, 331, 332, 336, 344, 351, 352*
Ohno, M. *78*
Okada, S. *357*
Okami, Y. 153, *321*
Okamoto, T. *352*
Okano, K. *354*
Okazaki, H. *73*
Oki, M. *78*
Okuda, R. 249, 260, *320, 344, 347, 350*
Okuda, S. *70*
Okumoto, T. *72*
Okumura, Y. 200, *331*
Oleinikova, G.K. 300, *357*
Olivera, B.M. 274, 275, *351, 352*
Olson, C.E. 209, *334*
Oltz, E.M. *360*
Omoto, S. *70*
Ooka, T. *75*
Ookuba, M. *361*
Orito, S. *341*
Oshima, Y. 147, 276, *353, 362*
Oshita, K. *71*
Ouvrier, D. *327*
Overal, W.L. *65*
Overeem, J.C. *73*
Oxford, A.E. 66, *73*
Ozawa, Y. *72*

Pagonis, S. 295, *356*
Palumbo, A. 309, *360*
Pan Q.-C. *147*
Pang, Z. *148*
Parham, W.E. *147*
Paris, R.P. *143*
Park, Y.H. *148*
Parthasarathy, P.C. *68*
Parthasathy, M.R. *70*
Patankar, S.S. *68*
Paterson, R.M.L. *144*
Patil, A.D. *338*
Patra, A. *326*
Patrick, V.A. *347*
Patterson, E.L. *74*
Patterson, G.W. *353, 354*
Paul, I.C. *328*
Paul, V.J. 277, *353*
Paull, K.D. *360*
Paulsch, W.E. *73*
Pavelka, L.A. *349*
Pawlik, J.R. *353*
Pedreira, G. *67*
Pelhate, M. *337*
Pelter, A. 65, 74, *362*
Perales, A. 270, *351*
Pereira, S.A. *67*
Perkinson, N.A. *67*
Perez, C. *325*
Perez, R. 270, *325, 351*
Pero, R.W. 71, *74*
Perzanowski, H.P. *324*
Peters, T. *337*
Pettit, G.R. 254, 270, *329, 339, 340, 349, 351, 360*
Petrovich, J.C. *341*
Pfaff, K.-P. *338, 339*
Phan, C.T. *74*
Phillipson, D.W. 204, *332*
Piccialli, V. 164, *325*
Piccinni-Leopardi, C. 195, *331*
Pichon, Y. *339*
Piers, E. *350*
Pietra, F. 157, 185, *324, 328, 340*
Pike, R. *63*
Piller, N.B. *74*
Pinho, S.L.V. *67*
Pistole, T.G. *363*
Pitcher, R.G. *77*
Pitout, M.J. *68*
Pizza, C. 283, 285, 287, 290, 295, *327, 354, 355, 356*

Plant, W.D. *65*
Plouvier, V. *74*
Pockl, E.E. *334*
Podojil, M. *71*
Poet, S.E. *343, 344*
Pohland, A.E. *76*
Pomerantz, M.W. *330*
Poonian, M. *330*
Pople, M. *68*
Popli, S.P. *74*
Popov, A.M. *357*
Popov, S.S. 212, 277, *332, 338, 340*
Posternak, T. *74*
Potenza, D. *66*
Potier, P. 239, *346*
Poupat, C. 239, *346*
Powell, D.R. *324*
Prasad, R.S. 269, *350*
Price, K.R. *67*
Prokofieva, N.G. 301, *354*
Prota, G. 309, *350, 360*
Proudfoot, J.R. *323*
Psaila, A.F. *362*
Puliti, R. 195, *330, 331*
Purushothaman, K.K. *148*
Pusset, J. 287, 290, 295, *354, 355, 356*
Pyrek, J.St. *143*

Qi-Tai, Z. 239, 276, *345, 353*
Quercia, V. *148*
Quinn, R.J. 203, *329, 330, 332*

Rack, M. *338*
Rácz, G. *145*
Rafii, S. *350, 351*
Rahman, A. 269, 271, *345, 350, 351*
Rai, P.P. *148*
Raistrick, H. *66, 74*
Raj, U. *362*
Rama, N.H. *63*
Ramachandran, K.S. *68*
Raman, P.V. *145*
Ramdahl, T. *327, 328*
Rao, C.B. *340*
Rao, J.V.L.N.S. *148*
Rao, R.V.K. *148*
Rapposch, M. *353*
Raston, C.L. *362*
Raszeja, W. *147*
Ratcliff, M.R. *332*
Rathmayer, W. *337*
Ratnagiriswaran, A.N. 148

Rau, G.H. *321*
Ravens, U. *337*
Ravi. B.N. 180, 239, 240, 251, *327, 333, 341, 344, 346, 347, 348*
Rebachuk, N.M. *361*
Reddy, G.S. *147*
Regan, A.C. *74*
Regan, J.D. *362*
Remaley, S. 269, *350*
Renaud, J.F. *336*
Renis, H.E. 309, *320, 360*
Renson, M. *74*
Renstrøm, B. *327*
Reshetova, O.S. 307, *359*
Reucroft, P.R. *360*
Reul, A. *336*
Riano-Martin, M. *65*
Riccio, R. 283, 285, 287, 290, 295, 296, 307, *327, 354, 355, 356, 359*
Rickets, E.F. *350*
Rideout, J.A. 308, *329, 359, 360*
Rieder, W. *69*
Rieker, A. *144*
Rifkin, J. 209, *334*
Rinehart, K.L. Jr. 185, 204, 309, *320, 328, 330, 332, 352, 360, 361*
Roberge, G. *148*
Roberts, J.C. *74*
Roberts, J.L. *148*
Robinson, W.T. *346*
Rodriguez, L.F. *145*
Rodriguez, M.L. 171, *326*
Roelofs, W.L. *64, 73*
Roeymans, H.J. *148*
Rogers, D. *74*
Rohmer, M. *322*
Roll, D.M. 185, *326, 329, 333*
Romey, G. *336*
Roschenthaler R. *69*
Rose, C.B. *324*
Roser, K. *69*
Rosett, T. *74*
Rosser, R.M. 155, *324, 340*
Rossi, C. *74*
Rotem, M. 173, *320, 327, 328*
Roush, W.R. *330*
Rovnykh, N.V. 300, *357*
Rozynov, B.V. 307, *359*
Rub, P. *78*
Ruble, J.R. 277, *353*
Rudloe, J.J. *349*
Russel, F.E. 211, *335, 337*

Russo, A.J. *334*
Rutledge, P.S. *148*

Sahara, M. *77*
Saibaeva, I.M. *71*
Sainsbury, M. *361*
Saitbaeva, I.M. *72*
Saitoh, S. *347*
Sakaguchi, S. *282, 354*
Sakai, K. *72*
Sakakibara, S. *273, 351*
Sakamoto, K. *355*
Sakamura, S. *74, 75*
Sakan, T. *77*
Sakandelidze, O.G. *307, 359*
Sakurai, S. *273, 351, 352*
Salazar, J. *146*
Saleki, S.M.H. *77*
Salem, T.M. *145*
Saluja, M.P. *74*
Samain, D. *320*
Samejima, Y. *335, 336*
Sammarco, P.W. *222, 342*
Sammes, P.G. *355*
Sanders, R.L. *251, 341*
Sanduja, R. *353*
Sanechika, Y. *298, 356*
San Feliciano, A. *74*
Sankawa, U. *74*
Sankhala, R.H. *74*
Sano, S. *72*
Santacroce, C. *323*
Santana, A.E.G. *67*
Santos, C.C. *67*
Santurbano, B. *64*
Saradambal, S. *148*
Sarel, S. *173, 327*
Sargent, J.R. *321*
Sargent, M.V. *328*
Sarkar, S.K. *74*
Sarma, N.S. *340*
Sasaki, K. *70*
Sassa, T. *73, 74*
Satake, M. *337, 339*
Sato, A. *74, 75, 349*
Sato, H. *75*
Sato, K. *78*
Sato, S. *352, 356*
Sato, Z. *70*
Satomura, Y. *71, 75*
Saunders, J.B. *362*
Saunders, S.J. *143*

Sawada, T. *75*
Sawai, Y. *72*
Scahill, T.A. *360, 361*
Schauer, R. *308, 360*
Scheffler, J.-J. *335*
Schembri, P. *318, 362*
Schettino, O. *145*
Scheuer, P.J. 157, 185, 206, 249, 262, 265,
   277, 312, *320, 321, 324, 326, 329, 332,*
   333, 340, 342, 345, 346, 347, 349, 350,
   351, 360, 362
Scheufler, E. *337*
Schiff, P.L. Jr. *66*
Schilcher, H. *148*
Schmalle, H.W. *75*
Schmid, H. *66*
Schmidt, O.T. *64, 75*
Schmidt, P.G. 223, 270, *331, 332, 342, 351*
Schmidt, U. *329, 351, 360*
Schmidt, J. *352*
Schmidt, J.M. *339, 340, 349*
Schmidtmayer, J. *337*
Schmitz, F.J. 155, 164, 217, 223, 269, 271,
   323, 324, 325, 330, 331, 332, 341, 342,
   345, 346, 350, 351
Schnehenburger, J. *70*
Schneider, P. *343*
Schöllhorn, E. *337*
Schönholzer, P. *345, 347*
Schroeder, H.W. *75*
Schulte, G.R. 157, *324, 349, 351*
Schulte, U. *148, 149*
Schulten, H.-R. *356*
Schulz, K.H. *75*
Schwartz, R.E. *320, 347*
Schwarz, W. *336*
Schweitz, H. *335, 338*
Scoft, A.I. *72*
Scognamiglio, G. *341*
Scolastico, C. *66*
Scott, F.E. *75*
Scott, P.M. *75, 77*
Sedmak, B. 212, *338*
Sedmera, P. *71, 75*
Seelkopf, C. *148*
Seldes, A.M. *356*
Self, R. 186, *328, 355*
Selover, S.J. *346*
Senatore, F. *354*
Senčič, L. *335*
SenGupta, P.K. 164, 269, *325, 350*
Seriguchi, D.G. *336*

Seshadri, T.R. *147*
Sesin, D.F. *360*
Sevcik, C. *321*
Sevenet, T. *354*
Shahripour, A. *324*
Sham, K.K. *64*
Shamoo, A.E. *334*
Shapiro, B.I. *352*
Sharma, G.M. *330*
Sharma, P. *70*
Sharypov, V.F. *358*
Shaw, D. *66*
Shaw, P.D. *320, 330*
Sheardown, M.J. *336*
Sheikh, Y.M. *344*
Shelly, J. *271, 351*
Sheumack, D.D. *282, 354*
Shibata, S. *73, 339*
Shibata, U. *70*
Shibata, Y. *76*
Shibatu, S. *74*
Shibita, M. *360*
Shibita, S. *73, 203, 332, 336, 337*
Shibkawa, M. *75*
Shida, Y. *362*
Shield, L.S. *320*
Shimada, H. *75*
Shimizu, I. *327, 347*
Shimizu, M. *340*
Shimizu, N. *333*
Shimizu, Y. *153, 282, 320, 321*
Shimojima, Y. *75*
Shimonaka, H. *63*
Shingu, T. *145*
Shiobara, Y. *146*
Shioiri, T. *351, 360*
Shiozaki, M. *75*
Shmueli, U. *327, 331*
Shochet, N.R. *200, 331*
Shomura, T. *70*
Shoolery, J.N. *185, 329, 331*
Shryock, J.C. *209, 334*
Shu, A.Y.L. *321*
Shubina, L.K. *155, 323*
Sibi, M.P. *75*
Sica, D. *164, 321, 322, 325, 332*
Sidyakin, G.P. *71, 72*
Sigg, H.P. *64*
Silva, M. *355*
Simard, J.-L. *304, 359*
Simonsen, J.L. *148*
Simpson, T.J. *270, 351*

Sims, J.J. *270, 351*
Singh, J. *148*
Singh, P. *70*
Singy, G. *344*
Sinha, N.D. *66*
Sivakumaron, S. *147*
Sjöstrand, U. *321, 322, 341*
Skajennikoff, M. *67*
Skelton, B.W. *173, 325, 327, 331, 343, 362*
Sklarz, B. *67*
Škrk, J. *212, 338*
Slates, H.L. *75*
Slathin, D.J. *66*
Sleeper, H.L. *277, 353*
Sleigh, R.W. *337*
Smalley, E.B. *77*
Smetanina, O.F. *307, 357, 358, 359*
Smith, G. *66*
Smith, L.R. *63*
Smith, N.B. *308, 359*
Snatzke, G. *63*
Snider, B.B. *67, 330*
Snieckus, V. *77*
Sodano, G. *154, 186, 205, 210, 258, 261, 262, 322, 323, 325, 328, 330, 332, 333, 334, 341, 343, 349, 350, 353*
Sokoloff, S. *173, 327*
Sokoloski, E.A. *65*
Solé-Cava, A.M. *340*
Son, B.W. *248, 341, 342, 346, 347*
Sorrentino, C. *295, 356*
Sotheeswaran, S. *65*
Souri, A. *77*
Spasor, S. *76*
Spaziani, E. *318, 363*
Spence, I. *282, 328, 354*
Spinella, A. *353*
Springer, J.P. *76*
Squillace-Greco, O. *290, 295, 355, 356*
Srocka, U. *148*
Stack, M.E. *76*
Stämpfli, R. *336*
Staunton, J. *68, 69, 74*
Stauton, J. *64*
Stauton, S. *63, 65*
Stefano, G.B. *354*
Steggles, A.W. *144*
Steglich, W. *71*
Steinhe, I. *63*
Steinrücken, H.C. *143*
Stengelin, S. *336, 337*

Author Index

Stepanenko, L.S. *71*
Stepanov, V.R. 304, *359*
Sterhell, S. *343*
Steudler, P.A. *346*
Stevenson, R. *64*
Steyn, P.S. *67, 70, 76, 77*
Stickings, C.E. *74*
Stierle, D.B. 158, 203, *324, 330, 332, 346*
Still, W.C. *339*
Stipanovic, R.D. *75*
Stockigt, J. *148*
Stodola, F.H. *76*
Stoessl, A. *76*
Stoilov, I. 212, 277, *338*
Stonard, R.J. *329, 341*
Stone, B.L. *351*
Stonik, V.A. 155, 294, 300, 301, 302, 304, 307, *323, 328, 354, 356, 357, 358, 359*
Stothers, J.B. *76*
Stoye-Herzog, M. *337*
Streamer, M. 222, *342*
Strichartz, G.B. *351*
Stringfellow, D.A. 309, *320, 360*
Strong, F.M. *77*
Sturaro, A. *340*
Suarez, E. *146*
Sudhakar, C.V. *148*
Sullivan, B.W. 158, *322, 324, 325, 327*
Sultanbawa, M.U.S. *65*
Sum, F.W. *350*
Sun, H.H. *351*
Sutherland, M.D. 308, *359*
Suvanto, P. *339*
Suzuki, H. *76, 148*
Suzuki, M. *68*
Suzuki, S. *341, 344, 345*
Swain, M. *343*
Swift, M.L. *354*
Switzner, D. *345*
Swynenberg, E.B. 309, *320, 360*
Szeto, S.K. *146*

Tachibana, K. *332, 341*
Tada, H. *327*
Tagle, B. *343, 346*
Taguchi, H. *74*
Taguchi, S. 288, *355*
Takagi, K. *339*
Takahashi, K. *147*, 200, *331, 362*
Takahashi, M. *339*
Takahashi, N. *71*
Takamatsu, H. *71*
Takashima, T. *75*
Takei, M. *359*
Takeuchi, N. *76*
Takeuchi, T. *68*
Takeya, K. 87, *146*
Tamkun, M.M. *333*
Tamm, Ch. *66*
Tanabe, M. 217, *341*
Tanaka, A.K. *76*
Tanaka, H. *70*
Tanaka, J. 200, *331, 338, 341*
Tanaka, N. *331*
Tanaka, O. *75*
Taneja, S. *70*
Tanenbaum, S.W. *77*
Taneyama, M. *77*
Tapiolas, D.M. 222, 223, 251, *342, 343, 344, 347, 348*
Tarasoff, P.G. *147*
Targett, N.M. *361*
Taschner, M.J. *324*
Tatsumaki, T. *77*
Taylor, A.R. *144*
Taylor, K.M. *328, 330*
Taylor, M.E.U. *74*
Taylor, W.C. *325*
Taylor, W.I. *71*
Tazieff-Depierre, F. *336*
Teshima, S. *354*
Tesseraux, I. *337, 339*
Tessier, A.M. 145, *148*
Thaller, V. *68, 70*
Thean, J.E. *76*
Thomas, B.R. *144*
Thomas, R. *74, 77*
Thompson, J.E. *321, 323, 326, 349*
Thompson, K. *334*
Thomson, R.H. 144, *148*, 223, *342*
Tidd, B.K. *69*
Tillekeratne, L.M.V. *330*
Tirard, P. 296, *356*
Tirodkar, R.B. *77*
Tischler, M. *350*
Tishchenko, L.Y. *358*
Tiwari, R.D. *148*
Tobias, J.D. *64*
Tobinaga, S. *76*
Todd, Lord *65*
Tokoroyama, T. *71, 77*
Tokutahe, N. *75*
Torimoto, N. *64*
Toth, J.A. *345*

Toube, T.B. *69*
Towers, G.H.N. *70*
Toyota, J. *344*
Tozawa, M. *349*
Traldi, P. *324, 328*
Trebilcock, M.J. *148*
Trimble, L.A. *75*
Trimmer, J.S. *341*
Trivellone, E. *353*
Troft, L. *222, 342*
Tschinkel, W.R. *71*
Tsuchiya, I. *68*
Tsugita, A. *335*
Tsuji, K. *352*
Tsukitani, Y. *327, 332, 343, 345, 348*
Tu, D. *148*
Tucker, D.J. *203, 332*
Turner, E.S. *63*
Turner, T.D. *148*
Turner, W.B. *63, 67, 68, 77*
Tursch, B. *163, 190, 225, 325, 327, 329, 342, 343, 344, 358*
Tutin, F. *143*
Tymiak, A.A. *185, 320, 328, 330, 352*
Tyszkiewics, J. *332*

Ubaldi, R. *74*
Uchio, Y. *231, 344, 345*
Uda, H. *333*
Ueda, K. *338, 339*
Ueda, S. *146*
Ueda, Y. *276, 353*
Uemura, D. *200, 331, 338, 339*
Uemura, M. *77*
Uhlenbruck, G. *193, 329*
Ulbricht, W. *337*
Ulubelen, A. *77*
Umezawa, H. *68, 77*
Unwin, C.H. *65*
Urakawa, N. *339*
Usgaonkar, R.N. *64, 72, 77*
Usieli, V. *173, 327*
Utz, R. *351*
Uyeo, S. *71*
Uzu, A. *362*

Vakula, T.R. *66*
Vanderah, D.J. *164, 325, 346*
van der Helm, D. *269, 271, 324, 331, 332, 345, 346, 350, 351*
van der Kreek, M. *147*
van der Merwe, K.J. *77*

van Egmond, H.P. *73*
van Eijk, G.W. *77, 89, 149*
van Engen, D. *330, 332, 343, 346*
van Gessel, M. *147*
Vanhaelen, M. *358*
van Meerssche, M. *195, 331, 342*
van Oeveren, B.C.J.A. *147*
van Rheenen, J.W.A. *295, 356*
van Walbeck, W. *75, 77*
Vanek, Z. *71*
Vanzanella, F. *324*
Varanda, W. *337*
Vavra, J.J. *320*
Vederas, J.C. *75*
Veera-Reddy, G.C. *148*
Venkatachalam, K. *148*
Venkateswarlu, Y. *340*
Vermes, B. *148*
Verpoorte, R. *147, 149*
Viltani, G. *258, 261, 262, 349, 350*
Vincent, J.-P. *335, 336, 338*
Vining, L.C. *72*
Viswanathan, N. *68*
Vleggar, R. *67, 77*
Vliegenthart, J.F.G. *308, 360*
Vokoun, J. *71, 75*
Volc, J. *75*
Volkwein, G. *301, 358*
Voogt, P.A. *295, 356*
Vora, V.C. *77*

Wachter, E. *325*
Wagatsuma, S. *72*
Wagner, H. *149*
Waitz, J.A. *77*
Walker, J. *67*
Walker, R.P. *158, 262, 267, 323, 324, 325, 326, 330, 350*
Walkup, R.D. *332*
Walser, A. *69*
Walts, A.E. *330*
Wang, G.K. *352*
Warashina, A. *337*
Ward, R.S. *74*
Warnick, J.E. *334*
Wassermann, O. *338*
Wasylyk, J.M. *309, 343, 360*
Watanabe, K. *314, 315, 361*
Watanabe, M. *77*
Watanabe, Y. *68, 353*
Watrous, J. *334*
Watson, R.D. *318, 363*

Watson, T.R. *66*
Watson, W.H. *347*
Webb, L. 158, *324, 342*
Weber, J.F. *330*
Weber, S. *75*
Webster, G.R.B. *72*
Wehole, H. *78*
Wei, R.-D. *77*
Weijer, J. *75*
Weinheimer, A.J. 277, *353*
Weinreich, D. *334*
Weische, A. 81, *149*
Weisleder, D. *78*
Weller, D.L. *320*
Wells, R.J. *71*, 153, 158, 172, 180, 190, 202, 239, 240, 251, *321, 324, 325, 327, 328, 329, 330, 331, 341, 342, 344, 345, 346, 347, 348, 361*
Wendler, N.L. *75*
Wetherington, J.B. *78*
Whalley, A.J.S. *63*
Whalley, W.B. *63*
White, A.H. 273, *325, 327, 331, 343, 347, 362*
White, D. *354*
White, P.Y. *144*
White, S. *320*
Wijekoon, W.M.D. *332*
Wijnsma, R. *149*
Wild, J. *329*
Wilkin, P.E. *144*
Williams, A.E. *144*
Williams, D.J. *74*
Williams, T. *77*
Williamson, J. *362*
Williard, P.G. 190, *329*
Willis, R.H. *342, 345*
Wilson, G. *146, 149*
Wilson, G.R. 309, *360*
Wilson, L. *320*
Wing, R.M. 270, *351*
Wirth, J.C. *78*
Withers, N.W. *321*
Woolard, F.X. *338*
Woolven, P. *74*
Wratten, S.J. *332, 349*
Wright, I.G. *72*
Wright, J.J. *65*
Wright, J.L.C. *348*
Wright, J.M. *342*
Wu, H. 165, *325, 326, 329*
Wulff, P. *348*

Wunderer, G. *337*

Xu, R. *147*

Yagi, A. *78*
Yagudaev, M.R. *72*
Yajima, H. *71*
Yalamanchili, G. *331*
Yalhi, B. *342*
Yamada, M. *73*
Yamada, S. *72*
Yamada, Y. 241, 248, *343, 344, 345, 347, 348*
Yamaguchi, K. *352*
Yamaki, T. *78*
Yamakado, T. *344*
Yamamoto, I. *73, 78*
Yamamoto, M. *78, 340*
Yamamoto, T. 200, *331*
Yamamoto, Y. *63, 73, 78*
Yamamura, S. *350*
Yamano, T. *78*
Yamashita, K. *76*
Yamatodani, S. *73, 78*
Yamauchi, T. *78*
Yamazaki, M. *71*
Yamazaki, Y. *78*
Yanagi, K. *341*
Yang, X.-P. *147*
Yashihara, M. 248, *347*
Yasuda, F. *327*
Yasumoto, T. 276, *353, 363*
Yasuzawa, T. 240, 248, *343, 347, 358*
Yee, C.W. *146*
Ying, B.-P. *147*
Yokotsuka, T. *71*
Yokoyama, N. *71*
Yonabaru, S. *362*
Yonaga, M. *338, 339*
Yoshiba, S. 273, 277, *351, 352*
Yoshida, S. *71, 77*
Yoshimura, T. *63*
Yoshioka, M. *78*
Young, D.N. *320*
Young, K. *69*
Yuda, Y. *70*

Zabel, V. *347*
Zahn, R.K. 193, *329*
Zahner, H. *78*
Zamir, L.O. *78*
Zanetti, L. *324*

Zdero, C. *65*, *68*
Zeeck, A. *78*
Zenk, M.H. 81, 89, *146*, *147*, *148*, *149*
Zewail, A.H. *145*
Zhongde, W. 225, *344*
Zhongnian, Y. 225, *344*
Zhuravlev, N.S. *144*, *149*

Zielinski, J. *321*, *322*
Zimmermann, D. *331*
Zollo, F. 283, 285, 287, 290, *327*, *354*, *355*, *356*
Zurenko, G.E. *330*
Zviely, M. *326*, *327*
Zwick, J. 213, *336*, *338*

Subject Index

By

R. BERNER, Wien

Aaptamine 195
Aaptos aaptos 195
Aaptos sp. 206
Acanthaglycoside 288
Acanthaster planci 283, 287, 288, 291, 295
Acanthella aurantiaca 195
− sp. 165
Acanthifolicin 203
Acanthodoral 263
Acanthodoris nanaimoensis 263
Acetate-polymalonate pathway 5
16β-Acetoxycholesta-7,24-dien-3β-ol 304
16α-Acetoxycholest-7-ene-3β-ol 304
20-Acetoxyclaviridone-b 248
20-Acetoxyclaviridone-c 249
20-Acetoxyclavulone 249
Acetoxycrenulide 270
12-Acetoxycyclosinularene 225
(2R,7S,11R)-7-Acetoxy-2-hydroxynardosin-1(10)-en-12-al 224
12α-Acetoxynakafuran-8 263
6β-Acetoxyolepupuane 263
(13Z,15E,17E,22E,24Z)3β-Acetoxy-12-oxomalabarica-13,15,17,22,24-pentaen-26-oic acid 180
(13Z,15E,17E,22E)3β-Acetoxy-12-oxomalabarica-13,15,17,22,24-pentaen-28-oic acid 180
12-Acetoxysinularene 225
Acetylcholine 154
Acetylenes 205, 258, 267, 268, 271
N^5-Acetylornithine 307
Actinia cari 211
− equina 211, 212
Actinobolin 4, 12, 52, 58
Actinodendron plumosum 211

Actinopyga agassizi 301, 304
− echinites 300
Adenosine 194
Adocia sp. 165
Aerophobin-1 186
Aerophobin-2 186
Aeroplysinin-1 184
Aeroplysinin-2 185, 186
Aerothionin 185
$\Delta^{9(15)}$-Africanene 223
Agassicin 263
Agelas nakamurai 164, 165
− sceptrum 195
− sp. 155, 164, 195
Agelasidine A 159
Agelasidine B 165
Agelasidine C 165
Agelasine A, B, C, D, E, F 164
Ageline A 164, 165
Ageline B 164
Aglaja depicta 277
Aglajne-1 277
Aglaophenia pluma 210
Aglykones 80, 88, 90, 218, 282, 283, 286, 287, 288, 289, 290, 291, 299, 301
Agrimolide 4, 36, 58
AI-77-B 54, 58
AI-77-C 54, 58
AI-77-D 55, 58
AI-77-F 53, 58
Ailanthus malamarica 180
Aiolochroia sp. 184
Albicanol 263
Albicanyl acetate 263
Alcaligenes faecalis 185
Alcyonidium gelatinosum 253

Alcyonium flaccidum 232
– *palmatum* 223
– *sp.* 218, 222, 240
– *utinomii* 232
Aldisa sanguinea cooperi 261, 268
β-Alectoronic acid 33, 58
Alizarin 80, 91, 131, 133, 134, 135, 136, 137, 138, 139, 140, 141, 142
Alizarin dimethyl ether 92, 142
Alizarin 1-methyl ether 92, 131, 133, 134, 135, 136, 137, 138, 139, 140, 142
Alizarin 2-methyl ether 93, 131, 134, 135, 136, 137, 139, 141, 142
Alkaloid 155, 190, 191, 194, 195, 199, 253
5-Alkylpyrrole-2-carboxaldehydes 194, 195
Allantoin 194
Aloe emodin 138
Altenuene 42, 58
Altenuisol 41, 58
Alternariol 41, 58
Alternariol methyl ether 42
Amarouciaxanthin A, B 316
Amaroucium pliciferum 316
Amathia convolute 254
Ambliofuran 163
Ambliol A, B, C 163
Amicoumacin A 55, 58
Amicoumacin B 54, 58
Amicoumacin C 54, 58
Amino acids 183, 211, 252, 272, 274, 298, 314, 315, 318
Amino acid sequence 211, 212, 213, 272, 274, 275, 319
4-Amino-5-bromopyrrolo-(2,3-d)-pyrimidine 194
AMPase 12
Amphimedon sp. 195
Anemonia sulcata 211, 212
Anhydrovalerenenol 224
Anisodoris nobilis 257
Anthelia glauca 217
Anthoarcuata graceae 261
Anthopleura elegantissima 211
– *xanthogrammica* 211
Anthragallol 93, 133, 137, 142
Anthragallol 1,2-dimethyl ether 94, 131, 132, 133, 138, 140
Anthragallol 1,3-dimethyl ether 94, 131, 133, 140
Anthragallol 2,3-dimethyl ether 95, 138, 142

Anthragallol 2-methyl ether 95, 132, 142
Anthragallol 1,2,3-trimethyl ether 96, 140
Anthraquinone-2-carbaldehyde 96, 138
Anthraquinones 85, 308
Antibiotic 153, 154, 171, 206, 251, 276
APA 211, 212
APC 211
Aplidium constellatum 316
Aplysia dactylomela 269, 270, 271
– *limacina* 210
– *oculifera* 271
– *vaccaria* 270
Aplysiatoxin 269
Aplysilla sulphurea 164
Aplysillin 164
Aplysina aerophoba 185, 186, 195
– *cavernicola* 185
– *fistularis* 185
– *sp.* 184
Aplysinopsin 183, 184
Aplysinopsis reticulata 183
Aplysistatin 269
Aplysulphurin 164
Arachidonic acid 248, 275
Archidoris montereyensis 262, 263, 265, 268
– *odhneri* 262, 265
Ardisic acid B 16, 58
Arenarol 158
Arenarone 158
(3S,4R,5S,6R,7S)-Aristol-9-en-3-ol 224
(4R,5S,6R,7S)-Aristol-9-en-3-one 224
Arsenobetain 318
Artemidin 14, 58
Artemidinal 14, 58
Artemidinol 15, 58
Artemidiol 15, 58
Artemisia dracunculus 15
Asbestinin 1–5 240
Asbestinin-5-acetate 240
Asbestinin epoxide 240
Ascochitine 8, 9
Ascidiaclamide 310
Ascidia nigra 310, 317
Aspena pubescens 12
Asperentin 6, 34, 58
Astaxanthin 316
Asterias amurensis 283
Asterias amurensis versicolor 289, 291
Asterias rubens 295
– *vulgaris* 290, 295
Asterina Pectinifera 283, 298
Asterogenol 295

Asterone 295
Asterosaponin A 283, 284
Asterosaponin P_1 294
Asterospicularia randalli 217
Astichopus multifidus 302, 304
Astropecten latespinosus 289, 298, 299
- *polyacanthus* 299
- *scoparius* 283
Atisane-3β,16α-diol 164
Atractylone 223
Atropine 275
ATX 211
ATX II 211, 212
ATX III 211
ATX IV 211
ATX V 211
Aulacomya ater 282
Axinella cannabina 154, 159, 257, 259
- *polycapella* 202
- *sp.* 259
- *verrucosa* 154, 195
Axisonitrile 261

Babylonia japonica 276
Bacillus subtilis 155, 158, 164, 165, 172, 185, 206, 253, 268, 275, 276, 313
Baciphelacin 12, 55, 58
Bacterium B-392 164
Bactobolin A 12, 52, 58
- B 53, 58
- C 52, 58
Balanoglossus carnosus 318
- *misakinesis* 318
Bastadin 1–7 185, 186
Bastaxanthin 181
Benzoic acid 277
2-Benzylxanthopurpurin 96, 133, 137
Bergenin 10, 58
Bicyclogermacrene 224
Biological activities
- antialgal activity 253
- antibacterial activity 152, 155
- antibiotic activity 206, 262
- anticonvulsant activity 240
- antifeedant activity 157, 200, 257, 261, 262, 263
- antifungal activity 172, 195, 200, 204, 253, 299
- antiinflammatory activity 172, 217, 251
- antileukemic activity 12

- antimicrobial activity 152, 157, 158, 164, 165, 172, 174, 183, 185, 193, 195, 202, 204, 210, 251, 253, 266, 268, 275, 276, 313
- antineoplastic activity 152, 215, 254, 272, 308
- antinicotinic activity 276
- antirheumatic activity 217
- antitumor activity 12, 200, 212, 215, 216, 217, 231, 249
- cardiotoxicity 208, 210
- cardiovascular effect 193
- chronotropic activity 194
- contractile response 273, 274, 276, 308
- convulsant response 231, 275
- cytotoxic activity 158, 164, 172, 185, 195, 200, 251, 270, 272, 299, 313
- defence effect 210, 317
- dermonecrosis 208
- deterrent activity 261
- effect on acetylcholin 154, 240, 274
- growth inhibition of
- - Hella cells 231
- - KB cells 203
- - L 1210 cells 202, 249, 275, 309
- - P 388 193, 195, 203, 215, 217, 308
- hemagglutination activity 253
- hemolytic effect 283, 291, 295, 299
- hypothermic effect 193
- inhibition of
- - cell division 152, 155, 157, 158, 171, 174, 194, 196, 239, 253, 268, 276, 277
- - development of fish embryo 193
- - gram negative bacteria 173, 195, 202
- - gram positive bacteria 165, 173, 195, 202
- - phospholipase A_2 172
- inotropic activity 194, 274
- mutagenic effect 239
- neurotoxic activity 208, 231, 273, 299, 318
- paralysis 231, 275
- predator repellent 263
- presynaptic effect 154
- sedative effect 231
- toxicity to corals 158
- - fish 155, 171, 172, 173, 174, 195, 196, 200, 206, 222, 223, 231, 261, 268, 270, 274, 276, 283, 299
- - shrimp 155, 171, 172
Biosynthesis 5, 81
3,4,3′,4′-Bisdehydroxyxanthomegnin 50

Bishomopalytoxin 213
Bispuupehenone 158
Bivittoside A, B, C, D 301
Bohadschia argus 301, 304
− *bivittata* 301
− *graeffei* 240
Bonella viridis 318
Bonellin 318
Bornträger reaction 89
Botrallin 43, 58
Brassicasterol 317
Brevifolincarboxylic acid 41, 58
Brianthein X, Y, Z 239
6-Bromoapsylinopsin 184
5-Bromo-7α-chlorocavernicolin 185
5-Bromo-7β-chlorocavernicolin 185
7α-Bromo-5-chlorocavernicolin 185
7β-Bromo-5-chlorocavernicolin 185
(+)3-Bromo-5-chloroverongiaquinol 185
6-Bromo-2′-de-N-methylaplysinopsin 183
6-Bromo-4′-N-demethylaplysinopsin 184
5-Bromo-N,N-dimethyltryptamine 183
7-Bromo-4-(2-ethoxyethyl)quinoline 253
1-Bromo-8-ketoamblio-A acetate 163
(±)-3-Bromoverongiaquinol 185
Bryostatin 1–6 254
Bryostatin 8 254

Ca-ATPase 186
Cacospongia mollior 259
− *scalaris* 171, 172
Cadlina luteomarginata 262, 263, 265
− *marginata* 171
Calyculone A, B, C 239
Candida albicans 155, 164, 165, 313
Canescin 7
Canescin A, B 30, 58
Canthaxanthin 307
Capillarin 14, 58
Capnella imbricata 225
Capnellane 225
4-O-Carbomethoxylamellicolic anhydride 7, 59
Carboxylic acids 203
Carcinoscorpius rotunda cauda 318
Cardisoma carnifex 318
Carotenoids 181, 282, 316
Carterospongia sp. 173
Casella atromarginata 265
Cassia sp. 80
Cavernicolin-1 185
Cavernicolin-2 185

Cavernoside 172
Cavernulina grandiflora 239
Cavernuline 239
Celenamide A, B, C, D 191
Cembranes 231
Cembranolides 231
Ceratocystis fimbriata 10
− *minor* 10
− *ulmi* 10
Cespitulari sp. 223, 225, 233, 239
Cerebratulus lacteus 317
Charonia lampas 289
− *sauliae* 276, 299
Cheilanthane 265
Chelyconus fulmen 275
Chironex fleckeri 208, 209
Chlamys tehuelcha 282
5-Chlorocavernicolin 185
Chloroisocumarin 6
Chlorovulone I 249
Cholest-4,14-diene-15,20ξ-diol-3,16-dione 217
(20R)-Cholest-5-ene-3β,21-diol 296
(20R)-Cholest-22-ene-3β,21-diol 296
Cholest-5-en-3β-ol 317
5α-Cholest-7-en-3β-ol 307
Cholest-4-en-3-one 154
Cholest-4-en-4,16β,18,22R-tetrol-3-one-16,18-diacetate 210
Cholesta-5,25-diene-3β-24ξ-diol 154
(20R,22S)-5α-Cholesta-9(11),24(25)-diene-3β,6α,20,22-tetraol 299
(20R)-Cholestane-3β,21-diol 296
5α-Cholestane-3β,4β,6α,8,15α,16β,26-heptol 296
5α-Cholestane-3β,5,6β,7α,15α,16β,26-heptol 296
5α-Cholestane-3β,6α,7α,8,15α,16β,26-heptol 296
5α-Cholestane-3β,5,6β,15α,16β,26-hexol 296
5α-Cholestane-3β,6β,7α,15α,16β,26-hexol 296
5α-Cholestane-3β,6α,8,15α,16β,26-hexol 296
5α-Cholestane-3β,4β,6α,8,15α,16β,26-octol 296
5α-Cholestane-3β,6β,15α,16β,26-pentol 296
Cholesterol 154, 317
Chorismic acid 81

Subject Index

Chromatography
- affinity 275
- droplet counter current 88, 90
- electrofocusing 275
- gel permeation 253, 263, 275
- GL 88, 89, 299, 308
- HPL 88, 89, 90, 283, 285, 287, 318
- ion exchange 253
- paper 88
- thinlayer 88, 89, 299, 308

Chromodoris albonotato 262
- *marisiae* 262
- *norrisi* 164
- *sedna* 266

Chrysaora quinquecirrha 208, 209
Cinachyra alloclada 205
Cinchona ledgeriana 90
Cinnamic acid 9
Ciona intestinalis 317
Citorellamine 313
Citrin 9
Citrinin 8
Cladosporin 4, 10, 34, 59
Cliona celata 191
Clionamide 191
Claviridenone a–d 248, 249
Clavukerin A, B 223
Clavukerin C 223, 251
Clavularia inflata 223, 225
- *koellikeri* 223, 225, 239, 251
- *sp.* 223
- *viridis* 217, 248, 249
Clavularin A, B 251
Clavulone 249
^{13}C-NMR-spectroscopy 3, 83, 87, 91ff., 164, 185, 186, 202, 212, 231, 283, 285, 287, 289, 298, 300, 310
Coelulatin 132
Collagenase 209
β-Collatolic acid 34, 59
Collisella limatula 277
Comasterias lurida 295
Comatula pectinata 308
Commiphora mukul 217
Condylactis aurantiaca 212
- *gigantea* 211
Conotoxin G I 273, 274
- G IA 273, 274
- G II 273, 274
- G III 273, 274
- G IV 274

Conus eburneus 273, 275
- *fulmen* 273
- *geographus* 257, 273, 274, 275
- *imperialis* 273
- *magus* 273, 274
- *striatus* 273, 274
- *tessulatus* 273, 275
- *textile* 273, 275
Copareolatin 132, 133
Copareolatin-1(or 5),6-dimethyl ether 97, 139
Copareolatin-6-methyl ether 97, 139
Corallium sp. 241
Corylopsin 16, 59
Coryphella lineata 261
Coscinasterias acutispina 283
Cotton effect 5
p-Coumaric acid 9
Crassostra virginica 282
Crenolide 270
Crinoids 307
Crispatane 277
Crispatene 277
Cryptoclauxin 46, 59
Cryptosporiopsin 8, 9
Cryptosporiopsinol 8, 9
(+)-Cubetol 225
Cubitanes 239
Cubitermes umbratus 239
Cucumaria fraudatix 304
- *frondosa* 304
- *japonica* 304
- *sp.* 304
Cucumarioside A$_2$ 304
Cucumarioside G$_1$ 304
Curcuma cedoaria 224
Cyanea capillata 209

(3E)-**Dactomelyne** 271
(3Z)-Dactomelyne 271
Damnacanthal 98, 133, 134, 139, 140
Damnacanthol 98, 132, 133, 138
Damnacanthol-ω-ethyl ether 98, 138, 140
Damnacanthol-ω-methyl ether 99
Dascyllus auranus 263
O-Deacetyl-12-O-benzoyl-12-pteroidin 239
12-Deacetyl-12,18-diepiscalaradial 172
12-Deacetyl-20-methyldeoxoscalarin 266
O-Deacetyl-propionylcavernulin 239
Dechlorogilmaniellin 45, 59
Dehydroaltenusin 4, 42, 59
Dehydroambliol-A 163

24-Dehydroechinoside A 300
25-Dehydrostichlorogenol 302
Dehydroustic acid 32, 59
3,4-Dehydroviomellein 49
3,4-Dehydroxanthomegnin 50
Demethyldysidenin 190
Demethylisodysidenin 190
6-Demethylkigelin 37
7-O-Demethylmonocerin 60
O-Demethylrenierone 196
Dendrilla sp. 163, 164, 259
Dendrillolide A, B, C 164
Dendrodoa grossularia 313
Dendrodoris grandiflora 262, 263, 266
– *limbata* 262
– *sp.* 263
Dendrogyra cylindrus 217
Dendrolasin 262
Denticulatin A, B 276
Denticulatolide 231
Deoxoscalarin 261
2-Deoxylemnacarnol 224
Deoxypalytoxin 213
Deoxyparguerol 270
9,10-Deoxytridachione 277
Depsipeptides 309
Dercitus sp. 183
Desacetylscalaradial 172
Deuterated compounds 5, 86, 87
(13Z,15E,17E)-3β,28-Diacetoxy-22-hydroxymalabarica-13,15,17,24-tetraen-12-one 180
15α,16α-Diacetoxyspongiane 164
Diaporthin 10, 32, 59
Diatoxanthin 316
Diaulula sandiegensis 258, 267, 268
5,6-Dibromo-N,N-dimethyltryptamine 183
Dichloroverongiaquinol 185
Didehydrofuranospongin-1 171
Didemnin A, B, C 309, 310
Didemnum sp. 312
– *ternatanum* 312
Diemensis A, B 276
Dihydrocitrone 30, 59
Dihydrohomalicine 16
Dihydromarthasterone 289, 290
Dihydrosiphonarin A, B 277
1,8-Dihydroxyanthraquinone 99, 131
Dihydroxydihydrocumarin 9, 10
1,3-Dihydroxy-2,5-dimethoxyanthraquinone 100, 131, 132

Dihydroxydeodactol 269
1,4-Dihydroxy-2-ethoxymethylanthraquinone 100, 134, 142
3,5-Dihydroxy-2-ethoxymethyl-4-methoxyanthraquinone 83, 100, 134
Dihydroxyflustramine C 253
1,4-Dihydroxy-2-hydroxymethylanthraquinone 101, 142
Dihydroxyisocumarin 3, 4, 10
5,6-Dihydroxylucidin 101, 138
1,6-Dihydroxy-4-methoxyanthrachinone 102, 134
1,3-Dihydroxy-4-methoxyanthraquinone 132
1,3-Dihydroxy-2-methoxymethylanthraquinone 134
1,4-Dihydroxy-5-(or 8)-methoxy-2-methylanthraquinone 102
4,5-Dihydroxy-7-methoxy-2-methylanthraquinone 102, 141
1,4-Dihydroxy-2-methylanthraquinone 103, 141
1,5-Dihydroxy-2-methylanthraquinone 103, 140, 141
1,4-Dihydroxy-6-methylanthraquinone 104, 141
1,4-Dihydroxynaphtoic acid 81, 82
1,4-Dihydroxy-3-prenyl-2-naphtoic acid 81, 82
2,5(or 3,5)-Dihydroxy-1,3,4(or 1,2,4)-trimethoxyanthraquinone 104, 132
Diketopiperazine 191
1,3-Dimethoxy-2-carboxyanthraquinone 141
5,6-Dimethoxy-1-(or 4)-hydroxy-2-(or 3)-hydroxymethylanthraquinone 104, 132
5,7-Dimethoxy-4-hydroxy-2-methylanthraquinone 105, 138
1,4-Dimethoxy-2,3-methylenedioxyanthraquinone 105, 132
N,N-Dimethylamino-3-guaiazulene 223
2-Dimethylamino-6-methyl-8-methylamino-14-1,3,7,9-tetrazacyclopentazulene 215
5-((3-N-Dimethylamino)-1,2,4-thiadiazolol)-3-indanyl-methanone 313
4α,14α-Dimethylcholest-9(11)-en-3β-ol 307
2,6-Dimethyl-5-heptenal 262
2,6-Dimethyl-5-heptenoic acid 262
1,4-Dimethyl-2,3,3α,4,5,6-hexahydroazulene 223

2,5-Dimethyl-6-methoxy-4,7-dihydroisoindole-4,7-dione 196
(13Z,15E,17E)-3,12-Dioxomalabarica-13,15,17,22, 24-pentaen-26-oic acid 180
(22E,24Z-3,12-Dioxomalabarica-13,15,17,22,24-pentaene-22,26-olide 180
N,N'-Diphenetylurea 312
Diplophyllum albicans 263
Discodermia kiiensis 193
Discodermin A, B, C, D 193
Diterpenes 157, 163, 165, 180, 222, 231, 239, 240, 265, 270
DNase 209
Dolabellanes 239, 270
Dolastatin 1 to 9 272
Dolatriol 270
Doridosine 257
Doriopsilla sp. 263
Dysidea amblia 158, 163, 259
- *arenaria* 158
- *aurea* 157
- *chlorea* 202
- *etheria* 158, 259
- *fragile* 157, 158, 258, 259
- *sp.* 155
Dysidenin 190
Dysidin 191

Eburnetoxin 275
Ecdysteroids 215, 318
Ecdysterone 215
Echinaster luzonicus 286
- *sepositus* 285, 295
Echinoside A, B 300
Ecteinascidia turbinata 317
ED_{50}-value
- Acanthifolicin 203
- Okadaic acid 203
Efflatournaria sp. 240, 241
Endothia parasitica 10
Ent-Valerenane 224
Epiactis prolifera 211, 212
12 Epideoxoscalarin 172
5α,8α-Epidioxysterol 154, 217, 317
5-Epi-ilimaquinone 158
(3E)-12-Epioptusenyne 271
12-Epioptusenyne 271
β-(−)-1,10-Epoxyaristolane 224
9α,11α-Epoxycholest-7-ene-3β,5α,6β,19-tetrol-6-acetate 155
16,23-Epoxycholesta-7,24-dien-3β-ol 304
Erythrolide A, B 239

Erythropodium caribaeorum 239
Escherichia coli 165
Esperiopsis digitata 154
2-Ethoxymethyl-3-hydroxy-4-methoxy-anthraquinone 106, 134
2-Ethoxymethyl-1-methoxy-3,5,6-trihydroxyanthraquinone 106, 140
2-Ethoxymethyl-3-methoxy-1,5,6-trihydroxyanthraquinone 106, 138
24-Ethylcholest-5-en-3β-ol 307
24-Ethylenecholest-5-en-3β-ol 307
Eudendrium glomeratum 210
- *spp.* 210, 261
Eudistoma olivaceum 313
Eudistomins 313
Eunicea calyculata 239
- *succinea* 233
Euplexaura erecta 223, 248
- *flara* 251
Euretaster insignis 296
Euryfuran 158, 261, 263
Euryspongia Sp. 157, 158, 259

Fasciaspongia cavernosa 172
Fasciculatin 266
Felicia wrightii 14
Fenestra spongia 158
Flabellina affinis 261
Flexibilide 231
Flustrabromine 253
Flustra foliacea 253
Flustramine A 253
Flustramine B 253
Flustramine C 253
Flustraminol A 253
Flustraminol B 253
Foliaspongin 172
Formamides 159, 206
N-Formyl-1,2-dihydrorenierone 196
5-Formylmellein 20, 59
(2R,7S)-7-Formyloxy-2-hydroxy-12-nornardosin-1(10)-en-11-one 224
Fulvoplumierin 277
Fungal isocumarins 2, 9, 10
Furanogermacradienes 224
Furanoquinol 223
Furanoquinone 223
Furanosesquiterpenes 158
Furanoterpenes 157, 171, 173, 265
Furodendin 171
Furodysinin 158, 263
Furodysinine 261

Furoscalarol 261
Furospinulosin 171
Furospongin-1-acetate 266
Furospongolide 171
Fusamarin 34, 59
Fusarentin-6,7-dimethylether 39
Fusarentin-6,8-dimethylether 39
Fusarentins 5

Galiprenylin 107, 134
Gallic acid 10
Gallium mollugo L. 81
Garveatin A 210
Garveia annulata 210
Geodia cydonium 193
- *mesotriaena* 193
Geodiastatin 1 193
Geodiastatin 2 193
Geodiatoxin 1 193
Geographutoxin I 274
Geographutoxin II 274
Geranylfarnesol 171
Geranylgeraniol 240
Gerardia savagli 215
Germacrene C 224
Gersemia rubiformis 217
Ghiselinin 263
Gibberellin 10
Gibbonsia elegans 277
Gilmaniellin 4, 44, 59
Gleophyllum sepiarium 17
Glossobalanus sp. 318
D-Glucuronic acid 318
Glycans 317
Glycerol ethers 205, 215
N-Glycoloylneuraminic acid 308
Glycoproteins 193, 274, 308
Glycoside 80, 88, 90, 283, 286, 295, 301, 304, 307
Gorgost-5-ene-3β,7α,11α,12β-tetraol-12-monoacetate 217
Gracilin A 164
Gracilin B 164
Grapholithia molesta 18
Grossularine 313
Guaiazulene 223
Guanidine 195
Gyrostoma helianthus 211

Hacelia attenuata 287, 288, 290, 295, 296
Halenaquinol 206
Halenaquinone 206

Halichondria cf. moorei 155
- *melanodoica* 203
- *okadai* 200, 203
- *sp.* 155, 159
Halichondrin B 200
Haliclona sp. 154, 206
Halocordyle disticha 210
Halocynine 314
Halocynthia aurantium 316
- *papillosa* 317
- *roretzi* 314, 315, 316
Halocynthiaxanthin 316
Halostanol sulfate 155
Hapalochlaena maculosa 282
Hervia peregrina 261
Heteronemin 172
1-0-Hexadecylglycerol 268
$2,2',4,4',5,5'$-Hexahydroxybiphenyl 202
Hippeastrine 53, 59
Hippuristanol 217
Histamine 209
H-NMR-spectroscopy 83, 85, 86, 91 ff., 164, 186, 217, 285, 287, 289, 295, 300, 310, 313
Holost-9(11)-ene-3β-ol 301
Holothuria edulis 300
- *floridana* 300
- *grisea* 300
- *leucospilota* 300
- *squamifera* 300
- *vagabunda* 300
Holothurin A, A_1, A_2 300, 301
- B, B_1 300, 301
Homalium laurifolium 16
- *zeylanicum* 16
Homalizine 16, 59
Homarus americanus 318
Homolycorine 53, 59
Homopalytoxin 213
Humiria balsamifera 16
Hydrangenol 4, 9, 23, 59
8-Hydroperoxy-24-vinylcholesterol 317
3-Hydroxyadipic acid 4
5(or-8)-Hydroxyalizarin-1-methyl ether 107, 133
2-Hydroxyanthraquinone 108, 131, 135, 136, 137, 139, 140, 142
3-Hydroxyanthraquinone-2-carbaldehyde 108, 138, 139
8-Hydroxyartemidin 23, 59
4'-Hydroxyasperentin 35, 59
5'-Hydroxyasperentin 35, 59

5′-Hydroxyasperentin-8-O-methyl ether 36
8-Hydroxycapillarin 23, 59
8-Hydroxydamnacanthol-ω-ethylether 109, 134
2-Hydroxyethyldimethylsulfoxonium ion 253
2α-Hydroxyhippuristanol 217
3β-Hydroxyholosta-7,25-diene-16-one 304
3β-Hydroxyholosta-9(11),25(26)-diene-16-one 304
1-Hydroxy-2-hydroxyethyl-3-methoxyanthraquinone 109, 134
1-Hydroxy-2-hydroxymethylanthraquinone 109, 132, 134, 139, 142
1-Hydroxy-6(or-7)-hydroxymethylanthraquinone 110, 139
5-Hydroxy-2-hydroxymethylanthraquinone 139
3-Hydroxy-4-hydroxymethyl-4-pentenoic acid 206
3-Hydroxyisocumarin 3
8-Hydroxyisocumarin 12
(13Z,15E,17E,22E)-3β-Hydroxymalabarica-13,15,17,22,24-pentaen-12-one 180
3-Hydroxymellein 21, 59
4-Hydroxymellein 21, 59
5-Hydroxymellein 22, 59
6-Hydroxymellein 26, 59
7-Hydroxymellein 23, 59
1-Hydroxy-5-methoxy-2-methylanthraquinone 110, 140
1-Hydroxy-2-methylanthraquinone 111, 131, 132, 135, 136, 137, 139, 140, 141, 142
2-Hydroxymethylanthraquinone 111, 134
3-Hydroxy-2-methylanthraquinone 112, 132, 133
23-Hydroxy-20-methyldeoxoscalarin 266
3β-Hydroxymethyl-A-norsteran 154
3β-Hydroxymethyl-A-norsterol 154
23-Hydroxy-20-methylscalarolide 266
9-Hydroxymonocerin 40, 59
3-Hydroxymorindone 112, 138
5-Hydroxynakafuran-8 263
12β-Hydroxy-24-norcholesta-1,4,22-trien-3-one 217
4-Hydroxyochratin 56, 59
(13Z,15E,17E,22E)-3β-Hydroxy-12-oxomalabarica-13,15,17,22,24-pentaen-22,26-olide 180
Hydroxyprolin 274
6-Hydroxyramulosin 26, 59

8-Hydroxysubspinosin 83
2-Hydroxy-1,3,4-trimethoxyanthraquinone 112, 132
(R)-2-Hydroxy-6-trimethylammoniohexanoate 314
3β-Hydroxy-4,4,14-trimethylpregna-9(11),16-dien-20-one 304
Hymeniacidon aldis 195
– *sanguinea* 195
– *sp.* 206
Hypotaurocyamine 165
Hypoxylon spp. 18
Hypselodoris agassizi 263
– *californensis* 262
– *ghiselini* 262, 263, 265
– *zebra* 158, 263
Hyrtios eubamma 158
– *sp.* 154, 257
Hystazarin monomethyl ether 140

Ianthella basta 181, 185, 186
IC_{50}-value
– -Chlorovulone I 249
– -Patellamide A 309
– -Punaglandin 3 249
– -Ulithiacyclamide 309
Idiadione 171, 261
Igernella notabilis 164
Ilikonapyrone 277
Ilimaquinone 158
Inflatene 223
Iotrochota 184
Ircinianin 173
Ircinia dendroides 171
– *fasciculata* 259
– *sp.* 174
– *variabilis* 170
– *wistarii* 173
IR-spectroscopy 83, 84, 85, 91ff., 298, 300
Isoacanthadoral 263
Isoagatholactone 164
Isochorismic acid 81, 82
Isocumarin 2
2-Isocyanopupukeanane 206
Isodeodactol 269
Isodysidenin 190
Isofistularin-3 185, 186
Isoparguerol 270
Isoparguerol-16-acetate 270
Isoprenylquinol 157
Isoptilocaulin 195
Isostichopus badionotus 307

Isothiocyanates 159, 206
Isovalerenenol 224

Jaspis stellifera 180
Jones reagent 180
Juzunal 113, 133, 134
Juzunol 133, 134

Kalihinol-A 165
- -B 165
- -C 165
- -D 165
- -E 165
- -F 165
Kelletia kelletii 276
- *sp.* 276
Kelletinin I 275
Kelletinin II 275
Keramadine 195
α-Ketoglutaric acid 81, 82
Kigelin 37

Laevigatoside 291
Laffariella variabilis 171
Lamellaria sp. 277
Lamellarin A–D 277
Lamellicolic anhydride 43, 59
Lamprometra palmata gyges 308
Latrunculia magnifica 200
Latrunculin A–D 200
Lemnacarnol 224
(2R,11S,12R)-Lemnal-1(10)-ene-2,12-diol 224
Lemnalia africana 224
- *humesi* 224
Lemnalol 225
Lendenfeldia sp. 172
Lenzitin 59
Leprocybin 30, 59
Leptogorgia sarmentosa 217
- *virgulata* 253
Lethal dose
- - Geodiatoxin 193
- - Lophotoxin 213
- - Okadaic acid 203
- - Petrosin 195
- - Sepositoside A 285
Leucetta microraphis 195
Leucettidine 195
Limatulone 277
Limulus polyphemus 318
Linckia laevigata 287, 290, 291

Lindera strychnifolia 223
Linderazulene 223
Linderene 223
Lissoclinum patella 309
LL-Z1640-5 33, 59
LL-Z1640-6 34, 59
LL-Z1640-7 33, 59
Lobohedleolide 231
(7Z)-Lobohedleolide 231
Lobophytum crassum 233
- *denticulatum* 231
- *depressum* 216, 248
- *hedleyi* 231
- *pauciflorum* 216, 232, 233
- *sp.* 232, 233, 240
(−)-Loliolide 270
Lophogorgia alba 233
- *sp.* 231
Lophotoxin 231
Lucidin 113, 131, 132, 133, 134, 135, 136, 137, 138, 139, 140, 141, 142
- 1-ethyl ether 141
- 3-ethyl ether 114
- ω-ethyl ether 114, 135, 138, 139, 141
- ω-methyl ether 115, 138, 139
Ludwigothuria grisca 317
Luidiaglycoside A–D 289
Luidia maculata 289, 290, 291, 295, 296
- *quinaria* 298
Lunatinin 33, 59
Luteone 265
Lutesporin 49, 59
Luzonicoside 286
- A 290
- B 290
- C 290
- D 290
Lytechinestatin 308
Lytechinus variegatus 308

Macrolides 199, 254
Macrophiotrix longipoda 307
Majoranal 115, 133
Malabarica triterpenes 180
Mallotus japonicus 16
Manoalide 171, 172
Mass-spectroscopy 85, 91 ff., 213, 285, 289, 290, 298, 308, 309, 310
Marginatafuran 265
Marislin 262
Marthasterias glacialis 290
Marthasterone 289

Marthasteroside A_1 290
Melibe leonina 262
Mellein 3, 4, 12, 18, 19, 59
Menaquinone 81
Methionine 172
2-Methoxyanthraquinone 116, 131, 134, 135, 136, 137, 140, 142
1-Methoxy-2-methylanthraquinone 116, 131, 134, 136, 137, 140, 142
1-Methoxy-2-methyl-3,5,6-trihydroxyanthraquinone 117, 140
5-Methoxy-2-(or-3)-methyl-1,4,6-trihydroxyanthraquinone 117, 132
4-Methoxy-1,3,5-trihydroxyanthraquinone 117, 132
9-Methyladenine 164
6-Methylalizarin 118, 133, 138
2-Methylanthraquinone 140
Methylaplysinopsin 183
2′-de-N-Methylaplysinopsin 183
24ξ-Methyl-26,27-bisnor-5α-cholest-22-ene-3β,4β,6β,8,15,16β,25-heptol 296
Methyl-(E)-3-(6-bromoindol-3-yl)-prop-3-enoate 184
(22 E)-24ξ-Methyl-5α-cholest-22-en-3β,4β,6α,8,15α,16β,26-heptol 296
24ξ-Methyl-5α-cholestane-3β,6α,8,15α-16β,26-hexol 296
24ξ-Methyl-5α-cholestane-3β,5,6β,22R,24-pentol-6-acetate 217
24-Methyl-5α-cholesten-3β-ol 307
24-Methyl-24,25-dioxoscalar-16-en-12β-yl-3-hydroxybutanoate 172
24-Methylenecholest-5-en-3β-ol 307
(20 R)-24-Methylenecholestane-3β-21-diol 296
24-Methylene-5α-cholesten-3β-ol 307
4α-Methyl-5α-ergosta-24(28)-en-3β-ol 215
1-Methylisoguanosine 193, 258
2-Methyl-2-(4-methylpent-3-enyl)-2H-chromen-6-ol 316
24ξ-Methyl-27-nor-5α-cholestane-3β,4β,6β,8,15α,16β,26-heptol 296
24ξ-Methyl-27-nor-5α-cholestane-3β,6β,8,15α,16β,26-hexol 269
Methyl nuapapuanoate 165
(Z)-7-Methyl-4-octen-3-one
Methylscalarene 172
24-Methylscalarin 266
2-Methyl-1,3,6-trihydroxyanthraquinone 118, 141

2-Methyl-3,5,6-trihydroxyanthraquinone 118, 138
6-Methylxanthopurin 140
Metridium senile 211, 212
Mevalonic acid 82
Mg-ATPase 304
Microciona prolifera 181
– *toxystila* 259
Microcionin-1,2,3,4 262
Microcosmus sulcatus 317
Monocerin 6, 40, 59
Monocerolide 39, 59
Monocerone 40, 59
Monoterpenes 210
Morinda citrifolia 89
– *lucida* 89
Morindaparvin A, B 119, 139
Morindone 120, 132, 133, 138, 139, 140
Morindone-5-methylester 119, 132, 139, 140
Munjistin 121, 131, 134, 135, 136, 137, 140, 141, 142
Munjistindimethylether 121
Munjistinmethylester 122, 139
Muqubilin 173, 174
Muricea california 216, 218
– *fruticosa* 218
Muricella sp. 240
Muricin 218
Mycale sp. 194
Mycalisine A, B 194
Myosin 186
Mytiloxanthinone 316
Mytilus edulis 282

Nakafuran-8 157
Nakafuran-9 157, 263
Na-K-ATPase 164, 165, 185, 186, 304
Nanimoal 263
Naphthalenes 308
Naphthalic anhydride 44, 59
Narcipoetine 53, 59
Navanax inermis 277
Navenone A 277
– B 277
– C 277
Neodolabelline 239
Neolemnane 225
2-Neomanoalide 172
Neopalytoxin 213
Neosurugatoxin 276

Nephthea brassica 232
- *chabrolii* 223, 224, 232, 241
- *sp.* 222, 232, 240, 241
Nerita albicilla 277
Nodososide 287
5-Nonylpyrrole-2-carbaldehyde 251
Norbergenin 15, 59
Nordamnacanthal 122, 132, 133, 138, 139, 141, 142
Norhalichondrin A 200
Norjuzunal 122, 133
Norpectinatone 276
Norrisolide 164
Norsesterterpene peroxide 173
Nucleosides 193, 298
Nudibranches 154, 157, 164, 193

Obscuronatin 240
(*E*)-Ocellenyne 271
(*Z*)-Ocellenyne 271
Ochracin 18, 59
Ochratoxin A 4, 7, 12, 56, 59
- B 12, 56, 59
- C 56, 59
3-Octadecyloxy-1,2-propanediol 307
Okadaic acid 203
Oktaketide 6
Olepupuane 263
Oligopeptides 309
Onchidium veruculatum 277
Oospoglycol 17, 59
Oospolactone 7, 10, 18, 59
Oosponol 7, 12, 17, 59
Oospora astringens 17
Ophidianoside B 290
- C 290
- F 291
Ophidiaster ophidianus 290, 291
Ophioxanthin 307
Ophirin 240
Ophiuroids 307
Oruwal 123, 139
Oruwalol 123, 139
22,25-Oxidoholothurinogenin 300
3-Oxochol-4-ene-24-oic acid 261

Pacifigorgia adamsii 224
- *media* 224
- *pulchra exilis* 224
Pacifigorgiol 224
Palauolide 172
Pallescensin 261

Pallescensolide 158
Palystatin A to D 215
Palythoa caribaeorum 213
- *liscia* 215, 268
- *sp.* 213
- *toxica* 213
- *tuberculosa* 213
Palytoxin 213
Pandaros acanthifolia 203
Panulirus cygnus 318
Paracentrotus lividus 309
Paragracine 215
Paralemnalia thyrsoides 224
Paramuricea chamaeleon 223
Parathyona sp. 304
Parathyonoside R 304
Parazoanthus gracilis 215
Parerythropodium fulvum 224
Parguerol 270
Parguerol-16-acetate 270
Patellamide A 309
- B 309
- C 309
Patiria pectinifera 294
Pectinatone 276
Peltaphorin 16, 59
Peltodoris atromaculata 204
Penicillium atrovenetum 204
- *chrysogenum* 205
Peniolactol 36, 59
Penlanfuran 157
Penticeraster alveolatus 290, 291
Peptide alkaloids 190, 272
Peptides 191, 193, 252, 257, 273, 274, 317, 318
Peronia peronii 277
Peroniatrol I 277
Peroniatrol II 277
Petrosia ficiformis 205, 259
- *seriata* 195
Petrosin 195
Petrosin A 196
Petrosin B 196
Phaeodactylum tricornutum 218
Phakellia flabellata 195
Phallusia mamillata 317
Phenalenone 7
Phenols 202, 318
Phenylalanine 9, 312
Phidolopin 253
Phidolopora pacifica 253
Philinopsis speciosa 277

Phyllodulcin 5, 9, 12, 23, 59
Phyllospongia dendyi 171, 172
- *foliascens* 171, 172, 202
- *radiata* 172
- *sp.* 172, 266
Physalia physalis 208, 209
Phytoalexin 10
Phytotoxins 10
Placogorgia sp. 223
Placopecten magellanicus 282
Placortic acid 204
Plakinamin A 155
Plakinamin B 155
Plakina sp. 155
Plakinic acid A 204
Plakinic acid B 204
Plakortin 204
Plakortis halichondrioides 204
- *zyggompha* 204, 206
Pleraplysilla spinifera 262
Pleraplysillin-2 262
Plexaura flava 241, 251
- *homomalla* 248
Plexaurella grisea 223
Plocamium cartilagineum 210
Plumeria sp. 277
Polyacetylenes 205
Polyandrocarpa sp. 314
Polyandrocarpidines A–D 314
Polycera tricolor 268
Polycitorella mariae 313
Polygodial 262
Polyketides 6
Polypeptides 152, 193, 211, 212, 252, 273, 317, 318
Polypropionates 276
Pregna-5,20-dien-3β-ol 218
Pregnedioside-A 218
Prianicin A 173
Prianicin B 173
Priano sp. 155
Prostaglandins 248
Prostanoids 249
Protease 209
Proteins 12, 208, 211, 215, 257, 275
Proteus vulgaris 185
Protoreaster nodosus 287, 288, 290, 291, 295, 296
Protoreasteroside 291
Psammaplysilla purea 186
- *purpurea* 185
Psammaplysin A 173

Psammaplysin B 173
Pseudactinia varia 212
Pseudocentrotus depressus 308
Pseudomonas aeruginosa 251
Pseudopterogorgia acerosa 239
Pseudopterolide 239
Pseudopurpurin 124, 131, 134, 135, 136, 137, 141, 142, 143
Psolothurin A 304
Psolusoside A 304
Pteroidine 239
Ptilocaulis spiculifer 195
Ptilocaulin 195
Ptychodera flava 318
Pukalide 231
Pulo'upone 277
Punaglandins 249
Purealin 186
Purine 164
Purpurin 124, 131, 132, 134, 135, 136, 137, 141, 142
Purpurin 1-methyl ether 125
Puupehenone 158
Pu'ulenal 262
Pyricularium oryzae 10
Pyrrol 195

Quinizarin 125, 142
Quinovose 304

Radianthus koseirensis 211
- *macrodactylus* 211
Ramulosin 18, 59
Renieramycin A 196
- B 196
- C 196
- D 196
Reniera sp. 196
Renierone 196
Reticulol 6, 12, 37, 59
Rhamnus sp. 80, 90
Rheum sp. 80, 90
Rhodeus ocellatus smithi 274, 275
Roboastra tigris 268
Rubiacea 81, 83, 88, 90
Rubiadin 126, 131, 132, 133, 134, 135, 136, 137, 138, 139, 140, 141, 142
- -1-methyl ether 126, 131, 132, 133, 134, 135, 138, 139, 140
- 3-methyl ether 127, 134
Rubia tinctorum L. 80, 81
Rubrosulphin 47, 59

Saccharomyces cerevisiae 173, 204, 313
(+)-α-Santalene 223
Saponins 283, 285, 286, 288, 290, 291, 294, 295, 299, 300, 301, 302, 304
Sarcina lutea 185
Sarcophyta elegans 232, 233
Sarcophyton crassocaule 232
– *decaryi* 232
– *glaucum* 217, 232, 233
– *sp.* 231, 232, 233
Sarcophytoxide 231
Saxitoxin 282
SC-28762 51, 59
SC-28763 51, 59
SC-30532 51, 59
Scalafuran 172
Scalaradial 172
Scalarane 172
Scalarbutenolide 173
Scalardysin A 173
Scalardysin B 173
Scalarin 172
Scalarolide 172
Sceptrin 195
Sclerin 8, 18, 59
Sclerotinia sclerotiorum 10
Sclerotinin A 10, 31, 59
Sclerotinin B 31, 59
Sclerotonin 8
Scrophulariaceae 81
Scytalium splendens 224
– *tentaculatum* 239
Semi-vioxanthin 47, 59
Senna sp. 90
Sepositoside A 285, 286
Serotonine 183
Sertularella classicaulis 210
Sesquiterpenes 157, 158, 163, 165, 180, 222, 224, 225, 262, 263, 270
Sessibugula translucens 268
Sesterterpenes 157, 163, 165, 170, 171, 172, 173, 180, 265, 266
Shikimate pathway 81
Shikimic acid 81
Sialic acid 308
Sialoglycoprotein 308
Sidnyaxanthin 316
Sidnyum argus 316
Sigmosceptrella laevis 173
Sigmosceptrellin A 173
– B 173
– C 173

Silphiperfdene 225
Sinularia capillosa 223
– *dissecta* 217
– *erecta* 223
– *facile* 232
– *firma* 223
– *flexibilis* 231, 251
– *mayi* 232
– *polydactyla* 223
– *sp.* 251
Sipholenol-A 180
– -B 180
– -C 180
– -D 180
– -E 180
Sipholenone A 180
Siphonaria atra 277
– *australis* 276
– *denticulata* 276
– *diemenensis* 276
– *laciniosa* 277
– *lessoni* 276
– *normalis* 277
– *pectinata* 276
– *sp.* 276
– *zelandica* 277
Siphonarin A 276
Siphonarin B 277
Siphonelliol 180
Siphonochalina siphonella 180
Siphonodictidine 158
Siphonodictyal-A 158
Siphonodictyal-B 158
Siphonodictyon coralliphagum 158
– *mucosa* 158
– *sp.* 157, 158
Smenospongia aurea 159
– *echina* 159
Sokotrasterol sulfate 155
Soranjidol 127, 132, 133, 138, 139, 140
– -1-methylether 128, 139
Sources of natural occurring
– Anthraquinones 92ff.
– Isocumarins 14ff., 60, 61, 62, 63
Spermidine derivatives 251
Spongia-13(16),14-dien-19-oic acid 163
Spongia-13(16),14-dien-19-al 163
Spongia-13(16),14-diene 163
Spongia idia 171, 172, 259
– *nitens* 173
– *officinalis* 163, 259
Spongionella gracilis 164

Subject Index

Staphylococcus aureus 155, 158, 164, 165, 172, 206, 262, 268, 276
Stellatin 25, 59
Stellatta sp. 180
Steroids 152, 153, 154, 155, 212, 215, 216, 217, 261, 271, 277, 282, 285, 295, 296, 307, 317
Stichaster striatus 290
Stichlorogenol 302
Stichloroside A, B, C 302
Stichodactyla haddoni 211
Stichoposide A, B, C, D, E 302
Stichopus chloronotus 302
– *japonicus* 307
– *variegatus* 302
Stoichactis helianthus 211
– *sp.* 211
Stoloniferone A, B, C, D 217
Streptomyces lavendulae 196
Striatoxin 274
Strongylocentrotus droebachiensis 308
Strongylostatin 1, 2 308
Stylocheilus longicauda 269
Subergorgia suberose 225
Subergorgic acid 225
o-Succinylbenzoic acid 81, 82
Surugatoxin 276
Suvanine 174
Swinholide A 200

Tambjamine A to D 268
Tambje abdere 268
– *eliora* 268
– *sp.* 268
Tavacbutenolide-1 157
– -2 158
Tavacfuran 157
Tavacpallescensin 158
Tealia felina 211
Tealia lofotensis 212
Tectoquinone 128, 139
Tedania digitata 193, 257, 259
Tedania ignis 154, 164, 191, 200
Tedanolide 200
Telesto riisei 249
Terpenes 153, 157, 184, 222, 224, 258, 261
Terrein 8
Tessulatoxin 275
Tethya aurantia 258
Tetrabenacine 183, 184
17 Z-Tetracosenylglycerol-1-ylether 205

Tetrahydropyrans 206
$1\beta,3\beta,5,6\beta$-Tetrahydroxy-5α-androstan-17-one 217
Tetrapyrrole 314
Tetrodotoxin 274, 275, 276, 282, 299
Tetrosia sp. 205
Thalysia junipertina 258
Thelenota ananas 302, 304
Thelenotoside A, B 302
Theloturin A, B 302
Theonella cf. swinhoei 159, 200
Theonellin 159
Thiamin pyrophosphate 81
Thiazoles 190, 272
Thornasterol A 290
Thornasterol B 283
Thornasteroside A 283, 288, 289, 290, 291
Thymin desoxyribonucleoside 298
Tinctomorone 129, 139
Toxadocia zumi 155
Toxin A III 318
Toxin B II 318
Toxin B IV 318
Toxopneustes pileolus 308
2,5,6-Tribromo-N-methylgramine 253
Tridachia crispata 277
Tridachiella diomedea 277
Tridachione 277
(S)-(+)-1-Tridecoxy-2,3-propanediol 206
Trididemnum sp. 309
Trigonelline 318
3,5,8-Trihydro-4-quinolone 195
1,2,4-Trihydroxybenzene 202
Trimethylsilylation 89
Triopha catalinae 268
Triophamine 268
Tripneustes gratilla 308
Trisnorsesquiterpenes 223
Triterpenes 180, 282
Tryptamine 183
Tryptophane 183, 313
Tubipora musica 224
d-Tubocurarine 274
Tunichrome B-1 310
Tutufa lissostoma 276
Tyrosine 183, 184, 185

Ulicylamide 309
Ulithiacyclamide 309
Ulosa ruetzleri 205
Upial 158

Uracil desoxyribonucleoside 298
Ustic acid 3, 32, 59
UV-spectroscopy 83, 84, 89, 91 ff.

Vakerin 16, 60
Valerenenol 224
Verapamil 275
Verbenaceae 81
Verongia see *Aplysina*
Versicoside A 289
Vibrio angiullarum 266
Viomellin 6, 48, 60
Viopurpurin 48, 60
Vioxanthin 48, 60
Virgularia sp. 217
Viriabilin 170
Viriditoxin 51, 60
Vitamin K 81, 83

Wax ester 153
Wistarin 173
Woodfordia fructicosa 15

Xanthomegnin 4, 6, 49, 60
Xanthopurpurin 129, 131, 134, 135, 136, 137, 140, 141, 142, 143
- dimethyl ether 129, 134, 135, 136, 137, 143
- 1-methyl ether 130, 141, 142
- 3-methyl ether 130, 141, 142, 143

Xanthoviridicatin D 47, 60
- G 47, 60
Xenia crassa 241
- *lilielae* 241
- *macro spiculata* 241
- *obscuronata* 240, 241
Xeniaphyllane 240
Xenia sp. 240
- *viridis* 241
Xenicin 240
Xeniolide 240
Xenoclauxin 45, 60
Xestospongia exigua 196, 206
- *sapra* 206
- *testudinaria* 203
Xestospongin A 196
- B 196
- C 196
- D 196
X-Ray-analysis 163, 173, 180, 185, 190, 195, 210, 224, 225, 265, 270, 276, 277, 313
Xylindein 50, 60
Xylose 304

Zoanthamide 215
Zoanthamine 215
Zoanthenamine 215
Zoanthus sociatus 215
Zoobotryon verticillatum 253

Fortschritte der Chemie organischer Naturstoffe
Progress in the Chemistry of Organic Natural Products

Volume 48:

1985. 33 figures. IX, 285 pages. Cloth DM 220,–. ISBN 3-211-81886-3

Contents: P. S. Steyn and R. Vleggaar: Tremorgenic Mycotoxins. – R. E. Moore: Structure of Palytoxin. – P. Crews and S. Naylor: Sesterterpenes: An Emerging Group of Metabolites from Marine and Terrestrial Organisms.

Volume 47:

1985. 16 figures. VIII, 290 pages. Cloth DM 198,–. ISBN 3-211-81864-2

Contents: R. Southgate and S. Elson: Naturally Occurring β-Lactams. – I. Howe and M. Jarman: New Techniques for the Mass Spectrometry of Natural Products. – P. G. McDougal and N. R. Schmuff: Chemical Synthesis of the Trichothecenes. – J. Polonsky: Quassinoid Bitter Principles II.

Volume 46:

1984. 7 figures. IX, 253 pages. Cloth DM 178,–. ISBN 3-211-81804-9

Contents: O. Tanaka and R. Kasai: Saponins of Ginseng and Related Plants. – E. Fujita, M. Node: Diterpenoids of *Rabdosia* Species. – S. Johne: The Quinazoline Alkaloids.

Volume 45:

1984. 2 figures. VIII, 288 pages. Cloth DM 194,–. ISBN 3-211-81755-7

Contents: D. A. H. Taylor: The Chemistry of the Limonoids from Meliaceae. – J. A. Elix, A. A. Whitton, and M. V. Sargent: Recent Progress in the Chemistry of Lichen Substances. – Y. Shimizu: Paralytic Shellfish Poisons.

Volume 44:

1983. 72 partly coloured figures. IX, 326 pages.
Cloth DM 208,—. ISBN 3-211-81754-9

Contents: F. J. Evans and S. E. Taylor: Pro-Inflammatory, Tumour-Promoting and Anti-Tumour Diterpenes of the Plant Families Euphorbiaceae and Thymelaeaceae. — A. Mondon and B. Epe: Bitter Principles of Cneoraceae. — S. Naylor, F. J. Hanke, L. V. Manes, and P. Crews: Chemical and Biological Aspects of Marine Monoterpenes. — J. G. Buchanan: The C-Nucleoside Antibiotics.

Volume 43:

1983. VIII, 383 pages. Cloth DM 208,—. ISBN 3-211-81741-7

Contents: J. L. Ingham: Naturally Occurring Isoflavonoids (1855–1981). — A. Koskinen and M. Lounasmaa: The Sarpagine-Ajmaline Group of Indole Alkaloids.

Volume 42:

1982. VII, 323 pages. Cloth DM 164,—. ISBN 3-211-81706-9

Contents: Y. Asakawa: Chemical Constituents of the Hepaticae. — M. Heidelberger: Cross-Reactions of Plant Polysaccharides in Antipneumococcal and Other Antisera, an Update.

All Volumes and Cumulative Index 1—20 available

Price reduction for subscribers: 10%

Special reduced price (20% reduction) for the complete Series Vols. 1—49 incl. the Cumulative Index to Vols. 1—20

Springer-Verlag Wien New York